STALLCUP'S®
ILLUSTRATED
CODE CHANGES

based on the *nec*® and related standards

2005

AN INTERNATIONAL CODES AND STANDARDS ORGANIZATION

Written by James Stallcup, Sr.
Edited by James Stallcup, Jr.
Design, graphics and layout by Billy G. Stallcup

Notice Concerning Liability: Publication of this work is for the purpose of circulating information and opinion among those concerned for fire and electrical safety and related subjects. While every effort has been made to achieve a work of high quality, neither the NFPA nor the authors and contributors to this work guarantee the accuracy or completeness of or assume any liability in connection with the information and opinions contained in this work. The NFPA and the authors and contributors shall in no event be liable for any personal injury, property, or other damages of any nature whatsoever, whether special, indirect, consequential, or compensatory directly or indirectly resulting from the publication, use of or reliance upon this work.

This work is published with the understanding that the NFPA and the authors and contributors to this work are supplying information and opinion but are not attempting to render engineering or other professional services. If such services are required the assistance of an appropriate professional should be sought.

The following are registered trademarks of the National Fire Protection Association: *National Electric Code*® and *NEC*®

NFPA No.: SICC05
ISBN: 0-87765-638-X
Library of Congress Card Catalog No.: 2004109209

Printed in the United States of America
05   06   07   08   5   4   3   2

Table of Contents

Foreword

The last two *NEC®* cycles have brought massive changes to the Code. The overwhelming amount of text that was either rewritten or reorganized, including redesigned and expanded tables as well as the introduction of new articles, has been done with the intention of making the NEC more user friendly. However, for a longtime user of the NEC, these changes didn't seem user friendly at first.

Now, once again, the 2005 NEC represents numerous changes. Some of these changes include new requirements for identification of ungrounded branch circuit and feeder conductors, revised rules for arc-fault circuit interrupters, and new zone hazarous area classification systems. Just learning what these changes are and the reasoning behind them shouldn't be the only reason for a user to learn them.

Correlation and interactivity of various codes and standards are mandatory in the electrical industry. Therefore, it is imperative that there be a code that other codes and standards can reference as the code of standard practice. The NEC is recognized as this code, and, therefore, for installation and maintenance of electrical systems, as well as for safety issues, keeping oneself aware of changes not only in the NEC, but in the electrical industry as a whole, is a must.

James Stallcup Sr.
James Stallcup Jr.

Introduction

The **2005 edition of the National Electrical Code® (NEC®)** contains many comprehensive revisions pertaining to specific NEC rules and regulations. Electrical personnel have an immediate and awesome task in not only learning but implementing these revisions in their everyday design, installation, and inspection of electrical systems.

The material in this book, if read and studied carefully in a continuous and enthusiastic manner, will provide a proper update on the revisions in the **2005 NEC**. However, even though it is true that only time and discussion among electrical personnel will provide the answers on how to interpret and apply some of these rules, one can use this book to get a headstart.

Illustrated Code Changes explains the major changes in the **2005 NEC** and can be used as a guide for fast and easy reference. These changes are presented in numerical order to correlate with the "Articles" and "Sections" as they appear in the **2005 NEC** and are also illustrated to give a more detailed description. Where appropriate, reasons for revisions and new articles are given and what kind of impact such changes will have on manufacturers, designers, installers, and inspectors.

For every proposal made, there was a reason, and this book will provide some of the reasons why a change resulted.

CORRELATING THE NEC WITH OTHER STANDARDS

Type of Change		Panel Action		UL	UL 508	API 500	API 505	OSHA
-		-		-	-	-	-	-
ROP		ROC		NFPA 70E	NFPA 70B	NFPA 79	NFPA	NEMA
pg. -	# -	pg. -	# -	-	-	-	-	-
log: -	CMP: -	log: -	Submitter: -			2002 NEC: -		IEC: -

With each change, references are made to well known standards that are used in the industry. However, the **ROP and ROC**, that are referenced to in the Table, are documents that are directly involved in the code-making process and are explained below along with the different types of changes and **Code Making Panel** actions.

NEC PROCESS

ROP (Report on Proposals)
The information found in this document not only specifies the proposed changes, but also includes who submitted the change and substantiation for such proposal.

ROC (Report on Comments)
All of the actions that were taken by the **Code Making Panels** and the **Technical Correlating Committee** at the meetings prior to the **NFPA** annual meeting and the adoption of the **2002 NEC** are found in this document.

Type of Change:
New Article, Section, Subsection, Subdivision, Exception and FPN
.... Revision
.... Deletion

Panel Action:
.... Accept
.... Accept in Principle
.... Accept in Part
.... Accept in Principle in Part
.... Reject

This Table was developed by the author with the intention of enhancing the perception of the change while correlating the NEC with other standards. Hopefully, the instructor and/or student of the code will find this information useful, if not interesting.

For seminar purposes the authors have listed 150 of what they feel will be the most important code changes. However, this book represents over 500 2005 NEC changes that are intended for the reader's review.

Administration and Enforcement, Introduction, Definitions and Requirements

Article 90 and **Chapter 1** of the NEC have always been referred to as the get acquainted material that every designer, installer, electrician, apprentice, inspector and maintenance person must review and understand before the other Chapters, Articles and Sections of the NEC can really be understood and applied.

The first Article in the NEC is **Article 90** which contains the Introduction. **Article 90** covers the purpose of the NEC along with other pertinent information that is applicable throughout each Chapter of the NEC.

Chapter 1 acquaints the user of the NEC with definitions and clearance rules that are mandatory to ensure the safety of the general public and personnel working in, near or on wiring methods and equipment.

Users as well as students of the NEC must review and become acquainted with **Article 90** and **Chapter 1** before attempting to study, learn and apply the other Articles and Chapters to a particular design or installation.

It is this concept of study that will make interpretations and applications of the many requirements in the NEC much easier to understand.

NEC Ch. – Article 90
Part – 90.1(D)

 1

Type of Change		Panel Action		UL	UL 508	API 500	API 505	OSHA
Revision		Accept		-	-	-	-	-
ROP		ROC		NFPA 70E	NFPA 70B	NFPA 79	NFPA	NEMA
pg. 53	# 1-16a	pg. -	# -	-	-	-	-	-
log: CP105	CMP: 1	log: -	Submitter: CMP 1			2002 NEC: 90.1(D)		IEC: 60364-1

2002 NEC – 90.1(D) Relation to International Standards

The requirements in this *Code* address the fundamental principles of protection for safety contained in Section 131 of International Electrotechnical Commission Standard 60364-1, *Electrical Installations of Buildings*.

2005 NEC – 90-1(D) Relation to <u>Other</u> International Standards

The requirements in this *Code* address the fundamental principles of protection for safety contained in Section 131 of International Electrotechnical Commission Standard 60364-1, *Electrical Installations of Buildings*.

Author's Substantiation. The word Other has been added in the title to clarify the NEC has been widely adopted in many countries. The NEC has been adopted in North, Central, and South America, Asia, and the Middle East and is also used by global companies in the implementation of their projects. This new term makes it clear that the NEC is compatible with international safety principles and calls attention to the fact that installations meeting the requirements of the NEC address the fundamental principles outlined in IEC 60364-1, *Electrical Installations of Buildings*, Section 131.

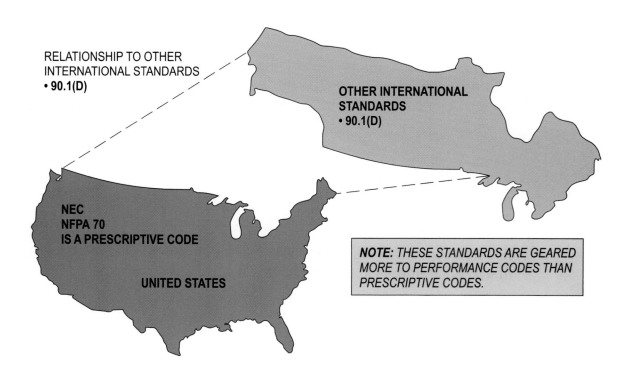

RELATIONSHIP TO OTHER INTERNATIONAL STANDARDS
• 90.1(D)

OTHER INTERNATIONAL STANDARDS
• 90.1(D)

NEC
NFPA 70
IS A PRESCRIPTIVE CODE

UNITED STATES

NOTE: THESE STANDARDS ARE GEARED MORE TO PERFORMANCE CODES THAN PRESCRIPTIVE CODES.

RELATION TO OTHER INTERNATIONAL STANDARDS
90.1(D)

Purpose of Change: The title head was revised to include other international standards.

NEC Ch. – Article 90
Part – 90.2(A)

Type of Change		Panel Action		UL	UL 508	API 500	API 505	OSHA
Revision		Accept in Principle		-	-	-	-	1910.302(a)(1)
ROP		ROC		NFPA 70E	NFPA 70B	NFPA 79	NFPA	NEMA
pg. 54	# 1-18	pg. 19	# 1-7	-	-	-	-	-
log: 1	CMP: 1	log: 1416	Submitter: James M. Daly			2002 NEC: 90.2(A)		IEC: -

2002 NEC – 90.2(A) Covered

This *Code* covers the installation of ~~electric~~ conductors, ~~electric~~ equipment, signaling and communications conductors and equipment, and fiber optic cables and raceways for the following:

2005 NEC – 90.2(A) Covered

This *Code* covers the installation of <u>electrical</u> conductors, <u>equipment, and raceways;</u> signaling and communications conductors, <u>equipment, and raceways;</u> and optical fiber cables and raceways for the following:

Author's Substantiation. The word raceways has been added to clarify what is covered. Repeating the word raceways clarifies that the Code applies to raceways for all three categories.

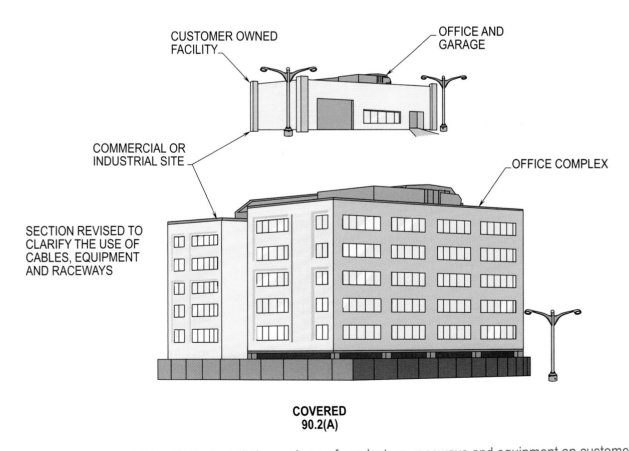

CUSTOMER OWNED FACILITY

OFFICE AND GARAGE

COMMERCIAL OR INDUSTRIAL SITE

OFFICE COMPLEX

SECTION REVISED TO CLARIFY THE USE OF CABLES, EQUIPMENT AND RACEWAYS

COVERED
90.2(A)

Purpose of Change: To identify the installation and use of conductors, raceways and equipment on customer owned facilities where the NEC applies.

NEC Ch. – Article 90
Part – 90.2(A)(2), FPN

Type of Change		Panel Action		UL	UL 508	API 500	API 505	OSHA
Revision		Accept		-	-	-	-	1910.302(a)(1)
ROP		ROC		NFPA 70E	NFPA 70B	NFPA 79	NFPA	NEMA
pg. 55	# 1-22a	pg. -	# -	-	-	-	-	-
log: CP100	CMP: 1	log: -	Submitter: CMP 1			2002 NEC: 90.2(A)(2), FPN		IEC: -

2002 NEC – 90.2(A)(2), FPN Covered

For additional information concerning such installations in an industrial or multibuilding complex, see ~~ANSI C2-1997~~, *National Electrical Safety Code.*

2005 NEC – 90.2(A)(2), FPN Covered

For additional information concerning such installations in an industrial or multibuilding complex, see ANSI C2-2002, *National Electrical Safety Code.*

Author's Substantiation. This revision is in accordance with Section 3.6.7.1.2 of the NFPA *Manual of Style* to revise the reference data in the fine print notes in **90.2(A)(2)** and **110.31(C)(1)** and in the note to **Table 110.31**.

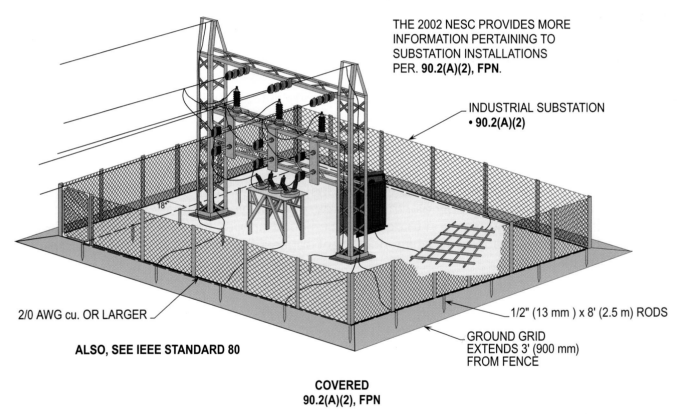

THE 2002 NESC PROVIDES MORE
INFORMATION PERTAINING TO
SUBSTATION INSTALLATIONS
PER. **90.2(A)(2), FPN**.

INDUSTRIAL SUBSTATION
• **90.2(A)(2)**

18"

2/0 AWG cu. OR LARGER

1/2" (13 mm) x 8' (2.5 m) RODS

GROUND GRID
EXTENDS 3' (900 mm)
FROM FENCE

ALSO, SEE IEEE STANDARD 80

**COVERED
90.2(A)(2), FPN**

Purpose of Change: To reference the latest publication of the *National Electrical Safety Code* (NESC), which is in
the 2002 edition.

NEC Ch. – Article 90
Part – 90.2(B), FPN to (4) and (5)

Type of Change		Panel Action		UL	UL 508	API 500	API 505	OSHA
New FPN		Accept		-	-	-	-	1910.269(a)
ROP		ROC		NFPA 70E	NFPA 70B	NFPA 79	NFPA	NEMA
pg. 57	# 1-25	pg. 19	# 1-9	-	-	-	-	-
log: 2003	CMP: 1	log: 1569	Submitter: Neil F. Labrake Jr.			2002 NEC: -		IEC: -

2005 NEC – 90.2(B), FPN to (4) and (5) Not Covered

Examples of utilities may include these entities, typically designated or recognized by governmental law or regulation by public service/utility commissions, that install, operate, and maintain electric supply (such as generation, transmission, or distribution systems) or communication systems (such as telephone, CATV, Internet, satellite, or data services). Utilities may be subject to compliance with codes and standards covering their regulated activities as adopted under governmental law or regulation. Additional information can be found through consultation with the appropriate governmental bodies, such as state regulatory commissions, Federal Energy Regulatory Commission, and Federal Communications Commission.

Author's Substantiation. The determination of what constitutes a utility is a question of federal and state law or regulation. This new Fine Print Note provides information reflecting this governmental determination and refers ultimate resolution of the matter in any particular instance to the appropriate governmental regulatory body.

Since both the NEC and the NESC can cover industrial plants, institutional facilities and electric generation as they function to supply electrical energy, it is always necessary to get clarification from the proper authority having jurisdiction to decide which code applies to these electrical systems. And once the determination is made as to which document applies, the document should continue in force for the life of the installation.

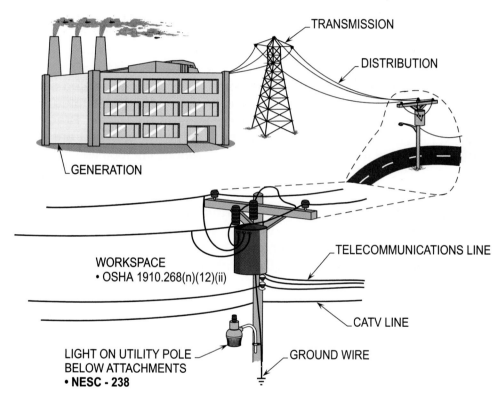

TRANSMISSION

DISTRIBUTION

GENERATION

WORKSPACE
• OSHA 1910.268(n)(12)(ii)

TELECOMMUNICATIONS LINE

CATV LINE

LIGHT ON UTILITY POLE
BELOW ATTACHMENTS
• NESC - 238

GROUND WIRE

NOT COVERED
90.2(B), FPN TO (4) AND (5)

Purpose of Change: To verify that utilities are typically recognized by public service/utility commissions or governmental law.

NEC Ch. – Article 90
Part – 90.5(C)

Type of Change		Panel Action		UL	UL 508	API 500	API 505	OSHA
New Sentence		Accept		-	-	-	-	1910.269(a)
ROP		ROC		NFPA 70E	NFPA 70B	NFPA 79	NFPA	NEMA
pg. 61	# 1-34	pg. -	# -	-	-	-	-	-
log: 1443	CMP: 1	log: -	Submitter: TCC			2002 NEC: 90.5(C)		IEC: -

2002 NEC – 90.5(C) Explanatory Material

Explanatory material, such as references to other standards, references to related sections of this *Code*, or information related to a *Code* rule, is included in this *Code* in the form of fine print notes (FPNs). Fine print notes are informational only and are not enforceable as requirements of this *Code*.

2005 NEC – 90.5(C) Explanatory Material

Explanatory material, such as references to other standards, references to related sections of this *Code*, or information related to a *Code* rule, is included in this *Code* in the form of fine print notes (FPNs). Fine print notes are informational only and are not enforceable as requirements of this *Code*. Brackets containing section references to another NFPA document are for informational purposes only and are provided as a guide to indicate the source of the extracted text. These bracketed references immediately follow the extracted text.

Author's Substantiation. A new second paragraph was added to prevent any misunderstanding about whether these bracketed references are for informational purposes only. These references were previously contained within Appendix A, which was very clearly a nonmandatory reference. A number of authorities having jurisdiction were interpreting the bracketed information as mandatory references and had questions about requiring the user of the Code to also comply with any requirements in the parent document. Adding this text makes it clear that these references remain for informational purposes only.

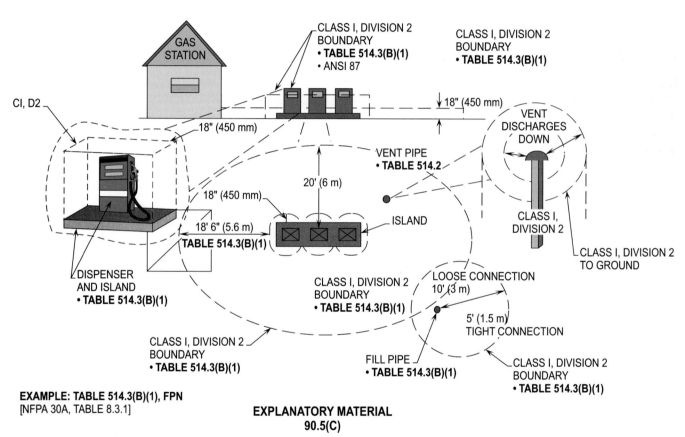

CLASS I, DIVISION 2
BOUNDARY
• **TABLE 514.3(B)(1)**
• ANSI 87

CLASS I, DIVISION 2
BOUNDARY
• **TABLE 514.3(B)(1)**

GAS
STATION

CI, D2

18" (450 mm)

18" (450 mm)

VENT
DISCHARGES
DOWN

VENT PIPE
• **TABLE 514.2**

20' (6 m)

18" (450 mm)

ISLAND

18' 6" (5.6 m)
TABLE 514.3(B)(1)

CLASS I,
DIVISION 2

CLASS I, DIVISION 2
TO GROUND

DISPENSER
AND ISLAND
• **TABLE 514.3(B)(1)**

LOOSE CONNECTION
10' (3 m)

CLASS I, DIVISION 2
BOUNDARY
• **TABLE 514.3(B)(1)**

5' (1.5 m)
TIGHT CONNECTION

CLASS I, DIVISION 2
BOUNDARY
• **TABLE 514.3(B)(1)**

FILL PIPE
• **TABLE 514.3(B)(1)**

CLASS I, DIVISION 2
BOUNDARY
• **TABLE 514.3(B)(1)**

EXAMPLE: TABLE 514.3(B)(1), FPN
[NFPA 30A, TABLE 8.3.1]

EXPLANATORY MATERIAL
90.5(C)

Purpose of Change:
To clarify that sections in the brackets referring to NFPA Standards are for informational material only.

NEC Ch. – Article 90
Part – 90.8(A)

Type of Change		Panel Action		UL	UL 508	API 500	API 505	OSHA
Revision		Accept in Principle		-	-	-	-	-
ROP		ROC		NFPA 70E	NFPA 70B	NFPA 79	NFPA	NEMA
pg. 61	# 1-35	pg. -	# -	-	-	-	-	-
log: 1228	CMP: 1	log: -	Submitter: David H. Kendall			2002 NEC: 90.8(A)		IEC: 132

2002 NEC – 90.8(A) Future Expansion and Convenience

Plans and specifications that provide ample space in raceways, spare raceways, and additional spaces allow for future increases in the use of electricity. Distribution centers located in readily accessible locations provide convenience and safety of operation.

2005 NEC – 90.8(A) Future Expansion and Convenience

Plans and specifications that provide ample space in raceways, spare raceways, and additional spaces allow for future increases in <u>electric power and communication circuits</u>. Distribution centers located in readily accessible locations provide convenience and safety of operation.

Author's Substantiation. This revision allows the addition of optical fiber and communications to the section on future expansion and convenience. Optical fiber and communication cabling technology is changing faster than electrical wiring. Within buildings, it is important to plan for future media additions or modifications.

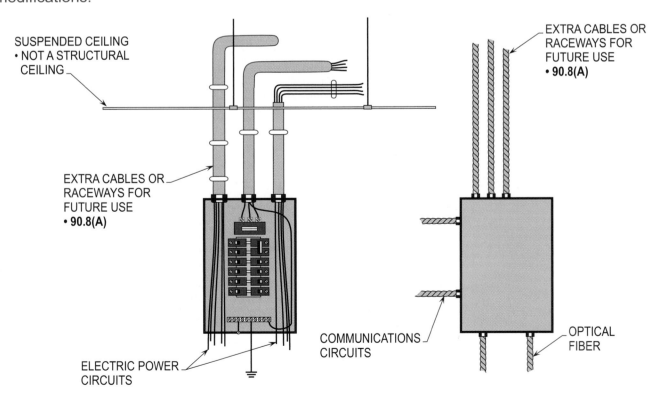

FUTURE EXPANSION AND CONVENIENCE
90.8(A)

Purpose of Change: To add the terms "optical fiber" and "communications" to the section.

NEC Ch. – Article 90
Part – 90.9(C)(4)

Type of Change		Panel Action		UL	UL 508	API 500	API 505	OSHA
Revision		Accept		-	-	-	-	-
ROP		ROC		NFPA 70E	NFPA 70B	NFPA 79	NFPA	NEMA
pg. 65	# 1-45	pg. -	# -	-	-	-	-	-
log: 1775	CMP: 1	log: -	Submitter: Tex Combs			2002 NEC: 90.9(C)(4)		IEC: -

2002 NEC – 90.9(C)(4) Safety

Where a negative impact on safety would result, ~~hard~~ conversion shall not be required.

2005 NEC – 90.9(C)(4) Safety

Where a negative impact on safety would result, <u>soft</u> conversion shall be used.

Author's Substantiation. Where safety is involved, the Code should give the correct conversion. Soft conversion is a direct mathematical conversion.

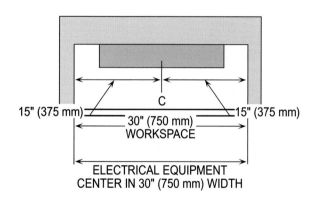

15" (375 mm)   C   15" (375 mm)

30" (750 mm)
WORKSPACE

ELECTRICAL EQUIPMENT
CENTER IN 30" (750 mm) WIDTH

EXAMPLE

Given: Using the "soft-conversion" method, determine the equivalent metric conversion for 30 in. where the calculation could have a negative impact on safety, such as a 30 in. minimum horizontal working space requirement in the rear of equipment that requires access to nonelectrical parts in 110.26(A)(1)(a).

Solution:

Step 1: 30 in. x 25.4 mm ÷ 1 in. = 762 mm

Step 2: Do not round off the calculation, because even a slight reduction in the original distance could have a negative impact on safety. The answer is 762 mm, and this metric dimension matches the minimum distance of 110.26(A)(1)(a) for a minimum horizontal working space.

SAFETY
90.9(C)(4)

Purpose of Change: To clarify when soft conversions shall be used.

NEC Ch. 1 – Article 100
Part I – Article 100

Type of Change	Panel Action	UL	UL 508	API 500	API 505	OSHA
New Definition	Accept in Principle	467	40.3	-	-	1910.399
ROP		ROC		NFPA 70E	NFPA 70B	NFPA 79	NFPA	NEMA

ROP		ROC		NFPA 70E	NFPA 70B	NFPA 79	NFPA	NEMA
pg. 78	# 1-63	pg. 26	# 1-38	410.10(B)	27.1.1.2	3.3.9	-	PB 22
log: 2454	CMP: 1	log: 1107	Submitter: Paul Dobrowsky			2002 NEC: -		IEC: Ch. 21

2005 NEC – Article 100 <u>Bonding Jumper, System</u>

<u>The connection between the grounded circuit conductor and the equipment grounding conductor at a separately derived system.</u>

Author's Substantiation. A new definition has been added to address the different connections used for bonding jumpers for a separately derived system or service.

Bonding Jumper, Main. The connection between the grounded circuit conductor and the equipment grounding conductor at the service.

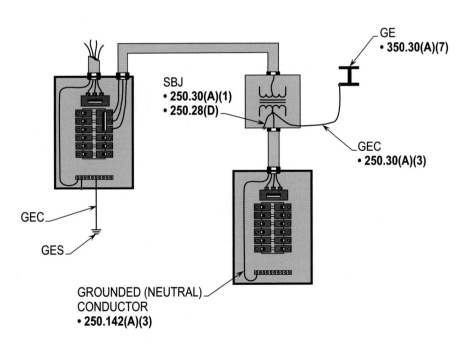

DEFINITIONS - PART I
BONDING JUMPER, SYSTEM
ARTICLE 100

Purpose of Change: To provide a definition for "bonding jumper, system," as used for bonding a separately derived system.

NEC Ch. 1 – Article 100
Part I – Article 100

Type of Change	Panel Action	UL	UL 508	API 500	API 505	OSHA
New Definition	Accept in Principle	489	18.4	-	-	1910.304(e)
ROP		ROC		NFPA 70E	NFPA 70B	NFPA 79	NFPA	NEMA

ROP		ROC		NFPA 70E	NFPA 70B	NFPA 79	NFPA	NEMA
pg. 86	# 1-76	pg. 29	# 1-55	410.9	13.4; 26.3	7.2.10	-	-
log: 1686	CMP: 1	log: 484	Submitter: James T. Dollard, Jr.		2002 NEC: -			IEC: 433.2

2005 NEC – Article 100 <u>Coordination (Selective)</u>

<u>Localization of an overcurrent condition to restrict outages to the circuit or equipment affected, accomplished by the choice of overcurrent-protective devices and their ratings or settings.</u>

Author's Substantiation. The new definition Coordination (Selective) has a global impact on the NEC. The term coordination applies to all overcurrent conditions, not just to fault conditions. By definition, overcurrent covers overload, short circuit, and ground fault.

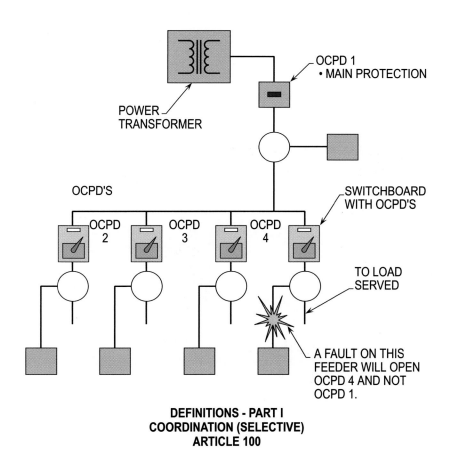

**DEFINITIONS - PART I
COORDINATION (SELECTIVE)
ARTICLE 100**

Purpose of Change: To provide a definition for the term "coordination (selective)" as used in **Article 100**.

NEC Ch. 1 – Article 100
Part I – Article 100

Type of Change		Panel Action		UL	UL 508	API 500	API 505	OSHA
Revision		Accept in Principle		-	-	-	-	-
ROP		ROC		NFPA 70E	NFPA 70B	NFPA 79	NFPA	NEMA
pg. 91	# 1-84	pg. -	# -	-	-	-	70A	-
log: 1398	CMP: 1	log: -	Submitter: Lanny G. McMahill			2002 NEC: Article 100		IEC: -

2002 NEC – Article 100 Dwelling Unit

~~One or more rooms for the use of one more persons as a housekeeping unit with space for eating, living, and sleeping, and permanent provisions for cooking and sanitation.~~

2005 NEC – Article 100 Dwelling Unit

A single unit, providing complete and independent living facilities for one or more persons, including permanent provisions for living, sleeping, cooking, and sanitation.

Author's Substantiation. This revision clarifies that not all dwellings have permanent provisions for cooking. A dwelling can be a dwelling with or without cooking provisions.

DWELLING UNIT WITH
• EATING AREA
• LIVING AREA
• COOKING FACILITIES
• SLEEPING AREA
• PERMANENT PROVISIONS
  FOR SANITATION
  (BATHROOM)

**DEFINITIONS - PART I
DWELLING UNIT
ARTICLE 100**

Purpose of Change: To delete cooking as one of the "permanent provisions for" requirement.

NEC Ch. 1 – Article 100
Part I – Article 100

Type of Change	Panel Action	UL	UL 508	API 500	API 505	OSHA
Revision	Accept in Principle	-	-	-	-	1910.399
ROP		ROC		NFPA 70E	NFPA 70B	NFPA 79	NFPA	NEMA

ROP		ROC		NFPA 70E	NFPA 70B	NFPA 79	NFPA	NEMA
pg. 93	# 1-89	pg. -	# -	Article 100	-	3.3.39	-	-
log: 3419	CMP: 1	log: -	Submitter: Louis A. Barrios, Jr.			2002 NEC: Article 100		IEC: -

2002 NEC – Article 100 Energized

Electrically connected to a source of voltage.

2005 NEC – Article 100 Energized

Electrically connected to, or is, a source of voltage.

Author's Substantiation. This sentence has been changed by adding the words "or is" and clarifies that equipment, such as capacitors, batteries, or conductors with induced voltages, is not literally connected to a source of voltage.

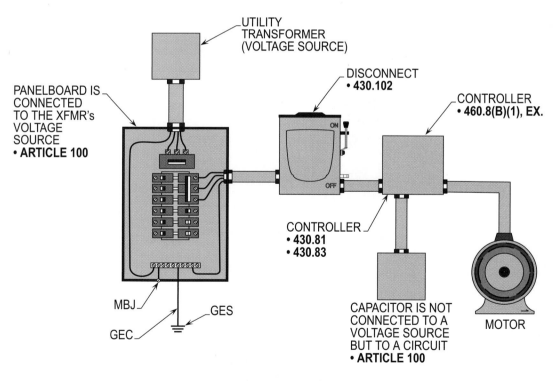

DEFINITIONS - PART I
ENERGIZED
ARTICLE 100

Purpose of Change: To properly define the term "energized" as used in **Article 100**.

NEC Ch. 1 – Article 100
Part I – Article 100

6

Type of Change	Panel Action	UL	UL 508	API 500	API 505	OSHA
New Definition	Accept in Principle	497	40.1	-	-	1910.304(f)(1)(v)(D)(3)
ROP		ROC	NFPA 70E	NFPA 70B	NFPA 79	NFPA	NEMA
pg. 126 # 1-136	pg. - # -		410.10(D)(1)	29.1	Ch. 8	-	-
log: 3345 CMP: 1	log: -	Submitter: Charles J. Palmieri		2002 NEC: -		IEC: 826

2005 NEC – Article 100 <u>Grounded Solidly</u>

<u>Connected to ground without inserting any resistor or impedance device.</u>

Author's Substantiation. This new definition complies with the style manual that definitions be centrally located in Article 100. This term is mentioned approximately 19 times in the Code.

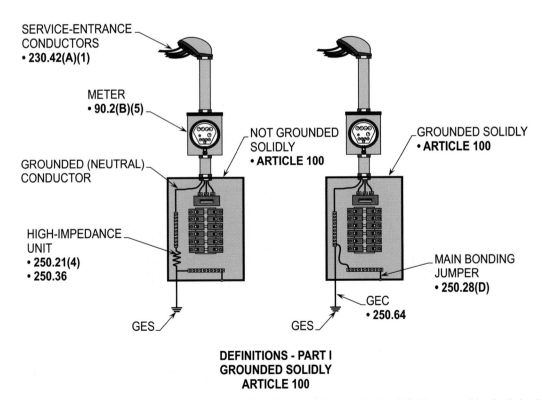

DEFINITIONS - PART I
GROUNDED SOLIDLY
ARTICLE 100

Purpose of Change: To provide a definition for the term "grounded solidly" as used in **Article 100**.

NEC Ch. 1 – Article 100
Part I – Article 100

Type of Change	Panel Action	UL	UL 508	API 500	API 505	OSHA
New Definition	Accept	-	18.4	-	-	1910.304(e)
ROP	**ROC**	**NFPA 70E**	**NFPA 70B**	**NFPA 79**	**NFPA**	**NEMA**
pg. 127 # 1-138	pg. 46 # 1-173	400.6	26.2	3.3.98	-	-
log: 3060 CMP: 1	log: 483 Submitter: Todd F. Lottmann			2002 NEC: -		IEC: Ch. 43

2005 NEC – Article 100 <u>Supplementary Overcurrent Protective Device</u>

<u>A device intended to provide limited overcurrent protection for specific applications and utilization equipment such as luminaires (lighting fixtures) and appliances. This limited protection is in addition to the protection provided in the required branch circuit by the branch circuit overcurrent protective device.</u>

Author's Substantiation. A new definition has been added to clarify the difference between overcurrent protective devices and supplementary overcurrent protective devices. Supplementary overcurrent protective devices are extremely application oriented, and prior to applying the devices, the differences and limitations for these devices must be investigated and found acceptable.

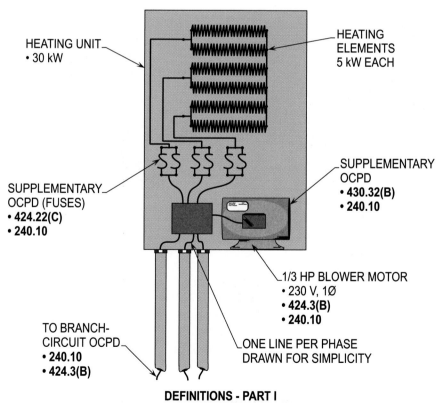

HEATING UNIT
• 30 kW

HEATING ELEMENTS
5 kW EACH

SUPPLEMENTARY OCPD (FUSES)
• **424.22(C)**
• **240.10**

SUPPLEMENTARY OCPD
• **430.32(B)**
• **240.10**

1/3 HP BLOWER MOTOR
• 230 V, 1Ø
• **424.3(B)**
• **240.10**

TO BRANCH-CIRCUIT OCPD
• **240.10**
• **424.3(B)**

ONE LINE PER PHASE DRAWN FOR SIMPLICITY

**DEFINITIONS - PART I
SUPPLEMENTARY OVERCURRENT PROTECTIVE DEVICE
ARTICLE 100**

Purpose of Change: To provide a definition for the term "supplementary overcurrent protective device" as used in **Article 100**.

NEC Ch. 1 – Article 110
Part I – 110.1

Type of Change		Panel Action		UL	UL 508	API 500	API 505	OSHA
Revision		Accept		-	-	-	-	1910.146(h)
ROP		ROC		NFPA 70E	NFPA 70B	NFPA 79	NFPA	NEMA
pg. 130	# 1-144a	pg. -	# -	-	-	-	-	-
log: CP104	CMP: 1	log: -	Submitter: CMP 1			2002 NEC: 110.1		IEC: -

2002 NEC – 110.1 Scope

This article covers general requirements for the examination and approval, installation and use, access to and spaces about electrical conductors and equipment, and tunnel installations.

2005 NEC – 110.1 Scope

This article covers general requirements for the examination and approval, installation and use, access to and spaces about electrical conductors and equipment; <u>enclosures intended for personnel entry</u>; and tunnel installations.

Author's Substantiation. The words "enclosures intended for personnel entry" correlate with the new Part V — Manholes and Other Electric Enclosures Intended for Personnel Entry, All Voltages, in Article 110.

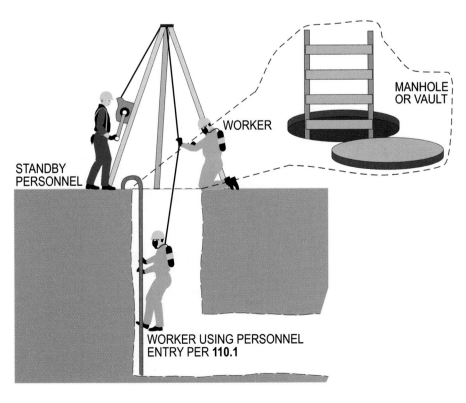

SCOPE
110.1

Purpose of Change: To include the words "enclosures intended for personnel entry" in the scope.

NEC Ch. 1 – Article 110
Part II – Table 110.26(A)(1), Condition 1

Type of Change		Panel Action		UL	UL 508	API 500	API 505	OSHA
Revision		Accept in Principle		-	-	-	-	Table S1
ROP		ROC		NFPA 70E	NFPA 70B	NFPA 79	NFPA	NEMA
pg. 164	# 1-208	pg. 60	# 1-240	T 400.15(A)(1)	-	T 12.5.1.1	-	-
log: 3255	CMP: 1	log: 550	Submitter: Gary J. Locke			2002 NEC: Table 110.26(A)(1), Condition 1		IEC: 134.12

2002 NEC – Table 110.26(A)(1), Condition 1

Exposed live parts on one side and no live or grounded parts on the other side of the working space, or exposed live parts on both sides effectively guarded by suitable wood or other insulating materials. ~~Insulated wire or insulated busbars operating at not over 300 volts to ground shall not be considered live parts.~~

2005 NEC – Table 110.26(A)(1), Condition 1

Exposed live parts on one side <u>of the working space</u> and no live or grounded parts on the other side of the working space, or exposed live parts on both sides <u>of the working space that are</u> effectively guarded by insulating materials.

Author's Substantiation. The last sentence has been deleted. The words "of the working space" have been added for clarification and parallel construction as required by 3.3.5 of the NEC *Style Manual*.

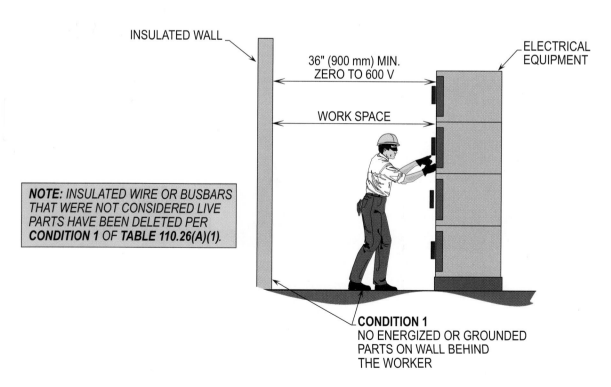

**DEPTH OF WORKING SPACE
TABLE 110.26(A)(1), CONDITION 1**

Purpose of Change: To more adequately clarify the intended enforcement issues of **Condition 1** to Table **110.26(A)(1)**.

NEC Ch. 1 – Article 110
Part II – Table 110.26(A)(1), Condition 3

Type of Change		Panel Action		UL	UL 508	API 500	API 505	OSHA
Revision		Accept in Principle		-	-	-	-	Table S1
ROP		ROC		NFPA 70E	NFPA 70B	NFPA 79	NFPA	NEMA
pg. 164	# 1-208	pg. 60	# 1-240	T 400.15(A)(1)	-	T 12.5.1.1	-	-
log: 3255	CMP: 1	log: 550	Submitter: Gary J. Locke			2002 NEC: Table 110.26(A)(1), Condition 3		IEC: 134.12

2002 NEC – Table 110.26(A)(1), Condition 3

Exposed live parts on both sides of the work space ~~(not guarded as provided in Condition 1) with the operator between.~~

2005 NEC – Table 110.26(A)(1), Condition 3

Exposed live parts on both sides of the <u>working</u> space.

Author's Substantiation. A portion of the sentence has been deleted. This change adds clarity and parallel construction as required by 3.3.5 of the NEC *Style Manual*.

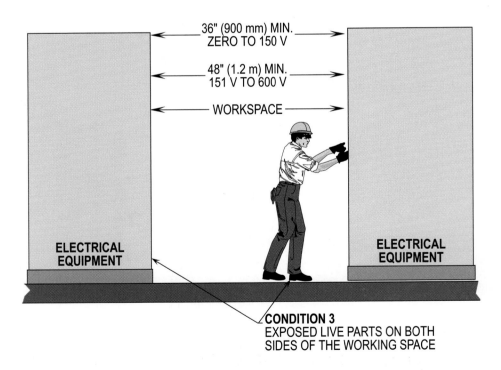

36" (900 mm) MIN.
ZERO TO 150 V

48" (1.2 m) MIN.
151 V TO 600 V

WORKSPACE

ELECTRICAL EQUIPMENT

ELECTRICAL EQUIPMENT

CONDITION 3
EXPOSED LIVE PARTS ON BOTH
SIDES OF THE WORKING SPACE

DEPTH OF WORKING SPACE
110.26(A)(1), CONDITION 3

Purpose of Change: To delete confusing wording in **Condition 3** and make the requirements more user friendly.

NEC Ch. 1 – Article 110
Part III – 110.31

Type of Change		Panel Action		UL	UL 508	API 500	API 505	OSHA
Revision		Accept		-	6.17	-	-	1910.303(g)(2)(i)
ROP		ROC		NFPA 70E	NFPA 70B	NFPA 79	NFPA	NEMA
pg. 174	# 1-225a	pg. -	# -	400.16(A)	B.1.17	14.3	-	-
log: CP103	CMP: 1	log: -	Submitter: CMP 1			2002 NEC: 110.31		IEC: 131.2.1

2002 NEC – 110.31 Enclosure for Electrical Installations

Electrical installations in a vault, room, or closet or in an area surrounded by a wall, screen, or fence, access to which is controlled by ~~lock and key~~ or other approved means, shall be considered to be accessible to qualified persons only. The type of enclosure used in a given case shall be designed and constructed according to the nature and degree of the hazard(s) associated with the installation.

2005 NEC – 110.31 Enclosure for Electrical Installations

Electrical installations in a vault, room, or closet or in an area surrounded by a wall, screen, or fence, access to which is controlled by a lock(s) or other approved means, shall be considered to be accessible to qualified persons only. The type of enclosure used in a given case shall be designed and constructed according to the nature and degree of the hazard(s) associated with the installation.

Author's Substantiation. The words "lock and key" were replaced by "a lock(s)" to clarify that the use of words "lock and key" were not meant to preclude the use of other types of locks.

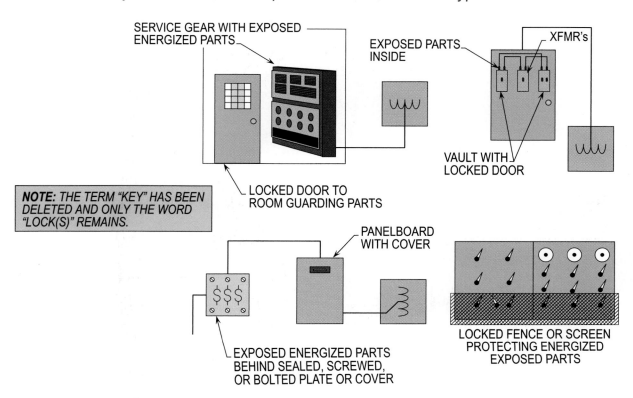

NOTE: THE TERM "KEY" HAS BEEN DELETED AND ONLY THE WORD "LOCK(S)" REMAINS.

SERVICE GEAR WITH EXPOSED ENERGIZED PARTS

LOCKED DOOR TO ROOM GUARDING PARTS

EXPOSED PARTS INSIDE

XFMR's

VAULT WITH LOCKED DOOR

PANELBOARD WITH COVER

EXPOSED ENERGIZED PARTS BEHIND SEALED, SCREWED, OR BOLTED PLATE OR COVER

LOCKED FENCE OR SCREEN PROTECTING ENERGIZED EXPOSED PARTS

ENCLOSURES FOR ELECTRICAL INSTALLATIONS
110.31

Purpose of Change: To delete the requirement for a key while emphasizing the use of a lock(s).

NEC Ch. 1 – Article 110
Part I – 110.34(A)

Type of Change		Panel Action		UL	UL 508	API 500	API 505	OSHA
Revision		Reject		-	-	-	-	-
ROP		ROC		NFPA 70E	NFPA 70B	NFPA 79	NFPA	NEMA
pg. 177	# 1-231	pg. 62	# 1-252	-	-	-	-	-
log: 2528	CMP: 1	log: 3260	Submitter: Alan Manche			2002 NEC: 110.34(A)		IEC: -

2002 NEC – 110.34(A) Working Space

Except as elsewhere required or permitted in this Code, the minimum clear working space in the direction of access to live parts of electrical equipment shall not be less than specified in Table 110.34(A). Distances shall be measured from the live parts, if such are exposed, or from the enclosure from opening if such are enclosed.

2005 NEC – 110.34(A) Working Space

Except as elsewhere required or permitted in this Code, equipment likely to require examination, adjustment, servicing, or maintenance while energized shall have clear working space in the direction of access to live parts of electrical equipment and shall not be less than specified in Table 110.34(A). Distances shall be measured from the live parts, if such are exposed, or from the enclosure front or opening if such are enclosed.

Author's Substantiation. This change correlates with 110.2 and the working space is required where the equipment is likely to be adjusted, maintained, examined or serviced while energized.

TESTING WITH HIGH-VOLTAGE TESTER

EXPOSED LIVE ENERGIZED PARTS

WORKING SPACE
110.34(A)

Purpose of Change: To clarify that electrical equipment with exposed live energized parts shall have appropriate working space provided for safety.

NEC Ch. 1 – Article 110
Part III – 110.36

Type of Change		Panel Action		UL	UL 508	API 500	API 505	OSHA
Revision		Accept		-	21.1	-	-	1910.308
ROP		ROC		NFPA 70E	NFPA 70B	NFPA 79	NFPA	NEMA
pg. 179	# 1-235	pg. -	# -	450.1(B)	33.18	Ch. 13	-	-
log: 96	CMP: 1	log: -	Submitter: James M. Daly			2002 NEC: 110.36		IEC: 133

2002 NEC – 110.36 Circuit Conductors

~~Open~~ runs of insulated wires and cables that have a bare lead sheath or a braided outer covering shall be supported in a manner designed to prevent physical damage to the braid or sheath. Supports for lead-covered cables shall be designed to prevent electrolysis of the sheath.

2005 NEC – 110.36 Circuit Conductors

<u>Exposed</u> runs of insulated wires and cables that have a bare lead sheath or a braided outer covering shall be supported in a manner designed to prevent physical damage to the braid or sheath. Supports for lead-covered cables shall be designed to prevent electrolysis of the sheath.

Author's Substantiation. The word "open" has been replaced with "exposed." This revision will limit the wording "open wiring" to open wiring on insulators, and the word "exposed" to open runs and open wiring not on insulators.

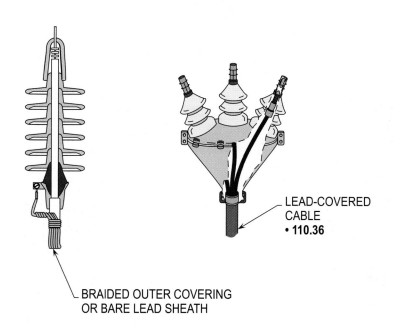

LEAD-COVERED CABLE
• **110.36**

BRAIDED OUTER COVERING OR BARE LEAD SHEATH

CIRCUIT CONDUCTORS
110.36

Purpose of Change: To provide requirements for exposed runs of wires and cables.

NEC Ch. 1 – Article 110
Part IV – 110.52

Type of Change		Panel Action		UL	UL 508	API 500	API 505	OSHA
Revision		Accept		-	16.13; 16.14	-	-	1910.305(j)(4)(i)
ROP		ROC		NFPA 70E	NFPA 70B	NFPA 79	NFPA	NEMA
pg. 179	# 1-236	pg. -	# -	420.10(E)	11.6	7.3	-	-
log: 225	CMP: 1	log: -	Submitter: James M. Daly			2002 NEC: 110.52		IEC: Ch. 43

2002 NEC – 110.52 Overcurrent Protection

Motor-operated equipment shall be protected from overcurrent in accordance with Article 430. Transformers shall be protected from overcurrent in accordance with Article 450.

2005 NEC – 110.52 Overcurrent Protection

Motor-operated equipment shall be protected from overcurrent in accordance with <u>Parts III, IV, and V of</u> Article 430. Transformers shall be protected from overcurrent in accordance with <u>450.3</u>.

Author's Substantiation. The wordind "Parts III, IV and V of" and "450.3" has been added to comply with 4.1.1 of the NEC *Style Manual,* which states "references shall not be made to an entire article unless additional conditions are specified."

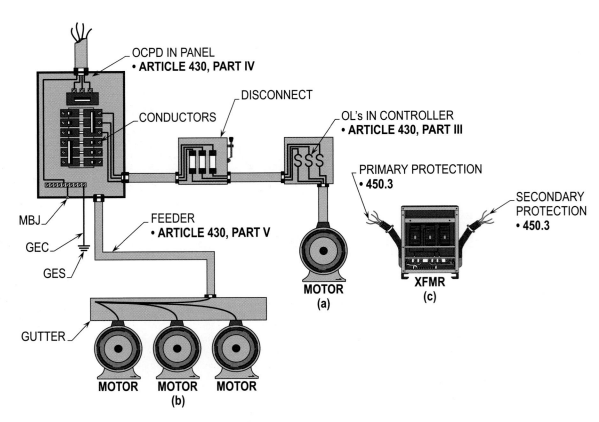

OVERCURRENT PROTECTION
110.52

Purpose of Change: To reference specific parts and sections of the NEC for motor and transformer protection.

NEC Ch. 1 – Article 110
Part IV – 110.59

Type of Change		Panel Action		UL	UL 508	API 500	API 505	OSHA
Revision		Accept		-	50.2	-	-	1910.304(f)(4)(i)
ROP		ROC		NFPA 70E	NFPA 70B	NFPA 79	NFPA	NEMA
pg. 181	# 1-239	pg. -	# -	400.12	B.1.17	6.2.2	-	-
log: 146	CMP: 1	log: -	Submitter: Joseph A. Tedesco			2002 NEC: 110.59		IEC: Ch. 43

2002 NEC – 110.59 Enclosures

Enclosures for use in tunnels shall be dripproof, weatherproof, or submersible as required by the environmental conditions. Switch or contactor enclosures shall not be used as junction boxes or as raceways for conductors feeding through or tapping off to other switches, unless ~~special designs are used to provide adequate space for this purpose~~.

2005 NEC – 110.59 Enclosures

Enclosures for use in tunnels shall be dripproof, weatherproof, or submersible as required by the environmental conditions. Switch or contactor enclosures shall not be used as junction boxes or as raceways for conductors feeding through or tapping off to other switches, unless <u>the enclosures comply with 312.8.</u>

Author's Substantiation. The words "special designs are used to provide adequate space for this purpose" were replaced with "the enclosures comply with 312.8." This revision removes code language that was vague, and complies with the NEC *Style Manual*.

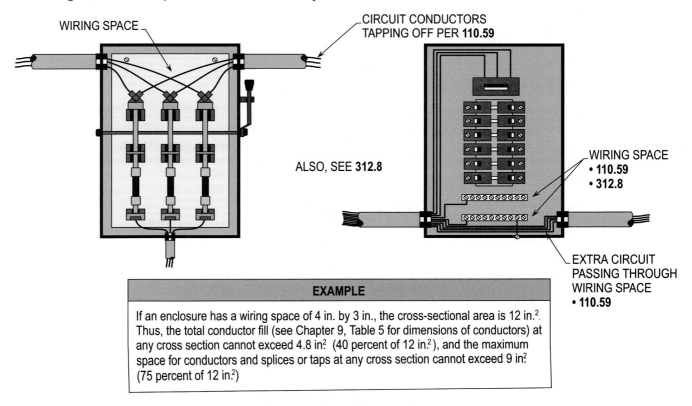

WIRING SPACE

CIRCUIT CONDUCTORS TAPPING OFF PER **110.59**

ALSO, SEE **312.8**

WIRING SPACE
• **110.59**
• **312.8**

EXTRA CIRCUIT PASSING THROUGH WIRING SPACE
• **110.59**

EXAMPLE

If an enclosure has a wiring space of 4 in. by 3 in., the cross-sectional area is 12 in.[2] Thus, the total conductor fill (see Chapter 9, Table 5 for dimensions of conductors) at any cross section cannot exceed 4.8 in[2] (40 percent of 12 in.[2]), and the maximum space for conductors and splices or taps at any cross section cannot exceed 9 in[2] (75 percent of 12 in.[2])

ENCLOSURES
110.59

Purpose of Change: To refer the user of the NEC to **312.8** for calculating wiring space.

Type of Change	Panel Action	UL	UL 508	API 500	API 505	OSHA
New Part	Accept in Principle	-	-	-	-	1910.956(b)
ROP	**ROC**	**NFPA 70E**	**NFPA 70B**	**NFPA 79**	**NFPA**	**NEMA**
pg. 182 # 1-240	pg. - # -	400.18(F)	10.2.2	-	-	-
log: 1048 CMP: 1	log: - Submitter: TCC			2002 NEC: 314.50		IEC: 481.2

Part IV Manholes and Other Electric Enclosures Intended for Personnel Entry
2002 NEC – ~~314.50~~ General

Electric enclosures intended for personnel entry and specifically fabricated for this purpose shall be of sufficient size to provide safe work space about electric equipment with live parts that is likely to require examination, adjustment, servicing, or maintenance while energized. ~~They~~ shall have sufficient size to permit ready installation or withdrawal of the conductors employed without damage to the conductors or to their insulation. They shall comply with the provisions of this part.

Exception: Where electric enclosures covered by Part IV of this article are part of an industrial wiring system operating under conditions of maintenance and supervision that ensure only qualified persons monitor and supervise the system, they shall be permitted to be designed and installed in accordance with appropriate engineering practice. If required by the authority having jurisdiction, design documentation shall be provided.

Part V Manholes and Other Electric Enclosures Intended for Personnel Entry, All Voltages
2005 NEC – 110.70 General

Electric enclosures intended for personnel entry and specifically fabricated for this purpose shall be of sufficient size to provide safe work space about electric equipment with live parts that is likely to require examination, adjustment, servicing, or maintenance while energized. Such enclosures shall have sufficient size to permit ready installation or withdrawal of the conductors employed without damage to the conductors or to their insulation. They shall comply with the provisions of this part.

Exception: Where electric enclosures covered by Part V of this article are part of an industrial wiring system operating under conditions of maintenance and supervision that ensure that only qualified persons monitor and supervise the system, they shall be permitted to be designed and installed in accordance with appropriate engineering practice. If required by the authority having jurisdiction, design documentation shall be provided.

Author's Substantiation. Part IV in Article 314 has been moved and renumbered as Part V in Article 110. This change is appropriate as the working clearance and safety requirements of electrical manholes and related fire resistivity will be contained with other relative information in Article 110. The title has been changed to be applicable to all voltages that will distinguish this new part in its location within Article 110 from requirements applicable to only 600 volts or greater.

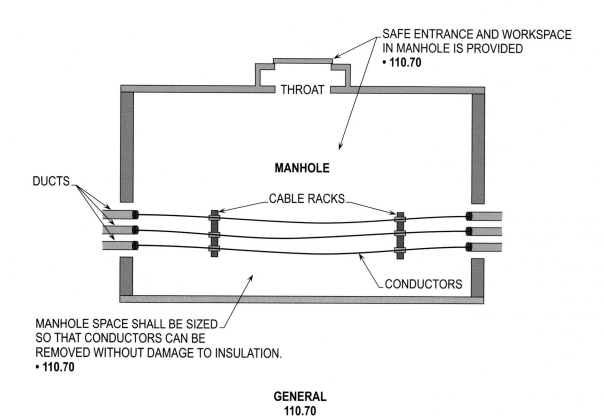

GENERAL
110.70

Purpose of Change: To relocate and renumber these sections in an effort to make the NEC more user friendly.

NEC Ch. 1 – Article 110
Part V – 110.71

Type of Change		Panel Action		UL	UL 508	API 500	API 505	OSHA
New Part		Accept in Principle		-	-	-	-	1910.956(b)
ROP		ROC		NFPA 70E	NFPA 70B	NFPA 79	NFPA	NEMA
pg. 182	# 1-240	pg. -	# -	400.18(F)	10.2.2	-	-	-
log: 1048	CMP: 1	log: -	Submitter: TCC			2002 NEC: 314.51		IEC: 481.2

2002 NEC – ~~314.51~~ Strength

Manholes, vaults, and their means of access shall be designed under qualified engineering supervision and shall withstand all loads likely to be imposed on the structures.

FPN: See ANSI C2-~~1997~~, *National Electrical Safety Code*, for additional information on the loading that can be expected to bear on underground enclosures.

2005 NEC – <u>110.71</u> Strength

Manholes, vaults, and their means of access shall be designed under qualified engineering supervision and shall withstand all loads likely to be imposed on the structures.

FPN: See ANSI C2-<u>2002,</u> *National Electrical Safety Code*, for additional information on the loading that can be expected to bear on underground enclosures.

Author's Substantiation. Part IV in Article 314 has been moved and renumbered as Part V in Article 110. This change is appropriate as the working clearance and safety requirements of electrical manholes and related fire resistivity will be contained with other relative information in Article 110.

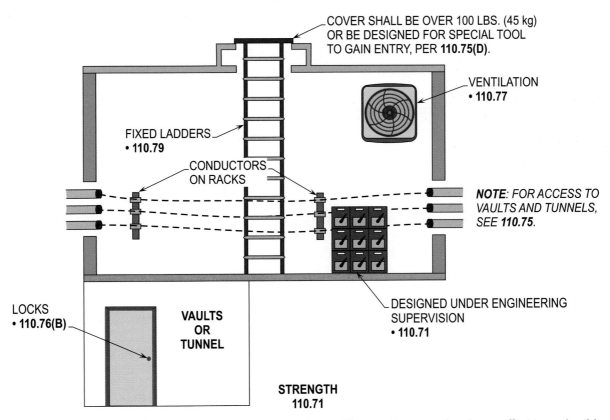

COVER SHALL BE OVER 100 LBS. (45 kg) OR BE DESIGNED FOR SPECIAL TOOL TO GAIN ENTRY, PER **110.75(D)**.

VENTILATION
• **110.77**

FIXED LADDERS
• **110.79**

CONDUCTORS ON RACKS

NOTE: FOR ACCESS TO VAULTS AND TUNNELS, SEE 110.75.

DESIGNED UNDER ENGINEERING SUPERVISION
• **110.71**

LOCKS
• **110.76(B)**

VAULTS OR TUNNEL

STRENGTH
110.71

Purpose of Change: To relocate to a more appropriate article and to renumber in an effort to make this part more user friendly.

NEC Ch. 1 – Article 110
Part V – 110.72

Type of Change	Panel Action	UL	UL 508	API 500	API 505	OSHA
New Part	Accept in Principle	-	-	-	-	1910.956(b)

ROP		ROC		NFPA 70E	NFPA 70B	NFPA 79	NFPA	NEMA
pg. 182	# 1-240	pg. -	# -	400.18(F)	10.2.2	-	-	-
log: 1048	CMP: 1	log: -	Submitter: TCC			2002 NEC: 314.52		IEC: 481.2

2002 NEC – 314.52 Cabling Work Space

A clear work space not less than 900 mm (3 ft) wide shall be provided where cables are located on both sides, and not less than 750 mm (2 1/2 ft) where cables are only on one side. The vertical headroom shall not be less than 1.8 m (6 ft) unless the opening is within 300 mm (1 ft), measured horizontally, of the adjacent interior side wall of the enclosure.

Exception: A manhole containing only one or more of the following shall be permitted to have one of the horizontal work space dimensions reduced to 600 mm (2 ft) where the other horizontal clear work space is increased so the sum of the two dimensions is not less than 1.8 m (6 ft):

   (1) Optical fiber cables as covered in Article 770

   (2) Power-limited fire alarm circuits supplied in accordance with 760.41

   (3) Class 2 or Class 3 remote-control and signaling circuits, or both, supplied in accordance with 725.41

2005 NEC – 110.72 Cabling Work Space

A clear work space not less than 900 mm (3 ft) wide shall be provided where cables are located on both sides, and not less than 750 mm (2 1/2 ft) where cables are only on one side. The vertical headroom shall not be less than 1.8 m (6 ft) unless the opening is within 300 mm (1 ft), measured horizontally, of the adjacent interior side wall of the enclosure.

Exception: A manhole containing only one or more of the following shall be permitted to have one of the horizontal work space dimensions reduced to 600 mm (2 ft) where the other horizontal clear work space is increased so the sum of the two dimensions is not less than 1.8 m (6 ft):

   (1) Optical fiber cables as covered in Article 770

   (2) Power-limited fire alarm circuits supplied in accordance with 760.41(A)

   (3) Class 2 or Class 3 remote-control and signaling circuits, or both, supplied in accordance with 725.41

Author's Substantiation. Part IV in Article 314 has been moved and renumbered as Part V in Article 110. This change is appropriate as the working clearance and safety requirements of electrical manholes and related fire resistivity will be contained with other relative information in Article 110.

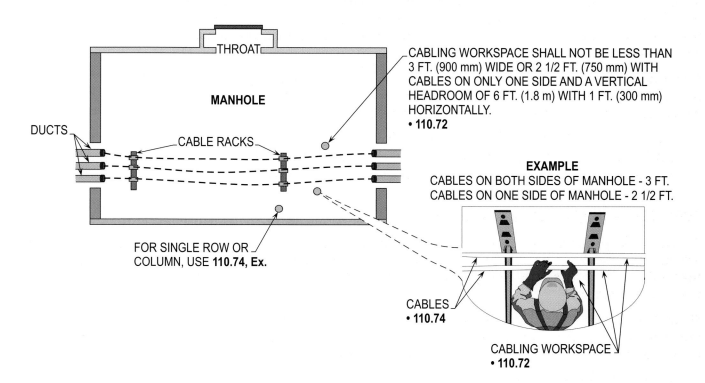

THROAT

MANHOLE

DUCTS

CABLE RACKS

CABLING WORKSPACE SHALL NOT BE LESS THAN
3 FT. (900 mm) WIDE OR 2 1/2 FT. (750 mm) WITH
CABLES ON ONLY ONE SIDE AND A VERTICAL
HEADROOM OF 6 FT. (1.8 m) WITH 1 FT. (300 mm)
HORIZONTALLY.
• **110.72**

EXAMPLE
CABLES ON BOTH SIDES OF MANHOLE - 3 FT.
CABLES ON ONE SIDE OF MANHOLE - 2 1/2 FT.

FOR SINGLE ROW OR
COLUMN, USE **110.74, Ex.**

CABLES
• **110.74**

CABLING WORKSPACE
• **110.72**

CABLING WORK SPACE
110.72

Purpose of Change: To relocate and renumber information to make the NEC more user friendly.

NEC Ch. 1 – Article 110
Part V – 110.73

Type of Change		Panel Action		UL	UL 508	API 500	API 505	OSHA
New Part		Accept in Principle		-	-	-	-	1910.956(b)
ROP		ROC		NFPA 70E	NFPA 70B	NFPA 79	NFPA	NEMA
pg. 182	# 1-240	pg. -	# -	400.18(F)	10.2.2	-	-	-
log: 1048	CMP: 1	log: -	Submitter: TCC			2002 NEC: 314.53		IEC: 481.2

2002 NEC – ~~314.53~~ Equipment Work Space

Where electric equipment with live parts that is likely to require examination, adjustment, servicing, or maintenance while energized is installed in a manhole, vault, or other enclosure designed for personnel access, the work space and associated requirements in 110.26 shall be met for installations operating at 600 volts or less. Where the installation is over 600 volts, the work space and associated requirements in 110.34 shall be met. A manhole access cover that weighs over 45 kg (100 lb) shall be considered as meeting the requirements of 110.34(C).

2005 NEC – <u>110.73</u> Equipment Work Space

Where electric equipment with live parts that is likely to require examination, adjustment, servicing, or maintenance while energized is installed in a manhole, vault, or other enclosure designed for personnel access, the work space and associated requirements in 110.26 shall be met for installations operating at 600 volts or less. Where the installation is over 600 volts, the work space and associated requirements in 110.34 shall be met. A manhole access cover that weighs over 45 kg (100 lb) shall be considered as meeting the requirements of 110.34(C).

Author's Substantiation. Part IV in Article 314 has been moved and renumbered as Part V in Article 110. This change is appropriate as the working clearance and safety requirements of electrical manholes and related fire resistivity will be contained with other relative information in Article 110.

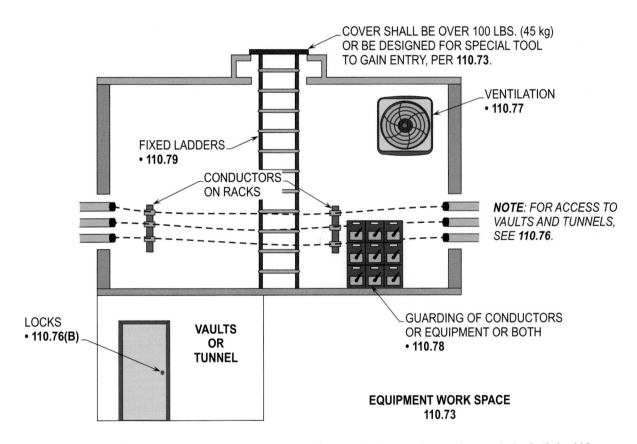

COVER SHALL BE OVER 100 LBS. (45 kg) OR BE DESIGNED FOR SPECIAL TOOL TO GAIN ENTRY, PER **110.73**.

VENTILATION
• **110.77**

FIXED LADDERS
• **110.79**

CONDUCTORS ON RACKS

*NOTE: FOR ACCESS TO VAULTS AND TUNNELS, SEE **110.76**.*

LOCKS
• **110.76(B)**

VAULTS OR TUNNEL

GUARDING OF CONDUCTORS OR EQUIPMENT OR BOTH
• **110.78**

EQUIPMENT WORK SPACE
110.73

Purpose of Change: To provide requirements for manholes, vaults, and tunnels in **Article 110**.

NEC Ch. 1 – Article 110
Part V – 110.74

Type of Change		Panel Action		UL	UL 508	API 500	API 505	OSHA
New Part		Accept in Principle		-	-	-	-	1910.956(b)
ROP		ROC		NFPA 70E	NFPA 70B	NFPA 79	NFPA	NEMA
pg. 182	# 1-240	pg. -	# -	400.18(F)	10.2.2	-	-	-
log: 1048	CMP: 1	log: -	Submitter: TCC			2002 NEC: 314.54		IEC: 481.2

2002 NEC – ~~314.54~~ Bending Space for Conductors

Bending space for conductors operating at 600 volts or below shall be provided in accordance with the requirements of 314.28~~(A)~~. Conductors operating over 600 volts shall be provided with bending space in accordance with 314.71(A) and 314.71(B), as applicable. All conductors shall be cabled, racked up, or arranged in an approved manner that provides ready and safe access for persons to enter for installation and maintenance.

Exception: Where 314.71(B) applies, each row or column of ducts on one wall of the enclosure shall be calculated individually, and the single row or column that provides the maximum distance shall be used.

2005 NEC – <u>110.74</u> Bending Space for Conductors

Bending space for conductors operating at 600 volts or below shall be provided in accordance with the requirements of 314.28. Conductors operating over 600 volts shall be provided with bending space in accordance with 314.71(A) and 314.71(B), as applicable. All conductors shall be cabled, racked up, or arranged in an approved manner that provides ready and safe access for persons to enter for installation and maintenance.

Exception: Where 314.71(B) applies, each row or column of ducts on one wall of the enclosure shall be calculated individually, and the single row or column that provides the maximum distance shall be used.

Author's Substantiation. Part IV in Article 314 has been moved and renumbered as Part V in Article 110. This change is appropriate as the working clearance and safety requirements of electrical manholes and related fire resistivity will be contained with other relative information in Article 110.

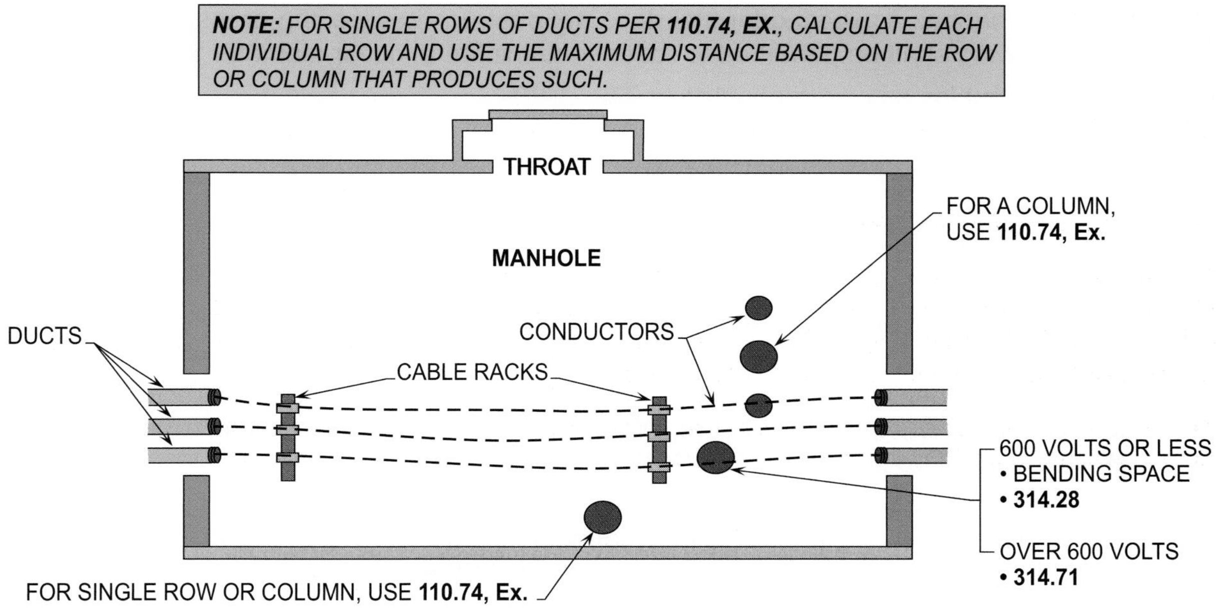

NOTE: *FOR SINGLE ROWS OF DUCTS PER **110.74, EX.**, CALCULATE EACH INDIVIDUAL ROW AND USE THE MAXIMUM DISTANCE BASED ON THE ROW OR COLUMN THAT PRODUCES SUCH.*

THROAT

MANHOLE

DUCTS

CONDUCTORS

CABLE RACKS

FOR A COLUMN,
USE **110.74, Ex.**

600 VOLTS OR LESS
• BENDING SPACE
• **314.28**

OVER 600 VOLTS
• **314.71**

FOR SINGLE ROW OR COLUMN, USE **110.74, Ex.**

BONDING SPACE FOR CONDUCTORS
110.74 AND Ex.

Purpose of Change: To provide rules for manholes in **Article 110**.

Type of Change		Panel Action		UL	UL 508	API 500	API 505	OSHA
New Part		Accept in Principle		-	-	-	-	1910.956(b)
ROP		ROC		NFPA 70E	NFPA 70B	NFPA 79	NFPA	NEMA
pg. 182	# 1-240	pg. -	# -	400.18(F)	10.2.2	-	-	-
log: 1048	CMP: 1	log: -	Submitter: TCC			2002 NEC: 314.55		IEC: 481.2

2002 NEC – 314.55 Access to Manholes

(A) Dimensions. Rectangular access openings shall not be less than 650 mm x 550 mm (26 in. x 22 in.). Round access openings in a manhole shall not be less than 650 mm (26 in.) in diameter.

Exception: A manhole that has a fixed ladder that does not obstruct the opening or that contains only one or more of the following shall be permitted to reduce the minimum cover diameter to 600 mm (2 ft):

(1) Optical fiber cables as covered in Article 770.

(2) Power-limited fire alarm circuits supplied in accordance with 760.41.

(3) Class 2 or Class 3 remote-control and signaling circuits, or both, supplied in accordance with 725.41.

(B) Obstructions. Manhole openings shall be free of protrusions that could injure personnel or prevent ready egress.

(C) Location. Manhole openings for personnel shall be located where they are not directly above electric equipment or conductors in the enclosure. Where this is not practicable, either a protective barrier or a fixed ladder shall be provided.

(D) Covers. Covers shall be over 45 kg (100 lb) or otherwise designed to require the use of tool to open. They shall be designed or restrained so they cannot fall into the manhole or protrude sufficiently to contact electrical conductors or equipment within the manhole.

(E) Marking. Manhole covers shall have an identifying mark or logo that prominently indicates their function, such as electric.

2005 NEC – 110.75 Access to Manholes

(A) Dimensions. Rectangular access openings shall not be less than 650 mm x 550 mm (26 in. x 22 in.). Round access openings in a manhole shall not be less than 650 mm (26 in.) in diameter.

Exception: A manhole that has a fixed ladder that does not obstruct the opening or that contains only one or more of the following shall be permitted to reduce the minimum cover diameter to 600 mm (2 ft):

(1) Optical fiber cables as covered in Article 770

(2) Power-limited fire alarm circuits supplied in accordance with 760.41

(3) Class 2 or Class 3 remote-control and signaling circuits, or both, supplied in accordance with 725.41

(B) Obstructions. Manhole openings shall be free of protrusions that could injure personnel or prevent ready egress.

(C) Location. Manhole openings for personnel shall be located where they are not directly above electric equipment or conductors in the enclosure. Where this is not practicable, either a protective barrier or a fixed ladder shall be provided.

(D) Covers. Covers shall be over 45 kg (100 lb) or otherwise designed to require the use of tools to open. They shall be designed or restrained so they cannot fall into the manhole or protrude sufficiently to contact electrical conductors or equipment within the manhole.

(E) Marking. Manhole covers shall have an identifying mark or logo that prominently indicates their function, such as electric.

Author's Substantiation. Part IV in Article 314 has been moved and renumbered as Part V in Article 110. This change is appropriate as the working clearance and safety requirements of electrical manholes and related fire resistivity will be contained with other relative information in Article 110.

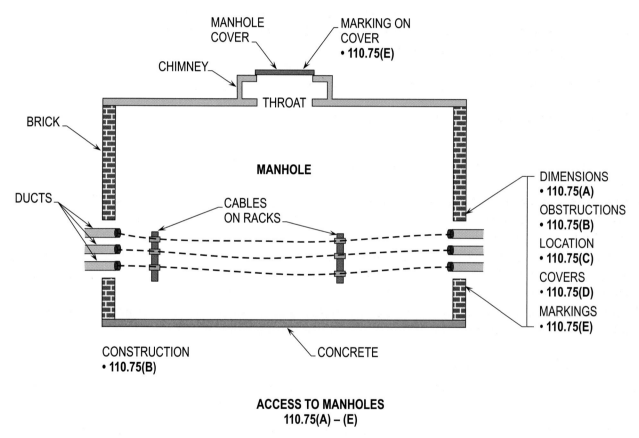

ACCESS TO MANHOLES
110.75(A) – (E)

Purpose of Change: To provide rules for manholes in **Article 110**.

NEC Ch. 1 – Article 110
Part V – 110.76

Type of Change		Panel Action		UL	UL 508	API 500	API 505	OSHA
New Part		Accept in Principle		-	-	-	-	1910.956(b)
ROP		ROC		NFPA 70E	NFPA 70B	NFPA 79	NFPA	NEMA
pg. 182	# 1-240	pg. -	# -	400.18(F)	10.2.2	-	-	-
log: 1048	CMP: 1	log: -	Submitter: TCC			2002 NEC: 314.56		IEC: 481.2

2002 NEC – ~~314.56~~ Access to Vaults and Tunnels

(A) Location. Access openings for personnel shall be located where they are not directly above electric equipment or conductors in the enclosure. Other openings shall be permitted over equipment to facilitate installation, maintenance, or replacement of equipment.

(B) Locks. In addition to compliance with the requirements of 110.34~~(C)~~, if applicable, access openings for personnel shall be arranged ~~so~~ that a person on the inside can exit when the access door is locked from the outside, or in the case of normally locking by padlock, the locking arrangement shall be such that the padlock can be closed on the locking system to prevent locking from the outside.

2005 NEC – <u>110.76</u> Access to Vaults and Tunnels

(A) Location. Access openings for personnel shall be located where they are not directly above electric equipment or conductors in the enclosure. Other openings shall be permitted over equipment to facilitate installation, maintenance, or replacement of equipment.

(B) Locks. In addition to compliance with the requirements of 110.34, if applicable, access openings for personnel shall be arranged <u>such</u> that a person on the inside can exit when the access door is locked from the outside, or in the case of normally locking by padlock, the locking arrangement shall be such that the padlock can be closed on the locking system to prevent locking from the outside.

Author's Substantiation. Part IV in Article 314 has been moved and renumbered as Part V in Article 110. This change is appropriate as the working clearance and safety requirements of electrical manholes and related fire resistivity will be contained with other relative information in Article 110.

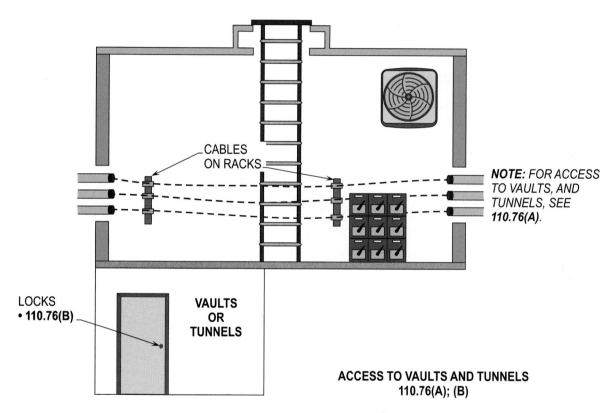

CABLES
ON RACKS

NOTE: *FOR ACCESS
TO VAULTS, AND
TUNNELS, SEE
110.76(A).*

LOCKS
• **110.76(B)**

**VAULTS
OR
TUNNELS**

**ACCESS TO VAULTS AND TUNNELS
110.76(A); (B)**

Purpose of Change: To provide requirements for manholes, vaults, and tunnels in **Article 110**.

NEC Ch. 1 – Article 110
Part V – 110.77

Type of Change		Panel Action		UL	UL 508	API 500	API 505	OSHA
New Part		Accept in Principle		-	-	-	-	1910.956(b)
ROP		ROC		NFPA 70E	NFPA 70B	NFPA 79	NFPA	NEMA
pg. 182	# 1-240	pg. -	# -	400.18(F)	10.2.2	-	-	-
log: 1048	CMP: 1	log: -	Submitter: TCC			2002 NEC: 314.57		IEC: 481.2

2002 NEC – ~~314.57~~ Ventilation

Where manholes, tunnels, and vaults have communicating openings into enclosed areas used by the public, ventilation to open air shall be provided wherever applicable.

2005 NEC – <u>110.77</u> Ventilation

Where manholes, tunnels, and vaults have communicating openings into enclosed areas used by the public, ventilation to open air shall be provided wherever applicable.

Author's Substantiation. Part IV in Article 314 has been moved and renumbered as Part V in Article 110. This change is appropriate as the working clearance and safety requirements of electrical manholes and related fire resistivity will be contained with other relative information in Article 110.

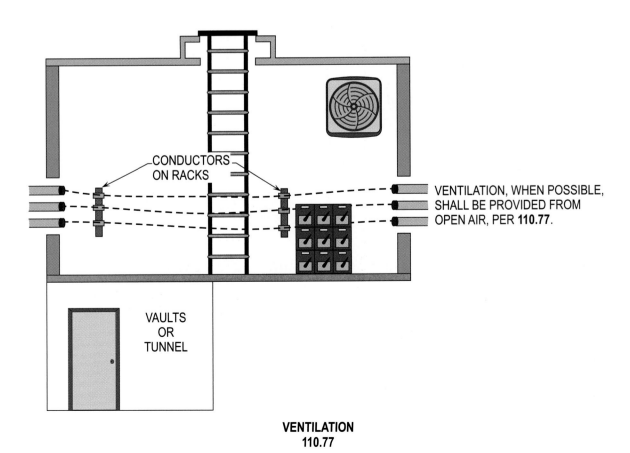

CONDUCTORS ON RACKS

VENTILATION, WHEN POSSIBLE, SHALL BE PROVIDED FROM OPEN AIR, PER **110.77**.

VAULTS OR TUNNEL

VENTILATION
110.77

Purpose of Change: To provide requirements for manholes, vaults, and tunnels, in **Article 110**.

NEC Ch. 1- Article 110
Part V – 110.78

Type of Change		Panel Action		UL	UL 508	API 500	API 505	OSHA
New Part		Accept in Principle		-	-	-	-	1910.956(b)
ROP		ROC		NFPA 70E	NFPA 70B	NFPA 79	NFPA	NEMA
pg. 182	# 1-240	pg. -	# -	400.18(F)	10.2.2	-	-	-
log: 1048	CMP: 1	log: -	Submitter: TCC			2002 NEC: 314.58		IEC: 481.2

2002 NEC – ~~314.58~~ Guarding

Where conductors or equipment, or both, could be contacted by objects falling or being pushed through a ventilating grating, both conductors and live parts shall be protected in accordance with the requirements of 110.27(A)(2) or 110.31~~(A)~~(1), depending on the voltage.

2005 NEC – 110.78 Guarding

Where conductors or equipment, or both, could be contacted by objects falling or being pushed through a ventilating grating, both conductors and live parts shall be protected in accordance with the requirements of 110.27(A)(2) or 110.31(B)(1), depending on the voltage.

Author's Substantiation. Part IV in Article 314 has been moved and renumbered as Part V in Article 110. This change is appropriate as the working clearance and safety requirements of electrical manholes and related fire resistivity will be contained with other relative information in Article 110.

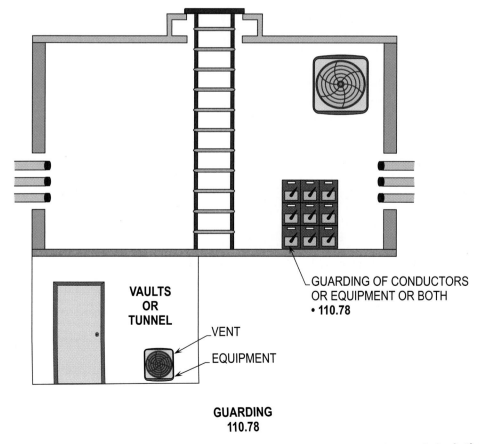

GUARDING OF CONDUCTORS
OR EQUIPMENT OR BOTH
• **110.78**

VAULTS
OR
TUNNEL

VENT

EQUIPMENT

**GUARDING
110.78**

Purpose of Change: To provide requirements for manholes, vaults, and tunnels in **Article 110**.

NEC Ch. 1 – Article 110
Part V – 110.79

Type of Change		Panel Action		UL	UL 508	API 500	API 505	OSHA
New Part		Accept in Principle		-	-	-	-	1910.956(b)
ROP		ROC		NFPA 70E	NFPA 70B	NFPA 79	NFPA	NEMA
pg. 182	# 1-240	pg. -	# -	400.18(F)	10.2.2	-	-	-
log: 1048	CMP: 1	log: -	Submitter: TCC			2002 NEC: 314.59		IEC: 481.2

2002 NEC – ~~314.59~~ Fixed Ladders

Fixed ladders shall be corrosion resistant.

2005 NEC – 110.79 Fixed Ladders

Fixed ladders shall be corrosion resistant.

Author's Substantiation. Part IV in Article 314 has been moved and renumbered as Part V in Article 110. This change is appropriate as the working clearance and safety requirements of electrical manholes and related fire resistivity will be contained with other relative information in Article 110.

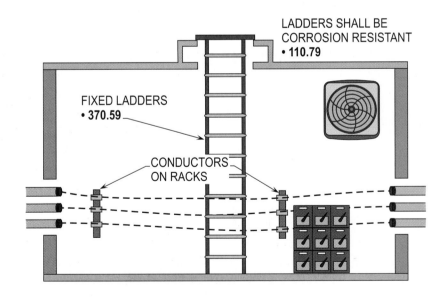

LADDERS SHALL BE
CORROSION RESISTANT
• **110.79**

FIXED LADDERS
• **370.59**

CONDUCTORS
ON RACKS

FIXED LADDERS
110.79

Purpose of Change: To provide requirements for manholes, vaults, and tunnels in **Article 110**.

2

Wiring and Protection

Chapter 2 of the NEC has always been referred to as the Designing Chapter and used by engineers, electrical contractors and electricians who have the responsibility of calculating loads and sizing the elements of the electrical system.

Chapter 2 is the starting point to begin calculating ampacities for branch-circuits and feeder-circuits. Even the ampacity for sizing service-entrance conductors is calculated by applying the rules of **Article 220**, which are found in **Parts I, II** and **III**, as well as **Part IV** of **Article 230**.

The key number for finding the requirements necessary for computing loads in **Chapter 2** is 200. In other words, all Articles and Sections will be identified by using a 200 series number. When designing electrical systems, **Chapter 2** and the 200 series is utilized with other pertinent Articles and Sections. It is nearly impossible for a designer to calculate loads and determine the size of various elements of the electrical system, if not properly acquainted with the calculation requirements of **Chapter 2**.

For example, if the user wanted to calculate the load in amps for a motor feeder-circuit, he or she would refer to **220.14(C)** and this section will reference **430.24** to find the rules, because **Article 220** does not list the rules for sizing loads for motor circuits. If **Article 220** does not contain the rules for calculating the load, it will refer the user to the needed section in other areas of the code.

NEC Ch. 2 – Article 200
Part – 200.6(B)(1) through (B)(3)

10

Type of Change		Panel Action		UL	UL 508	API 500	API 505	OSHA
Revision		Accept		497	40.2	-	-	1910.305(F)
ROP		ROC		NFPA 70E	NFPA 70B	NFPA 79	NFPA	NEMA
pg. 191	# 5-16	pg. 64	# 5-13	410.1(A)	27.7.3.1	14.2.3	-	-
log: 2644	CMP: 5	log: 1009	Submitter: Phil Simmons			2002 NEC: 200.6(B)		IEC: 514.3

2002 NEC – 200.6(B) Sizes Larger Than 6 AWG

An insulated grounded conductor larger than 6 AWG shall be identified either by a continuous white or gray outer finish or by three continuous white stripes on other than green insulation along its entire length or at the time of installation by a distinctive white marking at its terminations. This marking shall encircle the conductors or insulation.

2005 NEC – 200.6(B)(1) through (B)(3) Sizes Larger Than 6 AWG

An insulated grounded conductor larger than 6 AWG shall be identified by one of the following means:

(1) A continuous white or gray outer finish

(2) By three continuous white stripes along its entire length on other than green insulation

(3) At the time of installation, by a distinctive white or gray marking at its terminations. This marking shall encircle the conductors or insulation.

Author's Substantiation. This change is an editorial improvement and clarifies that gray tape or similar markings shall be permitted to be used to identify the grounded conductor(s) at terminations.

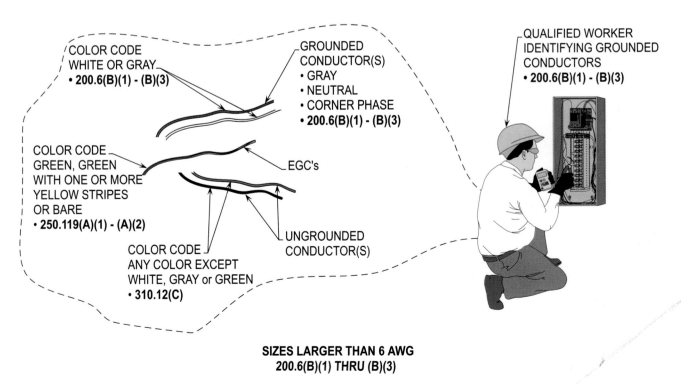

COLOR CODE
WHITE OR GRAY
• 200.6(B)(1) - (B)(3)

GROUNDED
CONDUCTOR(S)
• GRAY
• NEUTRAL
• CORNER PHASE
• 200.6(B)(1) - (B)(3)

QUALIFIED WORKER
IDENTIFYING GROUNDED
CONDUCTORS
• 200.6(B)(1) - (B)(3)

COLOR CODE
GREEN, GREEN
WITH ONE OR MORE
YELLOW STRIPES
OR BARE
• 250.119(A)(1) - (A)(2)

EGC's

COLOR CODE
ANY COLOR EXCEPT
WHITE, GRAY or GREEN
• 310.12(C)

UNGROUNDED
CONDUCTOR(S)

SIZES LARGER THAN 6 AWG
200.6(B)(1) THRU (B)(3)

Purpose of Change: To clarify that gray tape or similar markings shall be permitted to be used to identify the grounded conductor(s).

NEC Ch. 2 – Article 200
Part – 200.6(C)

Type of Change		Panel Action		UL	UL 508	API 500	API 505	OSHA
Revision		Accept in Principle		62	26	-	-	1910.305(f)
ROP		ROC		NFPA 70E	NFPA 70B	NFPA 79	NFPA	NEMA
pg. 192	# 5-18	pg. -	# -	410.6, 420.7	19.5	14.1.5	-	WC-5
log: 579	CMP: 5	log: -	Submitter: Vince Baclawski			2002 NEC: 200.6(C)		IEC: 522.7

2002 NEC – 200.6(C) Flexible Cords

An insulated conductor that is intended for use as a grounded conductor, where contained within a flexible cord, shall be identified by a white or gray outer finish ~~or by three continuous white stripes on other than green insulation~~ or by methods permitted by 400.22.

2005 NEC – 200.6(C) Flexible Cords

An insulated conductor that is intended for use as a grounded conductor, where contained within a flexible cord, shall be identified by a white or gray outer finish or by methods permitted by 400.22.

Author's Substantiation. This change clarifies that the grounded conductor in flexible cords does not need to be marked by three continuous white stripes on other than green insulation since flexible cord is a consumer product and is used by the general public, which may not know the difference between two black conductors, one of which has stripes on it.

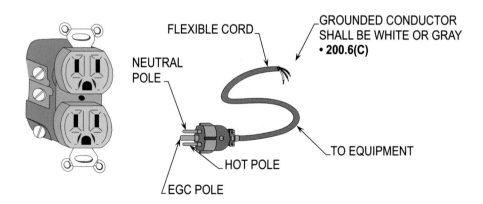

FLEXIBLE CORD
200.6(C)

Purpose of Change: To delete confusing language and make the section requirements more in line with 200.6 and 250.119 of the 2005 NEC.

NEC Ch. 2 – Article 200
Part – 200.6(D)(1) through (D)(3)

11

Type of Change		Panel Action		UL	UL 508	API 500	API 505	OSHA
Revision		Accept in Principle		62	26	-	-	1910.305(f)
ROP		ROC		NFPA 70E	NFPA 70B	NFPA 79	NFPA	NEMA
pg. 193	# 5-20	pg. 64	# 5-14	410.1(A)	20.14	33.47	-	-
log: 2783	CMP: 5	log: 1010	Submitter: Michael J. Callanan			2002 NEC: 200.6(D)		IEC: 514.3

2002 NEC – 200.6(D) Grounded Conductors of Different Systems

Where conductors of different systems are installed in the same raceway, cable, box, auxiliary gutter, or other type of enclosure, one system grounded conductor, if required, shall have an outer covering conforming to 200.6(A) or 200.6(B). Each other system grounded conductor shall have an outer covering of white with a readily distinguishable, different colored stripe other than green running along the insulation, or shall have other and different means of identification as allowed by 200.6(A) or (B) that will distinguish each system grounded conductor.

2005 NEC – 200.6(D)(1) through (D)(3) Grounded Conductors of Different Systems

Where grounded conductors of different systems are installed in the same raceway, cable, box, auxiliary gutter, or other type of enclosure, each grounded conductor shall be identified by system. Identification that distinguishes each system grounded conductor shall be permitted by one of the following means:

(1) One system grounded conductor shall have an outer covering conforming to 200.6(A) or 200.6(B).

(2) The grounded conductor(s) of other systems shall have a different outer covering conforming to 200.6(A) or (B) or by an outer covering of white or gray with a readily distinguishable colored stripe other than green running along the insulation.

(3) Other and different means of identification as allowed by 200.6(A) or (B) that will distinguish each system grounded conductor.

This means of identification shall be permanently posted at each branch-circuit panelboard.

Author's Substantiation. This change is an editorial improvement and clarifies all the recognized means of identification permitted in 200.6(A) and (B) for grounded conductors of different systems.

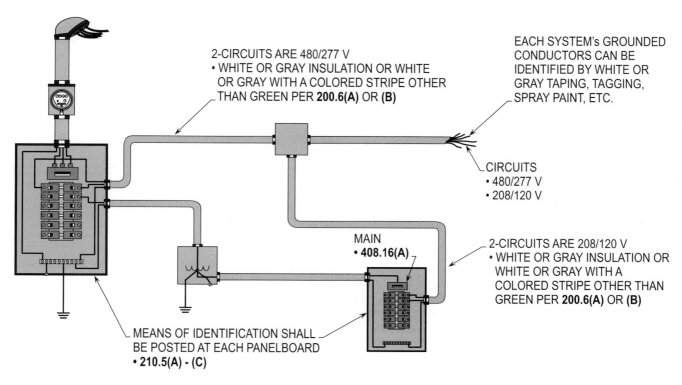

2-CIRCUITS ARE 480/277 V
• WHITE OR GRAY INSULATION OR WHITE OR GRAY WITH A COLORED STRIPE OTHER THAN GREEN PER **200.6(A)** OR **(B)**

EACH SYSTEM's GROUNDED CONDUCTORS CAN BE IDENTIFIED BY WHITE OR GRAY TAPING, TAGGING, SPRAY PAINT, ETC.

CIRCUITS
• 480/277 V
• 208/120 V

MAIN
• **408.16(A)**

2-CIRCUITS ARE 208/120 V
• WHITE OR GRAY INSULATION OR WHITE OR GRAY WITH A COLORED STRIPE OTHER THAN GREEN PER **200.6(A)** OR **(B)**

MEANS OF IDENTIFICATION SHALL BE POSTED AT EACH PANELBOARD
• **210.5(A) - (C)**

GROUNDED CONDUCTORS OF DIFFERENT SYSTEMS
200.6(D)(1) THRU (D)(3)

Purpose of Change: To rearrange the language in a systematic order and make the section more user friendly.

NEC Ch. 2 – Article 200
Part – 200.7(C)(1)

12

Type of Change	Panel Action	UL	UL 508	API 500	API 505	OSHA
New Sentence	Accept in Principle	1581	26	-	-	1910.304(a)
ROP	ROC	NFPA 70E	NFPA 70B	NFPA 79	NFPA	NEMA
pg. 196 # 5-26	pg. - # -	410.1(A)	20.14	33.47	-	WC-51
log: 2338 CMP: 5	log: -	Submitter: Brent Bitterman		2002 NEC: 200.7(C)(1)		IEC: 514.3

2002 NEC – 200.7(C)(1) Circuits of 50 Volts or More

The use of insulation that is white or gray or that has three continuous white stripes for other than a grounded conductor for circuits of 50 volts or more shall be permitted only as in (1) through (3).

   (1) If part of a cable assembly and where the insulation is permanently reidentified to indicate its use as an ungrounded conductor, by painting or other effective means at its termination, and at each location where the conductor is visible and accessible.

2005 NEC – 200.7(C)(1) Circuits of 50 Volts or More

The use of insulation that is white or gray or that has three continuous white stripes for other than a grounded conductor for circuits of 50 volts or more shall be permitted only as in (1) through (3).

   (1) If part of a cable assembly and where the insulation is permanently reidentified to indicate its use as an ungrounded conductor, by painting or other effective means at its termination, and at each location where the conductor is visible and accessible. <u>Identification shall encircle the insulation and shall be a color other than white, gray, or green.</u>

Author's Substantiation. This change clarifies that the use of insulation that is white or gray or that has three continuous white stripes for other than a grounded conductor shall be permitted if the identification of the insulation is reidentified by painting or other effective means at its terminations. This identification shall encircle the insulation and be of a color other than white, gray, or green.

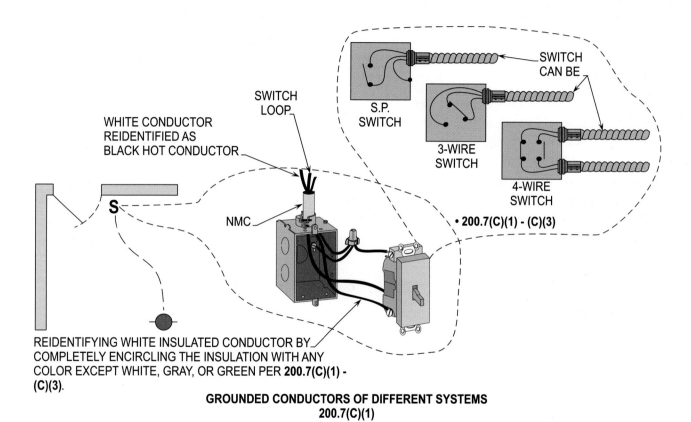

WHITE CONDUCTOR
REIDENTIFIED AS
BLACK HOT CONDUCTOR

SWITCH
LOOP

NMC

S

SWITCH
CAN BE

S.P.
SWITCH

3-WIRE
SWITCH

4-WIRE
SWITCH

• 200.7(C)(1) - (C)(3)

REIDENTIFYING WHITE INSULATED CONDUCTOR BY
COMPLETELY ENCIRCLING THE INSULATION WITH ANY
COLOR EXCEPT WHITE, GRAY, OR GREEN PER **200.7(C)(1) -
(C)(3)**.

**GROUNDED CONDUCTORS OF DIFFERENT SYSTEMS
200.7(C)(1)**

Purpose of Change: To revise this section by requiring the reidentification to completely encircle the conductor's existing insulation.

NEC Ch. 2 – Article 210
Part I – 210.4(B)

 13

Type of Change		Panel Action		UL	UL 508	API 500	API 505	OSHA
Revision		Accept in Principle		-	-	-	-	1910.304(d)(1)(ii)
ROP		ROC		NFPA 70E	NFPA 70B	NFPA 79	NFPA	NEMA
pg. 206	# 2-12	pg. -	# -	410.8(A)	11.5	5.33	-	-
log: 617	CMP: 2	log: -	Submitter: Dan Leaf			2002 NEC: 210.4(B)		IEC: 312.1

2002 NEC – 210.4(B) ~~Dwelling Units~~

~~In dwelling units,~~ a multiwire branch circuit supplying more than one device or equipment on the same yoke shall be provided with a means to disconnect simultaneously all ungrounded conductors at the panelboard where the branch circuit originated.

2005 NEC – 210.4(B) <u>Devices or Equipment</u>

Where a multiwire branch circuit supplies more than one device or equipment on the same yoke, <u>a means shall be provided to disconnect simultaneously all ungrounded conductors supplying those devices or equipment at the point</u> where the branch circuit originates.

Author's Substantiation. This change clarifies that all occupancies shall have a means provided to disconnect simultaneously all ungrounded conductors of multiwire branch circuits supplying devices or equipment at the point where the branch circuit originates.

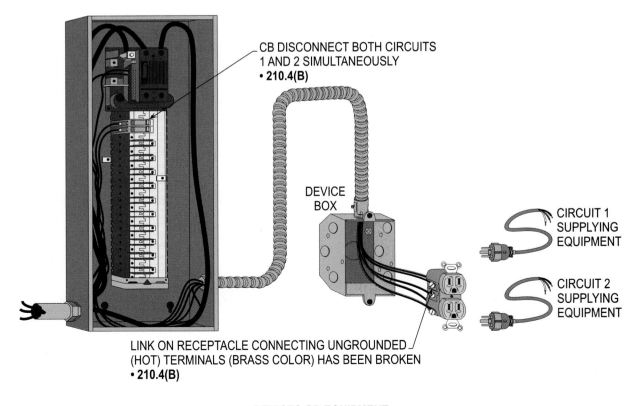

CB DISCONNECT BOTH CIRCUITS 1 AND 2 SIMULTANEOUSLY
• 210.4(B)

DEVICE BOX

CIRCUIT 1 SUPPLYING EQUIPMENT

CIRCUIT 2 SUPPLYING EQUIPMENT

LINK ON RECEPTACLE CONNECTING UNGROUNDED (HOT) TERMINALS (BRASS COLOR) HAS BEEN BROKEN
• 210.4(B)

DEVICES OR EQUIPMENT
210.4(B)

Purpose of Change: To delete the term "dwelling units" so that this requirement applies to all types of occupancies.

NEC Ch. 2 – Article 210
Part I – 210.5(C)

Type of Change		Panel Action		UL	UL 508	API 500	API 505	OSHA
Revision		Accept in Principle		-	-	-	-	1910.304(a)(1)
ROP		ROC		NFPA 70E	NFPA 70B	NFPA 79	NFPA	NEMA
pg. 213	# 2-30	pg. 68	# 2-16	410.2(A)	11.5	5.33	-	-
log: 2788	CMP: 2	log: 1574	Submitter: Michael J. Callanan			2002 NEC: 210.4(D)		IEC: 312.1

2002 NEC – ~~210.4(D) Identification of~~ Ungrounded Conductors

~~Where~~ more than one nominal voltage ~~exists in a building~~, each ungrounded conductor of a multwire branch circuit, where accessible, shall be identified by ~~phase and~~ system. ~~This~~ means of identification shall be permitted to be by separate color coding, marking tape, tagging, or other approved means and shall be permanently posted at each branch-circuit panelboard.

2005 NEC – <u>210.5(C) Ungrounded Conductors</u>

<u>Where the premises wiring system has branch circuits supplied from</u> more than one nominal voltage system, each ungrounded conductor of a branch circuit, where accessible, shall be identified by system. <u>The</u> means of identification shall be permitted to be by separate color coding, marking tape, tagging, or other approved means and shall be permanently posted at each branch-circuit panelboard <u>or similar branch-circuit distribution equipment</u>.

Author's Substantiation. This change clarifies that all ungrounded conductors for branch circuits supplied from more than one nominal system voltage shall be identified. This change will provide a greater degree of safety for occupancies that already have an established means of identification and the installation of new branch-circuits.

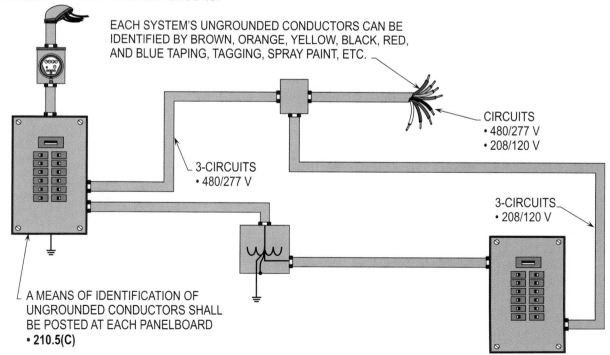

EACH SYSTEM'S UNGROUNDED CONDUCTORS CAN BE IDENTIFIED BY BROWN, ORANGE, YELLOW, BLACK, RED, AND BLUE TAPING, TAGGING, SPRAY PAINT, ETC.

CIRCUITS
• 480/277 V
• 208/120 V

3-CIRCUITS
• 480/277 V

3-CIRCUITS
• 208/120 V

A MEANS OF IDENTIFICATION OF UNGROUNDED CONDUCTORS SHALL BE POSTED AT EACH PANELBOARD
• 210.5(C)

UNGROUNDED CONDUCTORS
210.5(C)

Purpose of Change: To properly provide a means of identifying the ungrounded conductors of different systems.

NEC Ch. 2 – Article 210
Part I – 210.6(A)

15

Type of Change		Panel Action		UL	UL 508	API 500	API 505	OSHA
Revision		Accept		-	-	-	-	-
ROP		ROC		NFPA 70E	NFPA 70B	NFPA 79	NFPA	NEMA
pg. 201	# 2-5	pg. -	# -	-	-	-	-	-
log: 1032	CMP: 2	log: -	Submitter: TCC			2002 NEC: 210.6(A)		IEC: -

2002 NEC – 210.6(A) Occupancy Limitation

In dwelling units and guest rooms of hotels, motels, and similar occupancies, the voltage shall not exceed 120 volts, nominal, between conductors that supply the terminals of the following:

(1) Luminaires (lighting fixtures)

(2) Cord-and-plug-connected loads 1440 volt-amperes, nominal, or less or less than 1/4 hp

2005 NEC – 210.6(A) Occupancy Limitation

In dwelling units and guest rooms <u>or guest suites</u> of hotels, motels, and similar occupancies, the voltage shall not exceed 120 volts, nominal, between conductors that supply the terminals of the following:

(1) Luminaires (lighting fixtures)

(2) Cord-and-plug-connected loads 1440 volt-amperes, nominal, or less or less than 1/4 hp

Author's Substantiation. This change clarifies that occupancy limitations apply to guest suites of hotel, motels, and similar occupancies. The definition of an guest suite per NFPA *101* is an accommodation with two or more contiguous rooms comprising a compartment, with or without doors between such rooms, that provides living, sleeping, sanitary, and storage facilities.

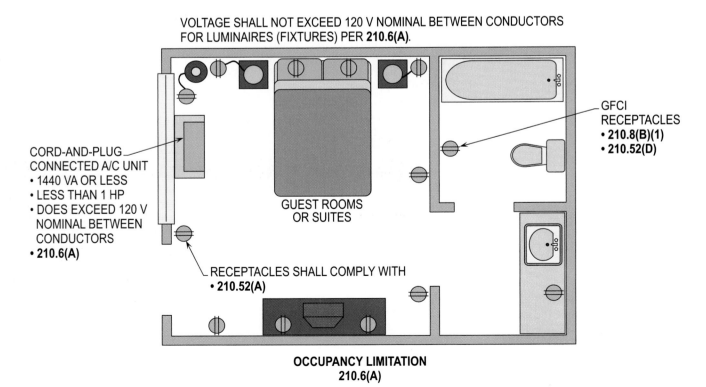

VOLTAGE SHALL NOT EXCEED 120 V NOMINAL BETWEEN CONDUCTORS FOR LUMINAIRES (FIXTURES) PER **210.6(A)**.

GFCI RECEPTACLES
• **210.8(B)(1)**
• **210.52(D)**

CORD-AND-PLUG CONNECTED A/C UNIT
• 1440 VA OR LESS
• LESS THAN 1 HP
• DOES EXCEED 120 V NOMINAL BETWEEN CONDUCTORS
• **210.6(A)**

GUEST ROOMS OR SUITES

RECEPTACLES SHALL COMPLY WITH
• **210.52(A)**

OCCUPANCY LIMITATION
210.6(A)

Purpose of Change: To add guest suites to this section and rearrange the requirements into a user friendly order.

NEC Ch. 2 – Article 210
Part I – 210.6(D)(2)

Type of Change		Panel Action		UL	UL 508	API 500	API 505	OSHA
Revision		Accept in Principle		-	-	-	-	-
ROP		ROC		NFPA 70E	NFPA 70B	NFPA 79	NFPA	NEMA
pg. 214	# 2-32	pg. -	# -	-	-	-	-	-
log: 3095	CMP: 2	log: -		Submitter: Thomas J. Garvey		2002 NEC: 210.6(D)(2)		IEC: -

2002 NEC – 210.6(D)(2) 600 Volts Between Conductors

Circuits exceeding 277 volts, nominal, to ground and not exceeding 600 volts, nominal, between conductors shall be permitted to supply the following:

(2) Cord-and-plug-connected or permanently connected utilization equipment

2005 NEC – 210.6(D)(2) 600 Volts Between Conductors

Circuits exceeding 277 volts, nominal, to ground and not exceeding 600 volts, nominal, between conductors shall be permitted to supply the following:

(2) Cord-and-plug-connected or permanently connected utilization equipment <u>other than luminaires (fixtures)</u>

Author's Substantiation. This change clarifies that luminaires (fixtures) are a type of utilization equipment and shall not be permitted on circuits exceeding 277 volts, nominal, to ground and exceeding 600 volts, nominal, between conductors.

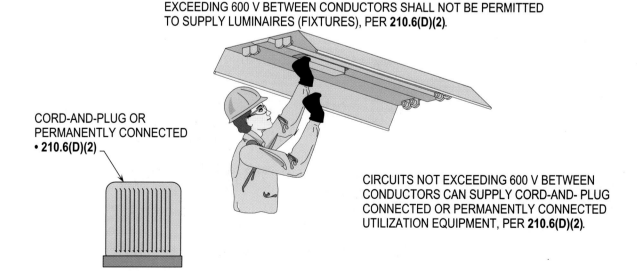

CIRCUITS EXCEEDING 277 VOLTS, NOMINAL TO GROUND AND NOT EXCEEDING 600 V BETWEEN CONDUCTORS SHALL NOT BE PERMITTED TO SUPPLY LUMINAIRES (FIXTURES), PER **210.6(D)(2)**.

CORD-AND-PLUG OR PERMANENTLY CONNECTED
• **210.6(D)(2)**

CIRCUITS NOT EXCEEDING 600 V BETWEEN CONDUCTORS CAN SUPPLY CORD-AND- PLUG CONNECTED OR PERMANENTLY CONNECTED UTILIZATION EQUIPMENT, PER **210.6(D)(2)**.

600 VOLTS BETWEEN CONDUCTORS
210.6(D)(2)

Purpose of Change: To clarify the equipment that can be supplied by circuits not exceeding 600 volts between conductors.

NEC Ch. 2 – Article 210
Part I – 210.7(B)

Type of Change		Panel Action		UL	UL 508	API 500	API 505	OSHA
Revision		Accept in Principle in Part		-	-	-	-	1910.304(b)
ROP		ROC		NFPA 70E	NFPA 70B	NFPA 79	NFPA	NEMA
pg. 216	# 2-35	pg. 69	# 2-20	-	-	-	-	-
log: 1876	CMP: 2	log: 1283	Submitter: Donald A. Ganiere			2002 NEC: 210.7(C)		IEC: 413.1.1.2

2002 NEC – ~~210.7(C)~~ Multiple Branch Circuits

Where more than one branch-circuit supplies ~~more than one receptacle~~ on the same yoke, a means to simultaneously disconnect the ungrounded conductors supplying those ~~receptacles~~ shall be provided at the ~~panelboard where~~ the branch circuits ~~originated~~.

2005 NEC – <u>210.7(B)</u> Multiple Branch Circuits

<u>Where two or more branch circuits supply devices or equipment</u> on the same yoke, a means to simultaneously disconnect the ungrounded conductors supplying those <u>devices</u> shall be provided at the <u>point at which</u> the branch circuits <u>originate</u>.

Author's Substantiation. This change clarifies that where devices on the same yoke are supplied by more than one branch circuit, a means shall be provided to simultaneously disconnect the ungrounded conductors at the point at which the branch circuits originate.

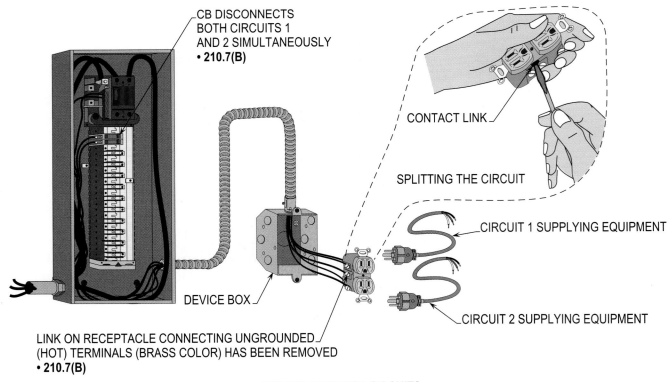

CB DISCONNECTS BOTH CIRCUITS 1 AND 2 SIMULTANEOUSLY
• 210.7(B)

CONTACT LINK

SPLITTING THE CIRCUIT

CIRCUIT 1 SUPPLYING EQUIPMENT

CIRCUIT 2 SUPPLYING EQUIPMENT

DEVICE BOX

LINK ON RECEPTACLE CONNECTING UNGROUNDED (HOT) TERMINALS (BRASS COLOR) HAS BEEN REMOVED
• 210.7(B)

MULTIPLE BRANCH CIRCUITS
210.7(B)

Purpose of Change: To provide requirements for disconnecting multiple circuits on the same yoke.

NEC Ch. 2 – Article 210
Part I – 210.8(A)(3), Ex. to (3)

Type of Change		Panel Action		UL	UL 508	API 500	API 505	OSHA
Revision		Accept in Principle		943	-	-	-	-
ROP		ROC		NFPA 70E	NFPA 70B	NFPA 79	NFPA	NEMA
pg. 220	# 2-43	pg. -	# -	410.4	14.3	-	-	-
log: 235	CMP: 2	log: -	Submitter: Lewis R. Wingert			2002 NEC: 210.8(A)(3), Ex.		IEC: -

2002 NEC – 210.8(A)(3), Ex. to (3) Outdoors

Receptacles that are not readily accessible and are supplied by a dedicated branch circuit for electric snow melting or deicing equipment shall be permitted to be installed in accordance with the applicable provisions of ~~Article 426~~.

2005 NEC – 210.8(A)(3), Ex. to (3) Outdoors

Receptacles that are not readily accessible and are supplied by a dedicated branch circuit for electric snow-melting or deicing equipment shall be permitted to be installed in accordance with 426.28.

Author's Substantiation. This change provides clarity for referencing the section in Article 426 for ground-fault protection of equipment for fixed outdoor electric deicing and snow-melting equipment.

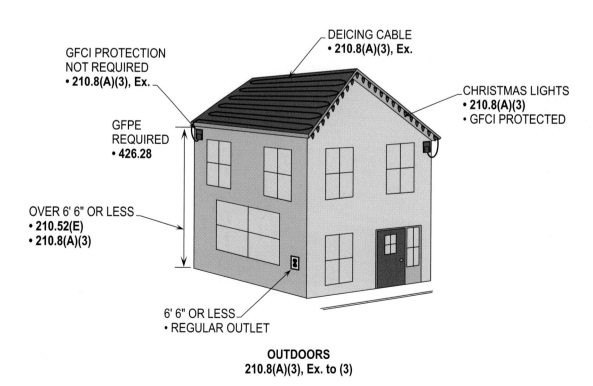

GFCI PROTECTION NOT REQUIRED
• **210.8(A)(3), Ex.**

DEICING CABLE
• **210.8(A)(3), Ex.**

CHRISTMAS LIGHTS
• **210.8(A)(3)**
• **GFCI PROTECTED**

GFPE REQUIRED
• **426.28**

OVER 6' 6" OR LESS
• **210.52(E)**
• **210.8(A)(3)**

6' 6" OR LESS
• REGULAR OUTLET

OUTDOORS
210.8(A)(3), Ex. to (3)

Purpose of Change: To require GFCI protection for all outdoor receptacles, except receptacles used to power snow and deicing cables.

NEC Ch. 2 – Article 210
Part I – 210.8(B)(2)

16

Type of Change		Panel Action		UL	UL 508	API 500	API 505	OSHA
Revision		Accept in Part		943	-	-	-	-
ROP		ROC		NFPA 70E	NFPA 70B	NFPA 79	NFPA	NEMA
pg. 237	# 2-85	pg. 77	# 2-58	410.4	14.3	-	-	280
log: 2761	CMP: 2	log: 757	Submitter: Douglas Hansen			2002 NEC: 210.8(B)(3)		IEC: 412.5

2002 NEC – 210.8(B) Other Than Dwelling Units

All 125-volt, single-phase, 15- and 20-ampere receptacles installed in the locations specified in (1), (2), and (3) shall have ground-fault circuit-interrupter protection for personnel:

(1) Bathrooms

~~(2)~~ Rooftops

~~(3)~~ ~~Kitchens~~

2005 NEC – 210.8(B) Other Than Dwelling Units

All 125-volt, single-phase, 15- and 20-ampere receptacles installed in the locations specified in (1) <u>through (5)</u> shall have ground-fault circuit-interrupter protection for personnel:

(2) <u>Commercial and institutional kitchens — for the purposes of this section, a kitchen is an area with a sink and permanent facilities for food preparation and cooking</u>

Author's Substantiation. This new subsection addresses the need for GFCI protection in commercial and institutional kitchens. Commercial and institutional kitchens are areas with a sink and permanent facilities for food preparation and cooking and not an area that might have a portable cooking appliance.

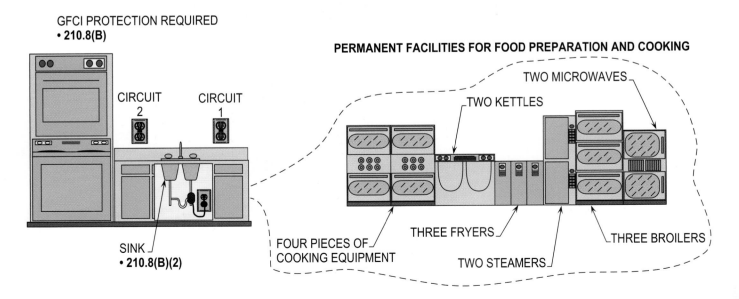

GFCI PROTECTION REQUIRED
• 210.8(B)

PERMANENT FACILITIES FOR FOOD PREPARATION AND COOKING

TWO MICROWAVES

TWO KETTLES

CIRCUIT 2   CIRCUIT 1

SINK
• 210.8(B)(2)

FOUR PIECES OF COOKING EQUIPMENT

THREE FRYERS

TWO STEAMERS

THREE BROILERS

**COMMERCIAL OR INSTITUTIONAL KITCHEN
210.8(B)(2)**

Purpose of Change: To define commercial and institutional kitchens.

NEC Ch. 2 – Article 210
Part I – 210.8(B)(4)

Type of Change		Panel Action		UL	UL 508	API 500	API 505	OSHA
New Item		Accept in Principle in Part		943	-	-	-	-
ROP		ROC		NFPA 70E	NFPA 70B	NFPA 79	NFPA	NEMA
pg. 231	# 2-70	pg. 78	# 2-64	410.4	14.3	-	-	280
log: 2446	CMP: 2	log: 2409	Submitter: William H. King, Jr.			2002 NEC: -		IEC: 412.5

2002 NEC – 210.8(B) Other Than Dwelling Units

All 125-volt, single-phase, 15- and 20-ampere receptacles installed in the locations specified in (1), (2), and (3) shall have ground-fault circuit-interrupter protection for personnel:

(1) Bathrooms

~~(2)~~ Rooftops

~~(3) Kitchens~~

2005 NEC – 210.8(B) Other Than Dwelling Units

All 125-volt, single-phase, 15- and 20-ampere receptacles installed in the locations specified in (1) <u>through (5)</u> shall have ground-fault circuit-interrupter protection for personnel:

(4) <u>Outdoors in public spaces — for the purpose of this section a public space is defined as any space that is for use by, or is accessible to, the public</u>

Author's Substantiation. This new subsection addresses the need for GFCI protection for any space that is for use by, or is accessible to, the public.

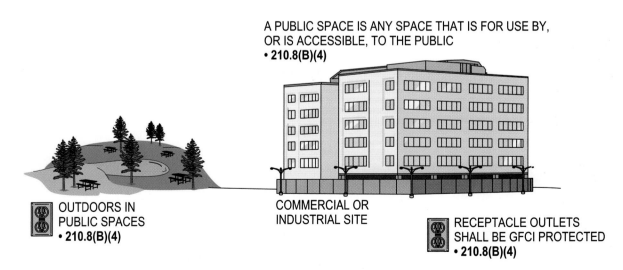

A PUBLIC SPACE IS ANY SPACE THAT IS FOR USE BY, OR IS ACCESSIBLE, TO THE PUBLIC
• 210.8(B)(4)

OUTDOORS IN PUBLIC SPACES
• 210.8(B)(4)

COMMERCIAL OR INDUSTRIAL SITE

RECEPTACLE OUTLETS SHALL BE GFCI PROTECTED
• 210.8(B)(4)

OTHER THAN DWELLING UNITS
210.8(B)(4)

Purpose of Change: To define a public space.

NEC Ch. 2 – Article 210
Part I – 210.11(A)

18

Type of Change		Panel Action		UL	UL 508	API 500	API 505	OSHA
Revision		Accept		-	-	-	-	-
ROP		ROC		NFPA 70E	NFPA 70B	NFPA 79	NFPA	NEMA
pg. 246	# 2-108	pg. -	# -	-	-	-	-	-
log: 1314	CMP: 2	log: -	Submitter: TCC			2002 NEC: 210.11(A)		IEC: -

2002 NEC – 210.11(A) Number of Branch Circuits

The minimum number of branch circuits shall be determined from the total ~~computed~~ load and the size or rating of the circuits used. In all installations, the number of circuits shall be sufficient to supply the load served. In no case shall the load on any circuit exceed the maximum specified in ~~220.4~~.

2005 NEC – 210.11(A) Number of Branch Circuits

The minimum number of branch circuits shall be determined from the total <u>calculated</u> load and the size or rating of the circuits used. In all installations, the number of circuits shall be sufficient to supply the load served. In no case shall the load on any circuit exceed the maximum specified in <u>220.18</u>.

Author's Substantiation. This change is the result of an effort by the NEC Usability Task Group to standardize the language throughout the NEC relative to the use of the terms "computed" and "calculated".

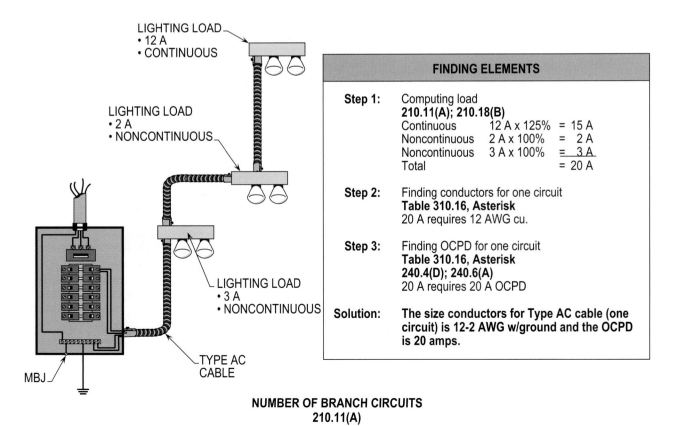

NUMBER OF BRANCH CIRCUITS
210.11(A)

Purpose of Change: To change the term "computed" to the term "calculated" and Section "220.4" to "220.18".

NEC Ch. 2 – Article 210
Part I – 210.11(C)(3), Ex.

Type of Change		Panel Action		UL	UL 508	API 500	API 505	OSHA
Revision		Accept in Principle		-	-	-	-	-
ROP		ROC		NFPA 70E	NFPA 70B	NFPA 79	NFPA	NEMA
pg. 249	# 2-115	pg. -	# -	-	-	-	-	-
log: 3330	CMP: 2	log: -	Submitter: Charles M. Trout			2002 NEC: 210.11(C)(3), Ex.		IEC: -

2002 NEC – 210.11(C)(3), Ex. Bathroom Branch Circuits

Where the 20-ampere circuit supplies a single bathroom, outlets for other equipment within the same bathroom shall be permitted to be supplied in accordance with 210.23(A).

2005 NEC – 210.11(C)(3), Ex. Bathroom Branch Circuits

Where the 20-ampere circuit supplies a single bathroom, outlets for other equipment within the same bathroom shall be permitted to be supplied in accordance with 210.23(A)(1) and (A)(2).

Author's Substantiation. This change clarifies that the exception pertains to cord-and-plug connected equipment not fastened in place and utilization equipment fastened in place within the same bathroom that is supplied by a 20-ampere circuit.

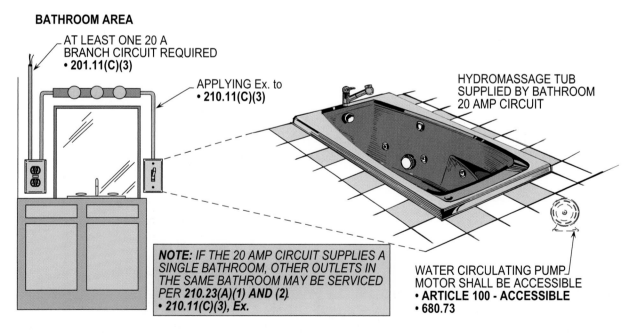

BATHROOM AREA

AT LEAST ONE 20 A BRANCH CIRCUIT REQUIRED
• 201.11(C)(3)

APPLYING Ex. to
• 210.11(C)(3)

HYDROMASSAGE TUB SUPPLIED BY BATHROOM 20 AMP CIRCUIT

NOTE: IF THE 20 AMP CIRCUIT SUPPLIES A SINGLE BATHROOM, OTHER OUTLETS IN THE SAME BATHROOM MAY BE SERVICED PER 210.23(A)(1) AND (2).
• 210.11(C)(3), Ex.

WATER CIRCULATING PUMP MOTOR SHALL BE ACCESSIBLE
• ARTICLE 100 - ACCESSIBLE
• 680.73

BATHROOM BRANCH CIRCUITS
210.11(C)(3), Ex.

Purpose of Change: To refer user to the appropriate section of the NEC for "other outlets served" rules.

NEC Ch. 2 – Article 210
Part I – 210.12(B)

19

Type of Change		Panel Action		UL	UL 508	API 500	API 505	OSHA
Revision		Accept		-	-	-	-	-
ROP		ROC		NFPA 70E	NFPA 70B	NFPA 79	NFPA	NEMA
pg. 275	# 2-134a	pg. -	# -	-	-	-	-	-
log: CP200	CMP: 2	log: CC200	Submitter: CMP 2			2002 NEC: 210.12(B)		IEC: -

2002 NEC – 210.12(B) Dwelling Unit Bedrooms

All branch circuits that supply ~~125~~ volt, single-phase, 15- and 20-ampere outlets installed in dwelling unit bedrooms shall be protected by an arc-fault circuit interrupter ~~listed~~ to provide protection of the ~~entire~~ branch circuit.

2005 NEC – 210.12(B) Dwelling Unit Bedrooms

All 120-volt, single phase, 15- and 20-ampere branch circuits supplying outlets installed in dwelling unit bedrooms shall be protected by a listed arc-fault circuit interrupter, combination type installed to provide protection of the branch circuit.

Branch/feeder AFCIs shall be permitted to be used to meet the requirements of 210.12(B) until January 1, 2008.

FPN: For information on type of arc-fault circuit interrupters, see UL 1699-1999, *Standard for Arc-Fault Circuit Interrupters.*

Author's Substantiation. This revision clarifies that a listed arc-fault circuit interrupter combination type shall protect all 120 volt, single-phase, 15 and 20 ampere branch-circuits supplying outlets installed in dwelling unit bedrooms.

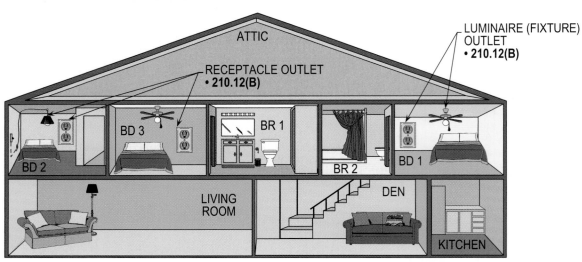

NOTE 1: *ALL 120 V, 1Ø, 15 AND 20 AMP OUTLETS SHALL HAVE AFCI PROTECTION IN BEDROOMS PER 210.12(B).*

NOTE 2: *AFCI COMBINATION TYPE SHALL BE LISTED AND INSTALLED TO PROTECT THE BRANCH-CIRCUIT PER 210.12(B).*

**ARC-FAULT CURCUIT-INTERRUPTER PROTECTION
210.12(B)**

Purpose of Change: To require an AFCI protection scheme of all 125 V, 1Ø, 15 and 20 amp outlets in bedrooms of dwelling units.

NEC Ch. 2 – Article 210
Part I – 210.12(B), Ex.

Type of Change		Panel Action		UL	UL 508	API 500	API 505	OSHA
Revision		Accept		-	-	-	-	-
ROP		ROC		NFPA 70E	NFPA 70B	NFPA 79	NFPA	NEMA
pg. 275	# 2-134a	pg. 84	# 2-87a	-	-	-	-	-
log: CP200	CMP: 2	log: CC200	Submitter: CMP 2			2002 NEC: 210.12(B)		IEC: -

2005 NEC – 210.12(B), Ex. Dwelling Unit Bedrooms

The location of the arc-fault circuit interrupter shall be permitted to be at other than the origination of the branch circuit in compliance with (a) and (b)

(a) The arc-fault circuit interrupter installed within 1.8 m (6 ft) of the branch circuit overcurrent device as measured along the branch circuit conductors.

(b) The circuit conductors between the branch circuit overcurrent device and the arc-fault circuit interrupter shall be installed in a metal raceway or a cable with a metallic sheath.

Author's Substantiation. This revision clarifies the location of the arc-fault circuit interrupter device between the branch-circuit conductors and the OCPD.

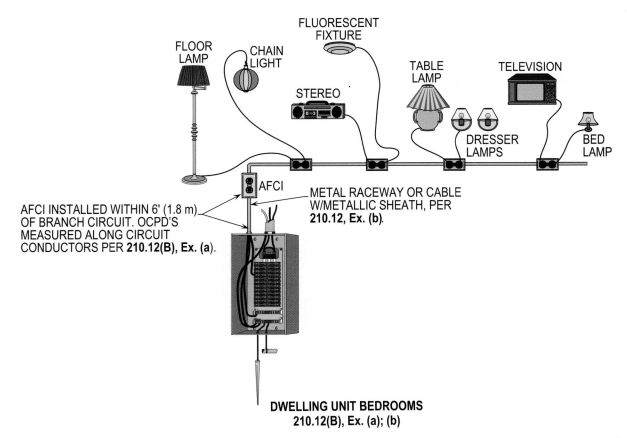

FLUORESCENT FIXTURE
CHAIN LIGHT
FLOOR LAMP
STEREO
TABLE LAMP
TELEVISION
DRESSER LAMPS
BED LAMP

AFCI

AFCI INSTALLED WITHIN 6' (1.8 m) OF BRANCH CIRCUIT. OCPD'S MEASURED ALONG CIRCUIT CONDUCTORS PER **210.12(B), Ex. (a)**.

METAL RACEWAY OR CABLE W/METALLIC SHEATH, PER **210.12, Ex. (b)**.

DWELLING UNIT BEDROOMS
210.12(B), Ex. (a); (b)

Purpose of Change: To appropriately locate AFCI protection device between the branch-circuit conductors and OCPD.

NEC Ch. 2 – Article 210
Part II – 210.19(A)(3), Ex. 1

Type of Change			Panel Action		UL	UL 508	API 500	API 505	OSHA
New Sentence			Accept		-	-	-	-	-
ROP			ROC		NFPA 70E	NFPA 70B	NFPA 79	NFPA	NEMA
pg. 287	# 2-183	pg. 93	# 2-118		-	-	-	-	-
log: 1229	CMP: 2	log: 2205	Submitter: David H. Kendall			2002 NEC: 210.19(A)(3), Ex. 1			IEC: -

2002 NEC – 210.19(A)(3), Ex. 1 Household Ranges and Cooking Appliances

Tap conductors supplying electric ranges, wall-mounted electric ovens, and counter-mounted electric cooking units from a 50-ampere branch circuit shall have an ampacity of not less than 20 and shall be sufficient for the load to be served. The taps shall not be longer than necessary for servicing the appliance.

2005 NEC – 210.19(A)(3), Ex. 1 Household Ranges and Cooking Appliances

Tap conductors supplying electric ranges, wall-mounted electric ovens, and counter-mounted electric cooking units from a 50-ampere branch circuit shall have an ampacity of not less than 20 and shall be sufficient for the load to be served. <u>These tap conductors include any conductors that are a part of the leads supplied with the appliance that are smaller than the branch circuit conductors.</u> The taps shall not be longer than necessary for servicing the appliance.

Author's Substantiation. This new sentence clarifies that the lead conductors supplied with range or oven are tap conductors and are smaller than the branch circuit conductors.

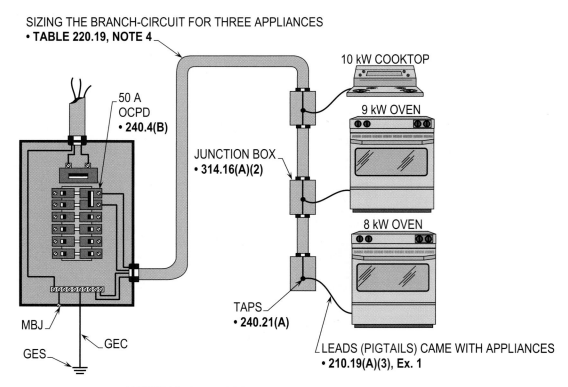

**HOUSEHOLD RANGES AND COOKING APPLIANCES
210.19(A)(3), Ex. 1**

Purpose of Change: To more accurately explain the rules for tap conductors that are smaller than the branch-circuit conductors.

NEC Ch. 2 – Article 210
Part II – 210.23(A)(1)

Type of Change		Panel Action		UL	UL 508	API 500	API 505	OSHA
Revision		Accept in Principle		-	-	-	-	-
ROP		ROC		NFPA 70E	NFPA 70B	NFPA 79	NFPA	NEMA
pg. 290	# 2-190	pg. -	# -	-	-	-	-	-
log: 824	CMP: 2	log: -	Submitter: Dan Leaf			2002 NEC: 210.23(A)(1)		IEC: -

2002 NEC – 210.23(A)(1) Cord-and-Plug-Connected Equipment

The rating of any one cord-and-plug-connected utilization equipment shall not exceed 80 percent of the branch-circuit ampere rating.

2005 NEC – 210.23(A)(1) Cord-and-Plug-Connected Equipment <u>Not Fastened in Place</u>

The rating of any one cord-and-plug-connected utilization equipment <u>not fastened in place</u> shall not exceed 80 percent of the branch-circuit ampere rating.

Author's Substantiation. This change clarifies that only cord-and-plug connected utilization equipment not fastened in place shall not exceed 80 percent of the branch-circuit ampere rating.

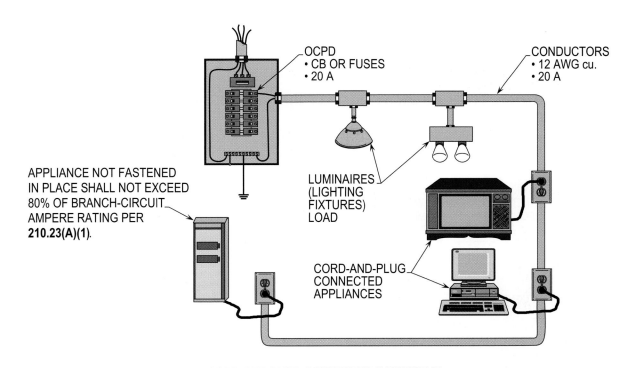

**CORD-AND-PLUG CONNECTED EQUIPMENT
NOT FASTENED IN PLACE
210.23(A)(1)**

Purpose of Change: To clarify that the appliance shall not be fastened in place.

NEC Ch. 2 – Article 210
Part III – 210.52(C)(5)

Type of Change		Panel Action		UL	UL 508	API 500	API 505	OSHA
Revision		Accept in Principle		-	-	-	-	-
ROP		ROC		NFPA 70E	NFPA 70B	NFPA 79	NFPA	NEMA
pg. 301	# 2-211a	pg. 94	# 2-125	-	-	-	-	-
log: CP201	CMP: 2	log: 2211	Submitter: CMP 2			2002 NEC: 210.52(C)(5)		IEC: -

2002 NEC – 210.52(C)(5) Receptacle Outlet Location

Receptacle outlets shall be located above, but not more than 500 mm (20 in.) above, the countertop. Receptacle outlets rendered not readily accessible by appliances fastened in place, appliance garages, or appliances occupying dedicated space shall not be considered as these required outlets.

2005 NEC – 210.52(C)(5) Receptacle Outlet Location

Receptacle outlets shall be located above, but not more than 500 mm (20 in.) above, the countertop. Receptacle outlets rendered not readily accessible by appliances fastened in place, appliance garages, <u>sinks, or rangetops as covered in 210.52(C)(1), Exception,</u> or appliances occupying dedicated space shall not be considered as these required outlets.

2005 NEC – <u>210.52(C)(1), Ex. Wall Counter Spaces</u>

<u>Receptacle outlets shall not be required on a wall directly behind a range or sink in the installation described in Figure 210.52.</u>

Author's Substantiation. This new exception and figure clarifies that receptacle outlets shall not be required on wall counter spaces behind a range or sink.

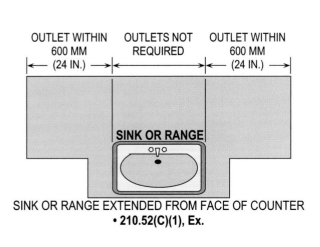

SINK OR RANGE EXTENDED FROM FACE OF COUNTER
• **210.52(C)(1), Ex.**

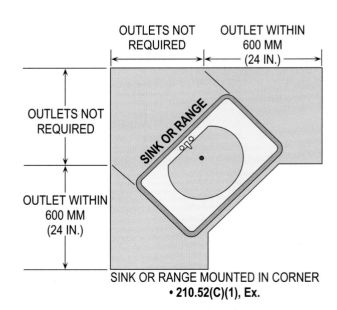

SINK OR RANGE MOUNTED IN CORNER
• **210.52(C)(1), Ex.**

RECEPTACLE OUTLET LOCATION
210.52(C)(5), Ex.

Purpose of Change: To determine the receptacle outlets that are <u>not</u> required in areas behind a sink or range.

NEC Ch. 2 – Article 210
Part III – 210.52(D) and Ex.

Type of Change		Panel Action		UL	UL 508	API 500	API 505	OSHA
New Exception		Accept in Principle		-	-	-	-	-
ROP		ROC		NFPA 70E	NFPA 70B	NFPA 79	NFPA	NEMA
pg. 308	# 2-229	pg. 96	# 2-130	-	-	-	-	-
log: 1831	CMP: 2	log: 3020	Submitter: Davidd Zinck			2002 NEC: -		IEC: -

2002 NEC – 210.52(D) Bathrooms

In dwelling units, at least one ~~wall~~ receptacle outlet shall be installed in bathrooms within 900 mm (3 ft) of the outside edge of each basin. The receptacle outlet shall be located on a wall or partition that is adjacent to the basin or basin countertop.

2005 NEC – 210.52(D) and <u>Ex.</u> Bathrooms

In dwelling units, at least one receptacle outlet shall be installed in bathrooms within 900 mm (3 ft) of the outside edge of each basin. The receptacle outlet shall be located on a wall or partition that is adjacent to the basin or basin countertop.

<u>**Exception.**</u> <u>The receptacle shall not be required to be mounted in the wall or partition where it is installed on the side or face of the basin cabinet not more than 300 mm (12 in.) below the countertop.</u>

Author's Substantiation. A new exception has been added to clarify that one receptacle outlet shall not be required to be mounted in the wall or partition where it is installed on the face of the basin cabinet not more than 300 mm (12 in.) below the countertop.

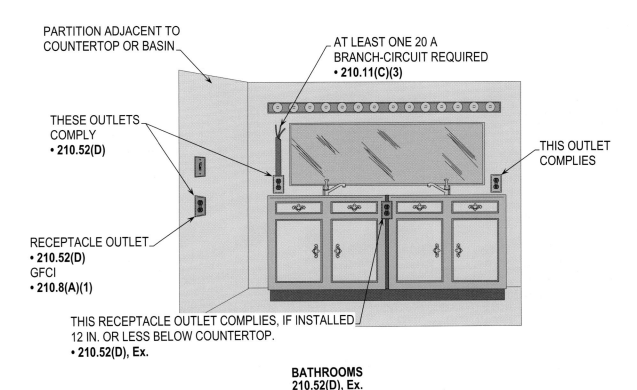

PARTITION ADJACENT TO COUNTERTOP OR BASIN

AT LEAST ONE 20 A BRANCH-CIRCUIT REQUIRED
• 210.11(C)(3)

THESE OUTLETS COMPLY
• 210.52(D)

THIS OUTLET COMPLIES

RECEPTACLE OUTLET
• 210.52(D)
GFCI
• 210.8(A)(1)

THIS RECEPTACLE OUTLET COMPLIES, IF INSTALLED 12 IN. OR LESS BELOW COUNTERTOP.
• 210.52(D), Ex.

BATHROOMS
210.52(D), Ex.

Purpose of Change: To permit a receptacle outlet to be installed on the face of the basin cabinet.

NEC Ch. 2 – Article 210
Part III – 210.60

Type of Change		Panel Action		UL	UL 508	API 500	API 505	OSHA
Revision		Accept		-	-	-	-	-
ROP		ROC		NFPA 70E	NFPA 70B	NFPA 79	NFPA	NEMA
pg. 312	# 2-242	pg. -	# -	-	-	-	-	-
log: 1033	CMP: 2	log: -	Submitter: TCC			2002 NEC: 210.60(A)		IEC: -

2002 NEC – 210.60 Guest Rooms

(A) General. Guest rooms in hotels, motels, and similar occupancies shall have receptacle outlets installed in accordance with 210.52(A) and 210.52(D). Guest rooms ~~meeting the definition of a dwelling unit~~ shall have receptacle outlets installed in accordance with all of the applicable rules in 210.52.

2005 NEC – 210.60 Guest Rooms <u>or Guest Suites</u>

(A) General. Guest rooms <u>or guest suites</u> in hotels, motels, and similar occupancies shall have receptacle outlets installed in accordance with 210.52(A) and 210.52(D). Guest rooms <u>or guest suites provided with permanent provisions for cooking</u> shall have receptacle outlets installed in accordance with all of the applicable rules in 210.52.

Author's Substantiation. This change clarifies that guest rooms or guest suites with permanent provisions for cooking in hotels, motels, and similar occupancies shall have receptacle outlets installed the same as for dwelling units.

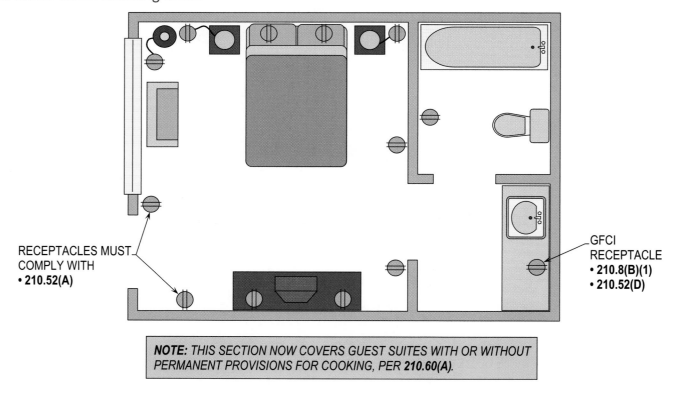

RECEPTACLES MUST COMPLY WITH
• 210.52(A)

GFCI RECEPTACLE
• 210.8(B)(1)
• 210.52(D)

NOTE: *THIS SECTION NOW COVERS GUEST SUITES WITH OR WITHOUT PERMANENT PROVISIONS FOR COOKING, PER 210.60(A).*

GUEST ROOMS OR GUEST SUITES
210.60(A)

Purpose of Change: To provide requirements for locating receptacle outlets in the guest rooms or guest suites of hotels and motels.

NEC Ch. 2 – Article 210
Part III – 210.63, Ex.

Type of Change		Panel Action		UL	UL 508	API 500	API 505	OSHA
New Exception		Accept in Principle		-	-	-	-	-
ROP		ROC		NFPA 70E	NFPA 70B	NFPA 79	NFPA	NEMA
pg. 316	# 2-250	pg. -	# -	-	-	-	-	-
log: 3445	CMP: 2	log: -	Submitter: Joseph Watson			2002 NEC: -		IEC: -

2005 NEC – 210.63, Ex. Heating, Air-Conditioning, and Refrigeration Equipment Outlet

A receptacle outlet shall not be required at one- and two-family dwellings for the service of evaporative coolers.

Author's Substantiation. A new exception has been added to clarify that a receptacle outlet for one- and two-family dwellings shall not be required for the service of evaporative cooler. An evaporative cooler does not have the same service and maintenance requirements as other types of air-conditioning equipment.

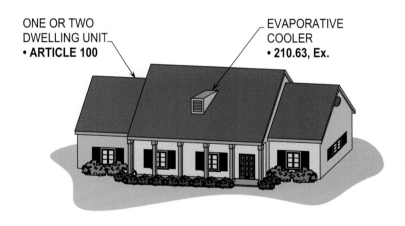

ONE OR TWO
DWELLING UNIT
• ARTICLE 100

EVAPORATIVE
COOLER
• 210.63, Ex.

NOTE: A RECEPTACLE OUTLET IS NOT REQUIRED FOR THE SERVICING OF AN EVAPORATIVE COOLER, PER 210.63, Ex.

HEATING, AIR-CONDITIONING, AND REFRIGERATION EQUIPMENT OUTLET
210.63, Ex.

Purpose of Change: To clarify the requirements for receptacle outlets at the location of evaporative coolers.

NEC Ch. 2 – Article 210
Part III – 210.70(B), Ex. 1 and 2

Type of Change		Panel Action		UL	UL 508	API 500	API 505	OSHA
New Exception		Accept		-	-	-	-	-
ROP		ROC		NFPA 70E	NFPA 70B	NFPA 79	NFPA	NEMA
pg. 321	# 2-264	pg.	# -	-	-	-	-	-
log: 1034	CMP: 2	log: -	Submitter: TCC			2002 NEC: 210.70(B)		IEC: -

2002 NEC – 210.70(B) Guest Rooms

At least one wall switch-controlled lighting outlet ~~or wall switch-controlled receptacle shall be installed in guest rooms in hotels, motels, or similar occupancies~~.

2005 NEC – 210.70(B) Guest Rooms <u>or Guest Suites</u>

<u>In hotels, motels, or similar occupancies, guest rooms or guest suites shall have</u> at least one wall switch-controlled lighting outlet <u>installed in every habitable room and bathroom.</u>

Exception No. 1: <u>In other than bathrooms and kitchens where provided, one or more receptacles controlled by a wall switch shall be permitted in lieu of lighting outlets.</u>

Exception No. 2: <u>Lighting outlets shall be permitted to be controlled by occupancy sensors that are (1) in addition to wall switches or (2) located at a customary wall switch location and equipped with a manual override that will allow the sensor to function as a wall switch.</u>

Author's Substantiation. This change clarifies that every habitable room and bathroom shall be provided with at least one wall switch-controlled lighting outlet. A new Exception 1 permits one or more receptacles controlled by a wall switch in lieu of lighting outlets in other than bathrooms and kitchens. A new Exception 2 permits lighting outlets to be controlled by occupancy sensors that are in addition to wall switches or located at customary wall switch location and equipped with a manual override that will allow the sensor to function as a wall switch.

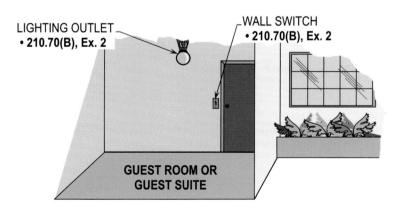

LIGHTING OUTLET
• 210.70(B), Ex. 2

WALL SWITCH
• 210.70(B), Ex. 2

GUEST ROOM OR GUEST SUITE

NOTE: *LIGHTING OUTLET CAN BE CONTROLLED BY OCCUPANCY SENSOR(S) PROVIDED THERE IS A MANUAL OVERRIDE THAT ALLOWS THE SENSOR TO OPERATE AS A WALL SWITCH, PER* **210.70(B), Ex. 2**

GUEST ROOMS OR GUEST SUITES
210.70(B), Ex. 2

Purpose of Change: To clarify the requirements for the operation of lighting outlets.

NEC Ch. 2 – Article 215
Part – 215.2(A)(1)

Type of Change		Panel Action		UL	UL 508	API 500	API 505	OSHA
New Sentence		Accept in Principle		467	40.2	-	-	1910.304(a)(1)
ROP		ROC		NFPA 70E	NFPA 70B	NFPA 79	NFPA	NEMA
pg. 324	# 2-270	pg. 100	# 2-148	410.1	20.14	3.3.48	-	-
log: 1106	CMP: 2	log: 2215	Submitter: Russell LeBlanc			2002 NEC: 215.2(A)(1)		IEC: 514.3

2002 NEC – 215.2(A)(1) General

Additional minimum sizes shall be as specified in (2), (3), and (4) under the conditions stipulated.

2005 NEC – 215.2(A)(1) General

The size of the feeder circuit grounded conductor shall not be smaller than that required by 250.122, except that 250.122(F) shall not apply where grounded conductors are run in parallel. Additional minimum sizes shall be as specified in 215.2(A)(2) and (A)(3) under the conditions stipulated.

Author's Substantiation. This new sentence clarifies that all grounded conductors of a feeder circuit, except those grounded conductors that are run in parallel per 250.122(F), shall not be smaller that that required by 250.122.

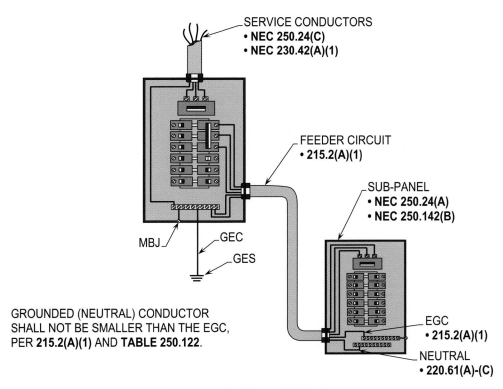

FEEDER CIRCUIT – GENERAL
215.2(A)(1)

Purpose of Change: To clarify that the grounded conductor (neutral) shall be as large as the equipment grounding conductor (EGC).

NEC Ch. 2 – Article 215
Part – 215.2(A)(2)

Type of Change	Panel Action	UL	UL 508	API 500	API 505	OSHA
Subsection Deleted	Accept in Principle in Part	-	-	-	-	-

ROP		ROC		NFPA 70E	NFPA 70B	NFPA 79	NFPA	NEMA
pg. 326	# 2-273	pg. -	# -	-	-		-	-
log: 832	CMP: 2	log: -	Submitter: Dan Leaf			2002 NEC: 215.2(A)(2)		IEC: -

2002 NEC – ~~215.2(A)(2) For Specified Circuits~~

~~The ampacity of feeder conductors shall not be less than 30 amperes where the load supplied consists of any of the following number and types of circuits:~~

~~(1) Two or more 2-wire branch circuits supplied by a 2-wire feeder~~

~~(2) More than two 2-wire branch circuits supplied by a 3-wire feeder~~

~~(3) Two or more 3-wire branch circuits supplied by a 3-wire feeder~~

~~(4) Two or more 4-wire branch circuits supplied by a 3-phase, 4-wire feeder~~

Author's Substantiation. This deletion clarifies that the limitations of the numbers of branch circuits supplied by a feeder are no longer warranted due to the limitations placed on the circuit size by other rules in the NEC.

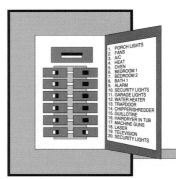

FEEDER CIRCUIT

THE AMPACITY OF FEEDER CONDUCTORS SHALL NOT BE LESS THAN 30 AMPS PER **215.2(A)(2)**.

• TWO OR MORE 2-WIRE BRANCH-CIRCUITS SUPPLIED BY A 2-WIRE FEEDER

• MORE THAN 2-WIRE BRANCH-CIRCUITS SUPPLIED BY A 3-WIRE FEEDER

• TWO OR MORE 3-WIRE BRANCH-CIRCUITS SUPPLIED BY A 3-WIRE FEEDER

• TWO OR MORE 4-WIRE BRANCH-CIRCUITS SUPPLIED A 3-PHASE, 4-WIRE FEEDER.

*NOTE: THIS SECTION HAS BEEN DELETED IN THE **2005 NEC**.*

FOR SPECIFIED CIRCUITS
215.2(A)(3)

Purpose of Change: To clarify that the number of branch-circuits supplied by a feeder is no longer limited by the NEC.

NEC Ch. 2 – Article 215
Part – 215.2(B)

Type of Change		Panel Action		UL	UL 508	API 500	API 505	OSHA
New Sentence		Accept in Principle		-	-	-	-	-
ROP		ROC		NFPA 70E	NFPA 70B	NFPA 79	NFPA	NEMA
pg. 328	# 2-276	pg. 101	# 2-152	-	-	-	-	-
log: 1107	CMP: 2	log: 2217	Submitter: Russell LeBlanc			2002 NEC: 215.2(B)		IEC: -

2002 NEC – 215.2(B) Feeder Over 600 Volts

The ampacity of conductors shall be in accordance with 310.15 and 310.60 as applicable. Feeder conductors over 600 volts shall be sized in accordance with 215.2(B)(1), (2), or (3).

2005 NEC – 215.2(B) Feeder Over 600 Volts

The ampacity of conductors shall be in accordance with 310.15 and 310.60 as applicable. <u>Where installed, the size of the feeder circuit grounded conductor shall not be smaller than that required by 250.122, except that 250.122(F) shall not apply where grounded conductors are run in parallel.</u> Feeder conductors over 600 volts shall be sized in accordance with 215.2(B)(1), (B)(2), or (B)(3).

Author's Substantiation. This new sentence clarifies that all grounded conductors of a feeder circuit, except those grounded conductors that are run in parallel per 250.122(F), shall not be smaller that that required by 250.122.

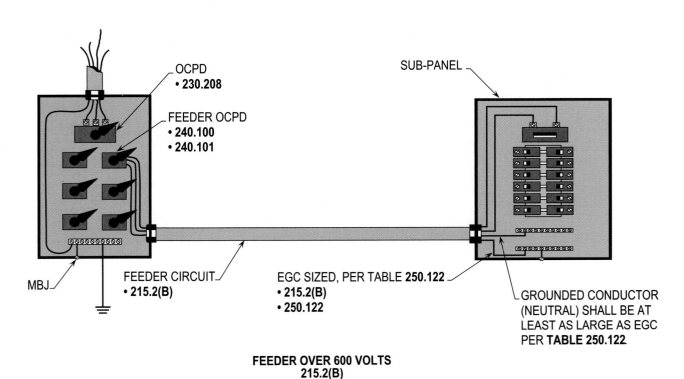

FEEDER OVER 600 VOLTS
215.2(B)

Purpose of Change: To clarify that the grounded conductor shall not be smaller than the EGC.

NEC Ch. 2 – Article 215
Part – 215.8

Type of Change		Panel Action		UL	UL 508	API 500	API 505	OSHA
Section Deleted		Accept in Principle		-	-	-	-	1910.304(f)(1)(iv)(C)
ROP		ROC		NFPA 70E	NFPA 70B	NFPA 79	NFPA	NEMA
pg. 330	# 2-280	pg. -	# -	410.10(A)(4)(3)	-	-	-	-
log: 606	CMP: 2	log: -	Submitter: Dan Leaf			2002 NEC: 215.8		IEC: 312.2

2002 NEC – ~~215.8 Means of Identifying Conductor with the Higher Voltage to Ground~~

~~On a 4-wire, delta-connected secondary where the midpoint of one phase winding is grounded to supply lighting and similar loads, the phase conductor having the higher voltage to ground shall be identified by an outer finish that is orange in color or by tagging or other effective means. Such identification shall be placed at each point where a connection is made if the grounded conductor is also present.~~

Author's Substantiation. This deletion clarifies that the means of identifying a conductor with the higher voltage to ground (high-leg) is covered in 110.15.

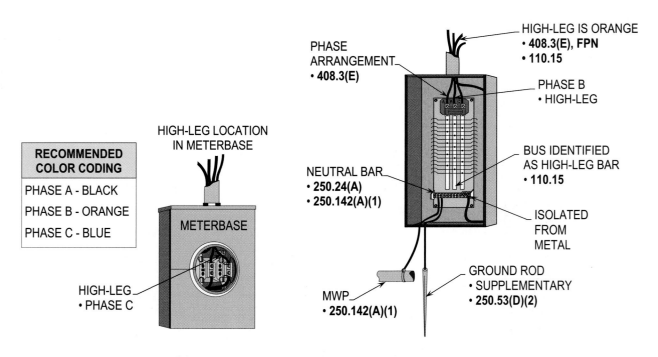

MEANS OF IDENTIFYING CONDUCTOR WITH THE HIGHER VOLTAGE TO GROUND
215.8

Purpose of Change: To delete this section and use the material in 110.15 because it contains the same requirements.

NEC Ch. 2 – Article 215
Part – 215.12

Type of Change		Panel Action		UL	UL 508	API 500	API 505	OSHA
New Section		Accept in Principle		467	40.2	-	-	1910.304(a)(1)
ROP		ROC		NFPA 70E	NFPA 70B	NFPA 79	NFPA	NEMA
pg. 335	# 2-289	pg. 102	# 2-159	410.1	20.14	3.3.48	-	-
log: 2789	CMP: 2	log: 1578	Submitter: Michael J. Callanan			2002 NEC: -		IEC: 514.3

2005 NEC – 215.12 Identification for Feeders

(A) Grounded Conductor. The grounded conductor of a feeder shall be identified in accordance with 200.6.

(B) Equipment Grounding Conductor. The equipment grounding conductor shall be identified in accordance with 250.119.

(C) Ungrounded Conductors. Where the premises wiring system has feeders supplied from more than one nominal system voltage, each ungrounded conductor of a feeder, where accessible, shall be identified by system. The means of identification shall be permitted to be by separate color coding, marking tape, tagging, or other approved means and shall be permanently posted at each panelboard or similar feeder distribution equipment.

Author's Substantiation. This new subsection clarifies the identification requirements for grounded conductors, equipment grounding conductors, and ungrounded conductors. See 210.5(C) for additional information.

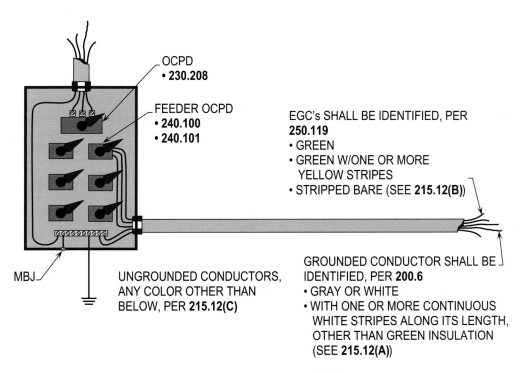

IDENTIFICATION OF FEEDERS
215.12(A) THRU (C)

Purpose of Change: To appropriately identify the circuit conductor of a feeder-circuit.

NEC Ch. 2 – Article 220
Parts I through V

26

Type of Change	Panel Action	UL	UL 508	API 500	API 505	OSHA
Renumbered	Accept in Principle	-	-	-	-	-
ROP	ROC	NFPA 70E	NFPA 70B	NFPA 79	NFPA	NEMA
pg. 337 # 2-292	pg. - # -	-	-	-	-	-
log: 1299 CMP: 2	log: -	Submitter: TCC		2002 NEC: Article 220		IEC: -

Author's Substantiation. Article 220 has been reorganized and renumbered under the guidance of the NFPA Task Group on Usability. See cross reference table for a better understanding of these new renumbered sections of Article 220.

CROSS REFERENCE		
PART I – GENERAL		
2005 NEC	**2002 NEC**	**SECTION HEADING**
220.1	220.1	Scope
220.3	——	Application of Other Articles
220.5	220.2	Calculations
220.5(A)	220.2(A)	Voltages
220.5(B)	220.2(B)	Fractions of an Ampere
PART II – BRANCH CIRCUIT LOAD CALCULATIONS		
220.10	——	General
220.12	220.3(A)	Lighting Load for Specified Occupancies
220.14	220.3(B)	Other Loads – All Occupancies
220.14(A)	220.3(B)(1)	Specific Appliances or Loads
220.14(B)	220.3(B)(2)	Electric Dryers and Household Electric Cooking Appliances
220.14(C)	220.3(B)(3)	Motor Loads
220.14(D)	220.3(B)(4)	Luminaires (Lighting Fixtures)
220.14(E)	220.3(B)(5)	Heavy-Duty Lampholders
220.14(F)	220.3(B)(6)	Sign and Outline Lighting
220.14(G)	220.3(B)(7)	Show Windows
220.14(H)	220.3(B)(8)	Fixed Multioutlet Assemblies
220.14(I)	220.3(B)(9)	Receptacle Outlets
220.14(J)	220.3(B)(10)	Dwelling Occupancies
220.14(K)	——	Banks and Office Buildings
220.14(L)	220.3(B)(11)	Other Outlets
220.16	220.3(C)	Loads for Additions to Existing Installations
220.16(A)	220.3(C)(1)	Dwelling Units
220.16(B)	220.3(C)(2)	Other Than Dwelling Units
220.18	220.4	Maximum Loads
220.18(A)	220.4(A)	Motor-Operated and Combination Loads
220.18(B)	220.4(B)	Inductive Lighting Loads
220.18(C)	220.4(C)	Range Loads

PART III – FEEDER AND SERVICE LOAD CALCULATIONS		
2005 NEC	**2002 NEC**	**SECTION HEADING**
220.40	220.10	General
220.42	220.11	General Lighting
220.43	220.12	Show Window and Track lighting
220.43(A)	220.12(A)	Show Windows
220.43(B)	220.12(B)	Track Lighting
220.44	220.13	Receptacle Loads – Other Than Dwelling Units
220.50	220.14	Motors
220.51	220.15	Fixed Electric Space Heating
220.52	220.16	Small Appliance and Laundry Loads - Dwelling Unit
220.52(A)	220.16(A)	Small Appliance Circuit Load
220.52(B)	220.16(B)	Laundry Circuit Load
220.53	220.17	Appliance Load – Dwelling Unit(s)
220.54	220.18	Electric Clothes Dryers – Dwelling Unit(s)
220.55	220.19	Electric Ranges and Other Cooking Appliances – Dwelling Unit(s)
220.56	220.20	Kitchen Equipment – Other Than Dwelling Unit(s)
220.60	220.21	Noncoincident Loads
220.61	220.22	Feeder or Service Neutral Load
220.61(A)	220.22	Basic Calculation
220.61(B)	220.22	Permitted Reductions
220.61(C)	220.22	Prohibited Reductions
PART IV – OPTIONAL FEEDER AND SERVICE LOAD CALCULATIONS		
220.80		General
220.82	220.30	Dwelling Unit
220.82(A)	220.30(A)	Feeder and Service Load
220.82(B)	220.30(B)	General Loads
220.82(C)	220.30(C)	Heating and Air-Conditioning Load
220.83	220.31	Existing Dwelling Unit
220.83(A)	220.31(A)	Where Additional Air-Conditioning Equipment or Electric Space-Heating Equipment Is Not to Be Installed
220.83(B)	220.31(B)	Where Additional Air-Conditioning Equipment or Electric Space-Heating Equipment Is to Be Installed
220.84	220.32	Multifamily Dwelling
220.84(A)	220.32(A)	Feeder or Service Load
220.84(B)	220.32(B)	House Loads
220.84(C)	220.32(C)	Connected Loads
220.85	220.33	Two Dwelling Units
220.86	220.34	Schools
220.87	220.35	Determining Existing Loads
220.88	220.36	New Restaurants
PART V – FARM LOAD CALCULATIONS		
220.100		General
220.102	220.40	Farm Loads – Buildings and Other Loads
220.102(A)	220.40(A)	Dwelling Unit
220.102(B)	220.40(B)	Other Than Dwelling Unit
220.103	220.41	Farm Loads – Total

Type of Change		Panel Action		UL	UL 508	API 500	API 505	OSHA
New FPN		Accept		-	-	-	-	-
ROP		ROC		NFPA 70E	NFPA 70B	NFPA 79	NFPA	NEMA
pg. 338	# 2-294	pg. -	# -	-	-	-	-	-
log: 1301	CMP: 2	log: -	Submitter: TCC			2002 NEC: 220.1		IEC: -

2002 NEC – 220.1 Scope

This article provides requirements for computing branch-circuit, feeder, and service loads.

~~Exception: Branch-circuit and feeder calculations for electrolytic cells as covered in 668.3(C)(1) and (4).~~

2005 NEC - 220.1 Scope

This article provides requirements for calculating branch-circuit, feeder, and service loads. Part I provides for general requirements for calculation methods. Part II provides calculation methods for branch circuit loads. Parts III and IV provide calculation methods for feeders and services. Part V provides calculation methods for farms.

FPN: See Figure 220.1 for information on the organization of Article 220.

Author's Substantiation. Article 220 has been reorganized and renumbered under the guidance of the NFPA Task Group on Usability. A new FPN has been added to refer the user to Figure 220.1 for information on the organization of Article 220.

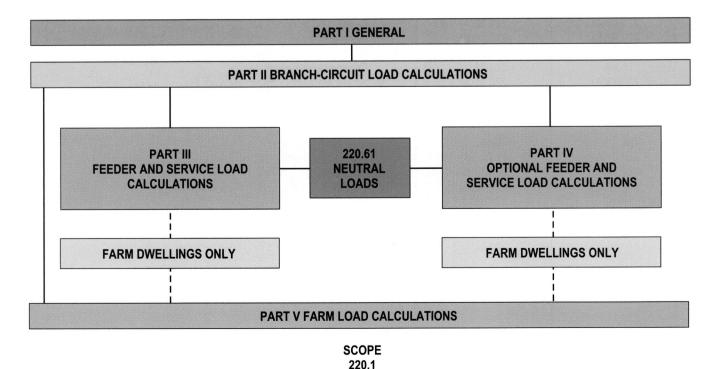

SCOPE
220.1

Purpose of Change: To arrange the calculation requirements for services, feeders, branch-circuits, and farm loads in a user friendly manner.

Type of Change		Panel Action		UL	UL 508	API 500	API 505	OSHA
New Section		Accept		-	-	-	-	-
ROP		ROC		NFPA 70E	NFPA 70B	NFPA 79	NFPA	NEMA
pg. 339	# 2-298	pg. -	# -	-	-	-	-	-
log: 1302	CMP: 2	log: -	Submitter: TCC			2002 NEC: -		IEC: -

2005 NEC – 220.3 Application of Other Articles

In other articles applying to the calculation of loads in specialized applications, there are requirements provided in Table 220.3 that are in addition to, or modifications of, those within this article.

Author's Substantiation. This new section has been added to provide additional load calculation references in other articles. Article 220 has been reorganized and renumbered under the guidance of the NFPA Task Group on Usability.

> NOTE: THIS TABLE IS AN EXAMPLE FOR SOME OF THE ADDITIONAL LOAD CALCULATIONS REFERENCED IN OTHER ARTICLES

CALCULATIONS	ARTICLE	SECTION (OR PART)
AIR-CONDITIONING AND REFRIGERATING EQUIPMENT, BRANCH-CIRCUIT CONDUCTOR SIZING	440	PART IV
MOTORS, FEEDER DEMAND FACTOR	430	430.26
MOTORS, MULTIMOTOR AND COMBINATION-LOAD EQUIPMENT	430	430.25
MOTORS, SEVERAL MOTORS OR A MOTOR(S) AND OTHER LOAD(S)	430	430.24
INDUSTRIAL MACHINERY, SUPPLY CONDUCTOR SIZING	670	670.4(A)
SENSITIVE ELECTRICAL EQUIPMENT, VOLTAGE DROP (MANDATORY CALCULATION)	647	647.4(D)

APPLICATION OF OTHER ARTICLES
220.3; TABLE 220.3

Purpose of Change: To reference other Articles, Parts, and Sections of the NEC for calculation loads.

NEC Ch. 2 – Article 220
Part I – 220.5

Type of Change		Panel Action		UL	UL 508	API 500	API 505	OSHA
Renumbered		Accept in Principle		-	-	-	-	-
ROP		ROC		NFPA 70E	NFPA 70B	NFPA 79	NFPA	NEMA
pg. 337	# 2-292	pg. -	# -	-	-	-	-	-
log: 1299	CMP: 2	log: -	Submitter: TCC			2002 NEC: 220.2		IEC: -

2002 NEC – ~~220.2 Computations~~

(A) **Voltages.** Unless other voltages are specified, for purposes of ~~computing~~ branch-circuit and feeder loads, nominal system voltages of 120, 120/240, 208Y/120, 240, 347, 480Y/277, 480, 600Y/347, and 600 volts shall be used.

(B) **Fraction of an Ampere.** Where ~~computations~~ result in a fraction of an ampere that is less than 0.5, such fractions shall be permitted to be dropped.

2005 NEC – 220.5 Calculations

(A) **Voltages.** Unless other voltages are specified, for purposes of calculating branch-circuit and feeder loads, nominal system voltages of 120, 120/240, 208Y/120, 240, 347, 480Y/277, 480, 600Y/347, and 600 volts shall be used.

(B) **Fraction of an Ampere.** Where calculations result in a fraction of an ampere that is less than 0.5, such fractions shall be permitted to be dropped.

Author's Substantiation. Article 220 has been reorganized and renumbered under the guidance of the NFPA Task Group on Usability.

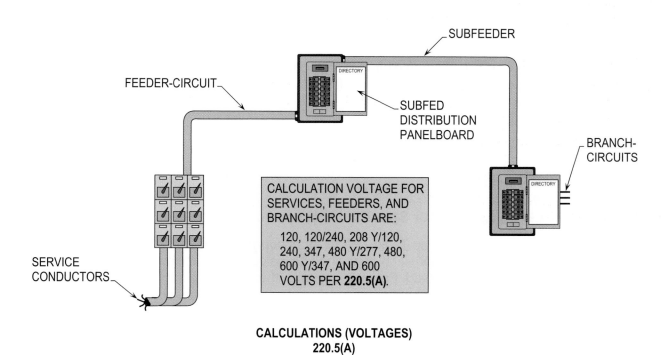

CALCULATION VOLTAGE FOR SERVICES, FEEDERS, AND BRANCH-CIRCUITS ARE:

120, 120/240, 208 Y/120, 240, 347, 480 Y/277, 480, 600 Y/347, AND 600 VOLTS PER **220.5(A)**.

CALCULATIONS (VOLTAGES)
220.5(A)

Purpose of Change: To relocate and renumber rules in **220.2(A)** to **220.5(A)**.

NEC Ch. 2 – Article 220
Part II – 220.10

Type of Change		Panel Action		UL	UL 508	API 500	API 505	OSHA
Renumbered		Accept in Principle		-	-	-	-	-
ROP		ROC		NFPA 70E	NFPA 70B	NFPA 79	NFPA	NEMA
pg. 337	# 2-292	pg. -	# -	-	-	-	-	-
log: 1299	CMP: 2	log: -	Submitter: TCC			2002 NEC: 220.3(A)		IEC: -

2002 NEC – ~~220.3 Computation of~~ Branch Circuit Load~~s~~

Branch-circuit loads shall be ~~computed~~ as shown in ~~220.3(A) through (C).~~

   ~~(A)~~ Lighting Load for Specified Occupancies. A unit load of not less than that specified in ~~Table 220.3(A)~~ for occupancies specified therein shall constitute the minimum lighting load. The floor area for each floor shall be ~~computed~~ from the outside dimension of the building, dwelling unit, or other area involved. For dwelling units, the ~~computed~~ floor area shall not include open porches, garages, or unused or unfinished spaces not adaptable for future use.

FPN: The unit values herein are based on minimum load conditions and 100 percent power factor and may not provide sufficient capacity for the installation contemplated.

2005 NEC – II. Branch Circuit Load <u>Calculations</u>
2005 NEC – <u>220.10 General</u>

Branch-circuit loads shall be <u>calculated</u> as shown in <u>220.12, 220.14, and 220.16</u>.

2005 NEC – <u>220.12</u> Lighting Load for Specified Occupancies

A unit load of not less than that specified in <u>Table 220.12</u> for occupancies specified therein shall constitute the minimum lighting load. The floor area for each floor shall be <u>calculated</u> from the outside dimensions of the building, dwelling unit, or other area involved. For dwelling units, the <u>calculated</u> floor area shall not include open porches, garages, or unused or unfinished spaces not adaptable for future use.

FPN: The unit values herein are based on minimum load conditions and 100 percent power factor and may not provide sufficient capacity for the installation contemplated.

Author's Substantiation. Article 220 has been reorganized and renumbered under the guidance of the NFPA Task Group on Usability.

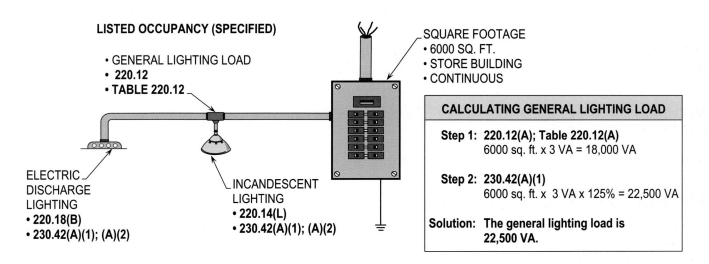

LISTED OCCUPANCY (SPECIFIED)

- GENERAL LIGHTING LOAD
- **220.12**
- **TABLE 220.12**

SQUARE FOOTAGE
- 6000 SQ. FT.
- STORE BUILDING
- CONTINUOUS

ELECTRIC
DISCHARGE
LIGHTING
- **220.18(B)**
- **230.42(A)(1); (A)(2)**

INCANDESCENT
LIGHTING
- **220.14(L)**
- **230.42(A)(1); (A)(2)**

CALCULATING GENERAL LIGHTING LOAD
Step 1: **220.12(A); Table 220.12(A)**   6000 sq. ft. x 3 VA = 18,000 VA
Step 2: **230.42(A)(1)**   6000 sq. ft. x 3 VA x 125% = 22,500 VA
Solution: **The general lighting load is**   **22,500 VA.**

LIGHTING LOAD FOR SPECIFIED OCCUPANCIES
220.12; TABLE 220.12

Purpose of Change: To relocate and renumber rules in **Table 220.3(A)** to **Table 220.12**.

NEC Ch. 2 – Article 220
Part II – 220.14(A)

Type of Change		Panel Action		UL	UL 508	API 500	API 505	OSHA
Renumbered		Accept in Principle		-	-	-	-	-
ROP		ROC		NFPA 70E	NFPA 70B	NFPA 79	NFPA	NEMA
pg. 337	# 2-292	pg. -	# -	-	-	-	-	-
log: 1299	CMP: 2	log: -	Submitter: TCC			2002 NEC: 220.3(B)(1)		IEC: -

2002 NEC – ~~220.3 Computation of~~ Branch Circuit Load~~s~~

Branch-circuit loads shall be ~~computed~~ as shown in ~~220.3(A) through (C).~~

    ~~(B)~~ Other Loads – All Occupancies. In all occupancies, the minimum load for each outlet for general-use receptacles and outlets not used for general illumination shall not be less than that ~~computed~~ in ~~220.3(B)(1) through (B)(11),~~ the loads shown being based on nominal branch-circuit voltages.

Exception: The loads of outlets serving switchboards and switching frames in telephone exchanges shall be waived from the ~~computations~~.

    ~~(1)~~ Specific Appliance or Loads. An outlet for a specific appliance or other load not covered in ~~(2) through (11)~~ shall be ~~computed~~ based on the ampere rating of the appliance or load served.

II. Branch Circuit Load <u>Calculations</u>
2005 NEC – <u>220.10 General</u>

Branch-circuit loads shall be <u>calculated</u> as shown in <u>220.12, 220.14, and 220.16</u>.

2005 NEC – <u>220.14 Other Loads – All Occupancies</u>

In all occupancies, the minimum load for each outlet for general-use receptacles and outlets not used for general illumination shall not be less than that <u>calculated</u> in <u>220.14(A) through (L)</u>, the loads shown being based on nominal branch-circuit voltages.

Exception: The loads of outlets serving switchboards and switching frames in telephone exchanges shall be waived from the <u>calculations</u>.

    <u>(A)</u> Specific Appliances or Loads. An outlet for a specific appliance or other load not covered in <u>220.14(B) through (L)</u> shall be <u>calculated</u> based on the ampere rating of the appliance or load served.

Author's Substantiation. Article 220 has been reorganized and renumbered under the guidance of the NFPA Task Group on Usability.

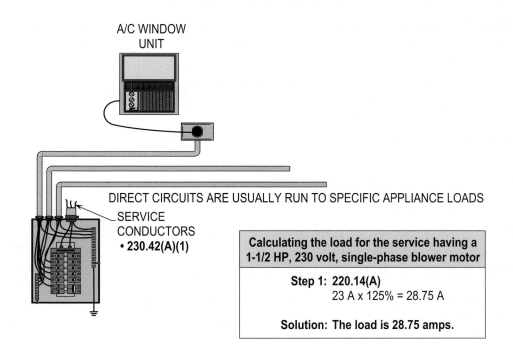

A/C WINDOW
UNIT

DIRECT CIRCUITS ARE USUALLY RUN TO SPECIFIC APPLIANCE LOADS

SERVICE
CONDUCTORS
• **230.42(A)(1)**

Calculating the load for the service having a 1-1/2 HP, 230 volt, single-phase blower motor
Step 1: 220.14(A)
23 A x 125% = 28.75 A
Solution: The load is 28.75 amps.

SPECIFIC APPLIANCES OR LOADS
220.14(A)

Purpose of Change: To relocate and renumber rules in **220.3(B)(1)** to **220.14(A)**.

NEC Ch. 2 – Article 220
Part II – 220.14(B)

Type of Change	Panel Action	UL	UL 508	API 500	API 505	OSHA
Renumbered	Accept in Principle	-	-	-	-	-
ROP		**ROC**	**NFPA 70E**	**NFPA 70B**	**NFPA 79**	**NFPA**	**NEMA**

pg. 337	# 2-292	pg. -	# -	-	-	-	-	-
log: 1299	CMP: 2	log: -	Submitter: TCC		**2002 NEC:** 220.3(B)(2)		IEC: -

2002 NEC – ~~220.3(B)(2)~~ Electric Dryers and Household Electric Cooking Appliances

Load ~~computations~~ shall be permitted as specified in ~~220.18~~ for electric dryers and in ~~220.19~~ for electric ranges and other cooking appliances.

2005 NEC – <u>220.14(B)</u> Electric Dryers and Household Electric Cooking Appliances

Load <u>calculations</u> shall be permitted as specified in <u>220.54</u> for electric dryers and in <u>220.55</u> for electric ranges and other cooking appliances.

Author's Substantiation. Article 220 has been reorganized and renumbered under the guidance of the NFPA Task Group on Usability.

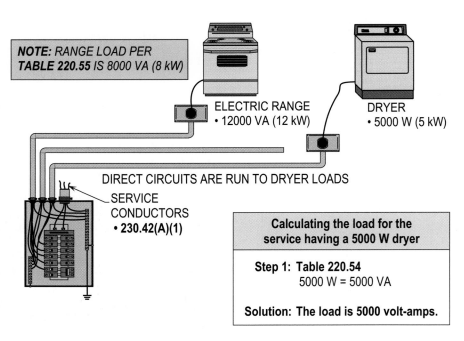

NOTE: *RANGE LOAD PER* **TABLE 220.55** *IS 8000 VA (8 kW)*

ELECTRIC RANGE
• 12000 VA (12 kW)

DRYER
• 5000 W (5 kW)

DIRECT CIRCUITS ARE RUN TO DRYER LOADS

SERVICE CONDUCTORS
• 230.42(A)(1)

Calculating the load for the service having a 5000 W dryer

Step 1: Table 220.54
5000 W = 5000 VA

Solution: The load is 5000 volt-amps.

ELECTRIC DRYERS AND HOUSEHOLD ELECTRIC COOKING APPLIANCES
220.14(B)

Purpose of Change: To relocate and renumber rules in **220.3(B)(2)** to **220.14(B)**.

NEC Ch. 2 – Article 220
Part II – 220.14(C)

Type of Change		Panel Action		UL	UL 508	API 500	API 505	OSHA
Renumbered		Accept in Principle		-	-	-	-	-
ROP		ROC		NFPA 70E	NFPA 70B	NFPA 79	NFPA	NEMA
pg. 337	# 2-292	pg. -	# -	-	-	-	-	-
log: 1299	CMP: 2	log: -	Submitter: TCC			2002 NEC: 220.3(B)(3)		IEC: -

2002 NEC – ~~220.3(B)(3)~~ Motor Loads

Outlets for motor loads shall be ~~computed~~ in accordance with the requirements in 430.22, 430.24, and 440.6.

2005 NEC – 220.14(C) Motor Loads

Outlets for motor loads shall be <u>calculated</u> in accordance with the requirements in 430.22, 430.24, and 440.6.

Author's Substantiation. Article 220 has been reorganized and renumbered under the guidance of the NFPA Task Group on Usability.

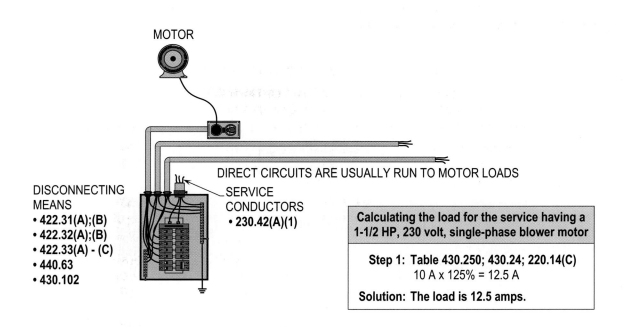

MOTOR

DIRECT CIRCUITS ARE USUALLY RUN TO MOTOR LOADS

DISCONNECTING MEANS
• 422.31(A);(B)
• 422.32(A);(B)
• 422.33(A) - (C)
• 440.63
• 430.102

SERVICE CONDUCTORS
• 230.42(A)(1)

Calculating the load for the service having a 1-1/2 HP, 230 volt, single-phase blower motor

Step 1: Table 430.250; 430.24; 220.14(C)
10 A x 125% = 12.5 A

Solution: The load is 12.5 amps.

MOTOR LOADS
220.14(C)

Purpose of Change: To relocate and renumber rules in **220.3(B)(3)** to **220.14(C)**.

NEC Ch. 2 – Article 220
Part II – 220.14(D)

Type of Change	Panel Action	UL	UL 508	API 500	API 505	OSHA
Revision	Accept in Principle	542	-	-	-	1910.304(e)(1)(i)
ROP	**ROC**	**NFPA 70E**	**NFPA 70B**	**NFPA 79**	**NFPA**	**NEMA**
pg. 343 # 2-307	pg. - # -	420.10(A)	21.3	7.2.6	-	OS 1
log: 1372 CMP: 2	log: -	Submitter: Jon Farren		2002 NEC: 220.3(B)(4)		IEC: Ch. 43

2002 NEC – ~~220.3(B)(4)~~ ~~Recessed~~ Luminaires (Lighting Fixtures)

An outlet supplying ~~recessed~~ luminaire(s) [lighting fixture(s)] shall be ~~computed~~ based on the maximum volt-ampere rating of the equipment and lamps for which the luminaire(s) [fixture(s)] is rated.

2005 NEC – <u>220.14(D)</u> Luminaires (Lighting Fixtures)

An outlet supplying luminaire(s) [lighting fixture(s)] shall be <u>calculated</u> based on the maximum volt-ampere rating of the equipment and lamps for which the luminaire(s) [fixture(s)] is rated.

Author's Substantiation. This revision clarifies that there are many outlets supplying luminaires with a load in excess of 180 volt-amperes. Article 220 has been reorganized and renumbered under the guidance of the NFPA Task Group on Usability.

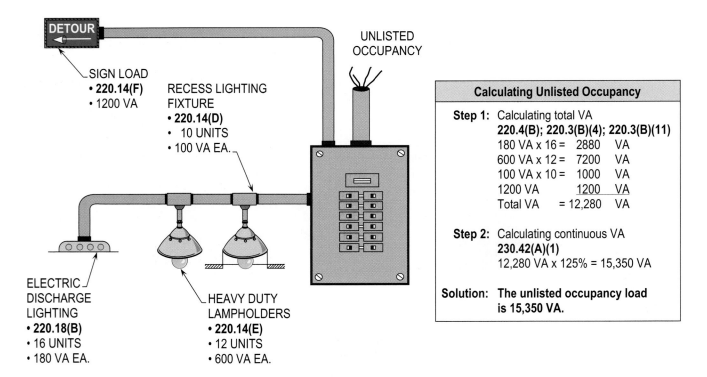

LUMINAIRES (LIGHTING FIXTURES)
220.14(D)

Purpose of Change: This revision clarifies that there are many outlets supplying luminaires with a load in excess of 180 volt-amperes.

NEC Ch. 2 – Article 220
Part II – 220.14(E)

Type of Change		Panel Action		UL	UL 508	API 500	API 505	OSHA
Renumbered		Accept in Principle		-	-	-	-	-
ROP		ROC		NFPA 70E	NFPA 70B	NFPA 79	NFPA	NEMA
pg. 337	# 2-292	pg. -	# -	-	-	-	-	-
log: 1299	CMP: 2	log: -	Submitter: TCC			2002 NEC: 220.3(B)(5)		IEC: -

2002 NEC – ~~220.3(B)(5)~~ Heavy-Duty Lampholders

Outlets for heavy-duty lampholders shall be ~~computed~~ at a minimum of 600 volt-amperes.

2005 NEC – 220.14(E) Heavy-Duty Lampholders

Outlets for heavy-duty lampholders shall be <u>calculated</u> at a minimum of 600 volt-amperes.

Author's Substantiation. Article 220 has been reorganized and renumbered under the guidance of the NFPA Task Group on Usability.

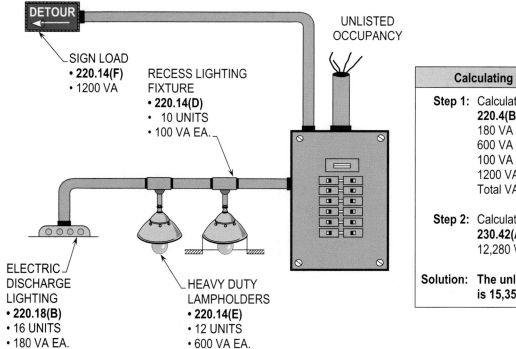

HEAVY-DUTY LAMPHOLDERS
220.14(E)

Purpose of Change: To relocate and renumber rules in **220.3(B)(5)** to **220.14(E)**.

NEC Ch. 2 – Article 220
Part II – 220.14(F)

Type of Change		Panel Action		UL	UL 508	API 500	API 505	OSHA
Renumbered		Accept in Principle		-	-	-	-	-
ROP		ROC		NFPA 70E	NFPA 70B	NFPA 79	NFPA	NEMA
pg. 337	# 2-292	pg. -	# -	-	-	-	-	-
log: 1299	CMP: 2	log: -	Submitter: TCC			2002 NEC: 220.3(B)(6)		IEC: -

2002 NEC – ~~220.3(B)(6)~~ Sign and Outline Lighting

Sign and outline lighting outlets shall be ~~computed~~ at a minimum of 1200 volt-amperes for each required branch circuit specified in 600.5(A).

2005 NEC – 220.14(F) Sign and Outline Lighting

Sign and outline lighting outlets shall be <u>calculated</u> at a minimum of 1200 volt-amperes for each required branch circuit specified in 600.5(A).

Author's Substantiation. Article 220 has been reorganized and renumbered under the guidance of the NFPA Task Group on Usability.

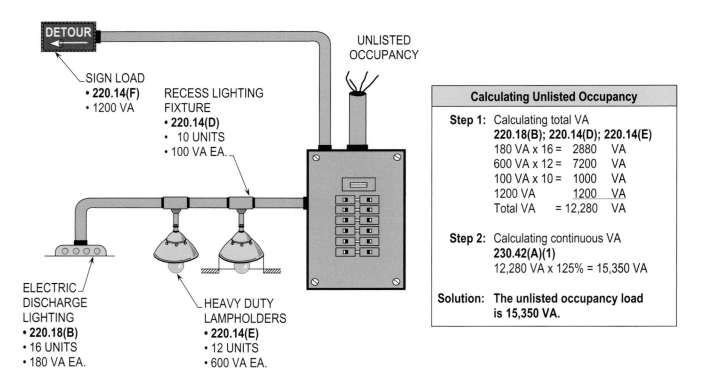

SIGN AND OUTLINE LIGHTING
220.14(F)

Purpose of Change: To relocate and renumber rules in **220.3(B)(6)** to **220.14(F)**.

NEC Ch. 2 – Article 220
Part II – 220.14(G)

Type of Change	Panel Action	UL	UL 508	API 500	API 505	OSHA
Renumbered	Accept in Principle	-	-	-	-	-
ROP	**ROC**	**NFPA 70E**	**NFPA 70B**	**NFPA 79**	**NFPA 20**	**NFPA 20**
pg. 337 # 2-292	pg. - # -	-	-	-	-	-
log: 1299 CMP: 2	log: -	Submitter: TCC		2002 NEC: 220.3(B)(7)		IEC: -

2002 NEC – ~~220.3(B)(7)~~ Show Windows

Show windows shall be ~~computed~~ in accordance with either of the following:

(1) The unit load per outlet as required in other provisions of this section.

(2) At 200 volt-amperes per 300 mm (1 ft) of show window.

2005 NEC – <u>220.14(G)</u> Show Windows

Show windows shall be <u>calculated</u> in accordance with either of the following:

(1) The unit load per outlet as required in other provisions of this section.

(2) At 200 volt-amperes per 300 mm (1 ft) of show window.

Author's Substantiation. Article 220 has been reorganized and renumbered under the guidance of the NFPA Task Group on Usability.

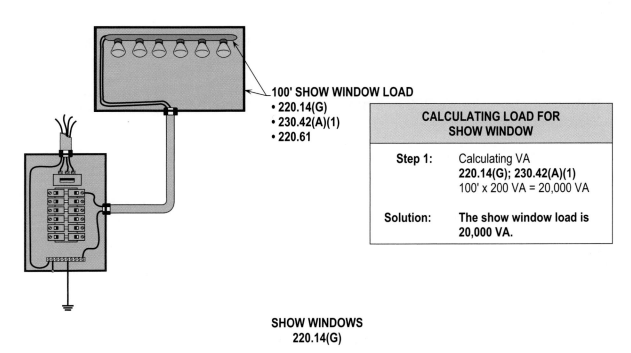

100' SHOW WINDOW LOAD
- 220.14(G)
- 230.42(A)(1)
- 220.61

CALCULATING LOAD FOR SHOW WINDOW	
Step 1:	Calculating VA **220.14(G); 230.42(A)(1)** 100' x 200 VA = 20,000 VA
Solution:	**The show window load is 20,000 VA.**

SHOW WINDOWS
220.14(G)

Purpose of Change: To relocate and renumber rules in **220.3(B)(7)** to **220.14(G)**.

NEC Ch. 2 – Article 220
Part II – 220.14(H)

Type of Change	Panel Action	UL	UL 508	API 500	API 505	OSHA
Renumbered	Accept in Principle	-	-	-	-	-
ROP		**ROC**	**NFPA 70E**	**NFPA 70B**	**NFPA 79**	**NFPA**	**NEMA**

pg. 337	# 2-292	pg. -	# -	-	-	-	-
log: 1299	CMP: 2	log: -	Submitter: TCC		2002 NEC: 220.3(B)(8)		IEC: -

2002 NEC – ~~220.3(B)(8)~~ Fixed Multioutlet Assemblies

Fixed multioutlet assemblies used in other than dwelling units or the guest room of hotels or motels shall be ~~computed~~ in accordance with (1) or (2). For the purposes of this section, the ~~computation~~ shall be permitted to be based on the portion that contains receptacle outlets.

(1) Where appliances are unlikely to be used simultaneously, each 1.5 m (5 ft) or fraction thereof of each separate and continuous length shall be considered as one outlet of not less than 180 volt-amperes.

(2) Where appliances are likely to be used simultaneously, each 300 mm (1 ft) or fraction thereof shall be considered as an outlet of not less than 180 volt-amperes.

2005 NEC – 220.14(H) Fixed Multioutlet Assemblies

Fixed multioutlet assemblies used in other than dwelling units or the guest rooms or guest suites of hotels or motels shall be calculated in accordance with (H)(1) or (H)(2). For the purposes of this section, the calculation shall be permitted to be based on the portion that contains receptacle outlets.

(1) Where appliances are unlikely to be used simultaneously, each 1.5 m (5 ft) or fraction thereof of each separate and continuous length shall be considered as one outlet of not less than 180 volt-amperes.

(2) Where appliances are likely to be used simultaneously, each 300 mm (1 ft) or fraction thereof shall be considered as an outlet of not less than 180 volt-amperes.

Author's Substantiation. Article 220 has been reorganized and renumbered under the guidance of the NFPA Task Group on Usability.

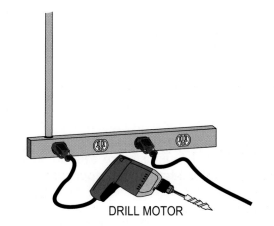

DRILL MOTOR

MULTIOUTLET ASSEMBLY
• 200 FT.
• **220.14(H)(1)**
• **220.14(H)(2)**
• APPLIANCES MAY OR MAY NOT BE
  USED SIMULTANEOUSLY

CALCULATING LOAD FOR MULTIOUTLET ASSEMBLY	
Step 1:	Calculating VA **220.14(H)(1)** 200' ÷ 5' x 180 VA = 7200 VA
Solution:	**The assembly load is 7200 VA.**

CALCULATING LOAD FOR MULTIOUTLET ASSEMBLY	
Step 1:	Calculating VA **220.14(H)(2)** 200' x 1' x 180 VA = 36,000 VA
Solution:	**The assembly load is 36,000 VA.**

OTHER LOADS – ALL OCCUPANCIES
220.14(H)(1); (H)(2)

Purpose of Change: To relocate and renumber rules in **220.3(B)(8)** to **220.14(H)**.

NEC Ch. 2 – Article 220
Part II – 220.14(I)

Type of Change		Panel Action		UL	UL 508	API 500	API 505	OSHA
Renumbered		Accept in Principle		498	-	-	-	1910.304(b)
ROP		ROC		NFPA 70E	NFPA 70B	NFPA 79	NFPA	NEMA
pg. 337	# 2-292	pg. -	# -	410.2(B)(2)	27.7.3.2	16.1.1	-	PR 4
log: 1299	CMP: 2	log: -	Submitter: TCC			2002 NEC: 220.3(B)(9)		IEC: 413.1.1.2

2002 NEC – ~~220.3(B)(9)~~ Receptacle Outlets

Except as covered in ~~220.3(B)(10),~~ receptacle outlets shall be ~~computed~~ at not less than 180 volt-amperes for each single or for each multiple receptacle on one yoke. A single piece of equipment consisting of a multiple receptacle comprised of four or more receptacles shall be ~~computed~~ at not less than 90 volt-amperes per receptacle.

This provision shall not be applicable to the receptacle outlets specified in 210.11(C)(1) and (2).

2005 NEC – 220.14(I) Receptacle Outlets

Except as covered in <u>220.14(J) and (K)</u>, receptacle outlets shall be <u>calculated</u> at not less than 180 volt-amperes for each single or for each multiple receptacle on one yoke. A single piece of equipment consisting of a multiple receptacle comprised of four or more receptacles shall be <u>calculated</u> at not less than 90 volt-amperes per receptacle. This provision shall not be applicable to the receptacle outlets specified in 210.11(C)(1) and <u>(C)</u>(2).

Author's Substantiation. Article 220 has been reorganized and renumbered under the guidance of the NFPA Task Group on Usability.

CALCULATING CONTINUOUS AND NONCONTINUOUS LOADS
Step 1: Finding VA (continuous) **220.14(I); 230.42(A)(1)** 180 VA x 40 x 125% = 9000 VA
Step 2: Finding VA (noncontinuous) **220.14(I); 230.42(A)(1)** 180 VA x 40 = 7200 VA
Solution: The noncontinuous load is 7200 VA, and the continuous load is 9000 VA.

CALCULATING RECEPTACLE LOAD AND APPLYING DEMAND FACTORS
Step 1: Calculating VA **220.14(I); 230.42(A)(1)** 150 x 180 VA = 27,000 VA
Step 2: Applying demand factors **Table 220.44** First 10,000 VA x 100% =10,000 VA Next 17,000 VA x 50% = 8,500 VA 18,500 VA
Solution: The demand load for the receptacles is 18,500 VA.

RECEPTACLE OUTLETS
220.14(I)

Purpose of Change: To relocate and renumber rules in **220.3(B)(9)** to **220.14(I)**.

NEC Ch. 2 – Article 220
Part II – 220.14(J)

Type of Change		Panel Action		UL	UL 508	API 500	API 505	OSHA
Renumbered		Accept in Principle		498	-	-	-	1910.304(b)
ROP		ROC		NFPA 70E	NFPA 70B	NFPA 79	NFPA	NEMA
pg. 337	# 2-292	pg. -	# -	410.2(B)(2)	27.7.3.2	16.1.1	-	-
log: 1299	CMP: 2	log: -	Submitter: TCC			2002 NEC: 220.3(B)(10)		IEC: 413.1.1.2

2002 NEC – ~~220.3(B)(10)~~ Dwelling Occupancies

In one-family, two-family, and multifamily dwellings and in guest rooms of hotels and motels, the outlets specified in (1), (2), and (3) are included in the general lighting load calculations of ~~220.3(A).~~ No additional load calculations shall be required for such outlets.

(1) All general-use receptacle outlets of 20-ampere rating or less, including receptacles connected to the circuits in 210.11(C)(3)

(2) The receptacle outlets specified in 210.52(E) and (G)

(3) The lighting outlets specified in 210.70(A) and (B)

2005 NEC – 220.14(J) Dwelling Occupancies

In one-family, two-family, and multifamily dwellings and in guest rooms <u>or guest suites</u> of hotels and motels, the outlets specified in <u>(J)</u>(1), <u>(J)</u>(2), and <u>(J)</u>(3) are included in the general lighting load calculations of <u>220.12</u>. No additional load calculations shall be required for such outlets.

(1) All general-use receptacle outlets of 20-ampere rating or less, including receptacles connected to the circuits in 210.11(C)(3)

(2) The receptacle outlets specified in 210.52(E) and (G)

(3) The lighting outlets specified in 210.70(A) and (B)

Author's Substantiation. Article 220 has been reorganized and renumbered under the guidance of the NFPA Task Group on Usability.

2700 SQ. FT. DWELLING UNIT

NOTE: INCLUDES OUTLETS IN **220.14(J)(1); (J)(2); AND (J)(3)**

Finding Number of 20 Amp Circuits
Step 1: Finding VA per sq. ft. **220.12; Table 220.12** The VA per sq. ft. is 3
Step 2: Finding total VA **Table 220.12** 2700 sq. ft. x 3 VA = 8100 VA
Step 3: Finding the number of circuits **210.11(A)** 8100 VA ÷ 20 A OCPD x 120 V = 3.375 (Rounded up to 4)
Solution: **The number of 20 amp, 2-wire circuits is 4.**

OTHER OUTLETS – ALL OCCUPANCIES
220.14(J)(1); (J)(2); AND (J)(3)

Purpose of Change: To relocate and renumber rules in **220.3(B)(10)** to **220.14(J)**.

NEC Ch. 2 – Article 220
Part II – 220.14(K)

29

Type of Change	Panel Action	UL	UL 508	API 500	API 505	OSHA
New Subsection	Accept in Principle	498	-	-	-	1910.304(b)
ROP		ROC		NFPA 70E	NFPA 70B	NFPA 79	NFPA	NEMA

ROP		ROC		NFPA 70E	NFPA 70B	NFPA 79	NFPA	NEMA
pg. 343	# 2-306	pg. -	# -	410.2(B)(2)	27.7.3.2	16.1.1	-	PR 4
log: 1510	CMP: 2	log: -	Submitter: Dan Strube			2002 NEC: -		IEC: -

2005 NEC – 220.14(K) Banks and Office Buildings

In banks or office buildings, the receptacle loads shall be calculated to be the larger of (1) or (2):

(1) The computed load from 220.14

(2) 11 volt-amperes/m^2 or 1 volt-ampere/ft^2

Author's Substantiation. A new section has been added to place the receptacle calculation for banks and office buildings in one location. Article 220 has been reorganized and renumbered under the guidance of the NFPA Task Group on Usability.

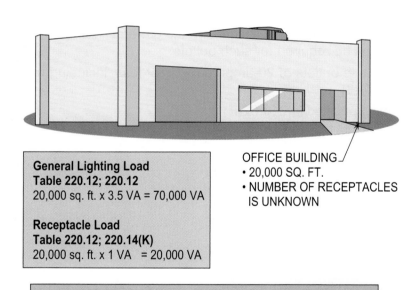

General Lighting Load
Table 220.12; 220.12
20,000 sq. ft. x 3.5 VA = 70,000 VA

Receptacle Load
Table 220.12; 220.14(K)
20,000 sq. ft. x 1 VA   = 20,000 VA

OFFICE BUILDING
• 20,000 SQ. FT.
• NUMBER OF RECEPTACLES IS UNKNOWN

NOTE: THE NONCONTINUOUS AND CONTINUOUS OPERATION RULES OF **220.14(I)** AND **230.42(A)(1)** HAVE NOT BEEN APPLIED.

BANKS AND OFFICE BUILDINGS
220.14(K)

Purpose of Change: To clarify the appropriate procedure for calculating an unknown number of receptacle outlets.

NEC Ch. 2 – Article 220
Part II – 220.14(L)

Type of Change		Panel Action		UL	UL 508	API 500	API 505	OSHA
Renumbered		Accept in Principle		498	-	-	-	1910.304(b)
ROP		ROC		NFPA 70E	NFPA 70B	NFPA 79	NFPA	NEMA
pg. 337	# 2-292	pg. -	# -	410.2(B)(2)	27.7.3.2	16.1.1	-	PR 4
log: 1299	CMP: 2	log: -	Submitter: TCC			2002 NEC: 220.3(B)(11)		IEC: -

2002 NEC – ~~220.3(B)(11)~~ Other Outlets

Other outlets not covered in ~~220.3(B)(1) through (10)~~ shall be computed based on 180 volt-amperes per outlet.

2005 NEC – 220.14(L) Other Outlets

Other outlets not covered in 220.14(A) through (K) shall be calculated based on 180 volt-amperes per outlet.

Author's Substantiation. Article 220 has been reorganized and renumbered under the guidance of the NFPA Task Group on Usability.

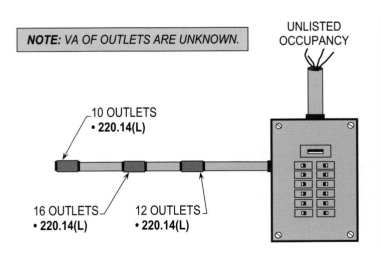

NOTE: VA OF OUTLETS ARE UNKNOWN.

UNLISTED OCCUPANCY

10 OUTLETS
• 220.14(L)

16 OUTLETS
• 220.14(L)

12 OUTLETS
• 220.14(L)

Calculating Load of Outlets

Step 1: Calculating total VA
220.14(L)
180 VA x 16 = 2880 VA
180 VA x 12 = 2160 VA
180 VA x 10 = 1800 VA
Total VA   = 6840 VA

Step 2: Calculating continuous VA
230.42(A)(1)
6840 VA x 125% = 8550 VA

Solution: **The unlisted occupancy load is 8550 VA.**

OTHER OUTLETS – ALL OCCUPANCIES
220.14(L)

Purpose of Change: To relocate and renumber rules in **220.3(B)(11)** to **220.14(L)**.

NEC Ch. 2 – Article 220
Part II – 220.16(A) and (B)

Type of Change	Panel Action	UL	UL 508	API 500	API 505	OSHA
Renumbered	Accept in Principle	1570	-	-	-	1910.305(j)(1)
ROP		ROC		NFPA 70E	NFPA 70B	NFPA 79	NFPA	NEMA

ROP		ROC		NFPA 70E	NFPA 70B	NFPA 79	NFPA	NEMA
pg. 337	# 2-292	pg. -	# -	420.10(A)	Ch. 17	16.2.4	-	270
log: 1299	CMP: 2	log: -	Submitter: TCC			2002 NEC: 220.3(C)		IEC: 714

2002 NEC – ~~220.3(C)~~ Loads for Additions to Existing Installations

~~(1)~~ **Dwelling Units.** Loads added to an existing dwelling unit(s) shall comply with the following as applicable:

(1) Loads for structural additions to an existing dwelling unit or for a previously unwired portion of an existing dwelling unit, either of which exceeds 46.5 m² (500 ft²), shall be ~~computed~~ in accordance with ~~220.3(A) and (B).~~

(2) Loads for new circuits or extended circuits in previously wired dwelling units shall be ~~computed~~ in accordance with either ~~220.3(A) or (B),~~ as applicable.

~~(2)~~ **Other Than Dwelling Units.** Loads for new circuits or extended circuits in other than dwelling units shall be ~~computed~~ in accordance with either ~~220.3(A) or (B),~~ as applicable.

2005 NEC – 220.16 Loads for Additions to Existing Installations

(A) **Dwelling Units.** Loads added to an existing dwelling unit(s) shall comply with the following as applicable:

(1) Loads for structural additions to an existing dwelling unit or for a previously unwired portion of an existing dwelling unit, either of which exceeds 46.5 m² (500 ft²), shall be <u>calculated</u> in accordance with <u>220.12 and 220.14.</u>

(2) Loads for new circuits or extended circuits in previously wired dwelling units shall be <u>calculated</u> in accordance with either <u>220.12 or 220.14,</u> as applicable.

(B) **Other Than Dwelling Units.** Loads for new circuits or extended circuits in other than dwelling units shall be <u>calculated</u> in accordance with either <u>220.12 or 220.14,</u> as applicable.

Author's Substantiation. Article 220 has been reorganized and renumbered under the guidance of the NFPA Task Group on Usability.

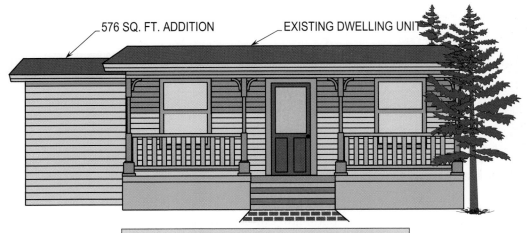

576 SQ. FT. ADDITION

EXISTING DWELLING UNIT

What is the general lighting and receptacle load for an addition to an existing dwelling?

Step 1: Calculating load
Table 220.12; 220.16(A)
sq. ft. x VA = Lighting Load
576 sq. ft. x 3 VA = 1728 VA

Solution: **The general lighting and receptacle load is 1728 volt-amps.**

DWELLING UNITS
220.16(A)(1)

Purpose of Change: To relocate and renumber rules in **220.3(C)(1)** to **220.16(A)**.

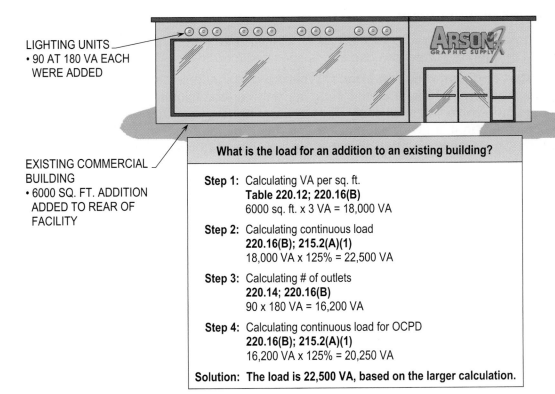

LIGHTING UNITS
• 90 AT 180 VA EACH
WERE ADDED

EXISTING COMMERCIAL
BUILDING
• 6000 SQ. FT. ADDITION
ADDED TO REAR OF
FACILITY

What is the load for an addition to an existing building?

Step 1: Calculating VA per sq. ft.
Table 220.12; 220.16(B)
6000 sq. ft. x 3 VA = 18,000 VA

Step 2: Calculating continuous load
220.16(B); 215.2(A)(1)
18,000 VA x 125% = 22,500 VA

Step 3: Calculating # of outlets
220.14; 220.16(B)
90 x 180 VA = 16,200 VA

Step 4: Calculating continuous load for OCPD
220.16(B); 215.2(A)(1)
16,200 VA x 125% = 20,250 VA

Solution: **The load is 22,500 VA, based on the larger calculation.**

OTHER THAN DWELLING UNITS
220.16(B)

Purpose of Change: To relocate and renumber rules in **220.3(C)(2)** to **220.16(B)**.

NEC Ch. 2 – Article 220
Part II – 220.18

Type of Change		Panel Action		UL	UL 508	API 500	API 505	OSHA
Renumbered		Accept in Principle		1004	68	-	-	1910.305(j)(4)
ROP		ROC		NFPA 70E	NFPA 70B	NFPA 79	NFPA	NEMA
pg. 337	# 2-292	pg. -	# -	420.10(E)	J.4	Ch. 15	-	MG 1
log: 1299	CMP: 2	log: -	Submitter: TCC			2002 NEC: 220.4		IEC: 533.2

2002 NEC – ~~220.4~~ Maximum Loads

The total load shall not exceed the rating of the branch circuit, and it shall not exceed the maximum loads specified in ~~220.4(A) through (C)~~ under the conditions specified therein.

(A) Motor-Operated and Combination Loads. Where a circuit supplies only motor-operated loads, Article 430 shall apply. Where a circuit supplies only air-conditioning equipment, refrigerating equipment, or both, Article 440 shall apply. For circuits supplying loads consisting of motor-operated utilization equipment that is fastened in place and has a motor larger than 1/8 hp in combination with other loads, the total computed load shall be based on 125 percent of the largest motor load plus the sum of the other loads.

(B) Inductive Lighting Loads. For circuits supplying lighting units that have ballasts, transformers, or autotransformers, the ~~computed~~ load shall be based on the total ampere ratings of such units and not on the total watts of the lamps.

(C) Range Loads. It shall be permissible to apply demand factors for range loads in accordance with ~~Table 220.19~~, including Note 4.

2005 NEC - 220.18 Maximum Loads

The total load shall not exceed the rating of the branch circuit, and it shall not exceed the maximum loads specified in 220.18(A) through (C) under the conditions specified therein.

(A) Motor-Operated and Combination Loads. Where a circuit supplies only motor-operated loads, Article 430 shall apply. Where a circuit supplies only air-conditioning equipment, refrigerating equipment, or both, Article 440 shall apply. For circuits supplying loads consisting of motor-operated utilization equipment that is fastened in place and has a motor larger than 1/8 hp in combination with other loads, the total calculated load shall be based on 125 percent of the largest motor load plus the sum of the other loads.

(B) Inductive Lighting Loads. For circuits supplying lighting units that have ballasts, transformers, or autotransformers, the calculated load shall be based on the total ampere ratings of such units and not on the total watts of the lamps.

(C) Range Loads. It shall be permissible to apply demand factors for range loads in accordance with Table 220.55, including Note 4.

Author's Substantiation. Article 220 has been reorganized and renumbered under the guidance of the NFPA Task Group on Usability.

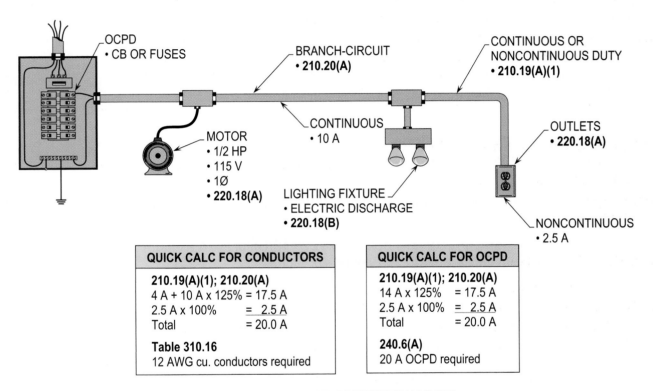

OCPD
• CB OR FUSES

BRANCH-CIRCUIT
• **210.20(A)**

CONTINUOUS OR
NONCONTINUOUS DUTY
• **210.19(A)(1)**

CONTINUOUS
• 10 A

MOTOR
• 1/2 HP
• 115 V
• 1Ø
• **220.18(A)**

LIGHTING FIXTURE
• ELECTRIC DISCHARGE
• **220.18(B)**

OUTLETS
• **220.18(A)**

NONCONTINUOUS
• 2.5 A

QUICK CALC FOR CONDUCTORS
210.19(A)(1); 210.20(A) 4 A + 10 A x 125% = 17.5 A 2.5 A x 100% = 2.5 A Total = 20.0 A
Table 310.16 12 AWG cu. conductors required

QUICK CALC FOR OCPD
210.19(A)(1); 210.20(A) 14 A x 125% = 17.5 A 2.5 A x 100% = 2.5 A Total = 20.0 A
240.6(A) 20 A OCPD required

MOTOR-OPERATED AND COMBINATION LOADS
220.18(A)

Purpose of Change: To relocate and renumber rules in **220.4(A)** to **220.18(A)**.

NEC Ch. 2 – Article 220
Part III – 220.40

Type of Change	Panel Action	UL	UL 508	API 500	API 505	OSHA
Renumbered	Accept in Principle	-	18.1	-	-	1910.304(e)
ROP		ROC		NFPA 70E	NFPA 70B	NFPA 79	NFPA	NEMA
pg. 337	# 2-292	pg. -	# -	410.9	3.3.18	13.5	-	WC 3
log: 1299	CMP: 2	log: -	Submitter: TCC			2002 NEC: 220.10		IEC: 314.2

Feeder and Services
2002 NEC – ~~220.10~~ General

The ~~computed~~ load of a feeder or service shall not be less than the sum of the loads on the branch circuits supplied, as determined by ~~Part I~~ of this article, after any applicable demand factors permitted by ~~Parts II, III, or IV~~ have been applied.

FPN: See Examples D1(A) through D10 in Annex D. See ~~220.4(B)~~ for the maximum load in amperes permitted for lighting units operating at less than 100 percent power factor.

Feeder and Service <u>Load Calculations</u>
2005 NEC – <u>220.40</u> General

The <u>calculated</u> load of a feeder or service shall not be less than the sum of the loads on the branch circuits supplied, as determined by <u>Part II</u> of this article, after any applicable demand factors permitted by <u>Parts III or IV or required by Part V</u> have been applied.

FPN: See Examples D1(A) through D10 in Annex D. See <u>220.18(B)</u> for the maximum load in amperes permitted for lighting units operating at less than 100 percent power factor.

Author's Substantiation. Article 220 has been reorganized and renumbered under the guidance of the NFPA Task Group on Usability.

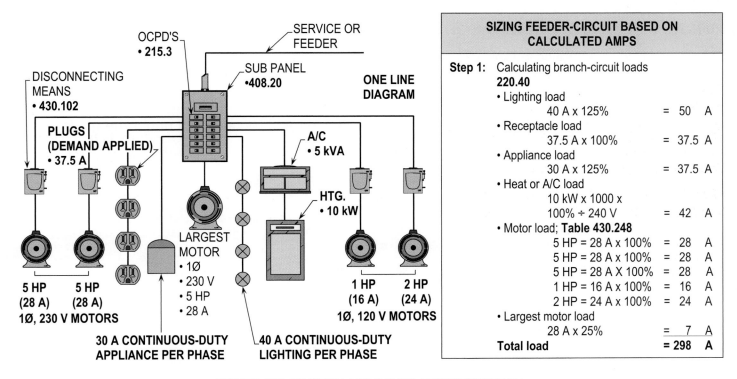

SIZING FEEDER-CIRCUIT BASED ON CALCULATED AMPS			
Step 1:	Calculating branch-circuit loads **220.40**		
	• Lighting load		
	40 A x 125%	= 50	A
	• Receptacle load		
	37.5 A x 100%	= 37.5	A
	• Appliance load		
	30 A x 125%	= 37.5	A
	• Heat or A/C load		
	10 kW x 1000 x		
	100% ÷ 240 V	= 42	A
	• Motor load; **Table 430.248**		
	5 HP = 28 A x 100%	= 28	A
	5 HP = 28 A x 100%	= 28	A
	5 HP = 28 A X 100%	= 28	A
	1 HP = 16 A x 100%	= 16	A
	2 HP = 24 A x 100%	= 24	A
	• Largest motor load		
	28 A x 25%	= 7	A
	Total load	**= 298**	**A**

OCPD'S
• 215.3

SERVICE OR FEEDER

SUB PANEL
•408.20

ONE LINE DIAGRAM

DISCONNECTING MEANS
• 430.102

PLUGS (DEMAND APPLIED)
• 37.5 A

A/C
• 5 kVA

HTG.
• 10 kW

LARGEST MOTOR
• 1Ø
• 230 V
• 5 HP
• 28 A

5 HP
(28 A)

5 HP
(28 A)

1Ø, 230 V MOTORS

1 HP
(16 A)

2 HP
(24 A)

1Ø, 120 V MOTORS

30 A CONTINUOUS-DUTY APPLIANCE PER PHASE

40 A CONTINUOUS-DUTY LIGHTING PER PHASE

**FEEDER AND SERVICE LOAD CALCULATIONS (GENERAL)
220.40**

Purpose of Change: To relocate and renumber rules in **220.10** to **220.40**.

NEC Ch. 2 – Article 220
Part III – 220.42

Type of Change		Panel Action		UL	UL 508	API 500	API 505	OSHA
Renumbered		Accept in Principle		542	-	-	-	1910.304(e)(1)(i)
ROP		ROC		NFPA 70E	NFPA 70B	NFPA 79	NFPA	NEMA
pg. 337	# 2-292	pg. -	# -	420.10(A)	21.3	7.2.6	-	OS 1
log: 1299	CMP: 2	log: -	Submitter: TCC			2002 NEC: 220.11		IEC: Ch. 43

2002 NEC – ~~220.11~~ General Lighting

The demand factors specified in ~~Table 220.11~~ shall apply to that portion of the total branch-circuit load ~~computed~~ for general illumination. They shall not be applied in determining the number of branch circuits for general illumination.

2005 NEC – <u>220.42</u> General Lighting

The demand factors specified in <u>Table 220.42</u> shall apply to that portion of the total branch-circuit load calculated for general illumination. They shall not be applied in determining the number of branch circuits for general illumination.

Author's Substantiation. Article 220 has been reorganized and renumbered under the guidance of the NFPA Task Group on Usability.

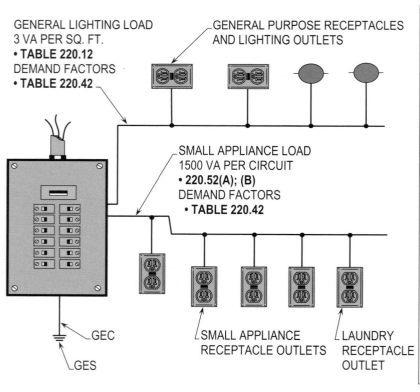

GENERAL LIGHTING LOAD 3 VA PER SQ. FT.
• TABLE 220.12
DEMAND FACTORS
• TABLE 220.42

GENERAL PURPOSE RECEPTACLES AND LIGHTING OUTLETS

SMALL APPLIANCE LOAD 1500 VA PER CIRCUIT
• 220.52(A); (B)
DEMAND FACTORS
• TABLE 220.42

GEC
GES
SMALL APPLIANCE RECEPTACLE OUTLETS
LAUNDRY RECEPTACLE OUTLET

What is the demand load for the general lighting load of a 3500 sq. ft. dwelling unit?

Step 1: Calculating the VA
Table 220.12
3500 sq. ft. x 3 VA = 10,500 VA

Step 2: Small appliance load
220.52(A)
1500 VA x 2 = 3,000 VA

Step 3: Laundry load
220.52(B)
1500 VA x 1 = 1,500 VA

Step 4: Total loads
General lighting load = 10,500 VA
Small appliance load = 3,000 VA
Laundry load = 1,500 VA
Total = 15,000 VA

Step 5: Applying demand factors
Table 220.42
First 3000 VA x 100% = 3,000 VA
Next 12,000 VA x 35% = 4,200 VA
Total = 7,200 VA

Solution: Demand load 1 is 7200 volt amps.

GENERAL LIGHTING
220.42; TABLE 220.42

Purpose of Change: To relocate and renumber rules in **Table 220.11** to **Table 220.42**.

NEC Ch. 2 – Article 220
Part III – 220.43

Type of Change	Panel Action	UL	UL 508	API 500	API 505	OSHA
Renumbered	Accept in Principle	2305	-	-	-	-
ROP	**ROC**	**NFPA 70E**	**NFPA 70B**	**NFPA 79**	**NFPA**	**NEMA**
pg. 337 / # 2-292	pg. - / # -	-	-	-	-	270
log: 1299 / CMP: 2	log: -	Submitter: TCC		2002 NEC: 220.12		IEC: Ch. 43

2002 NEC – ~~220.12~~ Show-Window and Track Lighting

(A) Show Windows. For show-window lighting, a load of not less than 660 volt-amperes/linear meter or 200 volt-amperes/linear foot shall be included for a show window, measured horizontally along its base.

FPN: See ~~220.3(B)(7)~~ for branch circuits supplying show windows.

(B) Track Lighting. For track lighting in other than dwelling units or guest rooms of hotels or motels, an additional load of 150 volt-amperes shall be included for every 600 mm (2 ft) of lighting track or fraction thereof. Where multicircuit track is installed, the total load shall be considered to be divided equally between the track circuits.

2005 NEC – 220.43 Show-Window and Track Lighting

(A) Show Windows. For show-window lighting, a load of not less than 660 volt-amperes/linear meter or 200 volt-amperes/linear foot shall be included for a show window, measured horizontally along its base.

FPN: See 220.14(G) for branch circuits supplying show windows.

(B) Track Lighting. For track lighting in other than dwelling units or guest rooms or guest suites of hotels or motels, an additional load of 150 volt-amperes shall be included for every 600 mm (2 ft) of lighting track or fraction thereof. Where multicircuit track is installed, the total load shall be considered to be divided equally between the track circuits.

Author's Substantiation. Article 220 has been reorganized and renumbered under the guidance of the NFPA Task Group on Usability.

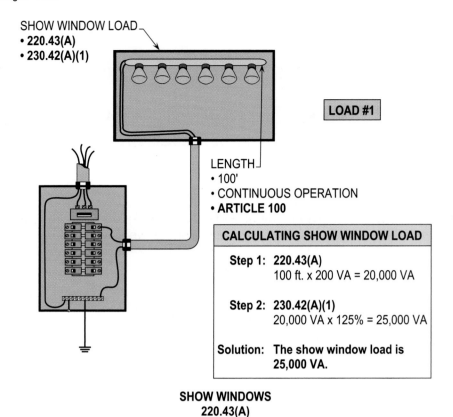

SHOW WINDOW LOAD
- **220.43(A)**
- **230.42(A)(1)**

LOAD #1

LENGTH
- 100'
- CONTINUOUS OPERATION
- **ARTICLE 100**

CALCULATING SHOW WINDOW LOAD

Step 1: **220.43(A)**
100 ft. x 200 VA = 20,000 VA

Step 2: **230.42(A)(1)**
20,000 VA x 125% = 25,000 VA

Solution: **The show window load is 25,000 VA.**

**SHOW WINDOWS
220.43(A)**

Purpose of Change: To relocate and renumber rules in **220.12(A)** to **220.43(A)**.

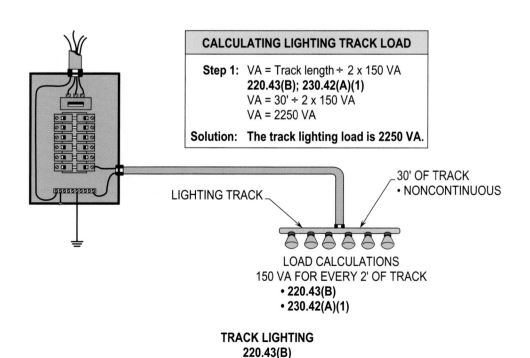

CALCULATING LIGHTING TRACK LOAD

Step 1: VA = Track length ÷ 2 x 150 VA
220.43(B); 230.42(A)(1)
VA = 30' ÷ 2 x 150 VA
VA = 2250 VA

Solution: **The track lighting load is 2250 VA.**

30' OF TRACK
- NONCONTINUOUS

LIGHTING TRACK

LOAD CALCULATIONS
150 VA FOR EVERY 2' OF TRACK
- **220.43(B)**
- **230.42(A)(1)**

**TRACK LIGHTING
220.43(B)**

Purpose of Change: To relocate and renumber rules in **220.12(B)** to **220.43(B)**.

NEC Ch. 2 – Article 220
Part III – 220.44

Type of Change		Panel Action		UL	UL 508	API 500	API 505	OSHA
Renumbered		Accept in Principle		1574	-	-	-	-
ROP		ROC		NFPA 70E	NFPA 70B	NFPA 79	NFPA	NEMA
pg. 337	# 2-292	pg. -	# -	-	-	-	-	270
log: 1299	CMP: 2	log: -	Submitter: TCC			2002 NEC: 220.13		IEC: -

2002 NEC – ~~220.13~~ Receptacle Loads – ~~Nondwelling~~ Units

~~In other than dwelling units,~~ receptacle loads ~~computed at not more than 180 volt-amperes per outlet in accordance with 220.3(B)(9) and fixed multioutlet assemblies computed in accordance with 220.3(B)(8)~~ shall be permitted to be ~~added to the lighting loads and~~ made subject to the demand factors given in ~~Table 220.11,~~ or ~~they shall be permitted to be made subject to the demand factors given in Table 220.13.~~

2005 NEC – 220.44 Receptacle Loads – Other Than Dwelling Units

Receptacle loads calculated in accordance with 220.14(H) and (I) shall be permitted to be made subject to the demand factors given in Table 220.42 or Table 220.44.

Author's Substantiation. Article 220 has been reorganized and renumbered under the guidance of the NFPA Task Group on Usability.

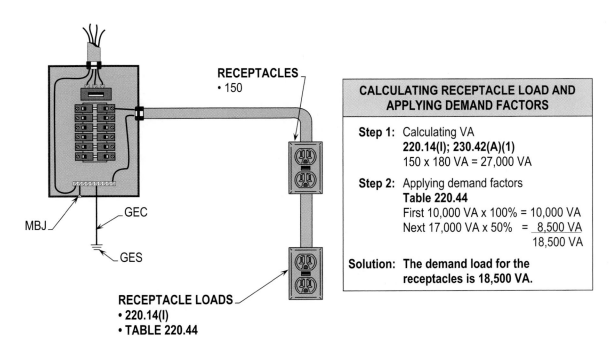

RECEPTACLES
• 150

RECEPTACLE LOADS
• 220.14(I)
• TABLE 220.44

MBJ
GEC
GES

CALCULATING RECEPTACLE LOAD AND APPLYING DEMAND FACTORS

Step 1: Calculating VA
220.14(I); 230.42(A)(1)
150 x 180 VA = 27,000 VA

Step 2: Applying demand factors
Table 220.44
First 10,000 VA x 100% = 10,000 VA
Next 17,000 VA x 50% = 8,500 VA
18,500 VA

Solution: The demand load for the receptacles is 18,500 VA.

RECEPTACLE LOADS – OTHER THAN DWELLING UNITS
220.44; TABLE 220.44

Purpose of Change: To relocate and renumber rules in **220.13** to **220.44**.

NEC Ch. 2 – Article 220
Part III – 220.50

Type of Change		Panel Action		UL	UL 508	API 500	API 505	OSHA
Renumbered		Accept in Principle		845	6.14	-	-	1910.305(j)(4)
ROP		ROC		NFPA 70E	NFPA 70B	NFPA 79	NFPA	NEMA
pg. 337	# 2-292	pg. -	# -	400.11	J.4	15.5	-	MG 1
log: 1299	CMP: 2	log: -	Submitter: TCC			2002 NEC: 220.14		IEC: 533.2

2002 NEC – ~~220.14~~ Motors

Motor loads shall be ~~computed~~ in accordance with 430.24, 430.25, and 430.26 and with 440.6 for hermetic refrigerant motor compressors.

2005 NEC – <u>220.50</u> Motors

Motor loads shall be <u>calculated</u> in accordance with 430.24, 430.25, and 430.26 and with 440.6 for hermetic refrigerant motor compressors.

Author's Substantiation. Article 220 has been reorganized and renumbered under the guidance of the NFPA Task Group on Usability.

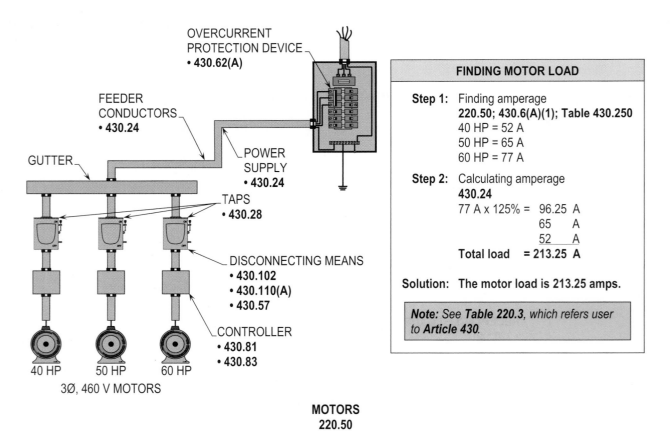

MOTORS
220.50

Purpose of Change: To relocate and renumber rules in **220.14** to **220.50**.

NEC Ch. 2 – Article 220
Part III – 220.51

Type of Change	Panel Action	UL	UL 508	API 500	API 505	OSHA
Renumbered	Accept in Principle	1042	-	-	-	1910.399
ROP	**ROC**	**NFPA 70E**	**NFPA 70B**	**NFPA 79**	**NFPA**	**NEMA**
pg. 337 # 2-292	pg. - # -	Article 100	21.3.12	-	-	DC 3
log: 1299 CMP: 2	log: -	Submitter: TCC		**2002 NEC:** 220.15		**IEC:** 463

2002 NEC – ~~220.15~~ Fixed Electric Space Heating

Fixed electric space heating loads shall be ~~computed~~ at 100 percent of the total connected load; however, in no case shall a feeder or service load current rating be less than the rating of the largest branch circuit supplied.

Exception: Where reduced loading of the conductors results from units operating on duty-cycle, intermittently, or from all units not operating at the same time, the authority having jurisdiction may grant permission for feeder and service conductors to have an ampacity less than 100 percent, provided the conductors have an ampacity for the load so determined.

2005 NEC – 220.51 Fixed Electric Space Heating

Fixed electric space heating loads shall be calculated at 100 percent of the total connected load. However, in no case shall a feeder or service load current rating be less than the rating of the largest branch circuit supplied.

Exception: Where reduced loading of the conductors results from units operating on duty-cycle, intermittently, or from all units not operating at the same time, the authority having jurisdiction may grant permission for feeder and service conductors to have an ampacity less than 100 percent, provided the conductors have an ampacity for the load so determined.

Author's Substantiation. Article 220 has been reorganized and renumbered under the guidance of the NFPA Task Group on Usability.

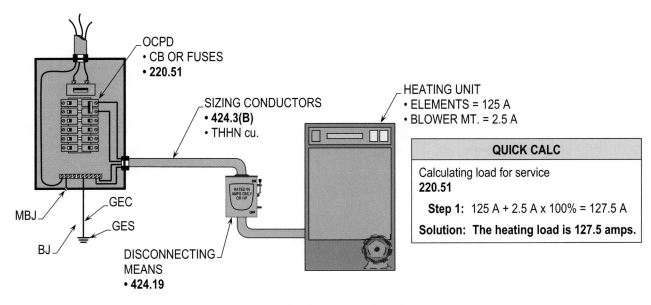

FIXED ELECTRIC SPACE HEATING
220.51

Purpose of Change: To relocate and renumber rules in **220.15** to **220.51**.

NEC Ch. 2 – Article 220
Part III – 220.52

Type of Change	Panel Action	UL	UL 508	API 500	API 505	OSHA
Renumbered	Accept in Principle	498	-	-	-	1910.304(b)
ROP	**ROC**	**NFPA 70E**	**NFPA 70B**	**NFPA 79**	**NFPA**	**NEMA**

pg. 337	# 2-292	pg. -	# -	410.2(B)(2)	27.7.3.2	16.1.1	-	PR 4
log: 1299	CMP: 2	log: -	Submitter: TCC			2002 NEC: 220.16		IEC: 413.1.1.2

2002 NEC – ~~220.16~~ Small Appliance and Laundry Loads – Dwelling Unit

(A) Small Appliance Circuit Load. In each dwelling unit, the load shall be ~~computed~~ at 1500 volt-amperes for each 2-wire small-appliance branch circuit required by 210.11(C)(1). Where the load is subdivided through two or more feeders, the ~~computed~~ load for each shall include not less than 1500 volt-amperes for each 2-wire small-appliance branch circuit. These loads shall be permitted to be included with the general lighting load and subjected to the demand factors provided in ~~Table 220.11~~.

Exception: The individual branch circuit permitted by 210.52(B)(1), Exception No. 2, shall be permitted to be excluded from the calculation required by ~~220.16~~.

(B) Laundry Circuit Load. A load of not less than 1500 volt-amperes shall be included for each 2-wire laundry branch circuit installed as required by 210.11(C)(2). This load shall be permitted to be included with the general lighting load and subjected to the demand factors provided in ~~Table 220.11~~.

2005 NEC – <u>220.52</u> Small Appliance and Laundry Loads – Dwelling Unit

(A) Small Appliance Circuit Load. In each dwelling unit, the load shall be <u>calculated</u> at 1500 volt-amperes for each 2-wire small-appliance branch circuit required by 210.11(C)(1). Where the load is subdivided through two or more feeders, the <u>calculated</u> load for each shall include not less than 1500 volt-amperes for each 2-wire small-appliance branch circuit. These loads shall be permitted to be included with the general lighting load and subjected to the demand factors provided in <u>Table 220.42</u>.

Exception: The individual branch circuit permitted by 210.52(B)(1), Exception No. 2, shall be permitted to be excluded from the calculation required by <u>220.52</u>.

(B) Laundry Circuit Load. A load of not less than 1500 volt-amperes shall be included for each 2-wire laundry branch circuit installed as required by 210.11(C)(2). This load shall be permitted to be included with the general lighting load and subjected to the demand factors provided in <u>Table 220.42</u>.

Author's Substantiation. Article 220 has been reorganized and renumbered under the guidance of the NFPA Task Group on Usability.

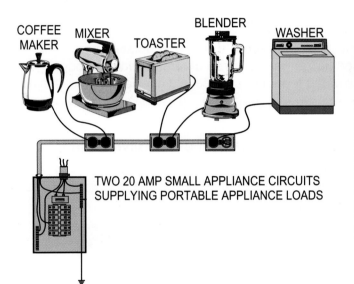

COFFEE MAKER
MIXER
TOASTER
BLENDER
WASHER

TWO 20 AMP SMALL APPLIANCE CIRCUITS
SUPPLYING PORTABLE APPLIANCE LOADS

Calculate the minimum VA rating for the small appliance and laundry loads required in a dwelling unit.

Step 1:   Calculating the loads in VA
          220.52(A)
          1500 VA x 2 = 3000 VA

Step 2:   **220.52(B)**
          1500 VA x 1 = 1500 VA

Solution: The load is **4500 VA.**

SMALL APPLIANCE AND LAUNDRY LOADS
220.52(A); (B)

Purpose of Change: To relocate and renumber rules in **220.16(A)** and **(B)** to **220.52(A)** and **(B)**.

NEC Ch. 2 – Article 220
Part III – 220.53

Type of Change	Panel Action	UL	UL 508	API 500	API 505	OSHA
Renumbered	Accept in Principle	1453	-	-	-	1910.305(j)(3)(i)
ROP	**ROC**	**NFPA 70E**	**NFPA 70B**	**NFPA 79**	**NFPA**	**NEMA**
pg. 337 # 2-292	pg. - # -	Article 100	B.1.5	-	-	-
log: 1299 CMP: 2	log: - Submitter: TCC			2002 NEC: 220.17		IEC: 424.2

2002 NEC – ~~220.17~~ Appliance Load – Dwelling Unit(s)

It shall be permissible to apply a demand factor of 75 percent to the nameplate rating load of four or more appliances fastened in place, other than electric ranges, clothes dryers, space-heating equipment, or air-conditioning equipment, that are served by the same feeder or service in a one-family, two-family, or multifamily dwelling.

2005 NEC – <u>220.53</u> Appliance Load – Dwelling Unit(s)

It shall be permissible to apply a demand factor of 75 percent to the nameplate rating load of four or more appliances fastened in place, other than electric ranges, clothes dryers, space-heating equipment, or air-conditioning equipment, that are served by the same feeder or service in a one-family, two-family, or multifamily dwelling.

Author's Substantiation. Article 220 has been reorganized and renumbered under the guidance of the NFPA Task Group on Usability.

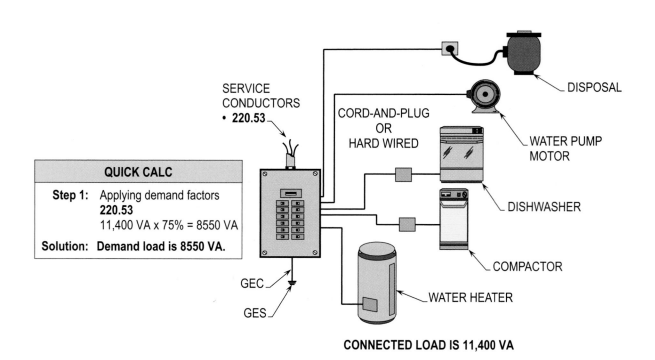

SERVICE CONDUCTORS
• 220.53

CORD-AND-PLUG OR HARD WIRED

DISPOSAL

WATER PUMP MOTOR

DISHWASHER

COMPACTOR

WATER HEATER

GEC

GES

QUICK CALC

Step 1: Applying demand factors
220.53
11,400 VA x 75% = 8550 VA

Solution: **Demand load is 8550 VA.**

CONNECTED LOAD IS 11,400 VA

APPLIANCE LOAD – DWELLING UNIT(S)
220.53

Purpose of Change: To relocate and renumber rules in **220.17** to **220.53**.

NEC Ch. 2 – Article 220
Part III – 220.54

Type of Change	Panel Action	UL	UL 508	API 500	API 505	OSHA
Renumbered	Accept in Principle	2158	-	-	-	-

ROP		ROC		NFPA 70E	NFPA 70B	NFPA 79	NFPA	NEMA
pg. 337	**#** 2-292	**pg.** -	**#** -	-	-	-	-	-
log: 1299	**CMP:** 2	**log:** -	**Submitter:** TCC			**2002 NEC:** 220.18		**IEC:** 311

2002 NEC – ~~220.18~~ Electric Clothes Dryers – Dwelling Unit(s)

The load for household electric clothes dryers in a dwelling unit(s) shall be 5000 watts (volt-amperes) or the nameplate rating, whichever is larger, for each dryer served. The use of the demand factors in ~~Table 220.18~~ shall be permitted. Where two or more single-phase dryers are supplied by a 3-phase, 4-wire feeder or service, the total load shall be ~~computed~~ on the basis of twice the maximum number connected between any two phases.

2005 NEC – <u>220.54</u> Electric Clothes Dryers – Dwelling Unit(s)

The load for household electric clothes dryers in a dwelling unit(s) shall be <u>either</u> 5000 watts (volt-amperes) or the nameplate rating, whichever is larger, for each dryer served. The use of the demand factors in <u>Table 220.54</u> shall be permitted. Where two or more single-phase dryers are supplied by a 3-phase, 4-wire feeder or service, the total load shall be <u>calculated</u> on the basis of twice the maximum number connected between any two phases.

Author's Substantiation. Article 220 has been reorganized and renumbered under the guidance of the NFPA Task Group on Usability.

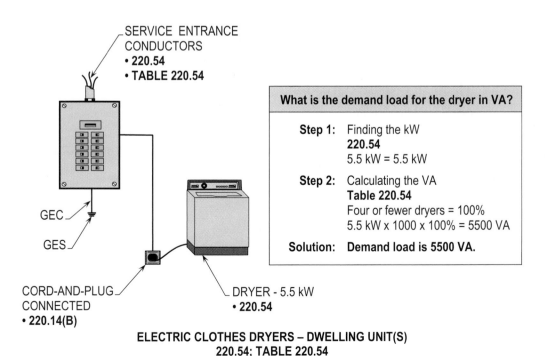

SERVICE ENTRANCE
CONDUCTORS
• **220.54**
• **TABLE 220.54**

GEC

GES

CORD-AND-PLUG
CONNECTED
• **220.14(B)**

DRYER - 5.5 kW
• **220.54**

What is the demand load for the dryer in VA?
Step 1: Finding the kW **220.54** 5.5 kW = 5.5 kW
Step 2: Calculating the VA **Table 220.54** Four or fewer dryers = 100% 5.5 kW x 1000 x 100% = 5500 VA
Solution: Demand load is 5500 VA.

ELECTRIC CLOTHES DRYERS – DWELLING UNIT(S)
220.54; TABLE 220.54

Purpose of Change: To relocate and renumber rules in **Table 220.18** to **Table 220.54**.

NEC Ch. 2 – Article 220
Part III – 220.55

Type of Change		Panel Action		UL	UL 508	API 500	API 505	OSHA
Renumbered		Accept in Principle		858	-	-	-	-
ROP		ROC		NFPA 70E	NFPA 70B	NFPA 79	NFPA	NEMA
pg. 337	# 2-292	pg. -	# -	-	-	-	-	-
log: 1299	CMP: 2	log: -	Submitter: TCC			2002 NEC: 220.19		IEC: 311

2002 NEC – ~~220.19~~ Electric Ranges and Other Cooking Appliances – Dwelling Unit(s)

The ~~demand~~ load for household electric ranges, wall-mounted ovens, counter-mounted cooking units, and other household cooking appliances individually rated in excess of 1 3/4 kW shall be permitted to be ~~computed~~ in accordance with ~~Table 220.19~~. Kilovolt-amperes (kVA) shall be considered equivalent to kilowatts (kW) for loads ~~computed~~ under this section.

Where two or more single-phase ranges are supplied by a 3-phase, 4-wire feeder or service, the total load shall be ~~computed~~ on the basis of twice the maximum number connected between any two phases.

FPN No. 1: See Example D5(A) in Annex D.

FPN No. 2: See ~~Table 220.20~~ for commercial cooking equipment.

FPN No. 3: See the examples in Annex D.

2005 NEC - <u>220.55</u> Electric Ranges and Other Cooking Appliances – Dwelling Unit(s)

The load for household electric ranges, wall-mounted ovens, counter-mounted cooking units, and other household cooking appliances individually rated in excess of 1 3/4 kW shall be permitted to be <u>calculated</u> in accordance with <u>Table 220.55</u>. Kilovolt-amperes (kVA) shall be considered equivalent to kilowatts (kW) for loads <u>calculated</u> under this section.

Where two or more single-phase ranges are supplied by a 3-phase, 4-wire feeder or service, the total load shall be <u>calculated</u> on the basis of twice the maximum number connected between any two phases.

FPN No. 1: See Example D5(A) in Annex D.

FPN No. 2: See <u>Table 220.56</u> for commercial cooking equipment.

FPN No. 3: See the examples in Annex D.

Author's Substantiation. Article 220 has been reorganized and renumbered under the guidance of the NFPA Task Group on Usability.

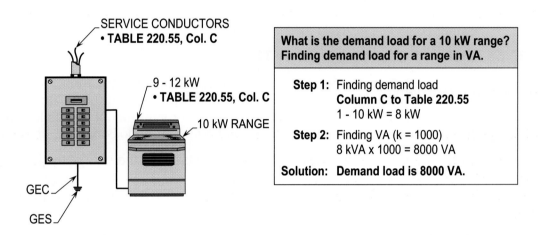

SERVICE CONDUCTORS
• **TABLE 220.55, Col. C**

9 - 12 kW
• **TABLE 220.55, Col. C**

10 kW RANGE

GEC

GES

**What is the demand load for a 10 kW range?
Finding demand load for a range in VA.**

Step 1: Finding demand load
Column C to Table 220.55
1 - 10 kW = 8 kW

Step 2: Finding VA (k = 1000)
8 kVA x 1000 = 8000 VA

Solution: Demand load is 8000 VA.

**ELECTRIC RANGES AND OTHER COOKING APPLIANCES – DWELLING UNITS
220.55; TABLE 220.55**

Purpose of Change: To relocate and renumber rules in **Table 220.19** to **Table 220.55**.

NEC Ch. 2 – Article 220
Part III – 220.56

Type of Change	Panel Action	UL	UL 508	API 500	API 505	OSHA
Renumbered	Accept in Principle	858	-	-	-	-
ROP	**ROC**	**NFPA 70E**	**NFPA 70B**	**NFPA 79**	**NFPA**	**NEMA**
pg. 337 # 2-292	pg. - # -	-	-	-	-	-
log: 1299 CMP: 2	log: -	Submitter: TCC		**2002 NEC:** 220.20		**IEC:** 311

2002 NEC – ~~220.20~~ Kitchen Equipment – Other Than Dwelling Unit(s)

It shall be permissible to ~~compute~~ the load for commercial electric cooking equipment, dishwasher booster heaters, water heaters, and other kitchen equipment in accordance with ~~Table 220.20~~. These demand factors shall be applied to all equipment that has either thermostatic control or intermittent use as kitchen equipment. ~~They~~ shall not apply to space-heating, ventilating, or air-conditioning equipment.

However, in on case shall the feeder or service ~~demand~~ be less than the sum of the largest two kitchen equipment loads.

2005 NEC – <u>220.56</u> Kitchen Equipment – Other Than Dwelling Unit(s)

It shall be permissible to <u>calculate</u> the load for commercial electric cooking equipment, dishwasher booster heaters, water heaters, and other kitchen equipment in accordance with <u>Table 220.56</u>. These demand factors shall be applied to all equipment that has either thermostatic control or intermittent use as kitchen equipment. <u>These demand factors</u> shall not apply to space-heating, ventilating, or air-conditioning equipment.

However, in on case shall the feeder or service <u>calculated load</u> be less than the sum of the largest two kitchen equipment loads.

Author's Substantiation. Article 220 has been reorganized and renumbered under the guidance of the NFPA Task Group on Usability.

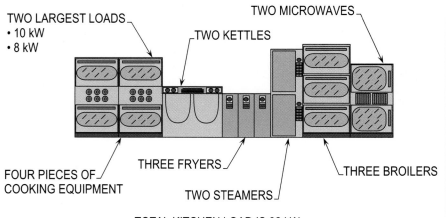

TWO LARGEST LOADS
• 10 kW
• 8 kW

TWO MICROWAVES

TWO KETTLES

FOUR PIECES OF COOKING EQUIPMENT

THREE FRYERS

TWO STEAMERS

THREE BROILERS

TOTAL KITCHEN LOAD IS 82 kW

CALCULATING LOAD FOR COOKING EQUIPMENT

Step 1: Calculating percentage
Table 220.56
16 pieces allowed 65%

Step 2: Applying demand factors
Table 220.56; 220.56
82 kW x 65% = 53.3 kVA

Solution: **The demand load of 53.3 kVA is greater than the two largest loads of 18 kVA.**

KITCHEN EQUIPMENT – OTHER THAN DWELLING UNIT(S)
220.56; TABLE 220.56

Purpose of Change: To relocate and renumber rules in **Table 220.20** to **Table 220.56**.

NEC Ch. 2 – Article 220
Part III – 220.60

Type of Change	Panel Action	UL	UL 508	API 500	API 505	OSHA
Renumbered	Accept in Principle	465	-	-	-	-
ROP	**ROC**	**NFPA 70E**	**NFPA 70B**	**NFPA 79**	**NFPA**	**NEMA**
pg. 337 # 2-292	pg. - # -	-	-	-	-	-
log: 1299 CMP: 2	log: - Submitter: TCC			**2002 NEC:** 220.21		**IEC:** Ch. 43

2002 NEC – ~~220.21~~ Noncoincident Loads

Where it is unlikely that two or more noncoincident loads will be in use simultaneously, it shall be permissible to use only the largest load(s) that will be used at one time, ~~in computing~~ the total load of a feeder or service.

2005 NEC – 220.60 Noncoincident Loads

Where it is unlikely that two or more noncoincident loads will be in use simultaneously, it shall be permissible to use only the largest load(s) that will be used at one time <u>for calculating</u> the total load of a feeder or service.

Author's Substantiation. Article 220 has been reorganized and renumbered under the guidance of the NFPA Task Group on Usability.

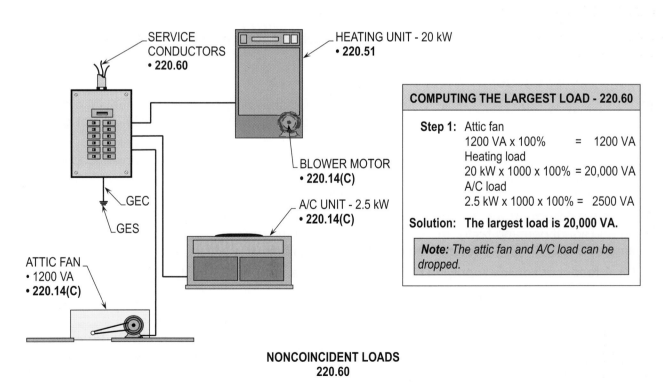

SERVICE CONDUCTORS
• **220.60**

HEATING UNIT - 20 kW
• **220.51**

BLOWER MOTOR
• **220.14(C)**

A/C UNIT - 2.5 kW
• **220.14(C)**

GEC

GES

ATTIC FAN
• 1200 VA
• **220.14(C)**

COMPUTING THE LARGEST LOAD - 220.60

Step 1: Attic fan
1200 VA x 100% = 1200 VA
Heating load
20 kW x 1000 x 100% = 20,000 VA
A/C load
2.5 kW x 1000 x 100% = 2500 VA

Solution: The largest load is 20,000 VA.

Note: The attic fan and A/C load can be dropped.

NONCOINCIDENT LOADS
220.60

Purpose of Change: To relocate and renumber rules in **220.21** to **220.60**.

Type of Change	Panel Action	UL	UL 508	API 500	API 505	OSHA
Revision	Accept in Principle	858	-	-	-	-

ROP		ROC		NFPA 70E	NFPA 70B	NFPA 79	NFPA	NEMA
pg. 357	# 2-336	pg. -	# -	-	-	-	-	-
log: 1306	CMP: 2	log: -	Submitter: TCC			2002 NEC: 220.22		IEC: 311

2002 NEC – ~~220.22~~ Feeder or Service Neutral Load

The feeder or service neutral load shall be the maximum unbalance of the load determined by this article. The maximum unbalanced load shall be the maximum net ~~computed~~ load between the neutral and any one ungrounded conductor, except that the load thus obtained shall be multiplied by 140 percent for 3-wire, 2-phase or 5-wire, 2-phase systems. For a feeder or service supplying household electric ranges, wall-mounted ovens, counter-mounted cooking units, and electric dryers, the maximum unbalanced load shall be considered as 70 percent of the load on the ungrounded conductors, as determined in accordance with ~~Table 220.19~~ for ranges and ~~Table 220.18~~ for dryers. For 3-wire dc or single-phase ac; 4-wire, 3-phase; 3-wire, 2-phase; or 5-wire, 2-phase systems, a further demand factor of 70 percent shall be permitted for that portion of the unbalanced load in excess of 200 amperes. There shall be no reduction of the neutral capacity for that portion of the load that consists of nonlinear loads supplied from a 4-wire, wye-connected, 3-phase system. There shall be no reduction in the capacity of the grounded conductor of a 3-wire circuit consisting of two phase wires and the neutral of a 4-wire, 3-phase, wye-connected system.

FPN No. 1: See Examples D1(A), D1(B), D2(B), D4(A), and D5(A) in Annex D.

FPN No. 2: A 3-phase, 4-wire, wye-connected system used to supply power to nonlinear loads may necessitate that the power system design allow for the possibility of high harmonic neutral currents.

2005 NEC – 220.61 Feeder or Service Neutral Load

(A) **Basic Calculation.** The feeder or service neutral load shall be the maximum unbalance of the load determined by this article. The maximum unbalanced load shall be the maximum net <u>calculated</u> load between the neutral and any one ungrounded conductor.

Exception: For 3-wire, 2-phase or 5-wire, 2-phase systems, the maximum unbalanced load shall be the maximum net calculated load between the neutral and any one ungrounded conductor multiplied by 140 percent.

(B) **Permitted Reductions.** A service or feeder supplying the following loads shall be permitted to have an additional demand factor of 70 percent applied to the amount in 220.61(B)(1) or portion of the amount in 220.61(B)(2) determined by the basic calculation:

(1) A feeder or service supplying household electric ranges, wall-mounted ovens, counter-mounted cooking units, and electric dryers, where the maximum unbalanced load has been determined in accordance with Table 220.55 for ranges and Table 220.54 for dryers

(2) That portion of the unbalanced load in excess of 200 amperes where the feeder or service is supplied from a 3-wire dc or single-phase ac system, or a 4-wire, 3-phase; 3-wire, 2-phase system, or a 5-wire, 2-phase system

(C) **Prohibited Reductions.** There shall be no reduction of the neutral or grounded conductor capacity applied to the amount in 220.61(C)(1), or portion of the amount in (C)(2), from that determined by the basic calculation:

(1) Any portion of a 3-wire circuit consisting of 2-phase wires and the neutral of a 4-wire, 3-phase, wye-connected system

(2) That portion consisting of nonlinear loads supplied from a 4-wire, wye-connected, 3-phase system

FPN No. 1: See Examples D1(A), D1(B), D2(B), D4(A), and D5(A) in Annex D.

FPN No. 2: A 3-phase, 4-wire, wye-connected <u>power</u> system used to supply power to nonlinear loads may necessitate that the power system design allow for the possibility of high harmonic neutral currents.

Author's Substantiation. The section has been restructured to enhance its usability and clarity. Article 220 has been reorganized and renumbered under the guidance of the NFPA Task Group on Usability.

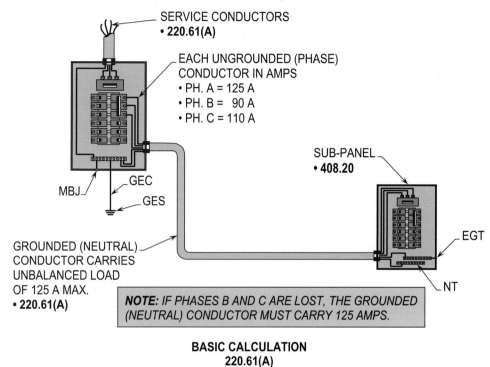

SERVICE CONDUCTORS
• **220.61(A)**

EACH UNGROUNDED (PHASE) CONDUCTOR IN AMPS
• PH. A = 125 A
• PH. B = 90 A
• PH. C = 110 A

SUB-PANEL
• **408.20**

MBJ
GEC
GES

GROUNDED (NEUTRAL) CONDUCTOR CARRIES UNBALANCED LOAD OF 125 A MAX.
• **220.61(A)**

EGT
NT

NOTE: IF PHASES B AND C ARE LOST, THE GROUNDED (NEUTRAL) CONDUCTOR MUST CARRY 125 AMPS.

BASIC CALCULATION
220.61(A)

Purpose of Change: To relocate and renumber rules in **220.22** to **220.61(A)**.

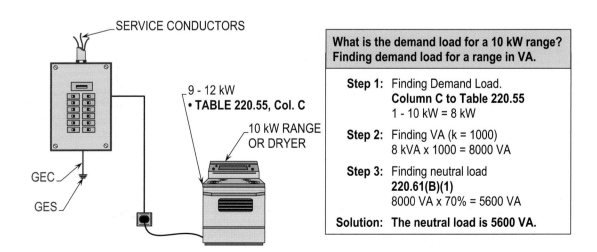

SERVICE CONDUCTORS

9 - 12 kW
• **TABLE 220.55, Col. C**

10 kW RANGE OR DRYER

GEC
GES

What is the demand load for a 10 kW range? Finding demand load for a range in VA.

Step 1: Finding Demand Load.
Column C to Table 220.55
1 - 10 kW = 8 kW

Step 2: Finding VA (k = 1000)
8 kVA x 1000 = 8000 VA

Step 3: Finding neutral load
220.61(B)(1)
8000 VA x 70% = 5600 VA

Solution: The neutral load is 5600 VA.

PERMITTED REDUCTIONS
220.61(B)(1)

Purpose of Change: To relocate and renumber revised rules in **220.22** to **220.61(B)(1)**.

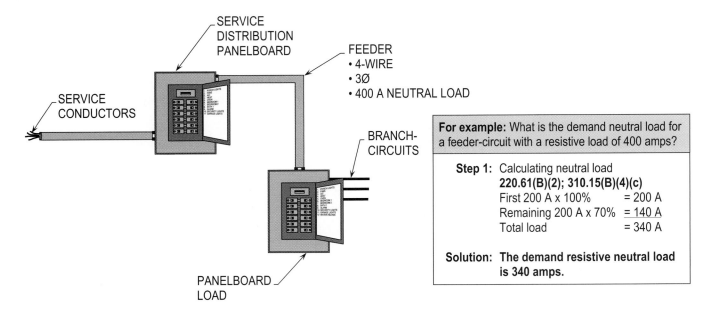

SERVICE
DISTRIBUTION
PANELBOARD

FEEDER
• 4-WIRE
• 3Ø
• 400 A NEUTRAL LOAD

SERVICE
CONDUCTORS

BRANCH-
CIRCUITS

PANELBOARD
LOAD

For example: What is the demand neutral load for a feeder-circuit with a resistive load of 400 amps?

Step 1: Calculating neutral load
220.61(B)(2); 310.15(B)(4)(c)
First 200 A x 100%   = 200 A
Remaining 200 A x 70% = 140 A
Total load   = 340 A

Solution: **The demand resistive neutral load is 340 amps.**

PERMITTED REDUCTIONS
220.61(B)(2)

Purpose of Change: To relocate and renumber revised rules in **220.22** to **220.61(B)(2)**.

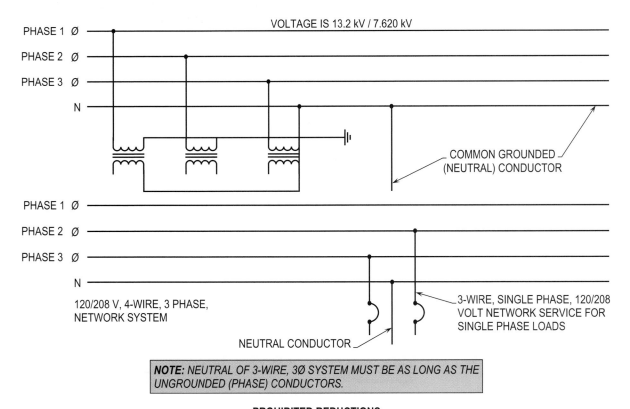

VOLTAGE IS 13.2 kV / 7.620 kV

PHASE 1 Ø
PHASE 2 Ø
PHASE 3 Ø
N

COMMON GROUNDED
(NEUTRAL) CONDUCTOR

PHASE 1 Ø
PHASE 2 Ø
PHASE 3 Ø
N

120/208 V, 4-WIRE, 3 PHASE,
NETWORK SYSTEM

3-WIRE, SINGLE PHASE, 120/208
VOLT NETWORK SERVICE FOR
SINGLE PHASE LOADS

NEUTRAL CONDUCTOR

NOTE: *NEUTRAL OF 3-WIRE, 3Ø SYSTEM MUST BE AS LONG AS THE UNGROUNDED (PHASE) CONDUCTORS.*

PROHIBITED REDUCTIONS
220.61(C)(1)

Purpose of Change: To relocate and renumber revised rules in **220.22** to **220.61(C)(1)**.

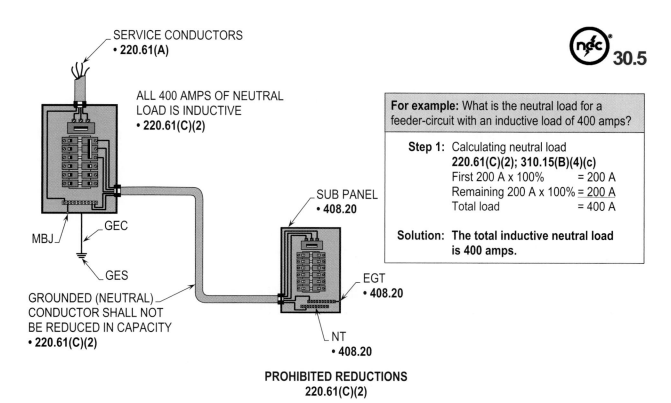

SERVICE CONDUCTORS
• **220.61(A)**

ALL 400 AMPS OF NEUTRAL
LOAD IS INDUCTIVE
• **220.61(C)(2)**

MBJ

GEC

GES

GROUNDED (NEUTRAL)
CONDUCTOR SHALL NOT
BE REDUCED IN CAPACITY
• **220.61(C)(2)**

SUB PANEL
• **408.20**

EGT
• **408.20**

NT
• **408.20**

For example: What is the neutral load for a
feeder-circuit with an inductive load of 400 amps?

Step 1: Calculating neutral load
220.61(C)(2); 310.15(B)(4)(c)
First 200 A x 100%   = 200 A
Remaining 200 A x 100% = 200 A
Total load   = 400 A

Solution: **The total inductive neutral load
is 400 amps.**

PROHIBITED REDUCTIONS
220.61(C)(2)

Purpose of Change: To relocate and renumber revised rules in **220.22** to **220.61(C)(2)**.

NEC Ch. 2 – Article 220
Part IV – 220.80

Type of Change		Panel Action		UL	UL 508	API 500	API 505	OSHA
Revision		Accept		-	-	-	-	-
ROP		ROC		NFPA 70E	NFPA 70B	NFPA 79	NFPA	NEMA
pg. 363	# 2-344	pg. -	# -	-	-	-	-	-
log: 2004	CMP: 2	log: -	Submitter: Neil F. LaBrake, Jr.			2002 NEC: 220.30		IEC: -

2002 NEC - ~~III. Optional Calculations for Computing Feeder and Service Loads~~
~~220.30 Optional Calculation –~~ Dwelling Unit

(A) Feeder and Service Load. For a dwelling unit having the total connected load served by a single ~~3-wire,~~ 120/240-volt or 208Y/120-volt set of service or feeder conductors with an ampacity of 100 ~~amperes~~ or greater, it shall be permissible to ~~compute~~ the feeder and service loads in accordance with this section instead of the method specified in ~~Part II~~ of this article. The calculated load shall be the result of adding the loads from ~~220.30(B) and (C)~~. Feeder and service-entrance conductors whose ~~demand~~ load is determined by this optional calculation shall be permitted to have the neutral load determined by ~~220.22~~.

(B) General Loads. The general calculated load shall be not less than 100 percent of the first 10 kVA plus 40 percent of the remainder of the following loads:

(1) 1500 volt-amperes for each 2-wire, 20-ampere small-appliance branch circuit and each laundry branch circuit specified in ~~220.16~~.

(2) 33 volt-amperes/m^2 or 3 volt-amperes/ft^2 for general lighting and general-use receptacles. The floor area for each floor shall be computed from the outside dimensions of the dwelling unit. The computed floor area shall not include open porches, garages, or unused or unfinished spaces not adaptable for future use.

(3) The nameplate rating of all appliances that are fastened in place, permanently connected, or located to be on a specific circuit, ranges, wall-mounted ovens, counter-mounted cooking units, clothes dryers, and water heaters.

(4) The nameplate ampere or kVA rating of all motors and of all low-power-factor loads.

(C) Heating and Air-Conditioning Load. The largest of the following six selections (load in kVA) shall be included:

(1) 100 percent of the nameplate rating(s) of the air conditioning and cooling.

(2) 100 percent of the nameplate ratings of the ~~heat pump compressors and supplemental heating unless the controller prevents the compressor and supplemental heating from operating at the same time.~~

(3) 100 percent of the nameplate ratings of electric thermal storage and other heating systems where the usual load is expected to be continuous at the full nameplate value. Systems qualifying under the selection shall not be calculated under any other selection in ~~220.30(C)~~.

(4) ~~65 percent of the nameplate(s) of the central electric space heating in heat pumps where the controller prevents the compressor and supplemental heating from operating at the same time.~~

(5) 65 percent of the nameplate rating(s) of electric space heating if less than four separately controlled units.

(6) 40 percent of the nameplate rating(s) of electric space heating if four or more separately controlled units.

2005 NEC – IV. Optional Feeder and Service Load Calculations
220.80 General

Optional feeder and service load calculations shall be permitted in accordance with Part IV.

220.82 Dwelling Unit

(A) Feeder and Service Load. This section applies to a dwelling unit having a total connected load served by a single 120/240-volt or 208Y/120-volt set of 3-wire service or feeder conductors with an ampacity of 100 or greater. It shall be permissible to calculate the feeder and service loads in accordance with this section instead of the method specified in Part III of this article. The calculated load shall be the result of adding the loads from 220.82(B) and (C). Feeder and service-entrance conductors whose calculated load is determined by this optional calculation shall be permitted to have the neutral load determined by 220.61.

(B) General Loads. The general calculated load shall be not less than 100 percent of the first 10 kVA plus 40 percent of the remainder of the following loads:

(1) 33 volt-amperes/m^2 or 3 volt-amperes/ft^2 for general lighting and general-use receptacles. The floor area for each floor shall be calculated from the outside dimensions of the dwelling unit. The calculated floor area shall not include open porches, garages, or unused or unfinished spaces not adaptable for future use.

(2) 1500 volt-amperes for each 2-wire, 20-ampere small-appliance branch circuit and each laundry branch circuit specified in 220.52.

(3) The nameplate rating of all appliances that are fastened in place, permanently connected, or located to be on a specific circuit, ranges, wall-mounted ovens, counter-mounted cooking units, clothes dryers, and water heaters.

(4) The nameplate ampere or kVA rating of all motors and of all low-power-factor loads.

(C) Heating and Air-Conditioning Load. The largest of the following six selections (load in kVA) shall be included:

(1) 100 percent of the nameplate rating(s) of the air conditioning and cooling.

(2) 100 percent of the nameplate rating(s) of the heating when a heat pump is used without any supplemental electric heating.

(3) 100 percent of the nameplate ratings of electric thermal storage and other heating systems where the usual load is expected to be continuous at the full nameplate value. Systems qualifying under this selection shall not be calculated under any other selection in 220.82(C).

(4) 100 percent of the nameplate rating(s) of the heat pump compressor and 65 percent of the supplemental electric heating for central electric space heating systems. If the heat pump compressor is prevented from operating at the same time as the supplementary heat, it does not need to be added to supplementary heat for the total central space heating load.

(5) 65 percent of the nameplate rating(s) of electric space heating if less than four separately controlled units.

(6) 40 percent of the nameplate rating(s) of electric space heating if four or more separately controlled units.

Author's Substantiation. A change has been made to recognize the disparity of calculating a larger service size for a heat pump compressor and supplemental heating system than the service size calculated for a total resistance type space heating system. Article 220 has been reorganized and renumbered under the guidance of the NFPA Task Group on Usability.

DWELLING UNIT
• 2500 SQ. FT.

HEATING UNIT
• 10,000 VA
• 2 UNITS

A/C UNIT
• 6000 VA

LOAD CALCULATED FOR GENERAL LIGHTING, SMALL APPLIANCE, AND SPECIAL APPLIANCE LOADS IS 50,200 VA PER PART III OF ARTICLE 220

Applying **220.82**	Applying **220.82**
First 10,000 VA X 100% = 10,000 VA	(Selecting largest load)
Next 40,200 VA X 40% = 16,080 VA	**220.82(C)(1); (2); (4); (5)**
Load 1 **= 26,080 VA**	Heating load
	10,000 VA x 2 x 65% = 13,000 VA
	A/C load
	6000 VA x 1 x 100% = 6,000 VA
	Load 2 (largest) **= 13,000 VA**
Total VA	Finding Amps
Load 1 = 26,080 VA	$I = VA \div V$
Load 2 = 13,000 VA	$I = 39,080 \text{ VA} \div 240 \text{ V}$
Total Load = 39,080 VA	$I = 163 \text{ A}$

DWELLING UNIT – OPTIONAL CALCULATION
220.82

Purpose of Change: To relocate and renumber rules in **220.30** to **220.82.**

NEC Ch. 2 – Article 220
Part IV – 220.83

Type of Change	Panel Action	UL	UL 508	API 500	API 505	OSHA
Renumbered	Accept in Principle	-	-	-	-	-
ROP	**ROC**	NFPA 70E	NFPA 70B	NFPA 79	NFPA	NEMA
pg. 337 # 2-292	pg. - # -	-	-	-	-	-
log: 1299 CMP: 2	log: - Submitter: TCC			2002 NEC: 220.31		IEC: -

2002 NEC – ~~220.31 Optional Calculations for Additional Loads in an~~ Existing Dwelling Unit

This section shall be permitted to be used to determine if the existing service or feeder is of sufficient capacity to serve additional loads. Where the dwelling unit is served by a 120/240-volt or 208Y/120-volt, 3-wire service, it shall be permissible to compute the total load in accordance with ~~220.31(A) or (B).~~

(A) Where Additional Air-Conditioning Equipment or Electric Space-Heating Equipment Is Not to Be Installed. The following formula shall be used for existing and additional new loads.

Load (~~kVa~~)	Percent of Load
First 8 ~~kVa~~ of load at	100
Remainder of load at	40

Load calculations shall include the following:

~~(1)~~ General lighting and general-use receptacles at 33 volt-amperes/m^2 or 3 volt-amperes/ft^2 as determined by ~~220.3(A)~~

(2) 1500 volt-amperes for each 2-wire, 20-ampere small-appliance branch circuit and each laundry branch circuit specified in ~~220.16~~

(3) Household range(s), wall-mounted oven(s), and counter-mounted cooking unit(s)

(4) All other appliances that are permanently connected, fastened in place, or connected to a dedicated circuit, at nameplate rating

(B) Where Additional Air-Conditioning Equipment or Electric Space-Heating Equipment Is to Be Installed. The following formula shall be used for existing and additional new loads. The larger connected load or air-conditioning or space-heating, but not both, shall be used.

Air-conditioning equipment	100
Central electric space heating	100
Less than four separately controlled space-heating units	100
First 8 kVA of all other loads	100
Remainder of all other loads	40

~~Other loads shall include the following:~~

(1) General lighting and general-use receptacles at 33 volt-ampere/m^2 or 3 volt-amperes/ft^2 as determined by ~~220.3(A)~~

(2) 1500 volt-amperes for each 2-wire, 20-ampere small-appliance branch circuit and each laundry branch circuit specified in ~~220.16~~

(3) Household range(s), wall-mounted oven(s), and counter-mounted cooking unit(s)

(4) All other appliances that are permanently connected, fastened in place, or connected to a dedicated circuit, including four or more separately controlled space-heating units, at nameplate rating

2005 NEC – 220.83 Existing Dwelling Unit

This section shall be permitted to be used to determine if the existing service or feeder is of sufficient capacity to serve additional loads. Where the dwelling unit is served by a 120/240-volt or 208Y/120-volt, 3-wire service, it shall be permissible to <u>calculate</u> the total load in accordance with <u>220.83(A) or (B)</u>.

(A) Where Additional Air-Conditioning Equipment or Electric Space-Heating Equipment Is Not to Be Installed. The following formula shall be used for existing and additional new loads.

Load <u>(kVA)</u>	Percent of Load
First 8 <u>kVA</u> of load at	100
Remainder of load at	40

Load calculations shall include the following:

(1) General lighting and general-use receptacles at 33 volt-amperes/m^2 or 3 volt-amperes/ft^2 as determined by <u>220.12</u>

(2) 1500 volt-amperes for each 2-wire, 20-ampere small-appliance branch circuit and each laundry branch circuit specified in <u>220.52</u>

(3) Household range(s), wall-mounted oven(s), and counter-mounted cooking unit(s)

(4) All other appliances that are permanently connected, fastened in place, or connected to a dedicated circuit, at nameplate rating

(B) Where Additional Air-Conditioning Equipment or Electric Space-Heating Equipment Is to Be Installed. The following formula shall be used for existing and additional new loads. The larger connected load or air-conditioning or space-heating, but not both, shall be used.

Load	Percent of Load
Air-conditioning equipment	100
Central electric space heating	100
Less than four separately controlled space-heating units	100
First 8 kVA of all other loads	100
Remainder of all other loads	40

<u>Other loads shall include the following:</u>

(1) General lighting and general-use receptacles at 33 volt-ampere/m^2 or 3 volt-amperes/ft^2 as determined by <u>220.12</u>

(2) 1500 volt-amperes for each 2-wire, 20-ampere small-appliance branch circuit and each laundry branch circuit specified in <u>220.52</u>

(3) Household range(s), wall-mounted oven(s), and counter-mounted cooking unit(s)

(4) All other appliances that are permanently connected, fastened in place, or connected to a dedicated circuit, including four or more separately controlled space-heating units, at nameplate rating

Author's Substantiation. Article 220 has been reorganized and renumbered under the guidance of the NFPA Task Group on Usability.

DWELLING UNIT
• 1800 SQ. FT.

LOAD CALCULATED FOR GENERAL LIGHTING, SMALL APPLIANCE, AND SPECIAL APPLIANCE LOADS IS 30,000 VA	
Applying **220.83(A)** First 8,000 VA X 100% = 8,000 VA Next 22,000 VA X 40% = 8,800 VA **Load 1** **= 16,800 VA**	Applying **220.83(B)** (Calculating added load) Fixed appliance load 5040 VA x 100% = 5,040 VA **Load 2** **= 5,040 VA**
Total VA Load 1 = 16,800 VA Load 2 = 5,040 VA **Total Load = 21,840 VA**	Can the additional load be added? I = VA ÷ V I = 21,840 VA ÷ 240 V **I = 91 A** Existing Service = 100 A **Newly Computed = 91 A** **The answer is yes.**

EXISTING DWELLING UNIT
220.83

Purpose of Change: To relocate and renumber revised rules in **220.31** to **220.83**.

Type of Change		Panel Action		UL	UL 508	API 500	API 505	OSHA
Renumbered		Accept in Principle		-	-	-	-	-
ROP			ROC	NFPA 70E	NFPA 70B	NFPA 79	NFPA	NEMA
pg. 337	# 2-292	pg. -	# -	-	-	-	-	-
log: 1299	CMP: 2	log: -	Submitter: TCC			2002 NEC: 220.32		IEC: -

2002 NEC – ~~220.32 Optional Calculation~~ – Multifamily Dwelling

(A) Feeder or Service Load. It shall be permissible to ~~compute~~ the load of a feeder or service that supplies ~~more than two~~ dwelling units of a multifamily dwelling in accordance with ~~Table 220.32~~ instead of ~~Part II~~ of this article ~~where~~ all the following conditions are met:

(1) No dwelling unit is supplied by more than one feeder.

(2) Each dwelling unit is equipped with electric cooking equipment.

> **Exception:** Where the ~~computed~~ load for multifamily dwellings without electric cooking in ~~Part II~~ of this article exceeds that ~~computed~~ under ~~Part III~~ for the identical load plus electric cooking (based on 8 kW per unit), the lesser of the two loads shall be permitted to be used.

(3) Each dwelling unit is equipped with either electric space heating, air conditioning, or both. Feeders and service conductors whose ~~demand~~ load is determined by this optional calculation shall be permitted to have the neutral load determined by ~~220.22~~.

(B) House Loads. House loads shall be ~~computed~~ in accordance with ~~Part II~~ of this article and shall be in addition to the dwelling unit loads ~~computed~~ in accordance with ~~Table 220.32~~.

(C) Connected loads. The ~~computed~~ load to which the demand factors of ~~Table 220.32~~ apply shall include the following:

(1) 1500 volt-amperes for each 2-wire, 20-ampere small appliance branch circuit and each laundry branch-circuit specified in ~~220.16~~.

(2) 33 volt-amperes/m^2 or 3 volt-amperes/ft^2 for general lighting and general-use receptacles.

(3) The nameplate rating of all appliances that are fastened in place, permanently connected or located to be on a specific circuit, ranges, wall-mounted ovens, counter-mounted cooking units, clothes dryers, water heaters, and space heaters. If water heater elements are interlocked so that all elements cannot be used at the same time, the maximum possible load shall be considered the nameplate load.

(4) The nameplate ampere or kilovolt-ampere rating of all motors and of all low-power-factor loads.

(5) The larger of the air-conditioning load or the space-heating load.

2005 NEC – __220.84__ Multifamily Dwelling

(A) Feeder or Service Load. It shall be permissible to __calculate__ the load of a feeder or service that supplies __three or more__ dwelling units of a multifamily dwelling in accordance with __Table 220.84__ instead of __Part III__ of this article if all the following conditions are met:

(1) No dwelling unit is supplied by more than one feeder.

(2) Each dwelling unit is equipped with electric cooking equipment.

Exception: Where the <u>calculated</u> load for multifamily dwellings without electric cooking in <u>Part III</u> of this article exceeds that <u>calculated</u> under <u>Part IV</u> for the identical load plus electric cooking (based on 8 kW per unit), the lesser of the two loads shall be permitted to be used.

(3) Each dwelling unit is equipped with either electric space heating, air conditioning, or both. Feeders and service conductors whose <u>calculated</u> load is determined by this optional calculation shall be permitted to have the neutral load determined by <u>220.61</u>.

(B) House Loads. House loads shall be <u>calculated</u> in accordance with <u>Part III</u> of this article and shall be in addition to the dwelling unit loads <u>calculated</u> in accordance with <u>Table 220.84</u>.

(C) Connected Loads. The <u>calculated</u> load to which the demand factors of <u>Table 220.84</u> apply shall include the following:

(1) 33 volt-amperes/m^2 or 3 volt-amperes/ft^2 for general lighting and general-use receptacles.

(2) 1500 volt-amperes for each 2-wire, 20-ampere small-appliance branch circuit and each laundry branch circuit specified in <u>220.52</u>.

(3) The nameplate rating of all appliances that are fastened in place, permanently connected or located to be on a specific circuit, ranges, wall-mounted ovens, counter-mounted cooking units, clothes dryers, water heaters, and space heaters. If water heater elements are interlocked so that all elements cannot be used at the same time, the maximum possible load shall be considered the nameplate load.

(4) The nameplate ampere or kilovolt-ampere rating of all motors and of all low-power-factor loads.

(5) The larger of the air-conditioning load or the space-heating load.

Author's Substantiation. Article 220 has been reorganized and renumbered under the guidance of the NFPA Task Group on Usability.

MULTIFAMILY DWELLING
• 25 UNITS

LOAD CALCULATED FOR MULTIFAMILY DWELLING, PER 220.84		
Calculating load **220.84(C)**		
General lighting load	=	187,500 VA
Cooking load	=	300,000 VA
Appliance load	=	205,000 VA
Heating load	=	500,000 VA
Total Load		**= 1,192,500 VA**
Applying Demand Factor **Table 220.84**		
1,192,500 VA x 35%	=	417,375 VA
Total Load	=	**417,375 VA**

MULTIFAMILY DWELLING
220.84; TABLE 220.84

Purpose of Change: To relocate and renumber rules in **220.32** to **220.84** and **Table 220.32** to **Table 220.84.**

NEC Ch. 2 – Article 220
Part IV – 220.85

Type of Change		Panel Action		UL	UL 508	API 500	API 505	OSHA
Renumbered		Accept in Principle		-	-	-	-	-
ROP		ROC		NFPA 70E	NFPA 70B	NFPA 79	NFPA	NEMA
pg. 337	# 2-292	pg. -	# -	-	-	-	-	-
log: 1299	CMP: 2	log: -	Submitter: TCC			2002 NEC: 220.33		IEC: -

2002 NEC – ~~220.33 Optional Calculation –~~ Two Dwelling Units

Where two dwelling units are supplied by a single feeder and the ~~computed~~ load under ~~Part II~~ of this article exceeds that for three identical units ~~computed~~ under ~~220.32~~, the lesser of the two loads shall permitted to be used.

2005 NEC – 220.85 Two Dwelling Units

Where two dwelling units are supplied by a single feeder and the calculated load under Part III of this article exceeds that for three identical units calculated under 220.84, the lesser of the two loads shall permitted to be used.

Author's Substantiation. Article 220 has been reorganized and renumbered under the guidance of the NFPA Task Group on Usability.

TWO DWELLING UNITS

GIVEN:
CALCULATED LOAD PER PART III OF ARTICLE 220 IS 36,000 VA FOR TWO DWELLING UNITS. · CALCULATED LOAD PER TABLE 220.84 OF ARTICLE 220 IS 28,000 VA FOR THREE DWELLING UNITS.
SOLUTION: THE TWO DWELLING UNITS CAN BE CALCULATED AT 28,000 VA.

TWO DWELLING UNITS
220.85

Purpose of Change: To relocate and renumber rules in **220.33** to **220.85**.

NEC Ch. 2 – Article 220
Part IV – 220.86

Type of Change		Panel Action		UL	UL 508	API 500	API 505	OSHA
Renumbered		Accept in Principle		-	-	-	-	-
ROP		ROC		NFPA 70E	NFPA 70B	NFPA 79	NFPA	NEMA
pg. 337	# 2-292	pg. -	# -	-	-	-	-	-
log: 1299	CMP: 2	log: -	Submitter: TCC			2002 NEC: 220.34		IEC: -

2002 NEC – ~~220.34 Optional Calculation –~~ Schools

The calculation of a feeder or service load for schools shall be permitted in accordance with ~~Table 220.34~~ in lieu of ~~Part II~~ of this article where equipped with electric space heating, air conditioning, or both. The connected load to which the demand factors of ~~Table 220.34~~ apply shall include all of the interior and exterior lighting, power, water heating, cooking, other loads, and the larger of the air-conditioning load or space-heating load within the building or structure.

Feeders and service-entrance conductors whose ~~demand~~ load is determined by this optional calculation shall be permitted to have the neutral load determined by ~~220.22~~. Where the building or structure load is calculated by this optional method, feeders within the building or structure shall have ampacity as permitted in ~~Part II~~ of this article; however, the ampacity of an individual feeder shall not be required to be larger than the ampacity for the entire building.

This section shall not apply to portable classroom buildings.

2005 NEC – 220.86 Schools

The calculation of a feeder or service load for schools shall be permitted in accordance with Table 220.86 in lieu of Part III of this article where equipped with electric space heating, air conditioning, or both. The connected load to which the demand factors of Table 220.86 apply shall include all of the interior and exterior lighting, power, water heating, cooking, other loads, and the larger of the air-conditioning load or space-heating load within the building or structure.

Feeders and service-entrance conductors whose calculated load is determined by this optional calculation shall be permitted to have the neutral load determined by 220.61. Where the building or structure load is calculated by this optional method, feeders within the building or structure shall have ampacity as permitted in Part III of this article; however, the ampacity of an individual feeder shall not be required to be larger than the ampacity for the entire building.

This section shall not apply to portable classroom buildings.

Author's Substantiation. Article 220 has been reorganized and renumbered under the guidance of the NFPA Task Group on Usability.

TOTAL SQUARE FOOTAGE OF SCHOOL

Classroom Area	=	30,000 sq. ft.
Auditorium Area	=	5,000 sq. ft.
Cafeteria Area	=	3,000 sq. ft.
Hall Area	=	1,000 sq. ft.
Total Area	=	39,000 sq. ft.

TOTAL VA OF SCHOOL

• 801,693 VA

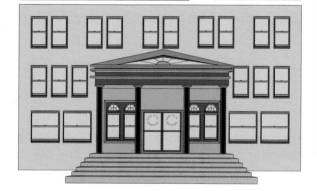

CALCULATING LOAD IN VA

Step 1: Calculating VA per sq. ft.
Table 220.86
VA per sq. ft. = Total VA ÷ Total sq. ft.
VA per sq. ft. = 801,693 VA ÷ 39,000 sq. ft.
VA per sq. ft. = 20.55623

Step 2: Applying demand factors
Table 220.86
First 3 VA per sq. ft. @ 100%
Next 17.55623 @ 75%

3 VA x 39,000 sq. ft. x 100%	= 117,000	VA
17.55623 VA x 39,000 sq. ft. x 75%	= 513,519.72	VA
Total VA	= 630,519.72	VA

Solution: **The demand load for the school is 630,519.72 VA.**

SCHOOLS
220.86; TABLE 220.86

Purpose of Change: To relocate and renumber rules in **220.34** to **220.86** and **Table 220.34** to **Table 220.86**.

NEC Ch. 2 – Article 220
Part IV – 220.87

Type of Change	Panel Action	UL	UL 508	API 500	API 505	OSHA
Renumbered	Accept in Principle	-	-	-	-	-
ROP		ROC		NFPA 70E	NFPA 70B	NFPA 79	NFPA	NEMA

ROP		ROC		NFPA 70E	NFPA 70B	NFPA 79	NFPA	NEMA
pg. 337	# 2-292	pg. -	# -	-	-	-	-	-
log: 1299	CMP: 2	log: -	Submitter: TCC			2002 NEC: 220.35		IEC: -

2002 NEC – ~~220.35 Optional Calculations for~~ Determining Existing Loads

The calculation of a feeder or service load for existing installations shall be permitted to use actual maximum demand to determine the existing load under the following conditions:

(1) The maximum demand data is available for a 1-year period.

Exception: If the maximum demand data for a 1-year period is not available, the calculated load shall be permitted to be based on the maximum demand (measure of average power demand over a 15-minute period) continuously recorded over a minimum 30-day period using a recording ammeter or power meter connected to the highest loaded phase of the feeder or service, based on the initial loading at the start of the recording. The recording shall reflect the maximum demand of the feeder or service by being taken when the building or space is occupied and shall include by measurement or calculation the larger of the heating or cooling equipment load, and other loads that may be periodic in nature due to seasonal or similar conditions.

(2) The maximum demand at 125 percent plus the new load does not exceed the ampacity of the feeder or rating of the service.

(3) The feeder has overcurrent protection in accordance with 240.4, and the service has overload protection in accordance with 230.90.

2005 NEC – 220.87 Determining Existing Loads

The calculation of a feeder or service load for existing installations shall be permitted to use actual maximum demand to determine the existing load under <u>all of</u> the following conditions:

(1) The maximum demand data is available for a 1-year period.

Exception: If the maximum demand data for a 1-year period is not available, the calculated load shall be permitted to be based on the maximum demand (measure of average power demand over a 15-minute period) continuously recorded over a minimum 30-day period using a recording ammeter or power meter connected to the highest loaded phase of the feeder or service, based on the initial loading at the start of the recording. The recording shall reflect the maximum demand of the feeder or service by being taken when the building or space is occupied and shall include by measurement or calculation the larger of the heating or cooling equipment load, and other loads that may be periodic in nature due to seasonal or similar conditions.

(2) The maximum demand at 125 percent plus the new load does not exceed the ampacity of the feeder or rating of the service.

(3) The feeder has overcurrent protection in accordance with 240.4, and the service has overload protection in accordance with 230.90.

Author's Substantiation. Article 220 has been reorganized and renumbered under the guidance of the NFPA Task Group on Usability.

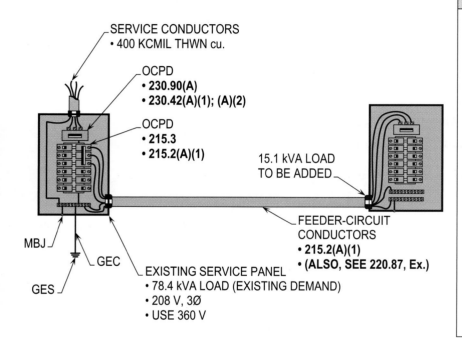

SERVICE CONDUCTORS
• 400 KCMIL THWN cu.

OCPD
• **230.90(A)**
• **230.42(A)(1); (A)(2)**

OCPD
• **215.3**
• **215.2(A)(1)**

15.1 kVA LOAD
TO BE ADDED

MBJ

GEC

GES

**FEEDER-CIRCUIT
CONDUCTORS**
• **215.2(A)(1)**
• **(ALSO, SEE 220.87, Ex.)**

EXISTING SERVICE PANEL
• 78.4 kVA LOAD (EXISTING DEMAND)
• 208 V, 3Ø
• USE 360 V

CALCULATING LOAD IN AMPS

Step 1: Finding demand
220.87
Maximum demand = 78.4 kVA

Step 2: Calculating existing demand
78.4 kVA x 125% = 98 kVA

Step 3: Calculating total kVA
230.42(A)(1)
98 kVA + 15.1 kVA = 113.1 kVA

Step 4: Calculating amperage
Table 310.16; 220.87
400 KCMIL THWN copper = 335 A
113.1 kVA x 1,000 = 113,100 VA
113,100 VA ÷ 208 x 1.732 = 314 A
314 A is less than 335 A

Solution: **The 15.1 kVA demand load can be
applied to the existing service
without upgrading the elements.**

**DETERMINING EXISTING LOADS
220.87**

Purpose of Change: To relocate and renumber rules in **220.35** to **220.87**.

NEC Ch. 2 – Article 220
Part IV – 220.88

Type of Change		Panel Action		UL	UL 508	API 500	API 505	OSHA
Renumbered		Accept in Principle		-	-	-	-	-
ROP		ROC		NFPA 70E	NFPA 70B	NFPA 79	NFPA	NEMA
pg. 337	# 2-292	pg. -	# -	-	-	-	-	-
log: 1299	CMP: 2	log: -	Submitter: TCC			2002 NEC: 220.36		IEC: -

2002 NEC – ~~220.36 Optional Calculation~~ – New Restaurants

Calculation of a service or feeder load, where the feeder serves the total load, for a new restaurant shall be permitted in accordance with ~~Table 220.36~~ in lieu of ~~Part II~~ of this article.

The overload protection of the service conductors shall be in accordance with 230.90 and 240.4.

Feeder conductors shall not be required to be of greater ampacity than the service conductors.

Service or feeder conductors whose ~~demand~~ load is determined by this optional calculation shall be permitted to have the neutral load determined by ~~220.22~~.

2005 NEC – 220.88 New Restaurants

Calculation of a service or feeder load, where the feeder serves the total load, for a new restaurant shall be permitted in accordance with Table 220.88 in lieu of Part III of this article.

The overload protection of the service conductors shall be in accordance with 230.90 and 240.4.

Feeder conductors shall not be required to be of greater ampacity than the service conductors.

Service or feeder conductors whose calculated load is determined by this optional calculation shall be permitted to have the neutral load determined by 220.61.

Author's Substantiation. Article 220 has been reorganized and renumbered under the guidance of the NFPA Task Group on Usability.

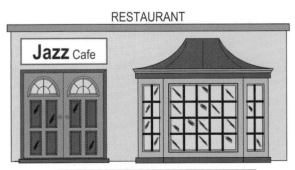

RESTAURANT

GIVEN LOADS

Lighting Loads	= 16,450 VA
Receptacle Loads	= 15,300 VA
Special Loads	= 95,015 VA
Heating or A/C Loads	= 40,000 VA
Motor Loads	= 6,809.9 VA

CALCULATING LOAD IN VA

Calculating total VA
220.88

Lighting loads	= 16,450 VA
Receptacle loads	= 15,300 VA
Special loads	= 95,015 VA
Heating or A/C loads	= 40,000 VA
Motor loads	= 6,809.9 VA
Total Loads	**= 173,574.9 VA**

Applying demand factors
Table 220.88
Demand load = VA x demand factor
Demand load = 173,574.9 VA x 80%
Demand Load = 138,860 VA

Solution: The demand load in VA for the restaurant is 138,860 VA.

NEW RESTAURANTS
220.88

Purpose of Change: To relocate and renumber rules in **220.36** to **220.88**.

NEC Ch. 2 – Article 220
Part V – 220.102

Type of Change	Panel Action	UL	UL 508	API 500	API 505	OSHA
Renumbered	Accept in Principle	-	-	-	-	-
ROP		ROC		NFPA 70E	NFPA 70B	NFPA 79	NFPA	NEMA

ROP		ROC		NFPA 70E	NFPA 70B	NFPA 79	NFPA	NEMA
pg. 337	# 2-292	pg. -	# -	-	-	-	-	-
log: 1299	CMP: 2	log: -	Submitter: TCC			2002 NEC: 220.40		IEC: -

2002 NEC – ~~220.40~~ Farm Loads – Buildings and Other Loads

(A) Dwelling Unit. The feeder or service load of a farm dwelling unit shall be ~~computed~~ in accordance with the provisions for dwellings in ~~Part II or III~~ of this article. Where the dwelling has electric heat and the farm has electric grain-drying systems, ~~Part III~~ of this article shall not be used to ~~compute~~ the dwelling load where the dwelling and farm load are supplied by a common service.

(B) Other Than Dwelling Unit. Where a feeder or service supplies a farm building or other load having two or more separate branch circuits, the load for feeders, service conductors, and service equipment shall be ~~computed~~ in accordance with demand factors not less than indicated in ~~Table 220.40~~.

2005 NEC – <u>220.102</u> Farm Loads – Buildings and Other Loads

(A) Dwelling Unit. The feeder or service load of a farm dwelling unit shall be <u>calculated</u> in accordance with the provisions for dwellings in <u>Part III or IV</u> of this article. Where the dwelling has electric heat and the farm has electric grain-drying systems, <u>Part IV</u> of this article shall not be used to <u>calculate</u> the dwelling load where the dwelling and farm load are supplied by a common service.

(B) Other Than Dwelling Unit. Where a feeder or service supplies a farm building or other load having two or more separate branch circuits, the load for feeders, service conductors, and service equipment shall be <u>calculated</u> in accordance with demand factors not less than indicated in <u>Table 220.102</u>.

Author's Substantiation. Article 220 has been reorganized and renumbered under the guidance of the NFPA Task Group on Usability.

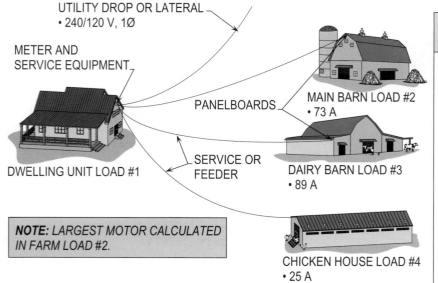

UTILITY DROP OR LATERAL
• 240/120 V, 1Ø

METER AND SERVICE EQUIPMENT

PANELBOARDS

DWELLING UNIT LOAD #1

MAIN BARN LOAD #2
• 73 A

SERVICE OR FEEDER

DAIRY BARN LOAD #3
• 89 A

NOTE: LARGEST MOTOR CALCULATED IN FARM LOAD #2.

CHICKEN HOUSE LOAD #4
• 25 A

What is the load in amps for the three farm loads?

Step 1:	Finding load for 3 farm loads **220.102; Table 220.102**	
	Load #2	= 73 A
	Load #3	= 89 A
	Load #4	= 25 A
	Total Load	= 187 A
Step 2:	Applying demand factors **Table 220.102**	
	First 60 A @ 100%	= 60 A
	Next 60 A @ 50%	= 30 A
	Remainder 67 A @ 25%	= 17 A
	Total load	= 107 A
Solution:	The total load for the three farm loads is 107 amps.	

FARM LOADS – BUILDING AND OTHER LOADS
220.102; TABLE 220.102

Purpose of Change: To relocate and renumber rules in **220.40** to **220.102**.

NEC Ch. 2 – Article 220
Part V – 220.103

Type of Change		Panel Action		UL	UL 508	API 500	API 505	OSHA
Renumbered		Accept in Principle		-	-	-	-	-
ROP		ROC		NFPA 70E	NFPA 70B	NFPA 79	NFPA	NEMA
pg. 337	# 2-292	pg. -	# -	-	-	-	-	-
log: 1299	CMP: 2	log: -	Submitter: TCC			2002 NEC: 220.41		IEC: -

2002 NEC – ~~220.41~~ Farm Loads – Total

Where supplied by a common service, the total load of the farm for service conductors and service equipment shall be ~~computed~~ in accordance with the farm dwelling unit load and demand factors specified in ~~Table 220.41~~. Where there is equipment in two or more farm equipment buildings or for loads having the same function, such loads shall be ~~computed~~ in accordance with ~~Table 220.40~~ and shall be permitted to be combined as a single load in ~~Table 220.41~~ for ~~computing~~ the total load.

2005 NEC – <u>220.103</u> Farm Loads – Total

Where supplied by a common service, the total load of the farm for service conductors and service equipment shall be <u>calculated</u> in accordance with the farm dwelling unit load and demand factors specified in <u>Table 220.103.</u> Where there is equipment in two or more farm equipment buildings or for loads having the same function, such loads shall be <u>calculated</u> in accordance with <u>Table 220.102</u> and shall be permitted to be combined as a single load in <u>Table 220.103</u> for <u>calculating</u> the total load.

Author's Substantiation. Article 220 has been reorganized and renumbered under the guidance of the NFPA Task Group on Usability.

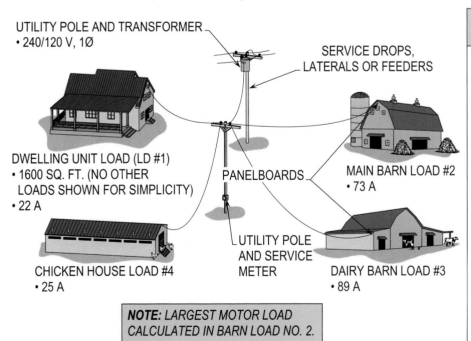

UTILITY POLE AND TRANSFORMER
• 240/120 V, 1Ø

SERVICE DROPS, LATERALS OR FEEDERS

DWELLING UNIT LOAD (LD #1)
• 1600 SQ. FT. (NO OTHER LOADS SHOWN FOR SIMPLICITY)
• 22 A

PANELBOARDS

MAIN BARN LOAD #2
• 73 A

UTILITY POLE AND SERVICE METER

CHICKEN HOUSE LOAD #4
• 25 A

DAIRY BARN LOAD #3
• 89 A

NOTE: LARGEST MOTOR LOAD CALCULATED IN BARN LOAD NO. 2.

What is the load in amps for four farm loads?

Step 1: Finding load for 3-farm loads
220.103; Table 220.103

Load #1	=	22 A
Load #2	=	73 A
Load #3	=	89 A
Load #4	=	25 A

Step 2: Applying demand factors
Largest demand load #1 - Ld. 3

89 A x 100%	=	89 A

Second largest load #2 - Ld. 2

73 A x 75%	=	55 A

Third largest load #3 - Ld. 4

25 A x 65%	=	16 A

Balance of load #4 - Ld. 1

22 A x 50%	=	11 A
Total load	=	171 A

Solution: **The total load for the four loads is 171 amps.**

FARM LOADS – TOTAL
220.103; TABLE 220.103

Purpose of Change: To relocate and renumber rules in **220.41** to **220.103** and **Table 220.41** to **Table 220.103**.

NEC Ch. 2 – Article 225
Part I – 225.10

Type of Change		Panel Action		UL	UL 508	API 500	API 505	OSHA
Revision		Accept		-	-	-	-	-
ROP		ROC		NFPA 70E	NFPA 70B	NFPA 79	NFPA	NEMA
pg. 375	# 4-86	pg. -	# -	-	-	-	-	-
log: CP403	CMP: 4	log: -	Submitter: CMP 4			2002 NEC: 225.10		IEC: -

2002 NEC – 225.10 Wiring on Buildings

The installation of outside wiring on surfaces of buildings shall be permitted for circuits of not over 600 volts, nominal, as open wiring on insulators, as multiconductor cable, as Type MC cable, as Type MI cable, as messenger supported wiring, in rigid metal conduit, in intermediate metal conduit, in rigid nonmetallic conduit, in cable trays, as cablebus, in wireways, in auxiliary gutters, in electrical metallic tubing, in flexible metal conduit, in liquidtight flexible metal conduit, in liquidtight flexible nonmetallic conduit, and in busways. Circuits of over 600 volts, nominal, shall be installed as provided in 300.37. ~~Circuits for signs and outline lighting shall be installed in accordance with Article 600.~~

2005 NEC – 225.10 Wiring on Buildings

The installation of outside wiring on surfaces of buildings shall be permitted for circuits of not over 600 volts, nominal, as open wiring on insulators, as multiconductor cable, as Type MC cable, as Type MI cable, as messenger supported wiring, in rigid metal conduit, in intermediate metal conduit, in rigid nonmetallic conduit, in cable trays, as cablebus, in wireways, in auxiliary gutters, in electrical metallic tubing, in flexible metal conduit, in liquidtight flexible metal conduit, in liquidtight flexible nonmetallic conduit, and in busways. Circuits of over 600 volts, nominal, shall be installed as provided in 300.37.

Author's Substantiation. The deletion of the last sentence improves usability since circuits for signs and outline lighting are covered in Article 600.

WIRING METHODS NOT OVER 600 V SHALL BE PERMITTED TO BE INSTALLED AS OPEN WIRING ON INSULATORS, MULTICONDUCTOR CABLE, MC CABLE, MI CABLE, MESSENGER SUPPORTED WIRING, IN RMC, IMC, RNC, CABLE TRAYS, AS CABLEBUS IN WIREWAYS, AUXILIARY GUTTERS, EMT, FMC, LTFM, LTNFC, AND BUSWAYS.

NOTE: *FOR WIRING METHODS CONNECTING SIGNS AND OUTLINE LIGHTING, SEE* **ARTICLE 600.**

WIRING ON BUILDINGS
225.10

Purpose of Change: To improve usability of the NEC by deleting reference to **Article 600**.

Type of Change	Panel Action	UL	UL 508	API 500	API 505	OSHA
New Section	Accept in Principle	-	-	-	-	-
ROP	ROC	NFPA 70E	NFPA 70B	NFPA 79	NFPA	NEMA
pg. 376 # 4-9a	pg. 105 # 4-6	-	-	-	-	-
log: CP402 CMP: 4	log: 2162 Submitter: CMP 4			2002 NEC: -		IEC: -

2005 NEC – 225.17 Masts as Supports

Where a mast is used for the support of final spans of feeders or branch circuits, it shall be of adequate strength or be supported by braces or guys to withstand safely the strain imposed by the overhead drop. Where raceway-type masts are used, all raceways fittings shall be identified for use with masts. Only the feeder or branch circuit conductors specified within this section shall be permitted to be attached to the feeder and/or branch circuit mast.

Author's Substantiation. This new section recognizes that overhead feeder and branch circuit final spans are sometimes attached to the building through the use of masts.

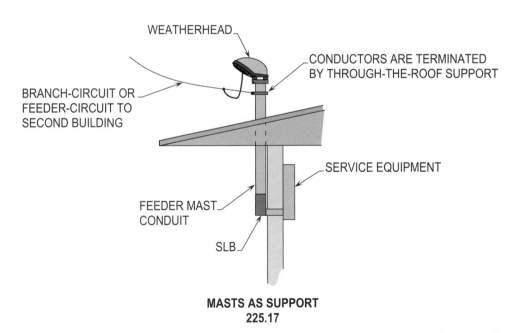

MASTS AS SUPPORT
225.17

Purpose of Change: To recognize a mast as a supporting means for branch-circuits or feeders.

NEC Ch. 2 – Article 225
Part I – 225.18

Type of Change		Panel Action		UL	UL 508	API 500	API 505	OSHA
Revision		Accept in Principle		-	-	-	-	-
ROP		ROC		NFPA 70E	NFPA 70B	NFPA 79	NFPA	NEMA
pg. 376	# 4-11	pg. -	# -	-	-	-	-	-
log: 826	CMP: 2	log: -	Submitter: Dan Leaf			2002 NEC: 225.18		IEC: -

2002 NEC – 225.18 Clearance from Ground

Overhead spans of open conductors and open multiconductor cables of not over 600 volts, nominal, shall ~~conform~~ to the following:

2005 NEC – 225.18 Clearance from Ground

Overhead spans of open conductors and open multiconductor cables of not over 600 volts, nominal, shall <u>have a clearance of not less than</u> the following:

Author's Substantiation. This revision clarifies that overhead spans of open conductors and open multiconductor cables, of not over 600 volts, nominal, shall be installed based on the minimum clearances from ground.

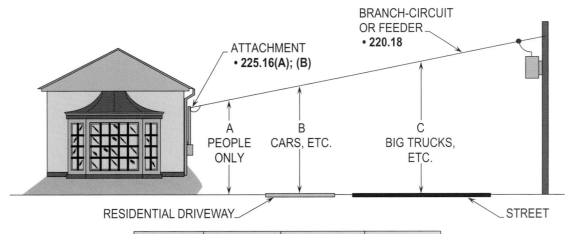

NOMINAL VOLTAGE	SERVICE ENTRANCE (A)	RESIDENTIAL PROPERTY (B)	PUBLIC STREETS (C)
0 - 150 V	10 FT. (3.0 m)	12 FT. (3.7 m)	18 FT. (5.5 m)
151 - 300 V	12 FT. (3.7 m)	12 FT. (3.7 m)	18 FT. (5.5 m)
301 - 600 V	15 FT. (4.5 m)	15 FT. (4.5 m)	18 FT. (5.5 m)

CLEARANCES FROM GROUND
225.18

Purpose of Change: This revision clarifies that overhead spans of open conductors and open multiconductor cables shall be based on the minimum clearances from ground.

Type of Change		Panel Action		UL	UL 508	API 500	API 505	OSHA
Revision		Accept		-	-	-	-	-
ROP		ROC		NFPA 70E	NFPA 70B	NFPA 79	NFPA	NEMA
pg. 382	# 4-20a	pg. -	# -	-	-	-	-	-
log: CP406	CMP: 2	log: -	Submitter: CMP 4			2002 NEC: 225.30		IEC: 60364

2002 NEC – 225.30 Number of Supplies

Where more than one building or other structure is on the same property and under single management, each additional building or other structure ~~served that is~~ on the load side of the service disconnecting means shall be supplied by one feeder or branch circuit unless permitted in 225.30(A) through (E). For the purpose of this section, a multiwire branch circuit shall be considered a single circuit.

2005 NEC – 225.30 Number of Supplies

Where more than one building or other structure is on the same property and under single management, each additional building or other structure <u>that is served by a branch circuit or feeder</u> on the load side of the service disconnecting means shall be supplied by <u>only</u> one feeder or branch circuit unless permitted in 225.30(A) through (E). For the purpose of this section, a multiwire branch circuit shall be considered a single circuit.

Author's Substantiation. This revision clarifies that a building or structure does not have to be fed by a branch circuit or feeder. If a building or structure is fed by a branch circuit or feeder, then compliance with 225.30(A) through (E) shall be followed.

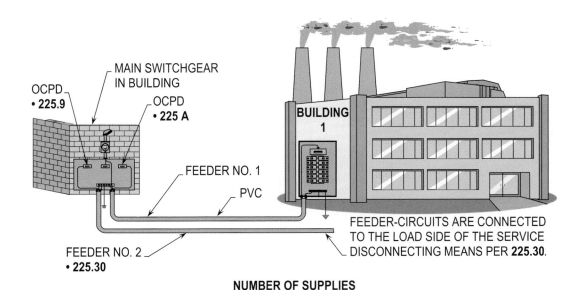

NUMBER OF SUPPLIES
225.30

Purpose of Change: To clarify rules when a building or structure is supplied by a branch-circuit or feeder.

NEC Ch. 2 – Article 225
Part II – 225.30(A)(6)

33

Type of Change		Panel Action		UL	UL 508	API 500	API 505	OSHA
New Item		Accept in Principle		-	-	-	-	-
ROP		ROC		NFPA 70E	NFPA 70B	NFPA 79	NFPA	NEMA
pg. 382	# 4-25a	pg. 107	# 4-19	-	-	-	-	-
log: CP400	CMP: 4	log: 1579	Submitter: CMP 4			2002 NEC: -		IEC: 60364

2005 NEC – 225.30(A)(6) Special Conditions

Additional feeders or branch circuits shall be permitted to supply the following:

(1) Fire pumps

(2) Emergency systems

(3) Legally required standby systems

(4) Optional standby systems

(5) Parallel power production systems

(6) <u>Systems designed for connection to multiple sources of supply for the purpose of enhanced reliability</u>

Author's Substantiation. A new item (A)(6) has been added to 225.30 to clarify that additional feeder or branch circuits shall be permitted for systems designed for connection to multiple sources of supply for the purpose of enhanced reliability. These systems are often necessary where supplying critical loads, such as medical laboratories that would not be normally covered under Article 517, banks, or computer loads, since the loads are not covered in Article 702. Article 702 applies only to onsite generation as indicated in the definition of optional standby systems.

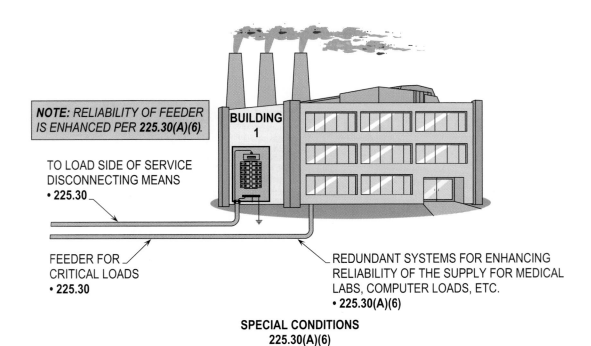

NOTE: *RELIABILITY OF FEEDER IS ENHANCED PER 225.30(A)(6).*

BUILDING 1

TO LOAD SIDE OF SERVICE DISCONNECTING MEANS
• 225.30

FEEDER FOR CRITICAL LOADS
• 225.30

REDUNDANT SYSTEMS FOR ENHANCING RELIABILITY OF THE SUPPLY FOR MEDICAL LABS, COMPUTER LOADS, ETC.
• 225.30(A)(6)

SPECIAL CONDITIONS
225.30(A)(6)

Purpose of Change: To make it clear that additional feeders for supplying critical loads are permitted.

Type of Change	Panel Action	UL	UL 508	API 500	API 505	OSHA
New Item	Accept in Principle	-	-	-	-	-
ROP	ROC	NFPA 70E	NFPA 70B	NFPA 79	NFPA	NEMA
pg. 394 # 4-44a	pg. 110 # 4-33	-	-	-	-	-
log: CP401 CMP: 4	log: 1580 Submitter: CMP 4			2002 NEC: -		IEC: -

2005 NEC – 230.2(A)(6) Special Conditions

Additional services shall be permitted to supply the following:

(1) Fire pumps

(2) Emergency systems

(3) Legally required standby systems

(4) Optional standby systems

(5) Parallel power production systems

(6) Systems designed for connection to multiple sources of supply for the purpose of enhanced reliability

Author's Substantiation. A new item (A)(6) has been added to 230.2 to clarify that additional service shall be permitted for systems designed for connection to multiple sources of supply for the purpose of enhanced reliability. These systems are often necessary where supplying critical loads, such as medical laboratories that would not be normally covered under Article 517, banks, or computer loads, since the loads are not covered in Article 702. Article 702 applies only to onsite generation as indicated in the definition of optional standby systems.

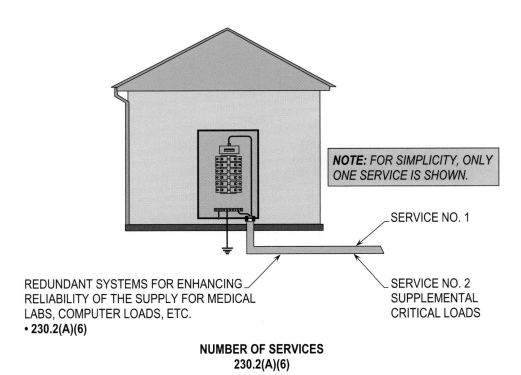

NOTE: FOR SIMPLICITY, ONLY ONE SERVICE IS SHOWN.

SERVICE NO. 1

REDUNDANT SYSTEMS FOR ENHANCING RELIABILITY OF THE SUPPLY FOR MEDICAL LABS, COMPUTER LOADS, ETC.
• 230.2(A)(6)

SERVICE NO. 2 SUPPLEMENTAL CRITICAL LOADS

NUMBER OF SERVICES
230.2(A)(6)

Purpose of Change: To clarify that two additional services to supply critical loads are permitted.

NEC Ch. 2 – Article 230
Part I – 230.9(A)

Type of Change		Panel Action		UL	UL 508	API 500	API 505	OSHA
Revision		Accept in Principle		-	-	-	-	1910.304(C)(3)
ROP		ROC		NFPA 70E	NFPA 70B	NFPA 79	NFPA	NEMA
pg. 399	# 4-55	pg. -	# -	410.7(C)	-	-	-	-
log: 2133	CMP: 4	log: -	Submitter: Joel Weisenberger			2002 NEC: 230.9(A)		IEC: -

2002 NEC – 230.9 Clearance ~~from~~ Building ~~Openings~~

Service conductors and final spans shall comply with 230.9(A), (B), and (C).

(A) Clearance ~~from Windows~~. Service conductors installed as open conductors or multiconductor cable without an overall outer jacket shall have a clearance of not less than 900 mm (3 ft) from windows that are designed to be opened, doors, porches, balconies, ladders, stairs, fire escapes, or similar locations.

2005 NEC – 230.9 Clearance<u>s</u> <u>on</u> Building<u>s</u>

Service conductors and final spans shall comply with 230.9(A), (B), and (C).

(A) Clearance<u>s</u>. Service conductors installed as open conductors or multiconductor cable without an overall outer jacket shall have a clearance of not less than 900 mm (3 ft) from windows that are designed to be opened, doors, porches, balconies, ladders, stairs, fire escapes, or similar locations.

Author's Substantiation. This revision in the title head clarifies that this subsection covers the clearances for service conductors on buildings and not only clearances from windows.

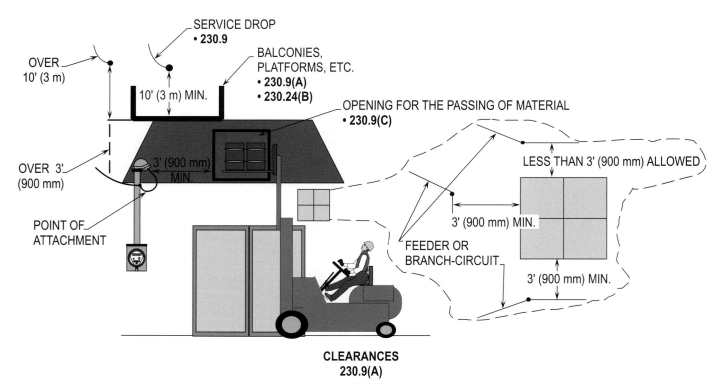

**CLEARANCES
230.9(A)**

Purpose of Change: To clarify that clearance rule applies to windows, balconies, and other areas.

NEC Ch. 2 – Article 230
Part II – 230.23(A)

Type of Change		Panel Action		UL	UL 508	API 500	API 505	OSHA
Revision		Accept		-	-	-	-	-
ROP		ROC		NFPA 70E	NFPA 70B	NFPA 79	NFPA	NEMA
pg. 403	# 4-62	pg. -	# -	-	-	-	-	-
log: 1316	CMP: 4	log: -	Submitter: TCC			2002 NEC: 230.23(A)		IEC: -

2002 NEC – 230.23(A) General

Conductors shall have sufficient ampacity to carry the current for the load as ~~computed~~ in accordance with Article 220 and shall have adequate mechanical strength.

2005 NEC – 230.23(A) General

Conductors shall have sufficient ampacity to carry the current for the load as <u>calculated</u> in accordance with Article 220 and shall have adequate mechanical strength.

Author's Substantiation. This revision clarifies that the term "calculate" will be the standardized language throughout the NEC for determining load values.

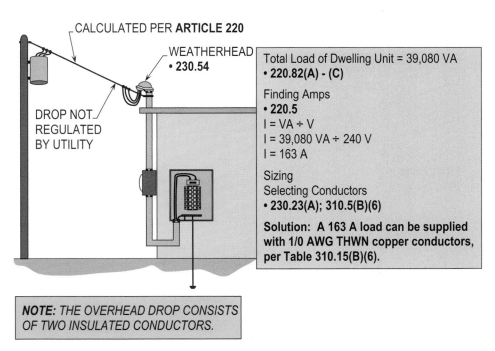

CALCULATED PER **ARTICLE 220**

WEATHERHEAD
• **230.54**

DROP NOT REGULATED BY UTILITY

Total Load of Dwelling Unit = 39,080 VA
• **220.82(A) - (C)**

Finding Amps
• **220.5**
$I = VA \div V$
$I = 39,080\ VA \div 240\ V$
$I = 163\ A$

Sizing
Selecting Conductors
• **230.23(A); 310.5(B)(6)**

Solution: A 163 A load can be supplied with 1/0 AWG THWN copper conductors, per Table 310.15(B)(6).

NOTE: THE OVERHEAD DROP CONSISTS OF TWO INSULATED CONDUCTORS.

SIZE AND RATING (GENERAL)
230.23(A)

Purpose of Change: The term "compute" was replaced with the term "calculated" because loads are usually determined by using a calculator.

NEC Ch. 2 – Article 230
Part II – 230.26

Type of Change		Panel Action		UL	UL 508	API 500	API 505	OSHA
Revision		Accept		-	-	-	-	1910.304(C)
ROP		ROC		NFPA 70E	NFPA 70B	NFPA 79	NFPA	NEMA
pg. 406	# 4-68	pg. -	# -	410.7(C)	-	-	-	-
log: 2006	CMP: 4	log: -	Submitter: Neil F. LaBrake, Jr.			2002 NEC: 230.26		IEC: -

2002 NEC – 230.26 Point of Attachment

The point of attachment of the service-drop conductors to a building or other structure shall provide the minimum clearances as specified in 230.24. In no case shall this point of attachment be less than 3.0 m (10 ft) above finished grade.

2005 NEC – 230.26 Point of Attachment

The point of attachment of the service-drop conductors to a building or other structure shall provide the minimum clearances as specified in <u>230.9</u> and 230.24. In no case shall this point of attachment be less than 3.0 m (10 ft) above finished grade.

Author's Substantiation. This revision clarifies that the clearances on buildings in 230.9 shall be considered for the point of attachment of the service-drop conductors to a building or other structure.

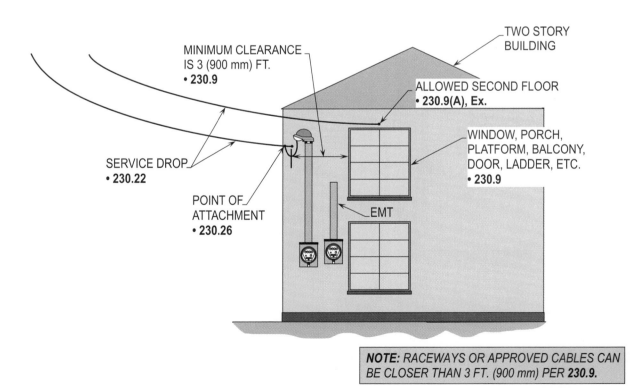

TWO STORY BUILDING

MINIMUM CLEARANCE IS 3 (900 mm) FT.
• **230.9**

ALLOWED SECOND FLOOR
• **230.9(A), Ex.**

WINDOW, PORCH, PLATFORM, BALCONY, DOOR, LADDER, ETC.
• **230.9**

SERVICE DROP
• **230.22**

POINT OF ATTACHMENT
• **230.26**

EMT

NOTE: *RACEWAYS OR APPROVED CABLES CAN BE CLOSER THAN 3 FT. (900 mm) PER* **230.9.**

POINT OF ATTACHMENT
230.26

Purpose of Change: To clarify that the vertical clearances in **230.9** are required.

NEC Ch. 2 – Article 230
Part III – 230.31(A)

Type of Change		Panel Action		UL	UL 508	API 500	API 505	OSHA
Revision		Accept		-	-	-	-	1910.304(e)
ROP		ROC		NFPA 70E	NFPA 70B	NFPA 79	NFPA	NEMA
pg. 403	**#** 4-62	**pg.** -	**#** -	410.9	-	-	-	TC 2
log: 1316	**CMP:** 4	**log:** -	**Submitter:** TCC			**2002 NEC:** 230.31(A)		**IEC:** Ch. 43

2002 NEC – 230.31 Size and Rating

(A) General. Service-lateral conductors shall have sufficient ampacity to carry the current for the load as ~~computed~~ in accordance with Article 220 and shall have adequate mechanical strength.

2005 NEC – 230.31 Size and Rating

(A) General. Service-lateral conductors shall have sufficient ampacity to carry the current for the load as <u>calculated</u> in accordance with Article 220 and shall have adequate mechanical strength.

Author's Substantiation. This revision clarifies that the term "calculate" will be standardized language for determining load values throughout the NEC.

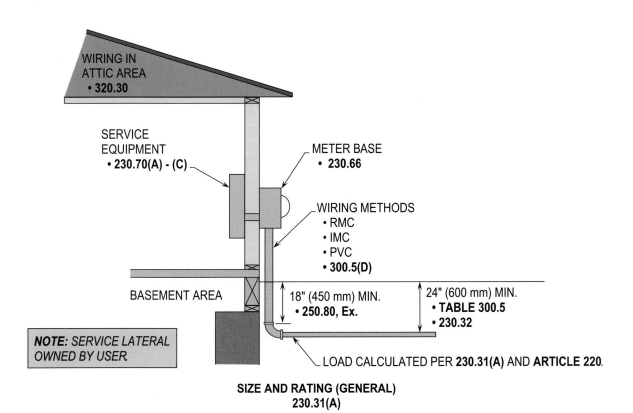

SIZE AND RATING (GENERAL)
230.31(A)

Purpose of Change: The term "computed" was replaced with the term "calculated" because loads are usually determined by using a calculator.

NEC Ch. 2 – Article 230
Part IV – 230.44, Ex.

Type of Change		Panel Action		UL	UL 508	API 500	API 505	OSHA
New Exception		Accept in Principle		White Book	-	-	-	1910.305(a)(2)(B)
ROP		ROC		NFPA 70E	NFPA 70B	NFPA 79	NFPA	NEMA
pg. 412	# 4-77	pg. -	# -	420.1(C)(2)	Ch. 24	14.5.10	-	FG 1
log: 455	CMP: 4	log: -	Submitter: Alan H. Nadon			2002 NEC: 230.44		IEC: 52

2002 NEC – 230.44 Cable Trays

Cable tray systems shall be permitted to support ~~cable used as~~ service-entrance conductors.

2005 NEC – 230.44 Cable Trays

Cable tray systems shall be permitted to support service-entrance conductors. Cable trays used to support service-entrance conductors shall contain only service-entrance conductors.

> **Exception:** Conductors other than service-entrance conductors shall be permitted to be installed in a cable tray with service-entrance conductors, provided a solid fixed barrier of a material compatible with the cable tray is installed to separate the service-entrance conductors from other conductors installed in the cable tray.

Author's Substantiation. A new sentence and exception has been added to clarify that service-entrance conductors shall be permitted to be installed in cable trays. Service-entrance conductors and other conductors shall be permitted in the same cable tray provided a solid fixed barrier is installed to separate the conductors.

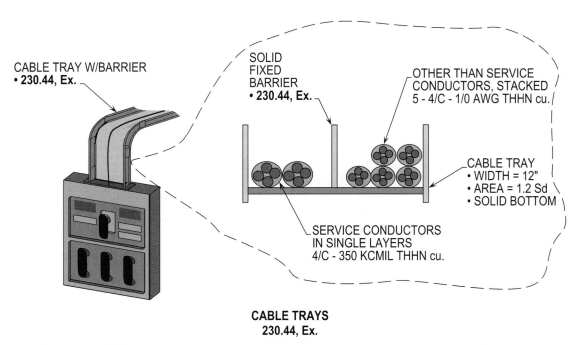

CABLE TRAY W/BARRIER
• **230.44, Ex.**

SOLID
FIXED
BARRIER
• **230.44, Ex.**

OTHER THAN SERVICE
CONDUCTORS, STACKED
5 - 4/C - 1/0 AWG THHN cu.

CABLE TRAY
• WIDTH = 12"
• AREA = 1.2 Sd
• SOLID BOTTOM

SERVICE CONDUCTORS
IN SINGLE LAYERS
4/C - 350 KCMIL THHN cu.

CABLE TRAYS
230.44, Ex.

Purpose of Change: To permit service conductors and other conductors in the same cable tray.

NEC Ch. 2 – Article 230
Part IV – 230.54(C), Ex.

Type of Change		Panel Action		UL	UL 508	API 500	API 505	OSHA
Revision		Accept in Principle		-	-	-	-	1910.304(C)(3)
ROP		ROC		NFPA 70E	NFPA 70B	NFPA 79	NFPA	NEMA
pg. 413	# 4-79	pg. -	# -	410.7(C)	-	-	-	-
log: 1854	CMP: 4	log: -	Submitter: Andre R. Cartal			2002 NEC: 230.54(C)		IEC: Ch. 43

2002 NEC – 230.54(C) Service Heads Above Service-Drop Attachment

Service heads and goosenecks in service-entrance cables shall be located above the point of attachment of the service-drop conductors to the building or other structure.

Exception: Where it is impracticable to locate the service head above the point of attachment, the service head location shall be permitted not farther than 600 mm (24 in.) from the point of attachment.

2005 NEC – 230.54(C) Service Heads <u>and Goosenecks</u> Above Service-Drop Attachment

Service heads and goosenecks in service-entrance cables shall be located above the point of attachment of the service-drop conductors to the building or other structure.

Exception: Where it is impracticable to locate the service head <u>or gooseneck</u> above the point of attachment, the service head <u>or gooseneck</u> location shall be permitted not farther than 600 mm (24 in.) from the point of attachment.

Author's Substantiation. This revision clarifies that goosenecks in service-entrance cables shall be permitted for the point of attachment of the service-drop conductors to the building or other structure.

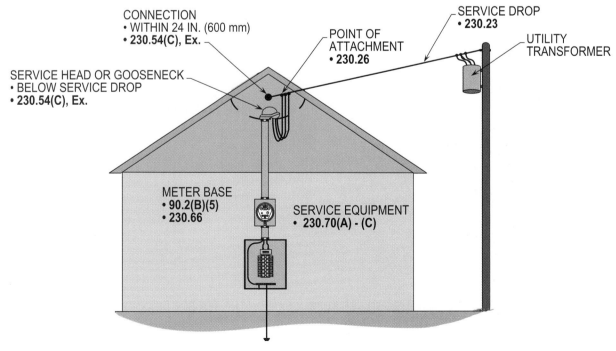

**SERVICE HEADS AND GOOSENECKS ABOVE
SERVICE-DROP ATTACHMENTS
230.54(C), Ex.**

Purpose of Change: To clarify that goosenecks are also included in this section and exception.

NEC Ch. 2 – Article 230
Part V – 230.82(3)

 36

Type of Change	Panel Action	UL	UL 508	API 500	API 505	OSHA
New Item	Accept in Principle in Part	414	-	-	-	-

ROP		ROC		NFPA 70E	NFPA 70B	NFPA 79	NFPA	NEMA
pg. 433	# 4-106	pg. -	# -	-	-	-	-	-
log: 2549	CMP: 4	log: -	Submitter: Jim Pauley			2002 NEC: -		IEC: Ch. 43

2005 NEC – 230.82 Equipment Connected on the Supply Side of Service Disconnect

Only the following equipment shall be permitted to be connected to the supply side of the service disconnecting means:

(3) Meter disconnect switches nominally rated not in excess of 600 volts that have a short-circuit current rating equal to or greater than the available short circuit current, provide all metal housings and service enclosures are grounded.

Author's Substantiation. This new item (3) permits meter disconnect switches nominally not rated in excess of 600 volts to be connected on the supply side of service disconnect. The meter disconnect switch shall have a short-circuit current rating equal to or greater than the available short circuit current.

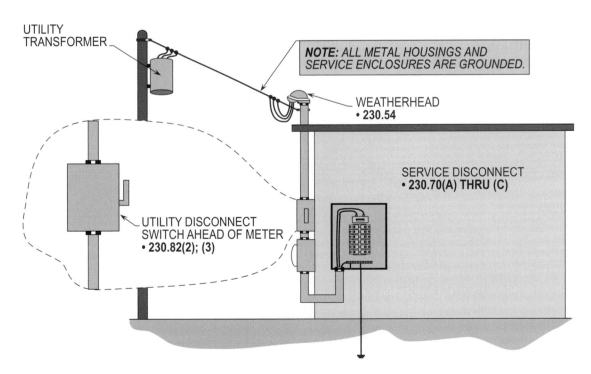

EQUIPMENT CONNECTED TO THE SUPPLY SIDE OF THE SERVICE DISCONNECT
230.82(3)

Purpose of Change: To verify that utility control disconnect switches shall have short-circuit capability.

NEC Ch. 2 – Article 230
Part VII – 230.94, Ex. 5

Type of Change		Panel Action		UL	UL 508	API 500	API 505	OSHA
Revision		Accept		-	-	-	-	1910.304(e)(L)(i)
ROP		ROC		NFPA 70E	NFPA 70B	NFPA 79	NFPA	NEMA
pg. 436	# 4-112a	pg. -	# -	410.8	3.3.18	7.2.8	-	-
log: CP405	CMP: 4	log: -	Submitter: CMP 4			2002 NEC: 230.94, Ex. 5		IEC: 533

2002 NEC – 230.94 Relative Location of Overcurrent Device and Other Service Equipment

The overcurrent device shall protect all circuits and devices.

Exception No. 5: Meters nominally rated not in excess of 600 volts, provided all meter housings and service enclosures are grounded ~~in accordance with Article 250.~~

2005 NEC – 230.94 Relative Location of Overcurrent Device and Other Service Equipment

The overcurrent device shall protect all circuits and devices.

Exception No. 5: Meters nominally rated not in excess of 600 volts <u>shall be permitted</u>, provided all meter housings and service enclosures are grounded.

Author's Substantiation. This revision provides better clarity and improves usability under the direction of the Technical Correlating Committee.

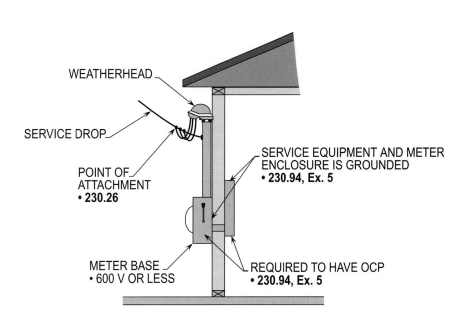

**RELATIVE LOCATION OF OVERCURRENT
DEVICE AND OTHER SERVICE EQUIPMENT
230.94, Ex. 5**

Purpose of Change: To clarify the requirements for protecting meter enclosures.

NEC Ch. 2 – Article 240
Part I – 240.2

Type of Change	Panel Action	UL	UL 508	API 500	API 505	OSHA
Revision	Accept	-	-	-	-	-
ROP	**ROC**	**NFPA 70E**	**NFPA 70B**	**NFPA 79**	**NFPA**	**NEMA**

pg. 445	# 10-9	pg. -	# -	-	-	-	-
log: 2906	CMP: 10	log: -	Submitter: Dorothy Kellogg		2002 NEC: 240.2		IEC: -

2002 NEC – 240.2 Supervised Industrial Installation

For the purposes of Part VIII, the industrial portions of a facility where all of the following conditions are met:

(1) Conditions of maintenance and engineering supervision ensure that only qualified persons monitor and service the system.

(2) The premises wiring system has 2500 kVA or greater of load used in industrial process(es), manufacturing activities, or both, as calculated in accordance with Article 220.

(3) The premises has at least one service that is more than 150 volts to ground and more than 300 volts phase-to-phase.

This definition excludes installations in buildings used by the industrial facility for offices, warehouses, garages, machine shops, and recreational facilities that are not an integral part of the industrial plant, substation, or control center.

2005 NEC – 240.2 Supervised Industrial Installation

For the purposes of Part VIII, the industrial portions of a facility where all of the following conditions are met:

(1) Conditions of maintenance and engineering supervision ensure that only qualified persons monitor and service the system.

(2) The premises wiring system has 2500 kVA or greater of load used in industrial process(es), manufacturing activities, or both, as calculated in accordance with Article 220.

(4) The premises has at least one service <u>or feeder</u> that is more than 150 volts to ground and more than 300 volts phase-to-phase.

This definition excludes installations in buildings used by the industrial facility for offices, warehouses, garages, machine shops, and recreational facilities that are not an integral part of the industrial plant, substation, or control center.

Author's Substantiation. This revision clarifies that not all supervised industrial installations are fed from services, but in some cases fed from one or more feeders.

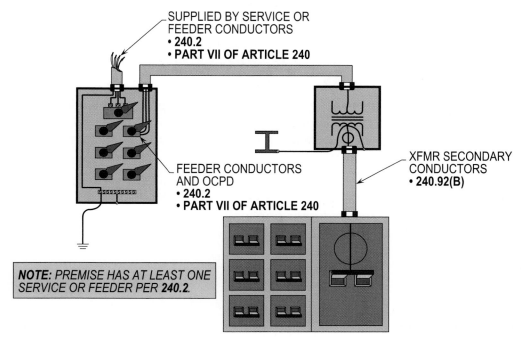

SUPPLIED BY SERVICE OR
FEEDER CONDUCTORS
• **240.2**
• **PART VII OF ARTICLE 240**

FEEDER CONDUCTORS
AND OCPD
• **240.2**
• **PART VII OF ARTICLE 240**

XFMR SECONDARY
CONDUCTORS
• **240.92(B)**

NOTE: *PREMISE HAS AT LEAST ONE
SERVICE OR FEEDER PER **240.2**.*

**DEFINITIONS
SUPERVISED INDUSTRIAL INSTALLATION
240.2**

Purpose of Change: To clarify that services or feeder conductors can be used in supervised industrial locations.

37

Type of Change		Panel Action		UL	UL 508	API 500	API 505	OSHA
Revision		Accept in Principle		845	-	-	-	1910.305(j)(4)
ROP		ROC		NFPA 70E	NFPA 70B	NFPA 79	NFPA	NEMA
pg. 454	# 10-23	pg. -	# -	420.10(E)	11.6	7.3	-	MG 1
log: 202	CMP: 10	log: -	Submitter: Nick Murphy			2002 NEC: 240.4(D)		IEC: 433.2

2002 NEC – 240.4(D) Small Conductors

Unless specifically permitted in 240.4(E) ~~through~~ (G), the overcurrent protection shall not exceed 15 amperes for 14 AWG, 20 amperes for 12 AWG, and 30 amperes for 10 AWG copper; or 15 amperes for 12 AWG and 25 amperes for 10 AWG aluminum or copper-clad aluminum after any correction factors for ambient temperature and number of conductors have been applied.

2005 NEC – 240.4(D) Small Conductors

Unless specifically permitted in 240.4(E) <u>or 240.4</u>(G), the overcurrent protection shall not exceed 15 amperes for 14 AWG, 20 amperes for 12 AWG, and 30 amperes for 10 AWG copper; or 15 amperes for 12 AWG and 25 amperes for 10 AWG aluminum or copper-clad aluminum after any correction factors for ambient temperature and number of conductors have been applied.

Author's Substantiation. This revision clarifies that small conductors shall not apply to tap conductors or overcurrent protection for specific conductor applications unless specifically permitted.

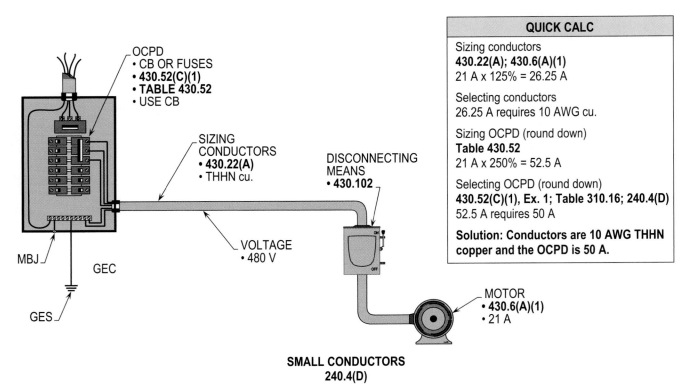

SMALL CONDUCTORS
240.4(D)

Purpose of Change: To clarify that **240.4(E)** or **(G)** can be used when OCPD is larger than the conductors.

NEC Ch. 2 – Article 240
Part I – 240.4(E)

Type of Change		Panel Action		UL	UL 508	API 500	API 505	OSHA
Revision		Accept in Principle		845	-	-	-	1910.305(j)(4)
ROP		ROC		NFPA 70E	NFPA 70B	NFPA 79	NFPA	NEMA
pg. 454	# 10-24	pg. -	# -	420.10(E)	11.6	7.3	-	MG 1
log: 1630	CMP: 10	log: -	Submitter: Allen Knickrehm			2002 NEC: 240.4(E)		IEC: 433.2

2002 NEC – 240.4(E) Tap Conductors

Tap conductors shall be permitted to be protected against overcurrent in accordance with 210.19(C) and (D), 240.5(B)(2), 240.21, 368.11, 368.12, and 430.53(D).

2005 NEC – 240.4(E) Tap Conductors

Tap conductors shall be permitted to be protected against overcurrent in accordance with <u>the following:</u>

(1) <u>210.19(A)(3) and (A)(4) Household Ranges and Cooking Appliances and Other Loads</u>

(2) 240.5(B)(2) <u>Fixture Wire</u>

(3) 240.21 <u>Location in Circuit</u>

(4) <u>368.17(B) Reduction in Ampacity Size of Busway</u>

(5) <u>368.17(C) Feeder or Branch Circuits (busway taps)</u>

(6) 430.53(D) <u>Single Motor Taps</u>

Author's Substantiation. This revision places the text in a list format that is more user friendly.

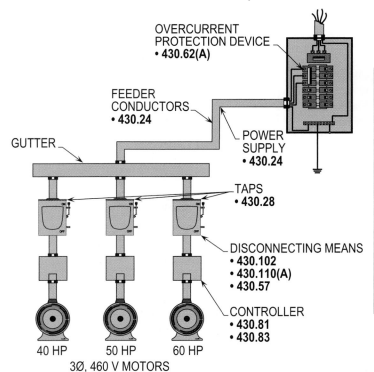

OVERCURRENT PROTECTION DEVICE
• **430.62(A)**

FEEDER CONDUCTORS
• **430.24**

GUTTER

POWER SUPPLY
• **430.24**

TAPS
• **430.28**

DISCONNECTING MEANS
• **430.102**
• **430.110(A)**
• **430.57**

CONTROLLER
• **430.81**
• **430.83**

40 HP   50 HP   60 HP
3Ø, 460 V MOTORS

FINDING THWN COPPER TAP CONDUCTORS

Step 1: Finding amperage
430.6(A)(1); Table 430.250
40 HP = 52 A
50 HP = 65 A
60 HP = 77 A

Step 2: Calculating size tap for 60 HP motor
430.28
77 A x 125% = 96.25 A

Step 3: Selecting conductors
310.10, FPN (2); Table 310.16
96.25 A requires 3 AWG THWN cu.

Solution: **The size THWN copper conductors are required to be 3 AWG.**

TAP CONDUCTORS
240.4(E)

Purpose of Change: To refer appropriate section and title head for a particular tap rule.

NEC Ch. 2 – Article 240
Part I – 240.10

Type of Change		Panel Action		UL	UL 508	API 500	API 505	OSHA
Revision		Accept		984	-	-	-	1910.304(e)(1)(i)
ROP		ROC		NFPA 70E	NFPA 70B	NFPA 79	NFPA	NEMA
pg. 462	# 10-32a	pg. -	# -	Article 100	13.2	3.3.98	-	-
log: CP1001	CMP: 10	log: -	Submitter: CMP 10			2002 NEC: 240.10		IEC: 533

2002 NEC – 240.10 Supplementary Overcurrent Protection

Where supplementary overcurrent protection is used for luminaires (lighting fixtures), appliances, and other equipment or for internal circuits and components of equipment, it shall not be used as a substitute for branch-circuit overcurrent devices or in place of the branch-circuit protection ~~specified in Article 210~~. Supplementary overcurrent devices shall not be required to be readily accessible.

2005 NEC – 240.10 Supplementary Overcurrent Protection

Where supplementary overcurrent protection is used for luminaires (lighting fixtures), appliances, and other equipment or for internal circuits and components of equipment, it shall not be used as a substitute for <u>required</u> branch-circuit overcurrent devices or in place of the <u>required</u> branch-circuit protection. Supplementary overcurrent devices shall not be required to be readily accessible.

Author's Substantiation. This revision provides better clarity and improves usability under the direction of the Technical Correlating Committee.

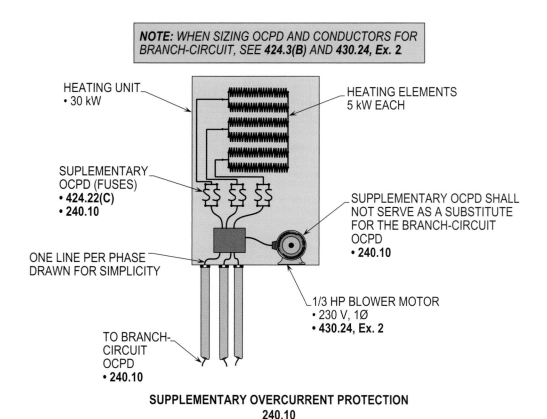

NOTE: WHEN SIZING OCPD AND CONDUCTORS FOR BRANCH-CIRCUIT, SEE **424.3(B)** AND **430.24, Ex. 2**

HEATING UNIT
• 30 kW

HEATING ELEMENTS
5 kW EACH

SUPLEMENTARY
OCPD (FUSES)
• **424.22(C)**
• **240.10**

SUPPLEMENTARY OCPD SHALL NOT SERVE AS A SUBSTITUTE FOR THE BRANCH-CIRCUIT OCPD
• **240.10**

ONE LINE PER PHASE DRAWN FOR SIMPLICITY

1/3 HP BLOWER MOTOR
• 230 V, 1Ø
• **430.24, Ex. 2**

TO BRANCH-CIRCUIT OCPD
• **240.10**

SUPPLEMENTARY OVERCURRENT PROTECTION
240.10

Purpose of Change: To revise the text and make this section easier to read and understand.

NEC Ch. 2 – Article 240
Part II – 240.21(B)(1)(1)a

Type of Change		Panel Action		UL	UL 508	API 500	API 505	OSHA
Revision		Accept		498	-	-	-	-
ROP		ROC		NFPA 70E	NFPA 70B	NFPA 79	NFPA	NEMA
pg. 469	# 10-43	pg. -	# -	-	B.1.4	-	-	SG 3
log: 1317	CMP: 10	log: -	Submitter: TCC			2002 NEC 240.21(B)(1)(1)		IEC: 473.1.1

2002 NEC – 240.21(B)(1) Taps Not Over 3 m (10 ft) Long

Where the length of the tap conductors does not exceed 3 m (10 ft) and the tap conductors comply with all of the following:

    (1) The ampacity of the tap conductors is

        a. Not less than the combined ~~computed~~ loads on the circuits supplied by the tap conductors, and

        b. Not less than the rating of the device supplied by the tap conductors or not less than the rating of the overcurrent-protective device at the termination of the tap conductors.

2005 NEC – 240.21(B)(1) Taps Not Over 3 m (10 ft) Long

Where the length of the tap conductors does not exceed 3 m (10 ft) and the tap conductors comply with all of the following:

    (1) The ampacity of the tap conductors is

        a. Not less than the combined <u>calculated</u> loads on the circuits supplied by the tap conductors, and

        b. Not less than the rating of the device supplied by the tap conductors or not less than the rating of the overcurrent-protective device at the termination of the tap conductors.

Author's Substantiation. This revision clarifies that the term "calculate" will be standardized language for determining load values throughout the NEC.

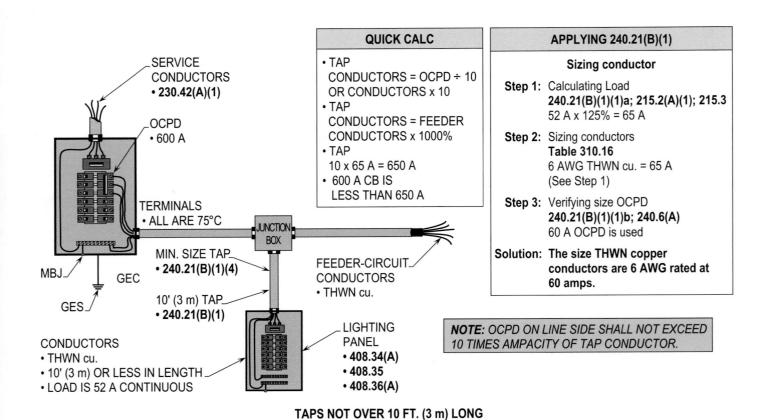

SERVICE CONDUCTORS
• **230.42(A)(1)**

OCPD
• 600 A

TERMINALS
• ALL ARE 75°C

MBJ

GEC

GES

CONDUCTORS
• THWN cu.
• 10' (3 m) OR LESS IN LENGTH
• LOAD IS 52 A CONTINUOUS

MIN. SIZE TAP
• **240.21(B)(1)(4)**

10' (3 m) TAP
• **240.21(B)(1)**

JUNCTION BOX

FEEDER-CIRCUIT CONDUCTORS
• THWN cu.

LIGHTING PANEL
• **408.34(A)**
• **408.35**
• **408.36(A)**

QUICK CALC
• TAP CONDUCTORS = OCPD ÷ 10 OR CONDUCTORS x 10 • TAP CONDUCTORS = FEEDER CONDUCTORS x 1000% • TAP 10 x 65 A = 650 A • 600 A CB IS LESS THAN 650 A

APPLYING 240.21(B)(1)
Sizing conductor

Step 1: Calculating Load
240.21(B)(1)(1)a; 215.2(A)(1); 215.3
52 A x 125% = 65 A

Step 2: Sizing conductors
Table 310.16
6 AWG THWN cu. = 65 A
(See Step 1)

Step 3: Verifying size OCPD
240.21(B)(1)(1)b; 240.6(A)
60 A OCPD is used

Solution: **The size THWN copper conductors are 6 AWG rated at 60 amps.**

NOTE: *OCPD ON LINE SIDE SHALL NOT EXCEED 10 TIMES AMPACITY OF TAP CONDUCTOR.*

TAPS NOT OVER 10 FT. (3 m) LONG
240.21(B)(1)(1)

Purpose of Change: To replace the term "compute" with the term "calculate" because loads are usually determined by using a calculator.

NEC Ch. 2 – Article 240
Part II – 240.21(B)(3)(2)

Type of Change		Panel Action		UL	UL 508	API 500	API 505	OSHA
Revision		Accept		489	-	-	-	-
ROP		ROC		NFPA 70E	NFPA 70B	NFPA 79	NFPA	NEMA
pg. 471	# 10-47	pg. -	# -	-	B.1.4	-	-	SG 3
log: 562	CMP: 10	log: -	Submitter: J. Kevin Vogel			2002 NEC: 240.21(B)(3)(2)		IEC: 473.1.1

2002 NEC – 240.21(B)(3) Taps Supplying a Transformer [Primary Plus Secondary Not Over 7.5 m (25 ft) Long]

Where the tap conductors supply a transformer and comply with all the following conditions:

(2) The conductors supplied by the secondary of the transformer shall have an ampacity that, when multiplied by the ratio of the secondary-to-primary voltage, is at least one-third of the rating of the overcurrent device protecting the feeder conductors.

2005 NEC – 240.21(B)(3) Taps Supplying a Transformer [Primary Plus Secondary Not Over 7.5 m (25 ft) Long]

Where the tap conductors supply a transformer and comply with all the following conditions:

(2) The conductors supplied by the secondary of the transformer shall have an ampacity that is not less than the value of the primary-to-secondary voltage ratio multiplied by one-third of the rating of the overcurrent device protecting the feeder conductors.

Author's Substantiation. This revision clarifies how calculations are to performed for taps supplying a transformer where primary plus secondary are not over 7.5 m (25 ft) long.

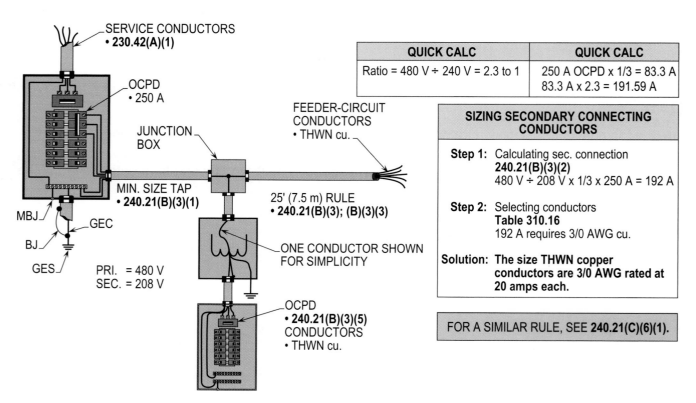

QUICK CALC	QUICK CALC
Ratio = 480 V ÷ 240 V = 2.3 to 1	250 A OCPD x 1/3 = 83.3 A 83.3 A x 2.3 = 191.59 A

SIZING SECONDARY CONNECTING CONDUCTORS

Step 1: Calculating sec. connection
240.21(B)(3)(2)
480 V ÷ 208 V x 1/3 x 250 A = 192 A

Step 2: Selecting conductors
Table 310.16
192 A requires 3/0 AWG cu.

Solution: The size THWN copper conductors are 3/0 AWG rated at 20 amps each.

FOR A SIMILAR RULE, SEE **240.21(C)(6)(1)**.

SERVICE CONDUCTORS
• **230.42(A)(1)**

OCPD
• 250 A

JUNCTION BOX

FEEDER-CIRCUIT CONDUCTORS
• THWN cu.

MIN. SIZE TAP
• **240.21(B)(3)(1)**

MBJ
GEC
BJ
GES

PRI. = 480 V
SEC. = 208 V

25' (7.5 m) RULE
• **240.21(B)(3); (B)(3)(3)**

ONE CONDUCTOR SHOWN FOR SIMPLICITY

OCPD
• **240.21(B)(3)(5)**
CONDUCTORS
• THWN cu.

TAPS SUPPLYING A TRANSFORMER
[PRIMARY PLUS SECONDARY NOT OVER 25 FT (7.5 m) LONG]
240.21(B)(3)(2)

Purpose of Change: To revise the wording and make the rules easier to read and understand.

NEC Ch. 2 – Article 240
Part II – 240.21(B)(4)

Type of Change		Panel Action		UL	UL 508	API 500	API 505	OSHA
Revision		Accept in Principle		489	-	-	-	-
ROP		ROC		NFPA 70E	NFPA 70B	NFPA 79	NFPA	NEMA
pg. 472	# 10-50	pg. 134	# 10-46	-	B.1.4	1.4		SG 3
log: 885	CMP: 10	log: 620	Submitter: James M. Daly			2002 NEC: 240.21(B)(4)(5)		IEC: 473.1.1

2002 NEC – 240.21(B)(4) Taps Over 7.5 m (25 ft) Long

Where the feeder is in a high bay manufacturing building over 11 m (35 ft) high at walls and the installation complies with all of the following conditions:

(5) The tap conductors are ~~suitably~~ protected from physical damage ~~or are~~ enclosed in ~~a~~ raceway.

2005 NEC – 240.21(B)(4) Taps Over 7.5 m (25 ft) Long

Where the feeder is in a high bay manufacturing building over 11 m (35 ft) high at walls and the installation complies with all of the following conditions:

(5) The tap conductors are protected from physical damage <u>by being</u> enclosed in <u>an approved</u> raceway <u>or by other approved means</u>.

Author's Substantiation. This revision clarifies that tap conductors are to be protected from physical damage by being enclosed in an approved raceway or by other approved means.

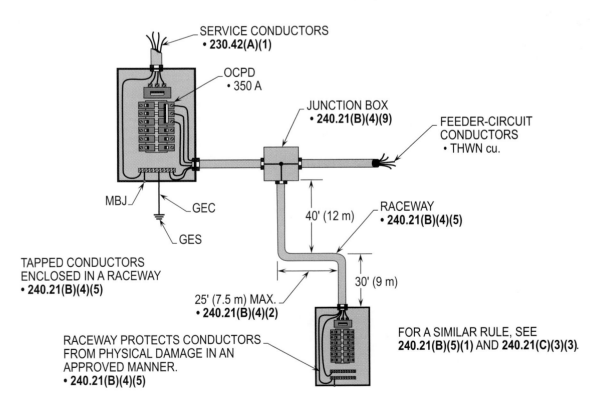

SERVICE CONDUCTORS
• 230.42(A)(1)

OCPD
• 350 A

JUNCTION BOX
• 240.21(B)(4)(9)

FEEDER-CIRCUIT
CONDUCTORS
• THWN cu.

MBJ

GEC

GES

RACEWAY
• 240.21(B)(4)(5)

40' (12 m)

30' (9 m)

TAPPED CONDUCTORS
ENCLOSED IN A RACEWAY
• 240.21(B)(4)(5)

25' (7.5 m) MAX.
• 240.21(B)(4)(2)

RACEWAY PROTECTS CONDUCTORS
FROM PHYSICAL DAMAGE IN AN
APPROVED MANNER.
• 240.21(B)(4)(5)

FOR A SIMILAR RULE, SEE
240.21(B)(5)(1) AND 240.21(C)(3)(3).

TAPS OVER 25 FT. (7.5 m) LONG
240.21(B)(4)

Purpose of Change: To ensure that tapped conductors are properly protected.

NEC Ch. 2 – Article 240
Part II – 240.21(C)

38

Type of Change		Panel Action		UL	UL 508	API 500	API 505	OSHA
Revision		Accept in Principle		489	-	-	-	-
ROP		ROC		NFPA 70E	NFPA 70B	NFPA 79	NFPA	NEMA
pg. 477	# 10-57	pg. 135	# 10-49	410.9(C)(2)	B.1.4	1.4	-	SG 3
log: 3474	CMP: 10	log: 1970	Submitter: Eric G. Schneier			2002 NEC: 240.21(C)		IEC: 473.1.1

2002 NEC – 240.21(C) Transformer Secondary Conductors

~~Conductors~~ shall be permitted to be connected to a transformer secondary, without overcurrent protection at the secondary, as specified in 240.21(C)(1) through (6).

2005 NEC – 240.21(C) Transformer Secondary Conductors

Each set of conductors feeding separate loads shall be permitted to be connected to a transformer secondary, without overcurrent protection at the secondary, as specified in 240.21(C)(1) through (C)(6). The provisions of 240.4(B) shall not be permitted for transformer secondary conductors.

Author's Substantiation. This revision clarifies some vague language that was unclear on whether or not it is referring to a single set of secondary conductors or multiple sets of secondary conductors to different overcurrent devices. A new sentence has been added to clearly state to the user that 240.4(B) shall not be permitted for transformer secondary conductors.

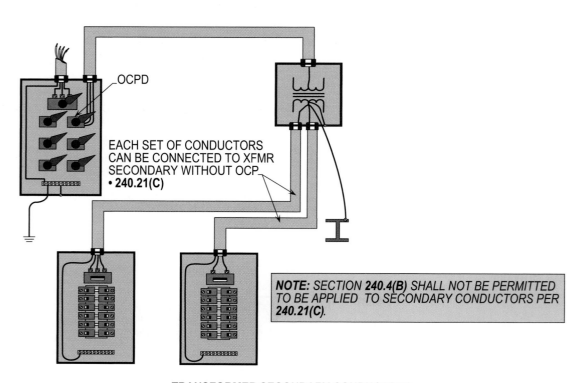

OCPD

EACH SET OF CONDUCTORS
CAN BE CONNECTED TO XFMR
SECONDARY WITHOUT OCP
• 240.21(C)

NOTE: SECTION **240.4(B)** SHALL NOT BE PERMITTED TO BE APPLIED TO SECONDARY CONDUCTORS PER **240.21(C)**.

TRANSFORMER SECONDARY CONDUCTORS
240.21(C)

Purpose of Change: To clarify rules for terminating conductors at transformer secondary.

NEC Ch. 2- Article 240
Part II – 240.24(B), Ex. 1(1) and (2)

Type of Change		Panel Action		UL	UL 508	API 500	API 505	OSHA
Revision		Accept		489	-	-	-	1910.304(e)(1)(iv)
ROP		ROC		NFPA 70E	NFPA 70B	NFPA 79	NFPA	NEMA
pg. 485	# 10-66	pg. -	# -	-	-	-	-	SG 3
log: 1035	CMP: 10	log: -	Submitter: TCC			2002 NEC: 240.21(B), Ex. 1(b)		IEC: 482.1.2

2002 NEC – 240.24(B) Occupancy

Each occupant shall have ready access to all overcurrent devices protecting the conductors supplying that occupancy.

Exception No. 1: Where electric service and electrical maintenance are provided by the building management and where these are under continuous building management supervision, the service overcurrent devices and feeder overcurrent devices supplying more than one occupancy shall be permitted to be accessible to only authorized management personnel in the following:

(a) Multiple-occupancy buildings

(b) Guest rooms of hotels and motels that are intended for transient occupancy

2005 NEC – 240.24(B) Occupancy

Each occupant shall have ready access to all overcurrent devices protecting the conductors supplying that occupancy.

Exception No. 1: Where electric service and electrical maintenance are provided by the building management and where these are under continuous building management supervision, the service overcurrent devices and feeder overcurrent devices supplying more than one occupancy shall be permitted to be accessible to only authorized management personnel in the following:

(1) Multiple-occupancy buildings

(2) Guest rooms or guest suites of hotels and motels that are intended for transient occupancy

Author's Substantiation. This revision clarifies that the service overcurrent devices and feeder overcurrent devices supplying guest rooms or guest suites of hotels and motels that are intended for transient occupancy shall be permitted to be accessible to only authorized management where under continuous supervision.

GUEST ROOMS OR SUITES OF
HOTEL/MOTEL WITH CONTINUOUS
BUILDING MANAGEMENT
• **240.24(B), Ex. 1(1); (2)**

ROOM OR SUITE IN A MULTIPLE-
OCCUPANCY BUILDING WHERE
OVERCURRENT DEVICES ARE NOT
ACCESSIBLE TO OCCUPANT
• **240.24(B), Ex. 1(1); (2)**
• **240.24(B), Ex. 2**

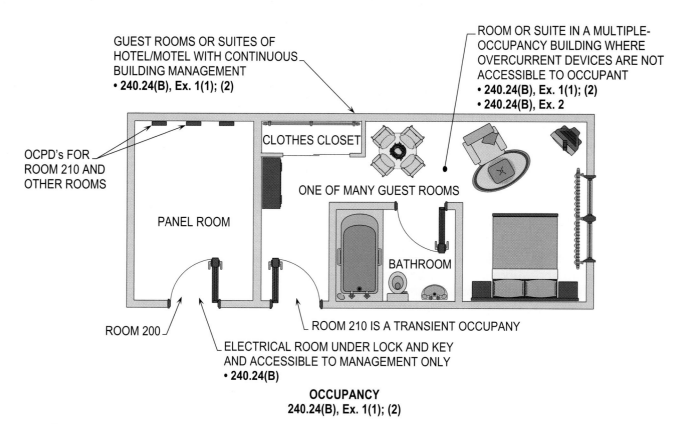

OCPD's FOR
ROOM 210 AND
OTHER ROOMS

CLOTHES CLOSET

ONE OF MANY GUEST ROOMS

PANEL ROOM

BATHROOM

ROOM 200

ROOM 210 IS A TRANSIENT OCCUPANY

ELECTRICAL ROOM UNDER LOCK AND KEY
AND ACCESSIBLE TO MANAGEMENT ONLY
• **240.24(B)**

OCCUPANCY
240.24(B), Ex. 1(1); (2)

Purpose of Change: To clarify that panels with OCPD's can be located behind locked doors with the supervision of management.

Type of Change		Panel Action		UL	UL 508	API 500	API 505	OSHA
Revision		Accept		-	-	-	-	1910.304(e)(1)(iv)
ROP		ROC		NFPA 70E	NFPA 70B	NFPA 79	NFPA	NEMA
pg. 482	# 10-66	pg. -	# -	-	-	-	-	-
log: 1035	CMP: 10	log: -	Submitter: TCC			2002 NEC: 240.24(E)		IEC: 482.1.2

2002 NEC – 240.24(E) Not Located in Bathrooms

In dwelling units and guest rooms of hotels and motels, overcurrent devices, other than supplementary overcurrent protection, shall not be located in bathrooms ~~as defined in Article 100.~~

2005 NEC – 240.24(E) Not Located in Bathrooms

In dwelling units and guest rooms <u>or guest suites</u> of hotels and motels, overcurrent devices, other than supplementary overcurrent protection, shall not be located in bathrooms.

Author's Substantiation. This revision clarifies that overcurrent devices, other than supplementary overcurrent protection, shall not be located in bathrooms of guest rooms or guest suites of hotels and motels.

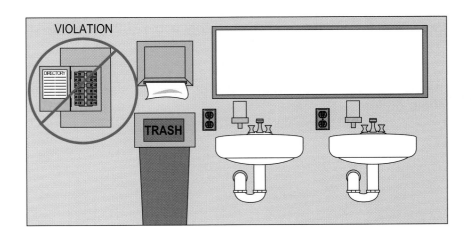

NOT LOCATED IN BATHROOMS
240.24(E)

Purpose of Change: To clarify that OCPD's are prohibited in the bathrooms of dwelling units or in the guest rooms or suites of hotels and motels.

NEC Ch. 2 – Article 240
Part IV – 240.40

Type of Change		Panel Action		UL	UL 508	API 500	API 505	OSHA
Revision		Accept		489	-	-	-	1910.304(e)(1)(iv)
ROP		ROC		NFPA 70E	NFPA 70B	NFPA 79	NFPA	NEMA
pg. 488	# 10-74	pg. -	# -	410.9(A)(3)	5.3	-	-	SG 3
log: 816	CMP: 10	log: -	Submitter: Dan Leaf			2002 NEC: 240.40		IEC: Ch. 43

2002 NEC – 240.40 Disconnecting Means for Fuses

A disconnecting means shall be provided on the supply side of all fuses in circuits over 150 volts to ground and cartridge fuses in circuits of any voltage where accessible to other than qualified persons so that each ~~individual~~ circuit containing fuses can be independently disconnected from the source of power. A current-limiting device without a disconnecting means shall be permitted on the supply side of the service disconnecting means as permitted by 230.82. A single disconnecting means shall be permitted on the supply side of more than one set of fuses as permitted by 430.112, Exception, for group operation of motors and 424.22(C) for fixed electric space-heating equipment.

2005 NEC – 240.40 Disconnecting Means for Fuses

A disconnecting means shall be provided on the supply side of all fuses in circuits over 150 volts to ground and cartridge fuses in circuits of any voltage where accessible to other than qualified persons, so that each circuit containing fuses can be independently disconnected from the source of power. A current-limiting device without a disconnecting means shall be permitted on the supply side of the service disconnecting means as permitted by 230.82. A single disconnecting means shall be permitted on the supply side of more than one set of fuses as permitted by 430.112, Exception, for group operation of motors and 424.22(C) for fixed electric space-heating equipment.

Author's Substantiation. The word "individual" was removed to clarify a circuit containing cartridge fuses may supply multiple outlets.

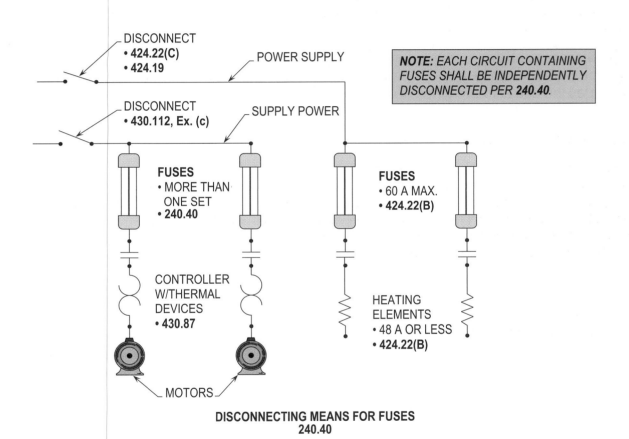

DISCONNECT
• **424.22(C)**
• **424.19**

POWER SUPPLY

NOTE: *EACH CIRCUIT CONTAINING FUSES SHALL BE INDEPENDENTLY DISCONNECTED PER* **240.40.**

DISCONNECT
• **430.112, Ex. (c)**

SUPPLY POWER

FUSES
• MORE THAN ONE SET
• **240.40**

FUSES
• 60 A MAX.
• **424.22(B)**

CONTROLLER W/THERMAL DEVICES
• **430.87**

HEATING ELEMENTS
• 48 A OR LESS
• **424.22(B)**

MOTORS

DISCONNECTING MEANS FOR FUSES
240.40

Purpose of Change: To make the requirements for the disconnecting of fuses easier to understand, the wording has been revised.

NEC Ch. 2 – Article 240
Part VI – 240.60(D)

Type of Change		Panel Action		UL	UL 508	API 500	API 505	OSHA
New Subsection		Accept in Principle		-	-	-	-	-
ROP		ROC		NFPA 70E	NFPA 70B	NFPA 79	NFPA	NEMA
pg. 491	# 10-78	pg. -	# -	-	-	-	-	-
log: 2532	CMP: 10	log: -	Submitter: Alan Manche			2002 NEC: -		IEC: -

2005 NEC – <u>240.60(D) Renewable Fuses</u>

<u>Class H cartridge fuses of the renewable type shall only be permitted to be used for replacement in existing installations where there is no evidence of overfusing or tampering.</u>

Author's Substantiation. A new subsection has been added to clarify that Class H cartridge fuses shall only be permitted to be used for replacement in existing installations. This reference to Class H only precludes the use of future types of current-limiting renewable fuses such as might be manufactured with polymers or solid state designs.

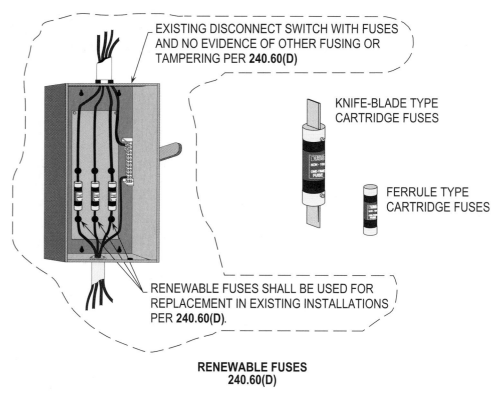

EXISTING DISCONNECT SWITCH WITH FUSES AND NO EVIDENCE OF OTHER FUSING OR TAMPERING PER **240.60(D)**

KNIFE-BLADE TYPE CARTRIDGE FUSES

FERRULE TYPE CARTRIDGE FUSES

RENEWABLE FUSES SHALL BE USED FOR REPLACEMENT IN EXISTING INSTALLATIONS PER **240.60(D)**.

RENEWABLE FUSES
240.60(D)

Purpose of Change: To verify where to use renewable Class H cartridge fuses.

NEC Ch. 2 – Article 240
Part VIII – 240.92(B)(1)(2), FPN

Type of Change		Panel Action		UL	UL 508	API 500	API 505	OSHA
New FPN		Accept		-	-	-	-	-
ROP		ROC		NFPA 70E	NFPA 70B	NFPA 79	NFPA	NEMA
pg. 495	# 10-84	pg. -	# -	410.9(B)(1)(a)	8.4.6.3	-	-	-
log: 2905	CMP: 10	log: -	-	Submitter: Dorothy Kellogg		2002 NEC: -		IEC: Ch. 43

2002 NEC – 240.92(B)(1) Short-Circuit and Ground-Fault Protection

The conductors shall be protected from short-circuit and ground-fault conditions by complying with one of the following conditions:

(2) The conductors are protected by a differential relay with a trip setting equal to or less than the conductor ampacity.

2005 NEC – 240.92(B)(1) Short-Circuit and Ground-Fault Protection

The conductors shall be protected from short-circuit and ground-fault conditions by complying with one of the following conditions:

(2) The conductors are protected by a differential relay with a trip setting equal to or less than the conductor ampacity.

> **FPN:** A differential relay is connected to be sensitive only to short-circuit or fault currents within the protected zone and is normally set much lower than the conductor ampacity. The differential relay is connected to trip protective devices that will de-energize the protected conductors if a short-circuit condition occurs.

Author's Substantiation. This new FPN has been added to clarify that a differential relay is connected to be sensitive only to short-circuit or fault conditions with the protected zone.

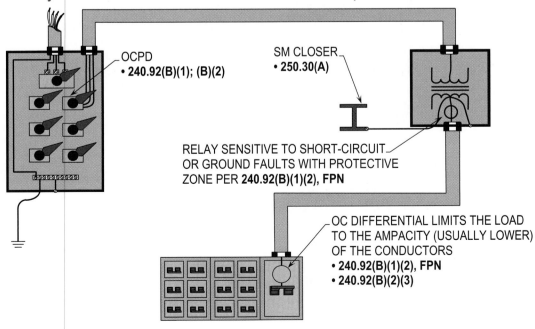

SHORT-CIRCUIT AND GROUND FAULT PROTECTION
240.92(B)(1)(2), FPN

Purpose of Change: To explain the benefits of using a differential relay.

NEC Ch. 2 – Article 250
Part I – 250.2

Type of Change		Panel Action		UL	UL 508	API 500	API 505	OSHA
Revision		Accept in Principle		467	40.1	-	-	1910.304(f)(3)
ROP		ROC		NFPA 70E	NFPA 70B	NFPA 79	NFPA	NEMA
pg. 504	# 5-48	pg. 142	# 5-34	Article 100	29.2.10	19.2	-	WD 1
log: 2955	CMP: 5	log: 2426	Submitter: Charles Mello			2002 NEC: 250.2		IEC: 542.1.2

2002 NEC – 250.2 Effective Ground-Fault Current Path

An intentionally constructed, permanent, low-impedance electrically conductive path designed and intended to carry current under ground-fault conditions from the point of a ground fault on a wiring system to the electrical supply source.

2005 NEC – 250.2 Effective Ground-Fault Current Path

An intentionally constructed, permanent, low-impedance electrically conductive path designed and intended to carry current under ground-fault conditions from the point of a ground fault on a wiring system to the electrical supply source <u>and that facilitates the operation of the overcurrent protective device or ground fault detectors on high-impedance grounded systems</u>.

Author's Substantiation. This revision clarifies that where a ground-fault condition occurs from the point of a ground fault on a wiring system to the electrical supply source, the ground fault shall facilitate the operation of the overcurrent protection device or ground fault detectors on high-impedance grounded systems.

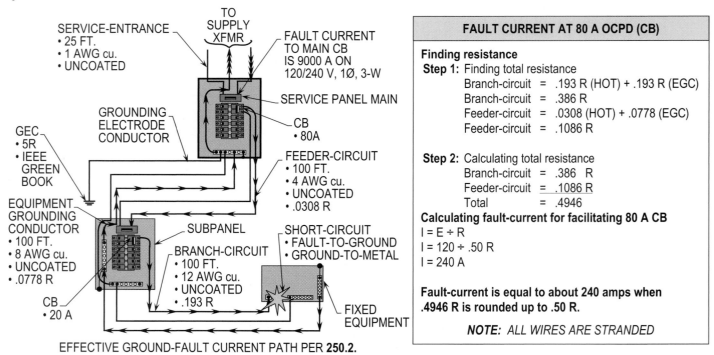

EFFECTIVE GROUND-FAULT CURRENT PATH PER **250.2**.

EFFECTIVE GROUND-FAULT CURRENT PATH
250.2

Purpose of Change: To clarify that the ground fault shall facilitate the operation of the overcurrent protection device or ground fault detections on high-impedance grounded systems.

NEC Ch. 2 – Article 250
Part I – 250.4(A)(5)

Type of Change		Panel Action		UL	UL 508	API 500	API 505	OSHA
Revision		Accept in Principle		467	40.1	-	-	1910.304(f)(3)
ROP		ROC		NFPA 70E	NFPA 70B	NFPA 79	NFPA	NEMA
pg. 506	# 5-52	pg. -	# -	Article 100	29.2.10	19.2	-	WD 1
log: 1557	CMP: 5	log: -	Submitter: Michael J. Johnston			2002 NEC: 250.4(A)(5)		IEC: 542.1.2

2002 NEC – 250.4(A)(5) Effective Ground-Fault Current Path

Electrical equipment and wiring and other electrically conductive material likely to become energized shall be installed in a manner that creates a permanent, low-impedance circuit capable of safely carrying the maximum ground-fault current likely to be imposed on it from any point on the wiring system where a ground fault may occur to the electrical supply source. The earth shall not be ~~used~~ as ~~the sole equipment grounding conductor or~~ effective ground-fault current path.

2005 NEC – 250.4(A)(5) Effective Ground-Fault Current Path

Electrical equipment and wiring and other electrically conductive material likely to become energized shall be installed in a manner that creates a permanent, low-impedance circuit <u>facilitating the operation of the overcurrent device or ground detector for high-impedance grounded systems. It shall be</u> capable of safely carrying the maximum ground-fault current likely to be imposed on it from any point on the wiring system where a ground fault may occur to the electrical supply source. The earth shall not be <u>considered</u> as <u>an</u> effective ground-fault current path.

Author's Substantiation. This revision clarifies that the earth shall not be used as a ground-fault current path.

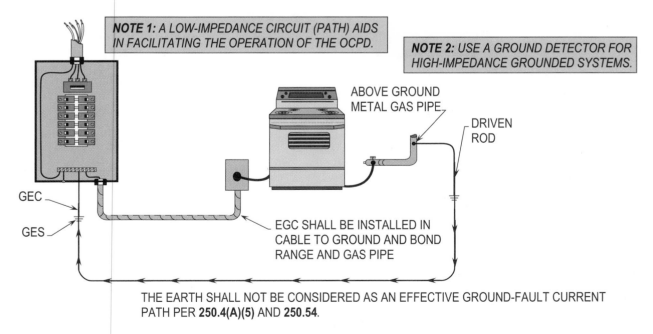

NOTE 1: A LOW-IMPEDANCE CIRCUIT (PATH) AIDS IN FACILITATING THE OPERATION OF THE OCPD.

NOTE 2: USE A GROUND DETECTOR FOR HIGH-IMPEDANCE GROUNDED SYSTEMS.

ABOVE GROUND METAL GAS PIPE

DRIVEN ROD

GEC

GES

EGC SHALL BE INSTALLED IN CABLE TO GROUND AND BOND RANGE AND GAS PIPE

THE EARTH SHALL NOT BE CONSIDERED AS AN EFFECTIVE GROUND-FAULT CURRENT PATH PER **250.4(A)(5)** AND **250.54**.

EFFECTIVE GROUND-FAULT CURRENT PATH
250.4(A)(5)

Purpose of Change: To clarify that the earth shall not be used as a ground-fault path.

NEC Ch. 2 – Article 250
Part I – 250.4(B)(4)

Type of Change		Panel Action		UL	UL 508	API 500	API 505	OSHA
Revision		Accept in Principle		467	40.1	-	-	1910.304(f)(3)
ROP		ROC		NFPA 70E	NFPA 70B	NFPA 79	NFPA	NEMA
pg. 507	# 5-54	pg. -	# -	Article 100	29.2.10	19.2	-	WD 1
log: 1558	CMP: 5	log: -	Submitter: Michael J. Johnston			2002 NEC: 250.4(B)(4)		IEC: 542.1.2

2002 NEC – 250.4(B)(4) Path for Fault Current

Electrical equipment, wiring, and other electrically conductive material likely to become energized shall be installed in a manner that creates a permanent, low-impedance circuit from any point on the wiring system to the electrical supply source to facilitate the operation of overcurrent devices should a second fault occur on the wiring system. The earth shall not be ~~used~~ as ~~the sole equipment grounding conductor or~~ effective fault-current path.

2005 NEC – 250.4(B)(4) Path for Fault Current

Electrical equipment, wiring, and other electrically conductive material likely to become energized shall be installed in a manner that creates a permanent, low-impedance circuit from any point on the wiring system to the electrical supply source to facilitate the operation of overcurrent devices should a second fault occur on the wiring system. The earth shall not be <u>considered</u> as <u>an</u> effective fault-current path.

Author's Substantiation. This revision clarifies that the earth shall not be considered as an effective fault-current path. By removing the reference and relation to the equipment grounding conductor and earth from this subsection, proper application of these performance requirements should be complied with.

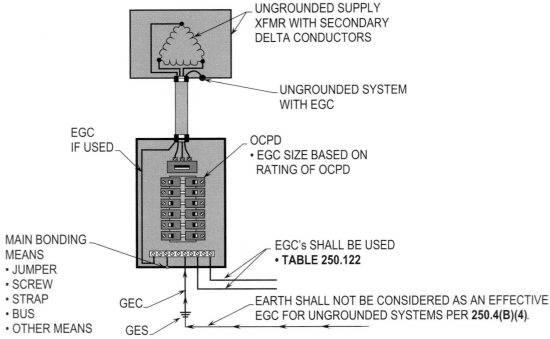

PATH FOR FAULT CURRENT
250.4(B)(4)

Purpose of Change: To verify that the earth shall not be used as an EGC.

NEC Ch. 2 – Article 250
Part II – 250.20

Type of Change		Panel Action		UL	UL 508	API 500	API 505	OSHA
Revision		Accept		467	40.1	-	-	1910.304(f)(1)
ROP		ROC		NFPA 70E	NFPA 70B	NFPA 79	NFPA	NEMA
pg. 509	# 5-58	pg. -	# -	410.10(C)(4)	29.2.2.1	-	-	WD 1
log: 2468	CMP: 5	log: -	Submitter: Paul Dobrowsky			2002 NEC: 250.20		IEC: 312.2

2002 NEC – 250.20 Alternating-Current ~~Circuits and~~ Systems to Be Grounded

Alternating-current ~~circuits and~~ systems shall be grounded as provided for in 250.20(A), (B), (C), or (D). Other ~~circuits and~~ systems shall be permitted to be grounded. If such systems are grounded, they shall comply with the applicable provisions of this article.

2005 NEC – 250.20 Alternating-Current Systems to Be Grounded

Alternating-current systems shall be grounded as provided for in 250.20(A), (B), (C), or (D). Other systems shall be permitted to be grounded. If such systems are grounded, they shall comply with the applicable provisions of this article.

Author's Substantiation. This revision clarifies that if the circuit is grounded, the system supplying it becomes grounded. The deletion of the term circuits should make this concept easier to understand and enforce.

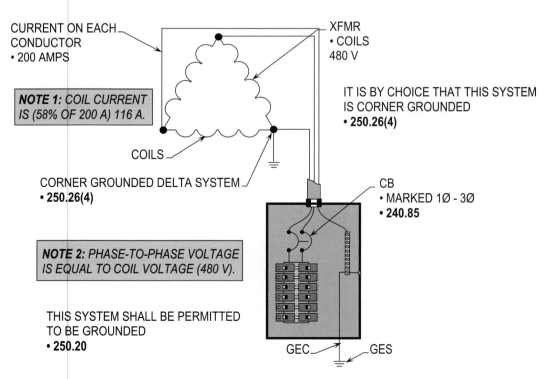

CURRENT ON EACH CONDUCTOR
• 200 AMPS

XFMR
• COILS
480 V

NOTE 1: *COIL CURRENT IS (58% OF 200 A) 116 A.*

IT IS BY CHOICE THAT THIS SYSTEM IS CORNER GROUNDED
• **250.26(4)**

COILS

CORNER GROUNDED DELTA SYSTEM
• **250.26(4)**

CB
• MARKED 1Ø - 3Ø
• **240.85**

NOTE 2: *PHASE-TO-PHASE VOLTAGE IS EQUAL TO COIL VOLTAGE (480 V).*

THIS SYSTEM SHALL BE PERMITTED TO BE GROUNDED
• **250.20**

GEC   GES

**ALTERNATING-CURRENT SYSTEMS TO BE GROUNDED
250.20**

Purpose of Change: To clarify that corner grounded systems are grounded by choice and not necessarily by requirement.

NEC Ch. 2 – Article 250
Part II – 250.20(A)

43

Type of Change		Panel Action		UL	UL 508	API 500	API 505	OSHA
Revision		Accept		467	32.3	-	-	1910.304(f)(1)
ROP		ROC		NFPA 70E	NFPA 70B	NFPA 79	NFPA	NEMA
pg. 509	# 5-58	pg. -	# -	410.10(C)(4)	-	8.3.1, Ex.	-	WD 1
log: 2468	CMP: 5	log: -	Submitter: Paul Dobrowsky			2002 NEC: 250.20(A)		IEC: 312.2

2002 NEC – 250.20(A) Alternating-Current ~~Circuits~~ of Less Than 50 Volts

Alternating-current ~~circuits~~ of less than 50 volts shall be grounded under any of the following conditions:

(1) Where supplied by transformers, if the transformer supply system exceeds 150 volts to ground

(2) Where supplied by transformers, if the transformer supply system is ungrounded

(3) Where installed as overhead conductors outside of buildings

2005 NEC – 250.20(A) Alternating-Current <u>Systems</u> of Less Than 50 Volts

Alternating-current <u>systems</u> of less than 50 volts shall be grounded under any of the following conditions:

(1) Where supplied by transformers, if the transformer supply system exceeds 150 volts to ground

(2) Where supplied by transformers, if the transformer supply system is ungrounded

(3) Where installed as overhead conductors outside of buildings

Author's Substantiation. This revision clarifies that if the circuit is grounded, the system supplying it becomes grounded. The deletion of the word "circuits" should make this concept easier to understand and enforce.

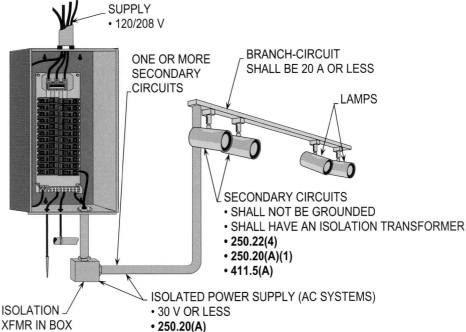

ALTERNATING-CURRENT SYSTEMS OF LESS THAN 50 VOLTS
250.20(A)

Purpose of Change: The word "circuits" was deleted and the term "systems" was added to this section to make this particular concept of grounding easier to understand.

NEC Ch. 2 – Article 250
Part II – 250.20(E)

Type of Change		Panel Action		UL	UL 508	API 500	API 505	OSHA
New Subsection		Accept in Principle		467	-	-	-	1910.304(f)(7)
ROP		ROC		NFPA 70E	NFPA 70B	NFPA 79	NFPA	NEMA
pg. 510	# 5-59	pg. -	# -	410.10(C)(5)	29.3.7	-	-	WD 1
log: 2469	CMP 5	log: -	Submitter: Paul Dobrowsky			2002 NEC: -		IEC: 312.2

2005 NEC – 250.20(E) Impedance Grounded Neutral Systems

Impedance grounded neutral systems shall be grounded in accordance with 250.36 or 250.186.

Author's Substantiation. This new subsection clarifies the impedance grounded neutral systems shall be grounded. Impedance grounded neutral systems are connected to earth but through an impedance device.

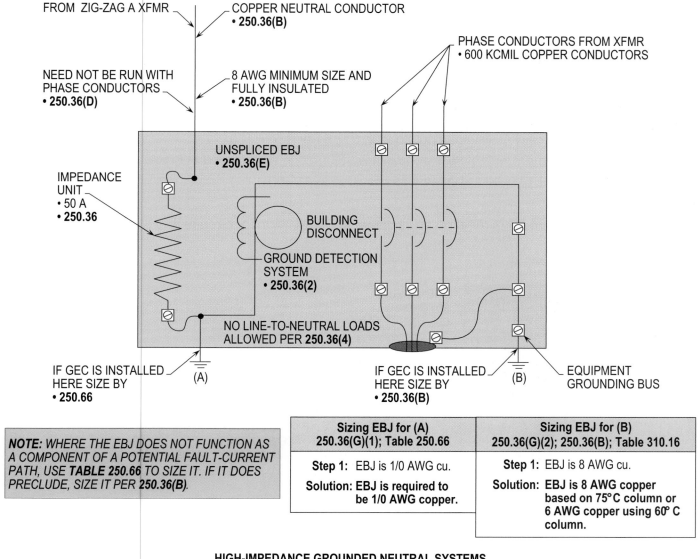

FROM ZIG-ZAG A XFMR
COPPER NEUTRAL CONDUCTOR
• 250.36(B)

PHASE CONDUCTORS FROM XFMR
• 600 KCMIL COPPER CONDUCTORS

NEED NOT BE RUN WITH PHASE CONDUCTORS
• 250.36(D)

8 AWG MINIMUM SIZE AND FULLY INSULATED
• 250.36(B)

UNSPLICED EBJ
• 250.36(E)

IMPEDANCE UNIT
• 50 A
• 250.36

BUILDING DISCONNECT

GROUND DETECTION SYSTEM
• 250.36(2)

NO LINE-TO-NEUTRAL LOADS ALLOWED PER 250.36(4)

IF GEC IS INSTALLED HERE SIZE BY
• 250.66
(A)

IF GEC IS INSTALLED HERE SIZE BY
• 250.36(B)
(B)

EQUIPMENT GROUNDING BUS

NOTE: *WHERE THE EBJ DOES NOT FUNCTION AS A COMPONENT OF A POTENTIAL FAULT-CURRENT PATH, USE **TABLE 250.66** TO SIZE IT. IF IT DOES PRECLUDE, SIZE IT PER **250.36(B)**.*

Sizing EBJ for (A) 250.36(G)(1); Table 250.66	Sizing EBJ for (B) 250.36(G)(2); 250.36(B); Table 310.16
Step 1: EBJ is 1/0 AWG cu.	**Step 1:** EBJ is 8 AWG cu.
Solution: EBJ is required to be 1/0 AWG copper.	Solution: EBJ is 8 AWG copper based on 75°C column or 6 AWG copper using 60° C column.

HIGH-IMPEDANCE GROUNDED NEUTRAL SYSTEMS
250.20(E)

Purpose of Change: To recognize a high-impedance grounded neutral system by adding a new subsection **(E)**.

NEC Ch. 2 – Article 250
Part II – 250.22(1) THRU (4)

Type of Change		Panel Action		UL	UL 508	API 500	API 505	OSHA
Revision		Accept		467	40.2	-	-	1910.304(e)(1)(ii)
ROP		ROC		NFPA 70E	NFPA 70B	NFPA 79	NFPA	NEMA
pg. 511	# 5-62a	pg. -	# -	410.9(A)(2)	29.2.17	7.2.2	-	WD 1
log: CP507	CMP: 5	log: -	Submitter: CMP 5			2002 NEC: 250.22(1) - (4)		IEC: 312.2

2002 NEC – 250.22 Circuit Not to Be Grounded

The following circuits shall not be grounded:

(1) ~~Cranes~~ (circuits for electric cranes operating over combustible fibers in Class III locations, as provided in ~~503.13~~)

(2) Health care facilities ~~(circuits~~ as provided in ~~Article 517)~~

(3) Electrolytic cells ~~(circuits~~ as provided in Article 668)

(4) Lighting systems ~~[secondary circuits~~ as provided in 411.5(A)]

2005 NEC – 250.22 Circuit Not to Be Grounded

The following circuits shall not be grounded:

(1) Circuits for electric cranes operating over combustible fibers in Class III locations, as provided in <u>503.155</u>

(2) <u>Circuits</u> in health care facilities as provided in <u>517.61 and 517.160</u>

(3) <u>Circuits for equipment within electrolytic cell working zone</u> as provided in Article 668

(4) <u>Secondary circuits of</u> lighting systems as provided in 411.5(A)

Author's Substantiation. This revision adds clarity to this section for circuits that shall not be grounded.

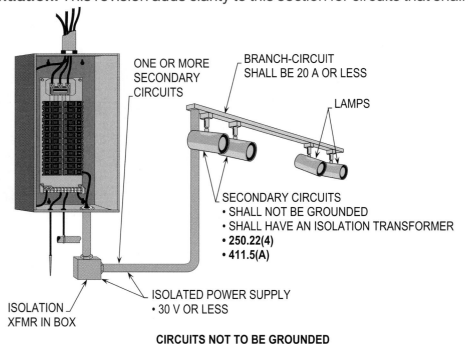

ONE OR MORE SECONDARY CIRCUITS

BRANCH-CIRCUIT SHALL BE 20 A OR LESS

LAMPS

SECONDARY CIRCUITS
• SHALL NOT BE GROUNDED
• SHALL HAVE AN ISOLATION TRANSFORMER
• **250.22(4)**
• **411.5(A)**

ISOLATION XFMR IN BOX

ISOLATED POWER SUPPLY
• 30 V OR LESS

CIRCUITS NOT TO BE GROUNDED
250.22(1) THRU (4)

Purpose of Change: For clarity, words were deleted and the language editorially revised.

Type of Change		Panel Action		UL	UL 508	API 500	API 505	OSHA
Revision		Accept in Principle		467	40.2	-	-	1910.304(f)
ROP		ROC		NFPA 70E	NFPA 70B	NFPA 79	NFPA	NEMA
pg. 515	# 5-67	pg. -	# -	410.10(B)	8.9.9.3	8.2	-	WD 1
log: 1543	CMP: 5	log: -	Submitter: Michael J. Johnston			2002 NEC: 250.24(A)(4)		IEC: 548

2002 NEC – 250.24(A)(4) Main Bonding Jumper as Wire or Busbar

Where the main bonding jumper specified in 250.28 is a wire or busbar and installed from the ~~neutral~~ bar or bus to the equipment grounding terminal bar or bus in the service equipment, the grounding electrode conductor shall be permitted to be connected to the equipment grounding terminal bar or bus to which the main bonding jumper is connected.

2005 NEC – 250.24(A)(4) Main Bonding Jumper as Wire or Busbar

Where the main bonding jumper specified in 250.28 is a wire or busbar and installed from the <u>grounded conductor terminal</u> bar or bus to the equipment grounding terminal bar or bus in the service equipment, the grounding electrode conductor shall be permitted to be connected to the equipment grounding terminal, bar, or bus to which the main bonding jumper is connected.

Author's Substantiation. This revision clarifies that all grounded conductors not just grounded conductors that are neutral conductors shall be connected to the grounded conductor terminal, bar or bus.

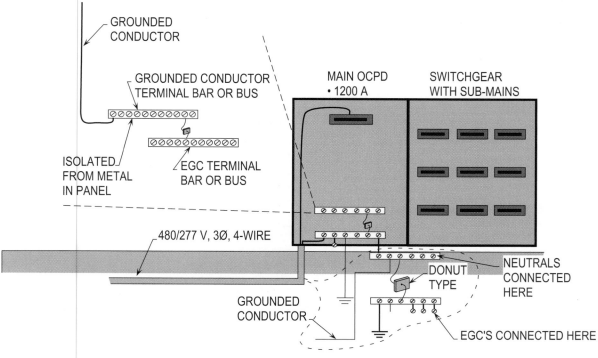

MAIN BONDING JUMPER AS WIRE OR BUSBAR
250.24(A)(4)

Purpose of Change: For clarity and application of rules, wording was deleted and the language editorially revised.

NEC Ch. 2 – Article 250
Part II – 250.24(B)

45

Type of Change		Panel Action		UL	UL 508	API 500	API 505	OSHA
New Section		Accept in Principle		467	40.2	-	-	1910.304(f)(4)
ROP		ROC		NFPA 70E	NFPA 70B	NFPA 79	NFPA	NEMA
pg. 516	# 5-69	pg. -	# -	410.10(B)	8.9.9.3	8.2	-	WD 1
log: 2473	CMP: 5	log: -	Submitter: Paul Dobrowsky			2002 NEC: -		IEC: 548

2005 NEC – 250.24(B) Main Bonding Jumper

For a grounded system, an unspliced main bonding jumper shall be used to connect the equipment grounding conductor(s) and the service-disconnect enclosure to the grounded conductor within the enclosure for each service disconnect in accordance with 250.28.

> **Exception No. 1:** Where more than one service disconnecting means is located in an assembly listed for use as service equipment, an unspliced main bonding jumper shall bond the grounded conductor(s) to the assembly enclosure.

> **Exception No. 2:** Impedance grounded neutral systems shall be permitted to be connected as provided in 250.36 and 250.186.

Author's Substantiation. This new section addresses the sizing and installation of the main bonding jumper for a grounded system.

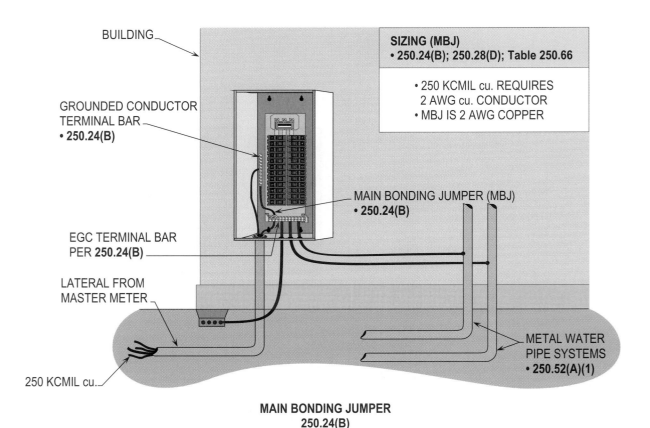

MAIN BONDING JUMPER
250.24(B)

Purpose of Change: To provide a section on how to size and install the main bonding jumper.

NEC Ch. 2 – Article 250
Part II – 250.24(C)

Type of Change	Panel Action	UL	UL 508	API 500	API 505	OSHA
Renumbered	Accept in Principle	467	-	-	-	1910.304(f)(1)(v)

ROP		ROC		NFPA 70E	NFPA 70B	NFPA 79	NFPA	NEMA
pg. 516	# 5-69	pg. -	# -	450.5(D)(2)	27.1.1.7	9.4.2.1	-	WD 1
log: 2473	CMP: 5	log: -	Submitter: Paul Dobrowsky			2002 NEC: 250.24(B)		IEC: 312.2

2002 NEC – ~~250.24(B)~~ Grounded Conductor Brought to Service Equipment

Where an AC system operating at less than 1000 volts is grounded at any point, the grounded conductor(s) shall be run to each service disconnecting means and shall be bonded to each disconnecting means enclosure. The grounded conductor(s) shall be installed in accordance with ~~250.24(B)(1) through (B)(3).~~

2005 NEC – 250.24(C) Grounded Conductor Brought to Service Equipment

Where an AC system operating at less than 1000 volts is grounded at any point, the grounded conductor(s) shall be run to each service disconnecting means and shall be bonded to each disconnecting means enclosure. The grounded conductor(s) shall be installed in accordance with 250.24(C)(1) through (C)(3).

Author's Substantiation. This subsection has been moved and renumbered due to 250.24(B) being added for main bonding jumper.

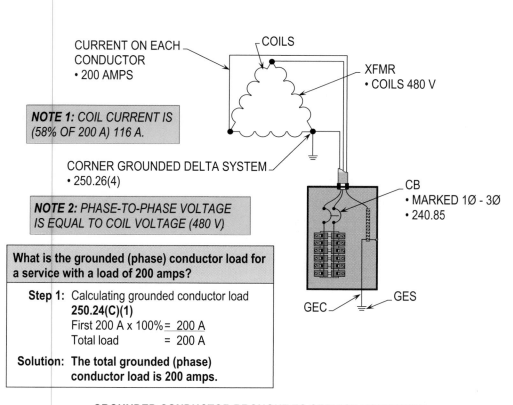

GROUNDED CONDUCTOR BROUGHT TO SERVICE EQUIPMENT
250.24(C)

Purpose of Change: For clarity of intent, the last sentence was revised and the title head relocated and renumbered.

NEC Ch. 2 – Article 250
Part II – 250.24(D)

47

Type of Change		Panel Action		UL	UL 508	API 500	API 505	OSHA
Renumbered		Accept in Principle		467	-	-	-	1910.304(f)(1)(v)
ROP		ROC		NFPA 70E	NFPA 70B	NFPA 79	NFPA	NEMA
pg. 516	# 5-69	pg. -	# -	450.5(D)(2)	27.1.1.7	9.4.2.1	-	WD 1
log: 2473	CMP: 5	log: -	Submitter: Paul Dobrowsky			2002 NEC: 250.24(C)		IEC: 312.2

2002 NEC – 250.24(C) Grounding Electrode Conductor

A grounding electrode conductor shall be used to connect the equipment grounding conductors, the service-equipment enclosures, and, where the system is grounded, the grounded service conductor to the grounding electrode(s) required by Part III of this article.

High-impedance grounded neutral system connections shall be made as covered in 250.36.

2005 NEC – 250.24(D) Grounding Electrode Conductor

A grounding electrode conductor shall be used to connect the equipment grounding conductors, the service-equipment enclosures, and, where the system is grounded, the grounded service conductor to the grounding electrode(s) required by Part III of this article.

High-impedance grounded neutral system connections shall be made as covered in 250.36.

Author's Substantiation. This subsection has been moved and renumbered due to 250.24(B) being added for main bonding jumper.

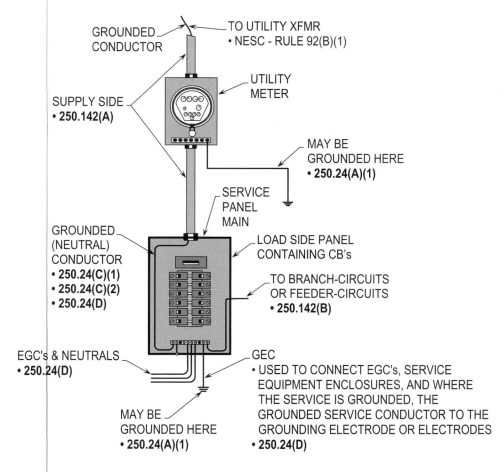

GROUNDED CONDUCTOR

TO UTILITY XFMR
• NESC - RULE 92(B)(1)

UTILITY METER

SUPPLY SIDE
• 250.142(A)

MAY BE GROUNDED HERE
• 250.24(A)(1)

SERVICE PANEL MAIN

GROUNDED (NEUTRAL) CONDUCTOR
• 250.24(C)(1)
• 250.24(C)(2)
• 250.24(D)

LOAD SIDE PANEL CONTAINING CB's

TO BRANCH-CIRCUITS OR FEEDER-CIRCUITS
• 250.142(B)

EGC's & NEUTRALS
• 250.24(D)

GEC
• USED TO CONNECT EGC's, SERVICE EQUIPMENT ENCLOSURES, AND WHERE THE SERVICE IS GROUNDED, THE GROUNDED SERVICE CONDUCTOR TO THE GROUNDING ELECTRODE OR ELECTRODES
• 250.24(D)

MAY BE GROUNDED HERE
• 250.24(A)(1)

GROUNDING ELECTRODE CONDUCTOR
250.24(D)

Purpose of Change: To improve the usability of the NEC, the requirements of existing **250.24(C)** were relocated to **250.24(D)**.

NEC Ch. 2 – Article 250
Part II – 250.28(D)

48

Type of Change	Panel Action	UL	UL 508	API 500	API 505	OSHA
Revision	Accept in Principle	467	40.2	-	-	1910.304(f)(4)
ROP	ROC	NFPA 70E	NFPA 70B	NFPA 79	NFPA	NEMA
pg. 518 # 5-74	pg. 145 # 5-46	410.10(B)	8.9.9.3	8.2	-	WD 1
log: 2474 CMP: 5	log: 1021 Submitter: Paul Dobrowsky			2002 NEC: 250.28(D)		IEC: 548

2002 NEC – 250.28(D) Size

~~The main bonding jumper~~ shall not be smaller than the sizes shown in Table 250.66 ~~for grounding electrode conductors~~. Where the ~~service-entrance phase~~ conductors are larger than 1100 kcmil copper or 1750 kcmil aluminum, the bonding jumper shall have an area that is not less than 12 1/2 percent of the area of the largest phase conductor except that, where the phase conductors and the bonding jumper are of different materials (copper or aluminum), the minimum size of the bonding jumper shall be based on the assumed use of phase conductors of the same material as the bonding jumper and with an ampacity equivalent to that of the installed phase conductors.

2005 NEC – 250.28(D) Size

<u>Main bonding jumpers and system bonding jumpers</u> shall not be smaller than the sizes shown in Table 250.66. Where the <u>supply</u> conductors are larger than 1100 kcmil copper or 1750 kcmil aluminum, the bonding jumper shall have an area that is not less than 12 1/2 percent of the area of the largest phase conductor except that, where the phase conductors and the bonding jumper are of different materials (copper or aluminum), the minimum size of the bonding jumper shall be based on the assumed use of phase conductors of the same material as the bonding jumper and with an ampacity equivalent to that of the installed phase conductors.

Author's Substantiation. This revision clarifies that the main bonding jumper and system bonding jumper for services and separately derived systems shall be sized not smaller than the sizes shown in Table 250.66 and where the supply conductors are larger than 1100 kcmil copper or 1750 kcmil, the bonding jumper shall have an area that is not less than 12 1/2 percent of the largest phase conductor.

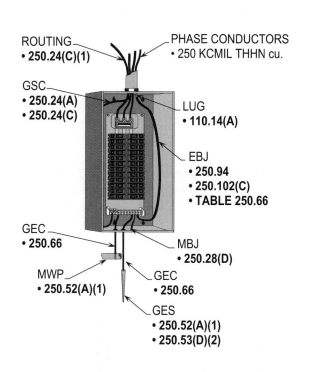

ROUTING
• 250.24(C)(1)

PHASE CONDUCTORS
• 250 KCMIL THHN cu.

GSC
• 250.24(A)
• 250.24(C)

LUG
• 110.14(A)

EBJ
• 250.94
• 250.102(C)
• TABLE 250.66

GEC
• 250.66

MBJ
• 250.28(D)

MWP
• 250.52(A)(1)

GEC
• 250.66

GES
• 250.52(A)(1)
• 250.53(D)(2)

SIZE
250.28(D)

CODE CHANGE LOOP

GROUNDED CONDUCTOR
• 250.24(C)(1)
• 250.142(A)(1)

MBJ
• 250.28(A) - (D)

EBJ
• 250.102(C)

GEC
• 250.66
• TABLE 250.66

GES
• 250.52(A)(1) - (A)(4)
• 250.52(A)(5)

Sizing Main Bonding Jumper
250.28(D)

Step 1: Calculating MBJ
250.28(D); Table 250.66
250 KCMIL cu. requires 2 AWG cu.

Solution: The size of the main bonding jumper is 2 AWG cu.

Purpose of Change: To clarify the sizing requirements for main bonding jumpers and system bonding jumpers.

Type of Change		Panel Action		UL	UL 508	API 500	API 505	OSHA
Revision		Accept in Principle		467	40.2	-	-	1910.304(f)(4)
ROP		ROC		NFPA 70E	NFPA 70B	NFPA 79	NFPA	NEMA
pg. 521	# 5-78	pg. -	# -	410.10(B)	8.9.9.3	8.2	-	WD 1
log: 1725	CMP: 5	log: -	Submitter: Michael J. Johnston			2002 NEC: 250.30(A)(1)		IEC: 548

2002 NEC – 250.30(A)(1) Bonding Jumper

A bonding jumper in compliance with 250.28(A) through (D) that is sized ~~for~~ the derived phase conductors shall be used to connect the equipment grounding conductors of the separately derived system to the grounded conductor. ~~Except as permitted by 250.24(A)(3),~~ this connection shall be made at any point on the separately derived system from the source to the first system disconnecting means or overcurrent device, or it shall be made at the source of a separately derived system that has no disconnecting means or overcurrent devices. ~~The point of connection shall be the same as the grounding electrode conductor as required in 250.30(A)(2).~~

> **Exception No. 1:** A bonding jumper at both the source and the first disconnecting means shall be permitted where doing so does not establish a parallel path for the grounded ~~circuit~~ conductor. Where a grounded conductor is used in this manner, it shall not be smaller than the size specified for the bonding jumper but shall not be required to be larger than the ungrounded conductor(s). For the purposes of this exception, connection through earth shall not be considered as providing a parallel path.

> **Exception No. 2:** The size of the bonding jumper for a system that supplies a Class 1, Class 2, or Class 3 circuit, and is derived from a transformer rated not more than 1000 volt-amperes, shall not be smaller than the derived phase conductors and shall not be smaller than 14 AWG copper or 12 AWG aluminum.

2005 NEC – 250.30(A)(1) System Bonding Jumper

An unspliced system bonding jumper in compliance with 250.28(A) through (D) that is sized based on the derived phase conductors shall be used to connect the equipment grounding conductors of the separately derived system to the grounded conductor. This connection shall be made at any single point on the separately derived system from the source to the first system disconnecting means or overcurrent device, or it shall be made at the source of a separately derived system that has no disconnecting means or overcurrent devices.

> **Exception No. 1:** For separately derived systems that are dual fed (double ended) in a common enclosure or grouped together in separate enclosures and employing a secondary tie, a single system bonding jumper connection to the tie point of the grounded circuit conductors from each power source shall be permitted.

> **Exception No. 2:** A system bonding jumper at both the source and the first disconnecting means shall be permitted where doing so does not establish a parallel path for the grounded conductor. Where a grounded conductor is used in this manner, it shall not be smaller than the size specified for the system bonding jumper but shall not be required to be larger than the ungrounded conductor(s). For the purposes of this exception, connection through earth shall not be considered as providing a parallel path.

> **Exception No. 3:** The size of the system bonding jumper for a system that supplies a Class 1, Class 2, or Class 3 circuit, and is derived from a transformer rated not more than 1000 volt-amperes, shall not be smaller than the derived phase conductors and shall not be smaller than 14 AWG copper or 12 AWG aluminum.

Author's Substantiation. The term "bonding jumper" has been changed to the term "system bonding jumper" to clear up any confusion between a bonding jumper and main bonding jumper.

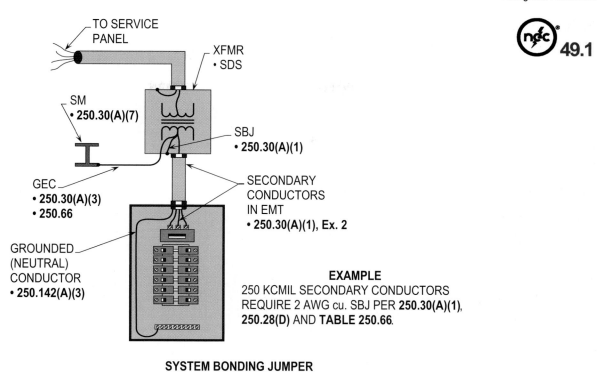

TO SERVICE PANEL

XFMR
• SDS

SM
• 250.30(A)(7)

SBJ
• 250.30(A)(1)

GEC
• 250.30(A)(3)
• 250.66

SECONDARY CONDUCTORS IN EMT
• 250.30(A)(1), Ex. 2

GROUNDED (NEUTRAL) CONDUCTOR
• 250.142(A)(3)

EXAMPLE
250 KCMIL SECONDARY CONDUCTORS REQUIRE 2 AWG cu. SBJ PER **250.30(A)(1)**, **250.28(D)** AND **TABLE 250.66**.

SYSTEM BONDING JUMPER
250.30(A)(1)

Purpose of Change: To change the term "bonding jumper" to the term "system bonding jumper" (SBJ) in an effort to make this section easier to read, understand, and enforce.

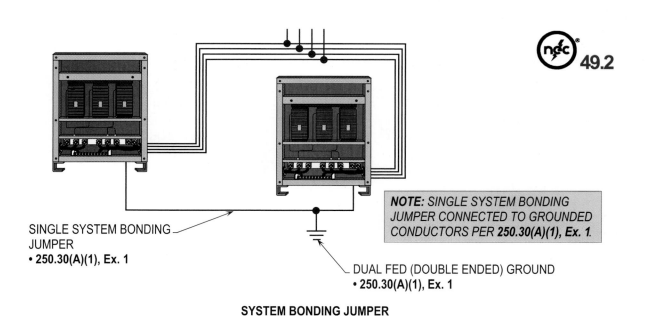

SINGLE SYSTEM BONDING JUMPER
• 250.30(A)(1), Ex. 1

NOTE: SINGLE SYSTEM BONDING JUMPER CONNECTED TO GROUNDED CONDUCTORS PER **250.30(A)(1), Ex. 1**.

DUAL FED (DOUBLE ENDED) GROUND
• 250.30(A)(1), Ex. 1

SYSTEM BONDING JUMPER
250.30(A)(1), Ex. 1

Purpose of Change: To include procedures for grounding dual fed separately derived systems employing a secondary tie.

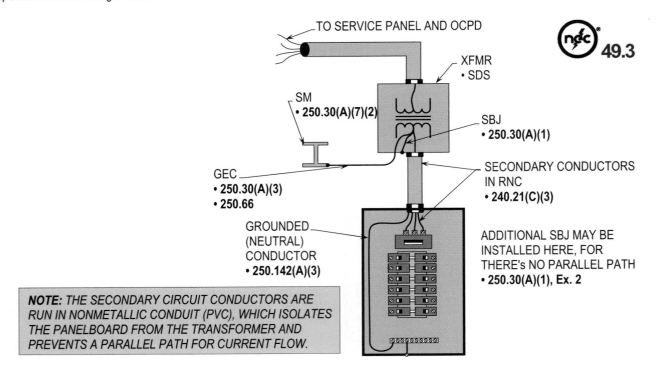

TO SERVICE PANEL AND OCPD

49.3

XFMR
• SDS

SM
• **250.30(A)(7)(2)**

SBJ
• **250.30(A)(1)**

GEC
• **250.30(A)(3)**
• **250.66**

SECONDARY CONDUCTORS
IN RNC
• **240.21(C)(3)**

GROUNDED
(NEUTRAL)
CONDUCTOR
• **250.142(A)(3)**

ADDITIONAL SBJ MAY BE
INSTALLED HERE, FOR
THERE's NO PARALLEL PATH
• **250.30(A)(1), Ex. 2**

*NOTE: THE SECONDARY CIRCUIT CONDUCTORS ARE
RUN IN NONMETALLIC CONDUIT (PVC), WHICH ISOLATES
THE PANELBOARD FROM THE TRANSFORMER AND
PREVENTS A PARALLEL PATH FOR CURRENT FLOW.*

SYSTEM BONDING JUMPER
250.30(A)(1), Ex. 2

Purpose of Change: To include requirements for installing a system bonding jumper at the power source
and OCPD in panelboard.

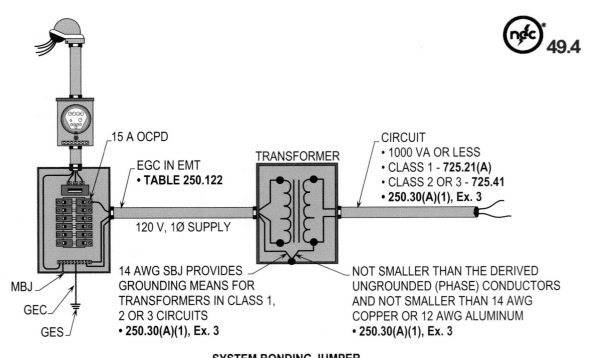

49.4

15 A OCPD

EGC IN EMT
• **TABLE 250.122**

TRANSFORMER

CIRCUIT
• 1000 VA OR LESS
• CLASS 1 - **725.21(A)**
• CLASS 2 OR 3 - **725.41**
• **250.30(A)(1), Ex. 3**

120 V, 1Ø SUPPLY

MBJ

GEC

GES

14 AWG SBJ PROVIDES
GROUNDING MEANS FOR
TRANSFORMERS IN CLASS 1,
2 OR 3 CIRCUITS
• **250.30(A)(1), Ex. 3**

NOT SMALLER THAN THE DERIVED
UNGROUNDED (PHASE) CONDUCTORS
AND NOT SMALLER THAN 14 AWG
COPPER OR 12 AWG ALUMINUM
• **250.30(A)(1), Ex. 3**

SYSTEM BONDING JUMPER
250.30(A)(1), Ex. 3

Purpose of Change: To provide requirements for the grounding of a VA rated transformer with a system
bonding jumper per the **2005 NEC**.

NEC Ch. 2 – Article 250
Part II – 250.30(A)(3)

50

Type of Change		Panel Action		UL	UL 508	API 500	API 505	OSHA
Revision		Accept in Principle		467	40.3	-	-	1910.304(f)(1)(v)
ROP		**ROC**		NFPA 70E	NFPA 70B	NFPA 79	NFPA	NEMA
pg. 521	# 5-78	pg. -	# -	410.10(C)(5)	Ch. 9	8.3	-	WD 1
log: 1725	CMP: 5	log: -	Submitter: Michael J. Johnston			2002 NEC: 250.30(A)(2)		IEC: 312.2

2002 NEC – ~~250.30(A)(2)~~ Grounding Electrode Conductor

~~The grounding electrode conductor shall be installed in accordance with (a) or (b). Where taps are connected to a common grounding electrode conductor, the installation shall comply with 250.30(A)(3).~~

~~(a)~~ **Single Separately Derived System.** A grounding electrode conductor for a single separately derived system shall be sized in accordance with 250.66 for the derived phase conductors and shall be used to connect the grounded conductor of the derived system to the grounding electrode as specified in ~~250.30(A)(4). Except as permitted by 250.24(A)(3) or (A)(4),~~ this connection shall be made at the same point on the separately derived system where the bonding jumper is installed.

> ~~Exception:~~ A grounding electrode conductor shall not be required for a system that supplies a Class 1, Class 2, or Class 3 circuit and is derived from a transformer rated not more than 1000 volt-amperes, provided the system grounded conductor is bonded to the transformer frame or enclosure by a jumper sized in accordance with ~~250.30(A)(1), Exception No. 2~~, and the transformer frame or enclosure is grounded by on the means specified in 250.134.

2005 NEC – 250.30(A)(3) Grounding Electrode Conductor, Single Separately Derived System

A grounding electrode conductor for a single separately derived system shall be sized in accordance with 250.66 for the derived phase conductors and shall be used to connect the grounded conductor of the derived system to the grounding electrode as specified in 250.30(A)(7). This connection shall be made at the same point on the separately derived system where the system bonding jumper is installed.

> **Exception No. 1:** Where the system bonding jumper specified in 250.30(A)(1) is a wire or busbar, it shall be permitted to connect the grounding electrode conductor to the equipment grounding terminal, bar, or bus, provided the equipment grounding terminal, bar, or bus is of sufficient size for the separately derived system.

> **Exception No. 2:** Where a separately derived system originates in listed equipment suitable as service equipment, the grounding electrode conductor from the service or feeder equipment to the grounding electrode shall be permitted as the grounding electrode conductor for the separately derived system, provided the grounding electrode conductor is of sufficient size for the separately derived system. Where the equipment ground bus internal to the equipment is not smaller than the required grounding electrode conductor for the separately derived system, the grounding electrode connection for the separately derived system shall be permitted to be made to the bus.

> **Exception No. 3:** A grounding electrode conductor shall not be required for a system that supplies a Class 1, Class 2, or Class 3 circuit and is derived from a transformer rated not more than 1000 volt-amperes, provided the grounded conductor is bonded to the transformer frame or enclosure by a jumper sized in accordance with 250.30(A)(1), Exception No. 3, and the transformer frame or enclosure is grounded by one of the means specified in 250.134.

Author's Substantiation. This section has been renumbered, with new exceptions added to clarify the requirements for the grounding electrode conductor for single separately derived systems.

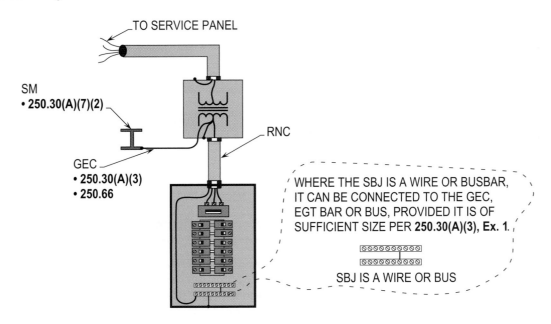

TO SERVICE PANEL

SM
• **250.30(A)(7)(2)**

GEC
• **250.30(A)(3)**
• **250.66**

RNC

WHERE THE SBJ IS A WIRE OR BUSBAR, IT CAN BE CONNECTED TO THE GEC, EGT BAR OR BUS, PROVIDED IT IS OF SUFFICIENT SIZE PER **250.30(A)(3), Ex. 1.**

SBJ IS A WIRE OR BUS

GROUNDING ELECTRODE CONDUCTOR, SINGLE SEPARATELY DERIVED SYSTEM
250.30(A)(3), Ex. 1

Purpose of Change: To include requirements for terminating the system bonding jumper for a separately derived system to the other components completing the grounding system.

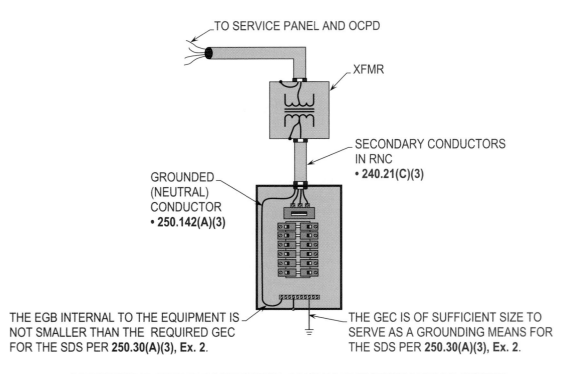

TO SERVICE PANEL AND OCPD

XFMR

SECONDARY CONDUCTORS IN RNC
• **240.21(C)(3)**

GROUNDED (NEUTRAL) CONDUCTOR
• **250.142(A)(3)**

THE EGB INTERNAL TO THE EQUIPMENT IS NOT SMALLER THAN THE REQUIRED GEC FOR THE SDS PER **250.30(A)(3), Ex. 2.**

THE GEC IS OF SUFFICIENT SIZE TO SERVE AS A GROUNDING MEANS FOR THE SDS PER **250.30(A)(3), Ex. 2.**

GROUNDING ELECTRODE CONDUCTOR, SINGLE SEPARATELY DERIVED SYSTEM
250.30(A)(3), Ex. 2

Purpose of Change: To determine the requirements for grounding to the equipment grounding bus in the equipment listed as service equipment.

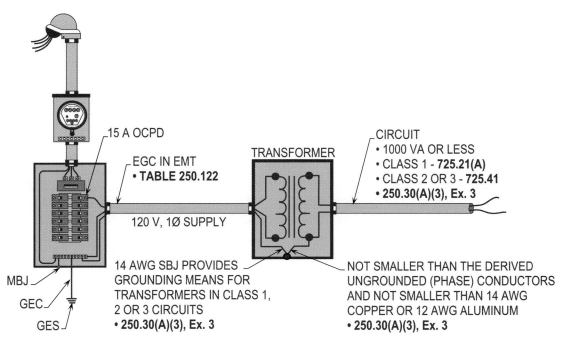

15 A OCPD

EGC IN EMT
• **TABLE 250.122**

TRANSFORMER

CIRCUIT
• 1000 VA OR LESS
• CLASS 1 - **725.21(A)**
• CLASS 2 OR 3 - **725.41**
• **250.30(A)(3), Ex. 3**

120 V, 1Ø SUPPLY

14 AWG SBJ PROVIDES
GROUNDING MEANS FOR
TRANSFORMERS IN CLASS 1,
2 OR 3 CIRCUITS
• **250.30(A)(3), Ex. 3**

NOT SMALLER THAN THE DERIVED
UNGROUNDED (PHASE) CONDUCTORS
AND NOT SMALLER THAN 14 AWG
COPPER OR 12 AWG ALUMINUM
• **250.30(A)(3), Ex. 3**

MBJ

GEC

GES

**GROUNDING ELECTRODE CONDUCTOR
250.30(A)(3), Ex. 3**

Purpose of Change: To provide requirements for the grounding of a 1000 VA transformer with a system bonding jumper per the **2005 NEC**.

51

Type of Change		Panel Action	UL	UL 508	API 500	API 505	OSHA
Revision		Accept in Principle	467	40.3	-	-	1910.304(f)(1)(v)
ROP		ROC	NFPA 70E	NFPA 70B	NFPA 79	NFPA	NEMA
pg. 521	# 5-78	pg. -	# -	410.10(C)(5)	Ch. 9	8.3	-	WD 1
log: 1725	CMP: 5	log: -	Submitter: Michael J. Johnston		2002 NEC: 250.30(A)(2)		IEC: 312.2

2002 NEC – ~~250.30(A)(2)~~ Grounding Electrode Conductor

~~The grounding electrode conductor shall be installed in accordance with (a) or (b). Where taps are connected to a common grounding electrode conductor, the installation shall comply with 250.30(A)(3).~~

(b) **Multiple Separately Derived Systems.** Where more than one separately derived system ~~is connected~~ to a common grounding electrode conductor ~~as provided in 250.30(A)(3),~~ the common grounding electrode conductor ~~shall be sized in accordance with 250.66, based on the total area of the largest derived phase conductor from each separately derived system.~~

2005 NEC – <u>250.30(A)(4)</u> Grounding Electrode Conductor, Multiple Separately Derived Systems

Where more than one separately derived system is <u>installed, it shall be permissible to connect a tap from each separately derived system</u> to a common grounding electrode conductor. <u>Each tap conductor shall connect the grounded conductor of the separately derived system to</u> the common grounding electrode conductor. <u>The grounding electrode conductors and taps shall comply with 250.30(A)(4)(a) through (A)(4)(c).</u>

<u>**Exception No. 1:** Where the system bonding jumper specified in 250.30(A)(1) is a wire or busbar, it shall be permitted to connect the grounding electrode conductor to the equipment grounding terminal, bar, or bus, provided the equipment grounding terminal, bar, or bus is of sufficient size for the separately derived system.</u>

<u>**Exception No. 2:** A grounding electrode conductor shall not be required for a system that supplies a Class 1, Class 2, or Class 3 circuit and is derived from a transformer rated not more than 1000 volt-amperes, provided the system grounded conductor is bonded to the transformer frame or enclosure by a jumper sized in accordance with 250.30(A)(1), Exception No. 3 and the transformer frame or enclosure is grounded by one the means specified in 250.134.</u>

<u>(a) **Common Grounding Electrode Conductor Size.** The common grounding electrode conductor shall not be smaller than 3/0 AWG copper or 250 kcmil aluminum.</u>

Author's Substantiation. This section has been renumbered, with new exceptions added to clarify the requirements for the grounding electrode conductor for multiple separately derived systems.

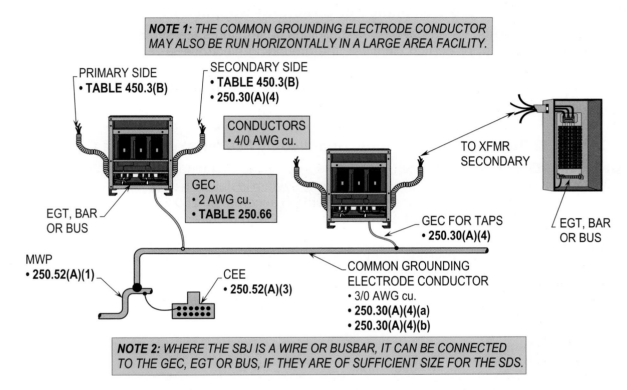

NOTE 1: *THE COMMON GROUNDING ELECTRODE CONDUCTOR MAY ALSO BE RUN HORIZONTALLY IN A LARGE AREA FACILITY.*

PRIMARY SIDE
• **TABLE 450.3(B)**

SECONDARY SIDE
• **TABLE 450.3(B)**
• **250.30(A)(4)**

CONDUCTORS
• 4/0 AWG cu.

TO XFMR SECONDARY

GEC
• 2 AWG cu.
• **TABLE 250.66**

EGT, BAR OR BUS

GEC FOR TAPS
• **250.30(A)(4)**

EGT, BAR OR BUS

MWP
• **250.52(A)(1)**

CEE
• **250.52(A)(3)**

COMMON GROUNDING ELECTRODE CONDUCTOR
• 3/0 AWG cu.
• **250.30(A)(4)(a)**
• **250.30(A)(4)(b)**

NOTE 2: *WHERE THE SBJ IS A WIRE OR BUSBAR, IT CAN BE CONNECTED TO THE GEC, EGT OR BUS, IF THEY ARE OF SUFFICIENT SIZE FOR THE SDS.*

GROUNDING ELECTRODE CONDUCTOR, MULTIPLE SEPARATELY DERIVED SYSTEMS
250.30(A)(4), Ex. 1

Purpose of Change: To provide requirements for connecting the system bonding jumper and grounding the separately derived systems to the common GEC.

NEC Ch. 2 – Article 250
Part II – 250.32(A)

52

Type of Change		Panel Action		UL	UL 508	API 500	API 505	OSHA
Revision		Accept in Principle		467	-	-	-	1910.304(f)(1)(v)
ROP		ROC		NFPA 70E	NFPA 70B	NFPA 79	NFPA	NEMA
pg. 541	# 5-109	pg. -	# -	450.5(D)(2)	27.1.1.7	9.4.2.1	-	WD 1
log: 1560	CMP: 5	log: -	Submitter: Michael J. Johnston			2002 NEC: 250.32(A)		IEC: 312.2

2002 NEC – 250.32 ~~Two or More~~ Buildings or Structures Supplied ~~from a Common Service~~

(A) Grounding Electrode. ~~When two or more~~ buildings or structures ~~are~~ supplied from a common ac service by ~~a~~ feeder(s) or branch circuit(s)~~,~~ the grounding electrode~~(s) required in Part III of this article at each building or structure~~ shall be connected ~~in the manner specified~~ in 250.32(B) or (C). Where there ~~are~~ no existing grounding ~~electrodes~~, the grounding electrode(s) required in ~~Part III of this article~~ shall be installed.

2005 NEC – 250.32 Buildings or Structures Supplied <u>by Feeder(s) or Branch Circuit(s)</u>

(A) Grounding Electrode. Building(s) or structure<u>(s)</u> supplied by feeder<u>(s)</u> or branch circuit(s) <u>shall have a grounding electrode or grounding electrode system installed in accordance with 250.50</u>. The grounding electrode <u>conductors</u> shall be connected in accordance 250.32(B) or (C). Where there is no existing grounding <u>electrode</u>, the grounding electrode(s) required in 250.50 shall be installed.

Author's Substantiation. This revision clarifies that building(s) or structure(s) supplied from a common ac service by feeder(s) or branch circuit(s) shall have a grounding electrode or grounding electrode system installed in accordance with 250.50. This change expands the installations beyond those limited to feeder(s) or branch circuit(s) that might not originate from a common service.

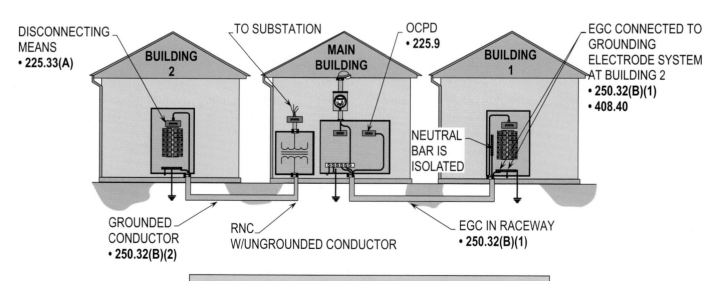

BUILDINGS OR STRUCTURES SUPPLIED BY FEEDER(S) OR BRANCH-CIRCUIT(S)
250.32(A)

Purpose of Change: Editorial changes were made for clarity, including the addition of a new title head.

NEC Ch. 2 – Article 250
Part II – 250.34(A)(1) and (A)(2)

Type of Change		Panel Action		UL	UL 508	API 500	API 505	OSHA
Revision		Accept in Principle		467	-	-	-	1926.404(f)(3)
ROP		ROC		NFPA 70E	NFPA 70B	NFPA 79	NFPA	NEMA
pg. 543	# 5-111	pg. 152	# 5-71	410.10(C)(7)	J.4	-	-	WD 1
log: 2478	CMP: 5	log: 3437	Submitter: -			2002 NEC: 250.34(A)(1) and (A)(2)		IEC: 551

2002 NEC – 250.34 Portable and Vehicle-Mounted Generators

(A) Portable Generators. The frame of a portable generator shall not be required to be ~~grounded and shall be permitted to serve as the grounding electrode~~ for a system supplied by the generator under the following conditions:

(1) The generator supplies only equipment mounted on the generator, cord-and-plug-connected equipment through receptacles mounted on the generator, or both, and

(2) The non-current-carrying metal parts of equipment and the equipment grounding conductor terminals of the receptacles are bonded to the generator frame.

2005 NEC – 250.34 Portable and Vehicle-Mounted Generators

(A) Portable Generators. The frame of a portable generator shall not be required to be <u>connected to a grounding electrode as defined in 250.52</u> for a system supplied by the generator under the following conditions:

(1) The generator supplies only equipment mounted on the generator, cord-and-plug-connected equipment through receptacles mounted on the generator, or both, and

(2) The non-current-carrying metal parts of equipment and the equipment grounding conductor terminals of the receptacles are bonded to the generator frame.

Author's Substantiation. This revision clarifies that the frame of a portable generator shall not be required to be connected to a grounding electrode.

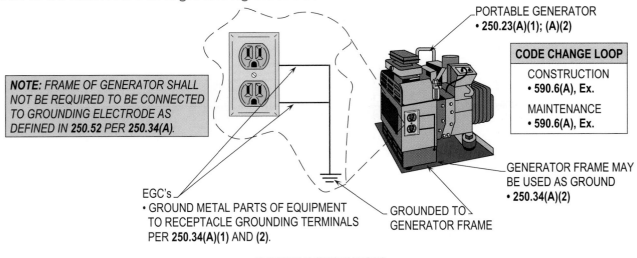

PORTABLE GENERATORS
250.34(A)(1) AND (A)(2)

Purpose of Change: To revise the text and provide clarity for not connecting generator frames to grounding electrode(s).

NEC Ch. 2- Article 250
Part III – 250.52(A)(7)

Type of Change		Panel Action		UL	UL 508	API 500	API 505	OSHA
Revision		Accept in Principle		467	-	-	-	1910.304(f)(1)(v)
ROP		ROC		NFPA 70E	NFPA 70B	NFPA 79	NFPA	NEMA
pg. 556	# 5-134	pg. 159	# 5-101	450.5(D)(2)	27.1.1.7	9.4.2.1	-	WD 1
log: 1617	CMP: 5	log: 2144	Submitter: Lanny Hughes			2002 NEC: 250.52(A)(7)		IEC: 312.2

2002 NEC – 250.52(A)(7) Other Local Metal Undergound Systems or Structures

Other local metal underground systems or structures such as piping systems and underground tanks.

2005 NEC – 250.52(A)(7) Other Local Metal Undergound Systems or Structures

Other local metal underground systems or structures such as piping systems, underground tanks, and underground metal well casings that are not effectively bonded to a metal water pipe.

Author's Substantiation. This revision clarifies that underground metal well casings are permitted to serve as grounding electrodes.

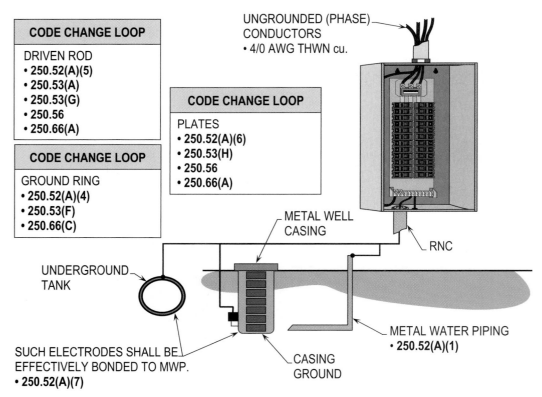

OTHER LOCAL METAL UNDERGROUND SYSTEMS OR STRUCTURES
250.52(A)(7)

Purpose of Change: To require electrodes outlined in **250.52(A)(7)** to be bonded to the metal water piping.

NEC Ch. 2 – Article 250
Part III – 250.54

Type of Change		Panel Action		UL	UL 508	API 500	API 505	OSHA
Revision		Accept		467	-	-	-	-
ROP		ROC		NFPA 70E	NFPA 70B	NFPA 79	NFPA	NEMA
pg. 558	# 5-140	pg. -	# -	-	-	-	-	WD 1
log: 1561	CMP: 5	log: -	Submitter: Michael J. Johnston			2002 NEC: 250.54		IEC: 312.2

2002 NEC – 250.54 Supplementary Grounding Electrodes

Supplementary grounding electrodes shall be permitted to be connected to the equipment grounding conductors specified in 250.118 and shall not be required to comply with the electrode bonding requirements in 250.50 or 250.53(C) or the resistance requirements of 250.56, but the earth shall not be used as ~~the sole equipment grounding conductor~~.

2005 NEC – 250.54 Supplementary Grounding Electrodes

Supplementary grounding electrodes shall be permitted to be connected to the equipment grounding conductors specified in 250.118 and shall not be required to comply with the electrode bonding requirements in 250.50 or 250.53(C) or the resistance requirements of 250.56, but the earth shall not be used as <u>an effective ground-fault current path as specified in 250.24(A)(5) and 250.4(B)(4)</u>.

Author's Substantiation. This revision clarifies that the earth shall not be used as an effective ground-fault current path.

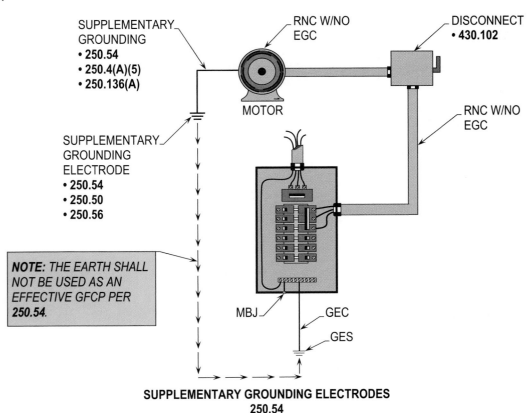

SUPPLEMENTARY GROUNDING ELECTRODES
250.54

Purpose of Change: To ensure the earth is not used as an effective ground-fault current path (GFCP).

NEC Ch. 2 – Article 250
Part III – 250.58

Type of Change		Panel Action		UL	UL 508	API 500	API 505	OSHA
Revision		Accept		-	-	-	-	-
ROP		ROC		NFPA 70E	NFPA 70B	NFPA 79	NFPA	NEMA
pg. 561	# 5-147	pg. -	# -	-	-	-	-	-
log: 2671	CMP: 5	log: -	Submitter: Phil Simmons			2002 NEC: 250.58		IEC: -

2002 NEC – 250.58 Common Grounding Electrode

Where an ac system is connected to a grounding electrode in or at a building ~~as specified in 250.24 and 250.32~~, the same electrode shall be used to ground conductor enclosures and equipment in or on that building. Where separate services supply a building and are required to be connected to a grounding electrode, the same grounding electrode shall be used.

Two or more grounding electrodes that are effectively bonded together shall be considered as a single grounding electrode system in this sense.

2005 NEC – 250.58 Common Grounding Electrode

Where an ac system is connected to a grounding electrode in or at a building or structure, the same electrode shall be used to ground conductor enclosures and equipment in or on that building or structure. Where separate services, feeders, or branch circuits supply a building and are required to be connected to a grounding electrode(s), the same grounding electrode(s) shall be used.

Two or more grounding electrodes that are effectively bonded together shall be considered as a single grounding electrode system in this sense.

Author's Substantiation. This revision clarifies that by adding the words or structure that some structures are not buildings but electrical systems may be required to be grounded at these structures. The words feeders or branch circuits was added to clarify that buildings or structures are permitted to be supplied by one or more of a service, feeder, or branch circuit.

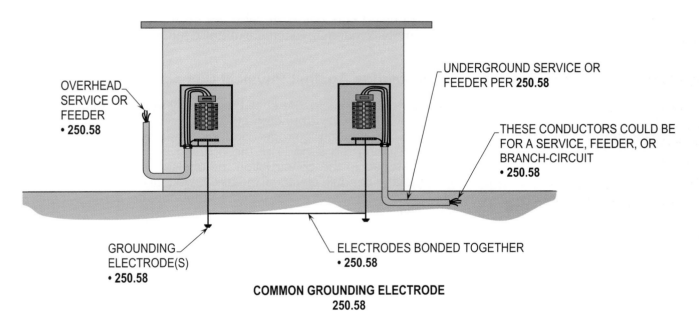

COMMON GROUNDING ELECTRODE
250.58

Purpose of Change: To improve usability of this section, services, feeders, and branch-circuits were added so that a common electrode will be provided for electrical supply systems.

NEC Ch. 2 – Article 250
Part III – 250.64(B)

Type of Change		Panel Action		UL	UL 508	API 500	API 505	OSHA
Revision		Accept in Principle		467	-	-	-	-
ROP		ROC		NFPA 70E	NFPA 70B	NFPA 79	NFPA	NEMA
pg. 562	# 5-150	pg. -	# -	-	-	-	-	-
log: 892	CMP: 5	log: -	Submitter: James M. Daly			2002 NEC: 250.64(B)		IEC: 522.6

2002 NEC – 250.64(B) Securing and Protection ~~from~~ Physical Damage

A grounding electrode conductor or its enclosure shall be securely fastened to the surface on which it is carried. A 4 AWG copper or aluminum ~~or larger~~ conductor shall be protected ~~if~~ exposed to ~~severe~~ physical damage. A 6 AWG grounding conductor that is free from exposure to physical damage shall be permitted to be run along the surface of the building construction without metal covering or protection where it is securely fastened to the construction; otherwise, it shall be in rigid metal conduit, intermediate metal conduit, rigid nonmetallic conduit, electrical metallic tubing, or cable armor. Grounding conductors smaller than 6 AWG shall be in rigid metal conduit, intermediate metal conduit, rigid nonmetallic conduit, electrical metallic tubing, or cable armor.

2005 NEC – 250.64(B) Securing and Protection <u>Against</u> Physical Damage

<u>Where exposed,</u> a grounding electrode conductor or its enclosure shall be securely fastened to the surface on which it is carried. A 4 AWG <u>or larger</u> copper or aluminum <u>grounding electrode</u> conductor shall be protected <u>where</u> exposed to physical damage. A 6 AWG grounding <u>electrode</u> conductor that is free from exposure to physical damage shall be permitted to be run along the surface of the building construction without metal covering or protection where it is securely fastened to the construction; otherwise, it shall be in rigid metal conduit, intermediate metal conduit, rigid nonmetallic conduit, electrical metallic tubing, or cable armor. Grounding <u>electrode</u> conductors smaller than 6 AWG shall be in rigid metal conduit, intermediate metal conduit, rigid nonmetallic conduit, electrical metallic tubing, or cable armor.

Author's Substantiation. This revision clarifies that a 4 AWG or larger copper or aluminum grounding electrode conductors shall be protected where exposed to physical damage. The word "severe" was deleted because it is difficult to distinguish between "physical damage" and "severe physical damage."

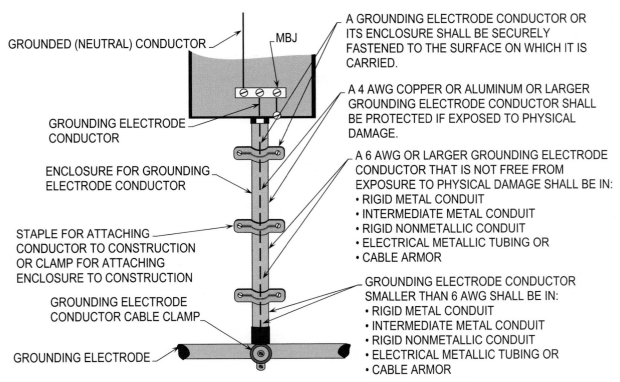

GROUNDED (NEUTRAL) CONDUCTOR

MBJ

A GROUNDING ELECTRODE CONDUCTOR OR ITS ENCLOSURE SHALL BE SECURELY FASTENED TO THE SURFACE ON WHICH IT IS CARRIED.

A 4 AWG COPPER OR ALUMINUM OR LARGER GROUNDING ELECTRODE CONDUCTOR SHALL BE PROTECTED IF EXPOSED TO PHYSICAL DAMAGE.

GROUNDING ELECTRODE CONDUCTOR

ENCLOSURE FOR GROUNDING ELECTRODE CONDUCTOR

A 6 AWG OR LARGER GROUNDING ELECTRODE CONDUCTOR THAT IS NOT FREE FROM EXPOSURE TO PHYSICAL DAMAGE SHALL BE IN:
• RIGID METAL CONDUIT
• INTERMEDIATE METAL CONDUIT
• RIGID NONMETALLIC CONDUIT
• ELECTRICAL METALLIC TUBING OR
• CABLE ARMOR

STAPLE FOR ATTACHING CONDUCTOR TO CONSTRUCTION OR CLAMP FOR ATTACHING ENCLOSURE TO CONSTRUCTION

GROUNDING ELECTRODE CONDUCTOR CABLE CLAMP

GROUNDING ELECTRODE CONDUCTOR SMALLER THAN 6 AWG SHALL BE IN:
• RIGID METAL CONDUIT
• INTERMEDIATE METAL CONDUIT
• RIGID NONMETALLIC CONDUIT
• ELECTRICAL METALLIC TUBING OR
• CABLE ARMOR

GROUNDING ELECTRODE

SECURING AND PROTECTION AGAINST PHYSICAL DAMAGE
250.64(B)

Purpose of Change: The word "severe" was deleted because of the difficulty in distinguishing between "physical damage" and "severe physical damage."

NEC Ch. 2 – Article 250
Part III – 250.64(C)

55

Type of Change		Panel Action		UL	UL 508	API 500	API 505	OSHA
Revision		Accept in Principle		-	-	-	-	-
ROP		ROC		NFPA 70E	NFPA 70B	NFPA 79	NFPA	NEMA
pg. 566	# 5-158	pg. 163	# 5-123	-	-	-	-	-
log: 2957	CMP: 5	log: 2407	Submitter: Charles Mello			2002 NEC: 250.64(C)		IEC: -

2002 NEC – 250.64(C) Continuous

~~The~~ grounding electrode conductor shall be installed in one continuous length without a splice or joint, ~~unless spliced~~ only by irreversible compression-type connectors ~~listed for the purpose~~ or by the exothermic welding process.

> ~~Exception:~~ Sections of busbars shall be permitted to be connected together to form a grounding electrode conductor.

2005 NEC – 250.64(C) Continuous

Grounding electrode conductor(s) shall be installed in one continuous length without a splice or joint except as permitted in (1) through (4).

(1) Splicing shall be permitted only by irreversible compression-type connectors listed as grounding and bonding equipment or by the exothermic welding process.

(2) Section of busbars shall be permitted to be connected together to form a grounding electrode conductor.

(3) Bonding jumper(s) from grounding electrode(s) and grounding electrode conductor(s) shall be permitted to be connected to an aluminum or copper busbar not less than 6 mm x 50 mm (1/4 in. x 2 in.). The busbar shall be securely fastened and shall be installed in an accessible location. Connections shall be made by a listed connector or by the exothermic welding process.

(4) Where aluminum busbars are used, the installation shall comply with 250.64(A).

Author's Substantiation. This revision clarifies the permitted methods for grounding electrode conductors that are not installed in one continuous length.

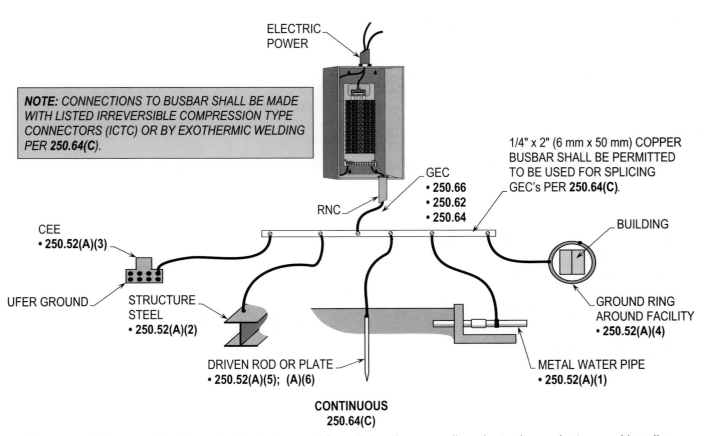

NOTE: *CONNECTIONS TO BUSBAR SHALL BE MADE WITH LISTED IRREVERSIBLE COMPRESSION TYPE CONNECTORS (ICTC) OR BY EXOTHERMIC WELDING PER **250.64(C)**.*

ELECTRIC POWER

1/4" x 2" (6 mm x 50 mm) COPPER BUSBAR SHALL BE PERMITTED TO BE USED FOR SPLICING GEC's PER **250.64(C)**.

GEC
• 250.66
• 250.62
• 250.64

RNC

BUILDING

CEE
• 250.52(A)(3)

UFER GROUND

STRUCTURE STEEL
• 250.52(A)(2)

GROUND RING AROUND FACILITY
• 250.52(A)(4)

DRIVEN ROD OR PLATE
• 250.52(A)(5); (A)(6)

METAL WATER PIPE
• 250.52(A)(1)

**CONTINUOUS
250.64(C)**

Purpose of Change: To allow a busbar to be used for splicing the grounding electrode conductors and bonding jumpers together.

NEC Ch. 2 – Article 250
Part III – 250.64(D)

Type of Change		Panel Action		UL	UL 508	API 500	API 505	OSHA
Revision		Accept in Principle		-	-	-	-	-
ROP		ROC		NFPA 70E	NFPA 70B	NFPA 79	NFPA	NEMA
pg. 568	# 5-161	pg. 164	# 5-125	-	-	-	-	-
log: 2652	CMP: 5	log: 3432	Submitter: Phil Simmons			2002 NEC: 250.64(D)		IEC: -

2002 NEC – 250.64(D) Grounding Electrode Conductor Taps

Where a service consists of more than a single enclosure as permitted in 230.40, Exception No. 2, it shall be permitted to connect taps to the grounding electrode conductor. Each such tap conductor shall extend to the inside of each such enclosure. The grounding electrode conductor shall be sized in accordance with 250.66, but the tap conductors shall be permitted to be sized in accordance with the grounding electrode conductors specified in 250.66 for the largest conductor serving the respective enclosures. The tap conductors shall be connected to the grounding electrode conductor in such a manner that the grounding electrode conductor remains without a splice.

2005 NEC – 250.64(D) Grounding Electrode Conductor Taps

Where a service consists of more than a single enclosure as permitted in 230.71(A), it shall be permitted to connect taps to the common grounding electrode conductor. Each such tap conductor shall extend to the inside of each such enclosure. The common grounding electrode conductor shall be sized in accordance with 250.66, based on the sum of the circular mil area of the largest ungrounded service entrance conductors. Where more than one set of service entrance conductors as permitted by 230.40, Exception No. 2 connect directly to a service drop or lateral, the common grounding electrode conductor shall be sized in accordance with Table 250.66, Note 1. The tap conductors shall be permitted to be sized in accordance with the grounding electrode conductors specified in 250.66 for the largest conductor serving the respective enclosures. The tap conductors shall be connected to the common grounding electrode conductor in such a manner that the common grounding electrode conductor remains without a splice or joint.

Author's Substantiation. This revision clarifies that a tap to the common grounding electrode conductor is permitted where multiple sets of service entrance conductors connect directly to a service drop or lateral where a service consists of more than a single enclosure.

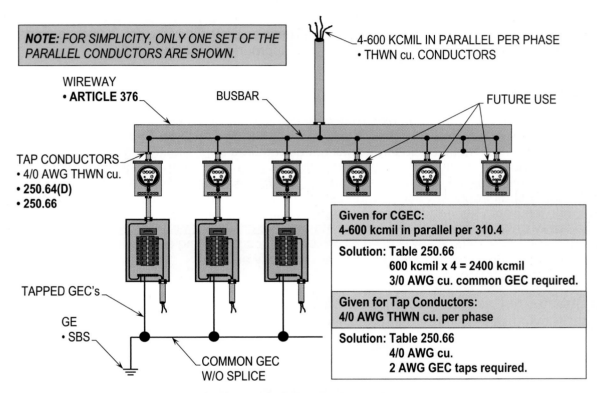

NOTE: *FOR SIMPLICITY, ONLY ONE SET OF THE PARALLEL CONDUCTORS ARE SHOWN.*

4-600 KCMIL IN PARALLEL PER PHASE
• THWN cu. CONDUCTORS

WIREWAY
• **ARTICLE 376**

BUSBAR

FUTURE USE

TAP CONDUCTORS
• 4/0 AWG THWN cu.
• **250.64(D)**
• **250.66**

TAPPED GEC's

GE
• SBS

COMMON GEC
W/O SPLICE

Given for CGEC:
4-600 kcmil in parallel per 310.4

Solution: Table 250.66
600 kcmil x 4 = 2400 kcmil
3/0 AWG cu. common GEC required.

Given for Tap Conductors:
4/0 AWG THWN cu. per phase

Solution: Table 250.66
4/0 AWG cu.
2 AWG GEC taps required.

GROUNDING ELECTRODE CONDUCTOR TAPS
250.64(D)

Purpose of Change: To clarify the procedure for sizing the common GEC and GEC taps according to the provision of this section.

NEC Ch. 2 – Article 250
Part III – 250.64(E)

Type of Change			Panel Action		UL	UL 508	API 500	API 505	OSHA
New Sentence			Accept in Principle		-	-	-	-	-
ROP			**ROC**		**NFPA 70E**	**NFPA 70B**	**NFPA 79**	**NFPA**	**NEMA**
pg. 569	# 5-162		pg. 164	# 5-126	-	-	-	-	-
log: 1564	CMP: 5		log: 58	Submitter: Michael J. Johnston		2002 NEC: 250.64(E)		IEC: -	

2002 NEC – 250.64(E) Enclosures for Grounding Electrode Conductors

Metal enclosures for grounding electrode conductors shall be electrically continuous from the point of attachment to cabinets or equipment to the grounding electrode shall be ~~made electrically continuous by bonding each end to the grounding electrode conductor~~. Where a raceway is used as protection for a grounding electrode conductor, the installation shall comply with the requirements of the appropriate raceway article.

2005 NEC – 250.64(E) Enclosures for Grounding Electrode Conductors

<u>Ferrous</u> metal enclosures for grounding electrode conductors shall be electrically continuous from the point of attachment to cabinets or equipment to the grounding electrode <u>and</u> shall be securely fastened to the ground clamp or fitting. Nonferrous metal enclosures shall not be required <u>to be electrically continuous. Ferrous metal enclosures that are not physically continuous from cabinets or equipment to the grounding electrode shall be made electrically continuous by bonding each end of the raceway or enclosure to the grounding electrode conductor. Bonding shall apply at each end and to all intervening ferrous raceways, boxes, and enclosures between the service equipment and the grounding electrode. The bonding jumper for a grounding electrode conductor raceway or cable armor shall be the same size as, or larger than, the required enclosed grounding electrode conductor.</u> Where a raceway is used as protection for a grounding electrode conductor, the installation shall comply with the requirements of the appropriate raceway article.

Author's Substantiation. Two sentences from 250.92(A)(3) and 250.102(C) have been relocated to 250.64(E) to provide better clarity and usability of the NEC.

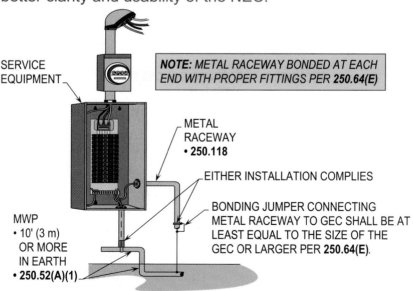

SERVICE EQUIPMENT

NOTE: *METAL RACEWAY BONDED AT EACH END WITH PROPER FITTINGS PER **250.64(E)***

METAL RACEWAY
• **250.118**

EITHER INSTALLATION COMPLIES

BONDING JUMPER CONNECTING METAL RACEWAY TO GEC SHALL BE AT LEAST EQUAL TO THE SIZE OF THE GEC OR LARGER PER **250.64(E)**.

MWP
• 10' (3 m) OR MORE IN EARTH
• **250.52(A)(1)**

ENCLOSURES FOR GROUNDING ELECTRODE CONDUCTORS
250.64(E)

Purpose of Change: For usability, two new sentences were taken from **250.92(A)(3)** and **250.102(C)** and added to this section.

NEC Ch. 2 – Article 250
Part III – 250.68(A), Ex. 2

Type of Change		Panel Action		UL	UL 508	API 500	API 505	OSHA
New Exception		Accept in Principle		-	-	-	-	-
ROP		ROC		NFPA 70E	NFPA 70B	NFPA 79	NFPA	NEMA
pg. 575	# 5-176	pg. -	# -	-	-	-	-	-
log: 1021	CMP: 5	log: -	Submitter: Lawrence A. Bey			2002 NEC: 250.68(A)		IEC: -

2002 NEC – 250.68(A) Accessibility

The connection of a grounding electrode conductor or bonding jumper to a grounding electrode shall be accessible.

> ~~Exception:~~ An encased or buried connection to a concrete-encased, driven, or buried grounding electrode shall not be required to be accessible.

2005 NEC – 250.68(A) Accessibility

The connection of a grounding electrode conductor or bonding jumper to a grounding electrode shall be accessible.

> **Exception No. 1:** An encased or buried connection to a concrete-encased, driven, or buried grounding electrode shall not be required to be accessible.

> **Exception No. 2:** An exothermic or irreversible compression connection to fire-proofed structural metal shall not be required to be accessible.

Author's Substantiation. A new Exception No. 2 has been added to 250.68(A) to clarify that exothermic or irreversible connections to fire-proofed structural metal shall not be required to be accessible.

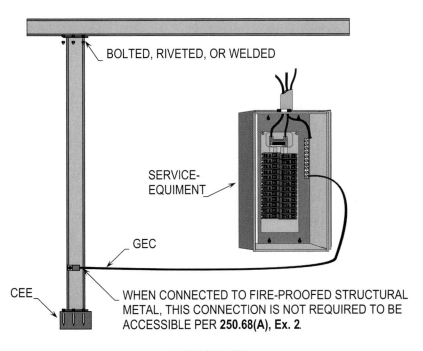

BOLTED, RIVETED, OR WELDED

SERVICE-EQUIMENT

GEC

CEE

WHEN CONNECTED TO FIRE-PROOFED STRUCTURAL METAL, THIS CONNECTION IS NOT REQUIRED TO BE ACCESSIBLE PER **250.68(A), Ex. 2**

ACCESSIBILITY
250.68(A), Ex. 2

Purpose of Change: To address the accessibility of the connection of the GEC to fire-proof structural metal.

NEC Ch. 2 – Article 250
Part IV – 250.84(A) and (B)

Type of Change	Panel Action	UL	UL 508	API 500	API 505	OSHA
Revision	Accept	-	-	-	-	-
ROP	ROC	NFPA 70E	NFPA 70B	NFPA 79	NFPA	NEMA
pg. 577 # 5-181	pg. - # -	-	-	-	-	-
log: 812 CMP: 5	log: -	Submitter: Dan Leaf		2002 NEC: 250.84(A) and (B)		IEC: -

2002 NEC – 250.84 Underground Service Cable or ~~Conduit~~

(A) **Underground Service Cable.** The sheath or armor of a continuous underground metal-sheathed ~~service~~ cable system that is ~~metallically connected~~ to the underground system shall not be required to be grounded at the building. The sheath or armor shall be permitted to be insulated from the interior conduit or piping.

(B) **Underground Service ~~Conduit~~ Containing Cable.** An underground service ~~conduit~~ that contains a metal-sheathed cable bonded to the underground system shall not be required to be grounded at the building. The sheath or armor shall be permitted to be insulated from the interior conduit or piping.

2005 NEC – 250.84 Underground Service Cable or <u>Raceway</u>

(A) **Underground Service Cable.** The sheath or armor of a continuous underground metal-sheathed <u>or armored service</u> cable system that is <u>bonded</u> to the <u>grounded</u> underground system shall not be required to be grounded at the building <u>or structure</u>. The sheath or armor shall be permitted to be insulated from the interior <u>metal raceway</u> conduit or piping.

(B) **Underground Service <u>Raceway</u> Containing Cable.** An underground <u>metal</u> service <u>raceway</u> that contains a metal-sheathed <u>or armored</u> cable bonded to the <u>grounded</u> underground system shall not be required to be grounded at the building <u>or structure</u>. The sheath or armor shall be permitted to be insulated from the interior <u>metal raceway</u> or piping.

Author's Substantiation. This revision clarifies that the sheath or armor shall be permitted to be insulated from the interior metal raceway. This revision clarifies that electrical metallic tubing (EMT) is permitted to be used.

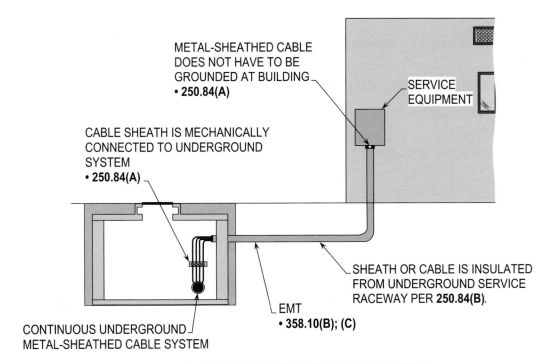

UNDERGROUND SERVICE CABLE OR RACEWAY
250.84(A); (B)

Purpose of Change: To verify that EMT shall be permitted to be used and to change the term "conduit" to the term "raceway" in title head and sections.

NEC Ch. 2 – Article 250
Part V – 250.92(B)(4)

Type of Change		Panel Action		UL	UL 508	API 500	API 505	OSHA
Revision		Accept in Principle		467	40.2	-	-	1910.304(f)(4)
ROP		ROC		NFPA 70E	NFPA 70B	NFPA 79	NFPA	NEMA
pg. 579	# 5-185	pg. 167	# 5-140	420.10(A)(1)	29.4.4	8.2.3.3	-	WD 1
log: 998	CMP: 5	log: 1023	Submitter: Noel Williams			2002 NEC: 250.92(B)(4)		IEC: 526

2002 NEC – 250.92(B) Method of Bonding at the Service

Electrical continuity at service equipment, service raceways, and service conductor enclosures shall be ensured by one of the following methods:

(4) Other ~~approved~~ devices, such as bonding-type locknuts and bushings

2005 NEC – 250.92(B) Method of Bonding at the Service

Electrical continuity at service equipment, service raceways, and service conductor enclosures shall be ensured by one of the following methods:

(4) Other <u>listed</u> devices, such as bonding-type locknuts, bushings, <u>or bushings with bonding jumpers</u>

Author's Substantiation. This revision clarifies that bonding-type locknuts, bushings, or bushings with bonding jumpers shall be listed devices.

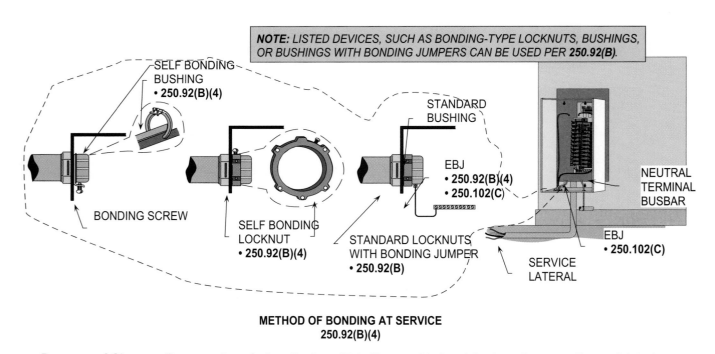

METHOD OF BONDING AT SERVICE
250.92(B)(4)

Purpose of Change: To prevent confusion, the term "listed" was added and the term "approved" was deleted.

NEC Ch. 2 – Article 250
Part V – 250.94

Type of Change		Panel Action		UL	UL 508	API 500	API 505	OSHA
Revision		Accept in Part		467	-	-	-	-
ROP		ROC		NFPA 70E	NFPA 70B	NFPA 79	NFPA	NEMA
pg. 579	# 5-186	pg. -	# -	450.5(D)(2)	-	-	-	-
log: 1571	CMP: 5	log: -		Submitter: Michael J. Johnston		2002 NEC: 250.94		IEC: 433; 528

2002 NEC – 250.94 Bonding for Other Systems

An accessible means external to enclosures for connecting intersystem bonding and grounding conductors shall be provided at the service equipment and at the disconnecting means for any additional buildings or structures by at least one of the following means:

2005 NEC – 250.94 Bonding for Other Systems

An accessible means external to enclosures for connecting intersystem bonding and grounding <u>electrode</u> conductors shall be provided at the service equipment and at the disconnecting means for any additional buildings or structures by at least one of the following means:

Author's Substantiation. This revision provides clarification and consistency with the NEC that the connection shall be grounding electrode conductors.

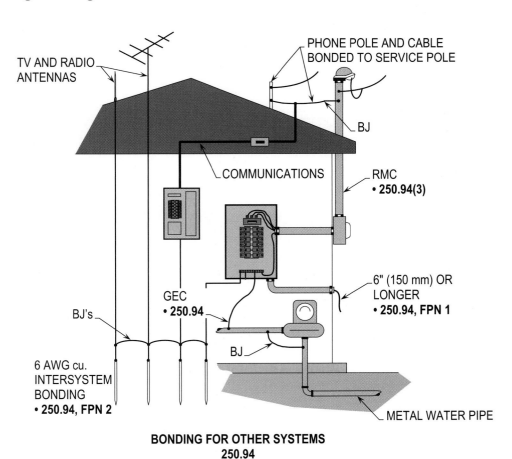

PHONE POLE AND CABLE BONDED TO SERVICE POLE

TV AND RADIO ANTENNAS

BJ

COMMUNICATIONS

RMC
• **250.94(3)**

6" (150 mm) OR LONGER
• **250.94, FPN 1**

GEC
• **250.94**

BJ's

BJ

6 AWG cu. INTERSYSTEM BONDING
• **250.94, FPN 2**

METAL WATER PIPE

BONDING FOR OTHER SYSTEMS
250.94

Purpose of Change: To revise and add words to more accurately represent the meaning of the text.

NEC Ch. 2 – Article 250
Part V – 250.97, Ex. (4)

Type of Change		Panel Action		UL	UL 508	API 500	API 505	OSHA
Revision		Accept		467	40.2	-	-	1910.304(f)(4)
ROP		ROC		NFPA 70E	NFPA 70B	NFPA 79	NFPA	NEMA
pg. 581	# 5-188a	pg. -	# -	420.1(A)(1)	29.4.4	8.2.3.3	-	WD 1
log: CP503	CMP: 5	log: -	Submitter: TCC			2002 NEC: 250.97, Ex. (4)		IEC: 526

2002 NEC – 250.97 Bonding for Over 250 Volts

For circuits of over 250 volts to ground, the electrical continuity of metal raceways or cables with metal sheaths that contain any conductor other than service conductors shall be ensured by one or more of the methods specified for services in 250.94(B), except for (1).

Exception: Where oversized, concentric, or eccentric knockouts are not encountered, or where a box or enclosure with concentric or eccentric knockouts is listed for the purpose, the following methods shall be permitted:

(d) Listed fittings that are identified for the purpose

2005 NEC – 250.97 Bonding for Over 250 Volts

For circuits of over 250 volts to ground, the electrical continuity of metal raceways or cables with metal sheaths that contain any conductor other than service conductors shall be ensured by one or more of the methods specified for services in 250.94(B), except for (B)(1).

Exception: Where oversized, concentric, or eccentric knockouts are not encountered, or where a box or enclosure with concentric or eccentric knockouts listed to provide a permanent, reliable electrical bond, the following methods shall be permitted:

(4) Listed fittings

Author's Substantiation. This revision improves clarity and removes any vague unenforceable terms.

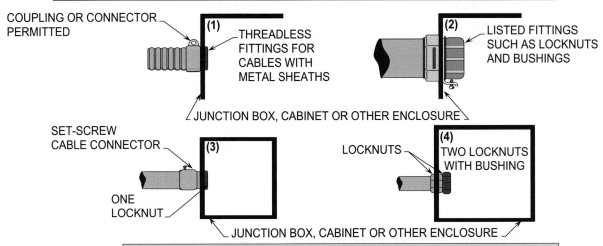

NOTE: *ITEMS LISTED PROVIDE A PERMANENT, RELIABLE ELECTRICAL BOND PER **250.94, Ex.***

COUPLING OR CONNECTOR
PERMITTED

(1) THREADLESS
FITTINGS FOR
CABLES WITH
METAL SHEATHS

(2) LISTED FITTINGS
SUCH AS LOCKNUTS
AND BUSHINGS

JUNCTION BOX, CABINET OR OTHER ENCLOSURE

SET-SCREW
CABLE CONNECTOR

(3)

ONE
LOCKNUT

LOCKNUTS

(4) TWO LOCKNUTS
WITH BUSHING

JUNCTION BOX, CABINET OR OTHER ENCLOSURE

BONDING FOR OVER 250 VOLTS-TO-GROUND PER 250.97, Ex.'s (1) - (4)

• GROUNDING BUSHING
• GROUNDING WEDGE [MAY BE USED, ONLY IF NO KNOCKOUT RINGS (LISTED
  OR NOT) ARE LEFT IN ENCLOSURE WALL AROUND THE CONDUIT]
• THREADLESS FITTINGS FOR CABLES W/METAL SHEATH
• TWO LOCKNUTS – ONE INSIDE, ONE OUTSIDE OF ENCLOSURE BUSHING USED
  OVER INSIDE LOCKNUT

BONDING FOR OVER 250 VOLTS
250.97, Ex. (4)

Purpose of Change: To revise the text and remove vague unenforceable terms which were confusing to the user.

NEC Ch. 2 – Article 250
Part V – 250.100

Type of Change	Panel Action	UL	UL 508	API 500	API 505	OSHA
New Sentence	Accept	-	-	-	-	-
ROP	**ROC**	NFPA 70E	NFPA 70B	NFPA 79	NFPA	NEMA
pg. 581 # 5-189	pg. - # -	-	-	-	-	-
log: 2482 CMP: 5	log: - Submitter: Paul Dobrowsky			2002 NEC: 250.100		IEC: -

2002 NEC – 250.100 Bonding in Hazardous (Classified) Locations

Regardless of the voltage of the electrical system, the electrical continuity of non-current-carrying metal parts of equipment, raceways, and other enclosures in any hazardous (classified) location as defined in Article 500 shall be ensured by any of the methods specified ~~for services~~ in ~~250.92(B)~~ that are approved for the wiring method used.

2005 NEC – 250.100 Bonding in Hazardous (Classified) Locations

Regardless of the voltage of the electrical system, the electrical continuity of non-current-carrying metal parts of equipment, raceways, and other enclosures in any hazardous (classified) location as defined in Article 500 shall be ensured by any of the methods specified in <u>250.92(B)(2) through (B)(4)</u> that are approved for the wiring method used. <u>One or more of these bonding methods shall be used whether or not supplementary equipment grounding conductors are installed.</u>

Author's Substantiation. This revision clarifies that bonding in hazardous (classified) locations shall be ensured by any of the methods in 250.92(B)(2) through (B)(4) that are approved for the wiring method used. This new sentence has been added to clarify that the supplementary equipment grounding conductor does not meet the intent of bonding in hazardous (classified) locations.

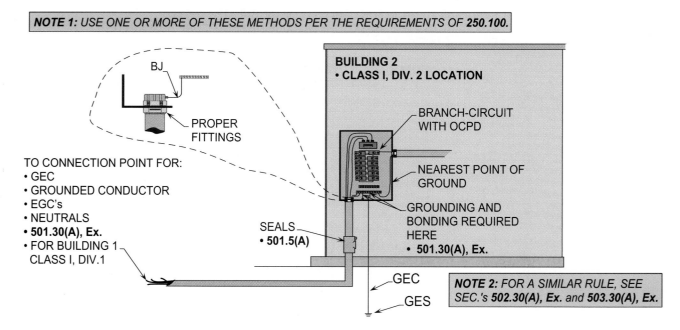

BONDING IN HAZARDOUS (CLASSIFIED) LOCATIONS
250.100

Purpose of Change: To delete the term "service" and revise the text to avoid confusion when bonding in hazardous locations.

NEC Ch. 2 – Article 250
Part V – 250.104(B)

Type of Change		Panel Action		UL	UL 508	API 500	API 505	OSHA
Revision		Accept		-	-	-	-	-
ROP		ROC		NFPA 70E	NFPA 70B	NFPA 79	NFPA	NEMA
pg. 585	# 5-197	pg. -	# -	-	-	-	-	-
log: 2484	CMP: 5	log: -	Submitter: Paul Dobrowsky			2002 NEC: 250.104(B)		IEC: -

2002 NEC – 250.104(B) Other Metal Piping

Where installed in or attached to a building or structure, metal piping system(s), including gas piping, that ~~may~~ become energized shall be bonded to the service equipment enclosure, the grounded conductor at the service, the grounding electrode conductor where of sufficient size, or to the one or more grounding electrodes used. The bonding jumper(s) shall be sized in accordance with 250.122 using the rating of the circuit that ~~may~~ energize the piping system(s). The equipment grounding conductor for the circuit that ~~may~~ energize the piping shall be permitted to serve as the bonding means. The points of attachment of the bonding jumper(s) shall be accessible.

2005 NEC – 250.104(B) Other Metal Piping

Where installed in or attached to a building or structure, metal piping system(s), including gas piping, that <u>is likely to</u> become energized shall be bonded to the service equipment enclosure, the grounded conductor at the service, the grounding electrode conductor where of sufficient size, or to the one or more grounding electrodes used. The bonding jumper(s) shall be sized in accordance with 250.122, using the rating of the circuit that <u>is likely to</u> energize the piping system(s). The equipment grounding conductor for the circuit that <u>is likely to</u> energize the piping shall be permitted to serve as the bonding means. The points of attachment of the bonding jumper(s) shall be accessible.

Author's Substantiation. The revision clarifies that other metal piping that is likely to become energized shall be bonded.

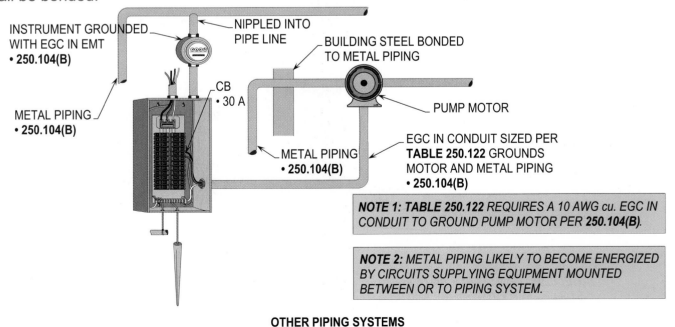

OTHER PIPING SYSTEMS
250.104(B)

Purpose of Change: To delete confusing terms and provide guidelines for bonding other metal piping system(s) that is likely to become energized by the power of a supplying circuit.

NEC Ch. 2 – Article 250
Part V – 250.104(C)

 58

Type of Change		Panel Action		UL	UL 508	API 500	API 505	OSHA
Revision		Accept in Principle		467	40.3	-	-	1910.304(f)(1)(v)
ROP		ROC		NFPA 70E	NFPA 70B	NFPA 79	NFPA	NEMA
pg. 586	# 5-200	pg. -	# -	410.10(C)(5)	Ch. 9	8.3	-	WD 1
log: 2485	CMP: 5	log: -	Submitter: Paul Dobrowsky			2002 NEC: 250.104(C)		IEC: 312.2

2002 NEC – 250.104(C) Structural ~~Steel~~

Exposed structural ~~steel~~ that is interconnected to form a ~~steel~~ building frame and is not intentionally grounded and ~~may~~ become energized shall be bonded to the service equipment enclosure, the grounded conductor at the service, the grounding electrode conductor where of sufficient size, or the one or more grounding electrodes used. The bonding jumper(s) shall be sized in accordance with Table 250.66 and installed in accordance with 250.64(A), (B), and (E). The points of attachment of the bonding jumper(s) shall be accessible.

2005 NEC – 250.104(C) Structural <u>Metal</u>

Exposed structural <u>metal</u> that is interconnected to form a <u>metal</u> building frame and is not intentionally grounded and <u>is likely to</u> become energized shall be bonded to the service equipment enclosure, the grounded conductor at the service, the grounding electrode conductor where of sufficient size, or the one or more grounding electrodes used. The bonding jumper(s) shall be sized in accordance with Table 250.66 and installed in accordance with 250.64(A), (B), and (E). The points of attachment of the bonding jumper(s) shall be accessible.

Author's Substantiation. This revision clarifies that all structural metal that is likely to become energized shall be bonded.

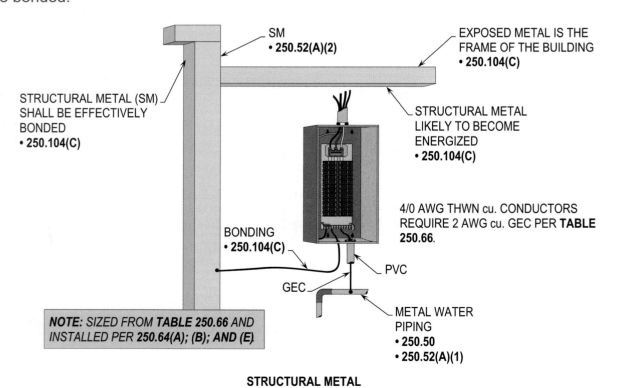

STRUCTURAL METAL
250.104(C)

Purpose of Change: To clarify that all structural metal that is likely to become energized shall be bonded.

NEC Ch. 2 – Article 250
Part V – 250.104(D)

Type of Change	Panel Action	UL	UL 508	API 500	API 505	OSHA
Relocated	Accept in Principle in Part	467	-	-	-	1910.304(f)(3)
ROP		**ROC**		**NFPA 70E**	**NFPA 70B**	**NFPA 79**	**NFPA**	**NEMA**

pg. 587	# 5-201	pg. 168	# 5-149	NFPA 70E -	NFPA 70B -	NFPA 79 -	NFPA -	NEMA -
log: 1540	CMP: 5	log: 763	Submitter: Michael J. Johnston			2002 NEC: 250.104(A)(4)		IEC: 312.2

2002 NEC – ~~250.104(A)(4)~~ Separately Derived Systems

The grounded conductor of each separately derived system shall be bonded to the nearest available point of the interior metal water piping system(s) in the area served by each separately derived system. This connection shall be made at the same point on the separately derived system where the grounding electrode conductor is connected. Each bonding jumper shall be sized in accordance with Table 250.66.

> **Exception:** A separate water piping bonding jumper to the metal water piping system shall not be required where the effectively grounded metal water piping system frame of a building or structure is used as the grounding electrode for a separately derived system and is bonded to the metallic water piping in the area served by the separately derived system.

2005 NEC – 250.104(D) Separately Derived Systems

Metal water piping systems and structural metal that is interconnected to form a building frame shall be bonded to separately derived systems in accordance with (D)(1) through (D)(3).

(1) Metal Water Piping System(s). The grounded conductor of each separately derived system shall be bonded to the nearest available point of the metal water piping system(s) in the area served by each separately derived system. This connection shall be made at the same point on the separately derived system where the grounding electrode conductor is connected. Each bonding jumper shall be sized in accordance with Table 250.66 based on the largest ungrounded conductor of the separately derived system.

> **Exception No. 1:** A separate bonding jumper to the metal water piping system shall not be required where the metal water piping system is used as the grounding electrode for the separately derived system.

> **Exception No. 2:** A separate water piping bonding jumper shall not be required where the metal frame of a building or structure is used as the grounding electrode for a separately derived system and is bonded to the metal water piping in the area served by the separately derived system.

(2) Structural Metal. Where exposed structural metal that is interconnected to form the building frame exists in the area served by the separately derived system, it shall be bonded to the grounded conductor of each separately derived system. This connection shall be made at the same point on the separately derived system where the grounding electrode conductor is connected. Each bonding jumper shall be sized in accordance with Table 250.66 based on the largest ungrounded conductor of the separately derived system.

> **Exception No. 1:** A separate bonding jumper to the building structural metal shall not be required where the metal frame of a building or structure is used as the grounding electrode for the separately derived system.

> **Exception No. 2:** A separate bonding jumper to the building structural metal shall not be required where the water piping of a building or structure is used as the grounding electrode for a separately

derived system and is bonded to the <u>building structural metal</u> in the area served by the separately derived system.

(3) Common Grounding Electrode Conductor. Where a common grounding electrode conductor <u>is installed for multiple separately derived systems as permitted by 250.30(A)(4), and</u> exposed structural metal that is interconnected to form the building frame or interior metal piping exists in the area served by the separately derived system, <u>the metal piping and the structural metal member</u> shall be bonded to the <u>common</u> grounding electrode conductor.

> **Exception:** <u>A separate bonding jumper from each derived system to metal water piping and to structural metal members shall not be required where the metal water piping and the structural metal members in the area served by the separately derived system are bonded to the common grounding electrode conductor.</u>

Author's Substantiation. This new section places the bonding requirements for metal water piping system(s), structural metal, and common grounding electrode conductors for separately derived systems into one location.

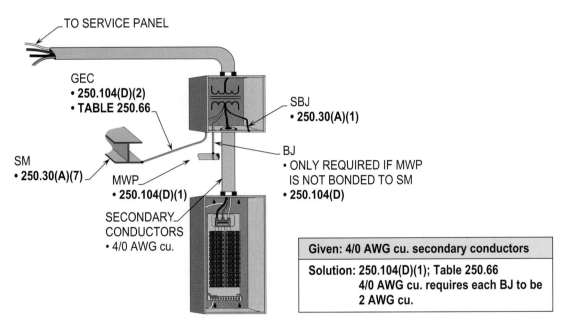

METAL WATER PIPING SYSTEMS
250.104(D)(1)

Purpose of Change: To ensure that the section rules are specific on how to determine the correct size of the bonding jumper to bond to the metal water pipe.

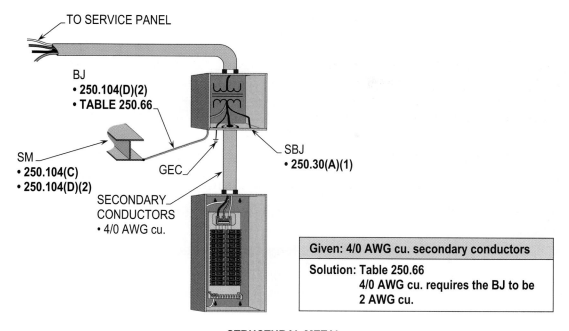

BJ
- **250.104(D)(2)**
- **TABLE 250.66**

TO SERVICE PANEL

SM
- **250.104(C)**
- **250.104(D)(2)**

GEC

SBJ
- **250.30(A)(1)**

SECONDARY
CONDUCTORS
- 4/0 AWG cu.

Given: 4/0 AWG cu. secondary conductors
Solution: Table 250.66 4/0 AWG cu. requires the BJ to be 2 AWG cu.

STRUCTURAL METAL
250.104(D)(2)

Purpose of Change: To ensure that the section rules are specific on how to determine the correct size of the bonding jumper to bond the grounded conductor and XFMR enclosure to the structural metal of the building.

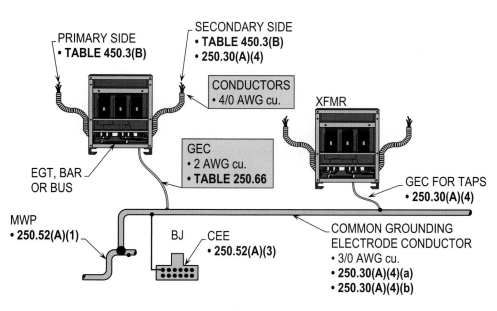

PRIMARY SIDE
- **TABLE 450.3(B)**

SECONDARY SIDE
- **TABLE 450.3(B)**
- **250.30(A)(4)**

CONDUCTORS
- 4/0 AWG cu.

XFMR

EGT, BAR
OR BUS

GEC
- 2 AWG cu.
- **TABLE 250.66**

GEC FOR TAPS
- **250.30(A)(4)**

MWP
- **250.52(A)(1)**

BJ   CEE
- **250.52(A)(3)**

COMMON GROUNDING
ELECTRODE CONDUCTOR
- 3/0 AWG cu.
- **250.30(A)(4)(a)**
- **250.30(A)(4)(b)**

COMMON GROUNDING ELECTRODE CONDUCTOR
250.104(D)(3)

Purpose of Change: To provide requirements for bonding the structural metal and metal piping to the separately derived systems and to the common GEC.

NEC Ch. 2 – Article 250
Part VI – 250.118(5)d

Type of Change		Panel Action		UL	UL 508	API 500	API 505	OSHA
Revision		Accept in Principle		467	40.1	-	-	-
ROP		ROC		NFPA 70E	NFPA 70B	NFPA 79	NFPA	NEMA
pg. 597	# 5-216	pg. 177	# 5-194	-	14.2.2	-	-	WD 1
log: 2768	CMP: 5	log: 3593	Submitter: Douglas Hansen			2002 NEC: 250.118(6)d		IEC: Ch. 43; 526

2002 NEC – 250.118 Types of Equipment Grounding Conductors

The equipment grounding conductor run with or enclosing the circuit conductors shall be one or more or a combination of the following:

(6) Listed flexible metal conduit ~~that is not listed for grounding,~~ meeting all the following conditions:

a. The conduit is terminated in fittings listed for grounding.

b. The circuit conductors contained in the conduit are protected by overcurrent devices rated at 20 amperes or less.

c. The combined length of flexible metal conduit and flexible metallic tubing and liquidtight flexible metal conduit in the same ground return path does not exceed 1.8 m (6 ft).

d. ~~The conduit is not installed for flexibility.~~

2005 NEC – 250.118 Types of Equipment Grounding Conductors

The equipment grounding conductor run with or enclosing the circuit conductors shall be one or more or a combination of the following:

(5) Listed flexible metal conduit meeting all the following conditions:

a. The conduit is terminated in fittings listed for grounding.

b. The circuit conductors contained in the conduit are protected by overcurrent devices rated at 20 amperes or less.

c. The combined length of flexible metal conduit and flexible metallic tubing and liquidtight flexible metal conduit in the same ground return path does not exceed 1.8 m (6 ft).

d. Where used to connect equipment where flexibility is necessary after installation, an equipment grounding conductor shall be installed.

Author's Substantiation. This revision clarifies that listed flexible metal conduit shall have an equipment grounding conductor installed where flexibility is necessary to connect equipment.

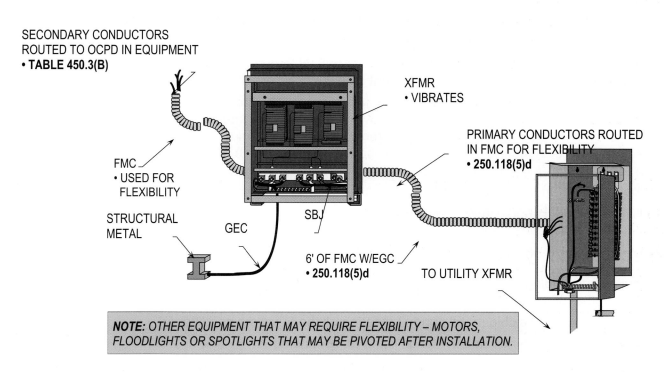

SECONDARY CONDUCTORS
ROUTED TO OCPD IN EQUIPMENT
• **TABLE 450.3(B)**

XFMR
• VIBRATES

PRIMARY CONDUCTORS ROUTED
IN FMC FOR FLEXIBILITY
• **250.118(5)d**

FMC
• USED FOR
  FLEXIBILITY

STRUCTURAL
METAL

GEC

SBJ

6' OF FMC W/EGC
• **250.118(5)d**

TO UTILITY XFMR

NOTE: *OTHER EQUIPMENT THAT MAY REQUIRE FLEXIBILITY – MOTORS, FLOODLIGHTS OR SPOTLIGHTS THAT MAY BE PIVOTED AFTER INSTALLATION.*

TYPES OF EQUIPMENT GROUNDING CONDUCTOR (FMC)
250.118(5)d

Purpose of Change: To clarify the rules in this section when installing a 6 ft. (1.8 m) piece of FMC for flexibility between equipment.

NEC Ch. 2 – Article 250
Part VI – 250.118(6)e

Type of Change		Panel Action		UL	UL 508	API 500	API 505	OSHA
Revision		Accept in Principle		467	40.2	-	-	1910.304(a)(1)
ROP		ROC		NFPA 70E	NFPA 70B	NFPA 79	NFPA 1	NEMA
pg. 598	# 5-218	pg. 177	# 5-192	Article 100	29.2.2.3	14.2.2	360	WD 1
log: 3185	CMP: 5	log: 3582	Submitter: -			2002 NEC: 250.118(5)e		IEC: 544.1

2002 NEC – 250.118 Types of Equipment Grounding Conductors

The equipment grounding conductor run with or enclosing the circuit conductors shall be one or more or a combination of the following:

(5) Listed liquidtight flexible metal conduit meeting all the following conditions:

a. The conduit is terminated in fittings listed for grounding.

b. For metric designators 12 through 16 (trade sizes 3/8 through 1/2), the circuit conductors contained in the conduit are protected by overcurrent devices rated at 20 amperes or less.

c. For metric designators 21 through 35 (trade sizes 3/4 through 1 1/4), the circuit conductors contained in the conduit are protected by overcurrent devices rated not more than 60 amperes and there is no flexible metal conduit, flexible metallic tubing, or liquidtight flexible metal conduit in trade sizes metric designators 12 through 16 (trade sizes 3/8 through 1/2) in the grounding path.

d. The combined length of flexible metal conduit and flexible metallic tubing and liquidtight flexible metal conduit in the same ground return path does not exceed 1.8 m (6 ft).

e. The conduit is not installed for flexibility.

2005 NEC – 250.118 Types of Equipment Grounding Conductors

The equipment grounding conductor run with or enclosing the circuit conductors shall be one or more or a combination of the following:

(6) Listed liquidtight flexible metal conduit meeting all the following conditions:

a. The conduit is terminated in fittings listed for grounding.

b. For metric designators 12 through 16 (trade sizes 3/8 through 1/2), the circuit conductors contained in the conduit are protected by overcurrent devices rated at 20 amperes or less.

c. For metric designators 21 through 35 (trade sizes 3/4 through 1 1/4), the circuit conductors contained in the conduit are protected by overcurrent devices rated not more than 60 amperes and there is no flexible metal conduit, flexible metallic tubing, or liquidtight flexible metal conduit in trade sizes metric designators 12 through 16 (trade sizes 3/8 through 1/2) in the grounding path.

d. The combined length of flexible metal conduit and flexible metallic tubing and liquidtight flexible metal conduit in the same ground return path does not exceed 1.8 m (6 ft).

e. Where used to connect equipment where flexibility is necessary after installation, an equipment grounding conductor shall be installed.

Author's Substantiation. This revision clarifies that liquidtight flexible metal conduit shall have an equipment grounding conductor installed where flexibility is necessary to connect equipment.

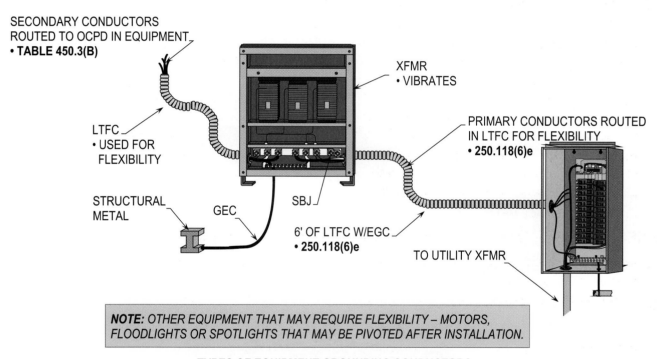

SECONDARY CONDUCTORS
ROUTED TO OCPD IN EQUIPMENT
• **TABLE 450.3(B)**

LTFC
• USED FOR
  FLEXIBILITY

STRUCTURAL
METAL

GEC

SBJ

6' OF LTFC W/EGC
• **250.118(6)e**

XFMR
• VIBRATES

PRIMARY CONDUCTORS ROUTED
IN LTFC FOR FLEXIBILITY
• **250.118(6)e**

TO UTILITY XFMR

NOTE: *OTHER EQUIPMENT THAT MAY REQUIRE FLEXIBILITY – MOTORS, FLOODLIGHTS OR SPOTLIGHTS THAT MAY BE PIVOTED AFTER INSTALLATION.*

**TYPES OF EQUIPMENT GROUNDING CONDUCTORS
250.118(6)e**

Purpose of Change: To clarify the rules in this section when installing a 6 ft. piece of LTFC for flexibility between equipment.

NEC Ch. 2 – Article 250
Part VI – 250.118(14)

Type of Change			Panel Action		UL	UL 508	API 500	API 505	OSHA
New Item			Accept		467	40.2	-	-	1910.304(a)(1)
ROP			ROC		NFPA 70E	NFPA 70B	NFPA 79	NFPA	NEMA
pg. 593	#	5-210	pg. 174	# 5-180	Article 100	29.2.23	14.2.2	5	WD 1
log: 192	CMP: 5		log: 565	Submitter: TCC			2002 NEC: 250.118(14)		IEC: 544.1

2005 NEC – 250.118 Types of Equipment Grounding Conductors

The equipment grounding conductor run with or enclosing the circuit conductors shall be one or more or a combination of the following:

(14) Surface metal raceways listed for grounding.

Author's Substantiation. This new item permits surface metal raceways to be used as equipment grounding conductors if listed for such use.

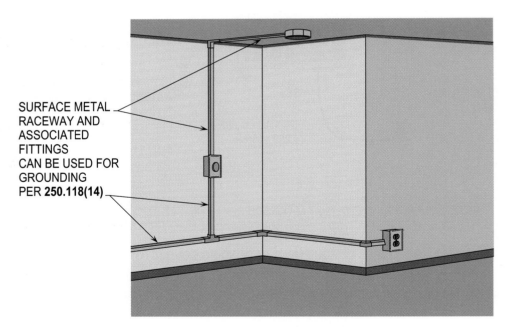

SURFACE METAL RACEWAY AND ASSOCIATED FITTINGS CAN BE USED FOR GROUNDING PER **250.118(14)**

TYPES OF EQUIPMENT GROUNDING CONDUCTORS
250.118(14)

Purpose of Change: To recognize surface metal raceways as a grounding means.

NEC Ch. 2 – Article 250
Part VI – 250.119(A)

Type of Change	Panel Action	UL	UL 508	API 500	API 505	OSHA
Revision	Accept in Part	467	40.2	-	-	1910.304(a)(1)

ROP		ROC		NFPA 70E	NFPA 70B	NFPA 79	NFPA	NEMA
pg. 598	# 5-219	pg. -	# -	Article 100	29.2.23	14.2.2	-	WD 1
log: 807	CMP: 5	log: -	Submitter: Dan Leaf			2002 NEC: 250.119(A)		IEC: 544.1

2002 NEC – 250.119(A) Conductors Larger Than 6 AWG

An insulated or covered conductor larger than 6 AWG copper or aluminum shall be permitted, at the time of installation, to be permanently identified as an equipment grounding conductor at each end and at every point where the conductor is accessible. Identification shall encircle the conductor and shall be accomplished by one of the following:

(1) Stripping the insulation or covering from the entire exposed length

(2) Coloring the exposed insulation or covering green

(3) Marking the exposed insulation or covering with green tape or green adhesive labels

2005 NEC – 250.119(A) Conductors Larger Than 6 AWG

Equipment grounding conductors larger than 6 AWG shall comply with 250.119(A)(1) and (A)(2).

(1) An insulated or covered conductor larger than 6 AWG shall be permitted, at the time of installation, to be permanently identified as an equipment grounding conductor at each end and at every point where the conductor is accessible.

Exception: Conductors larger than 6 AWG shall not be required to be marked in conduit bodies that contain no splices or unused hubs.

(2) Identification shall encircle the conductor and shall be accomplished by one of the following:

a. Stripping the insulation or covering from the entire exposed length

b. Coloring the exposed insulation or covering green

c. Marking the exposed insulation or covering with green tape or green adhesive labels

Author's Substantiation. This revision clarifies the marking requirements for equipment grounding conductors larger than 6 AWG and equipment grounding conductors in conduit bodies that contain no splices or unused hubs shall not be required to be marked.

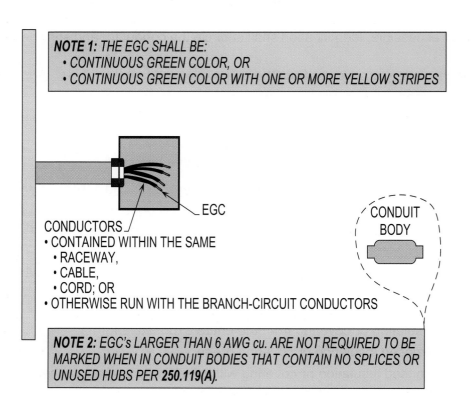

NOTE 1: *THE EGC SHALL BE:*
- *CONTINUOUS GREEN COLOR, OR*
- *CONTINUOUS GREEN COLOR WITH ONE OR MORE YELLOW STRIPES*

EGC

CONDUIT
BODY

CONDUCTORS
- CONTAINED WITHIN THE SAME
  - RACEWAY,
  - CABLE,
  - CORD; OR
- OTHERWISE RUN WITH THE BRANCH-CIRCUIT CONDUCTORS

NOTE 2: *EGC's LARGER THAN 6 AWG cu. ARE NOT REQUIRED TO BE MARKED WHEN IN CONDUIT BODIES THAT CONTAIN NO SPLICES OR UNUSED HUBS PER **250.119(A)**.*

CONDUCTORS LARGER THAN 6 AWG
250.119(A)

Purpose of Change: To appropriately address the color coding and marking of an EGC.

NEC Ch. 2 – Article 250
Part VI – 250.122(F)(2)(3)

Type of Change	Panel Action	UL	UL 508	API 500	API 505	OSHA
Revision	Accept	467	40.2	-	-	-
ROP	**ROC**	**NFPA 70E**	**NFPA 70B**	**NFPA 79**	**NFPA**	**NEMA**
pg. 606 # 5-231a	pg. - # -	Article 100	29.2.23	14.2.2	-	WD 1
log: CP504 CMP: 5	log: -	Submitter: CMP 5		2002 NEC: 250.122(F)(2)(3)		IEC: 544.1

2002 NEC – 250.122(F) Conductors in Parallel

Where conductors are run in parallel in multiple raceways or cables as permitted in 310.4, the equipment grounding conductors, where used, shall be run in parallel in each raceway or cable. One of the methods in 250.122(F)(1) or (2) shall be used to ensure the equipment grounding conductors are protected.

    (2) Where ground-fault protection of equipment is installed, each parallel equipment grounding conductor in a multiconductor cable shall be permitted to be sized in accordance with Table 250.122 on the basis of the trip rating of the ground-fault protection where the following conditions are met:

        (1) Conditions of maintenance and supervision ensure that only qualified persons will service the installation.

        (2) The ground-fault protection equipment is set to trip at not more than the ampacity of a single ungrounded conductor of one of the cables in parallel.

        (3) The ground-fault protection is listed for the purpose.

2005 NEC – 250.122(F) Conductors in Parallel

Where conductors are run in parallel in multiple raceways or cables as permitted in 310.4, the equipment grounding conductors, where used, shall be run in parallel in each raceway or cable. One of the methods in 250.122(F)(1) or (F)(2) shall be used to ensure the equipment grounding conductors are protected.

    (2) Ground-Fault Protection of Equipment Installed. Where ground-fault protection of equipment is installed, each parallel equipment grounding conductor in a multiconductor cable shall be permitted to be sized in accordance with Table 250.122 on the basis of the trip rating of the ground-fault protection where the following conditions are met:

        (1) Conditions of maintenance and supervision ensure that only qualified persons will service the installation.

        (2) The ground-fault protection equipment is set to trip at not more than the ampacity of a single ungrounded conductor of one of the cables in parallel.

        (3) The ground-fault protection is listed for the purpose of protecting the equipment grounding conductor.

Author's Substantiation. This revision clarifies that the equipment grounding conductors shall be protected by ground-fault protection that is listed for the purpose.

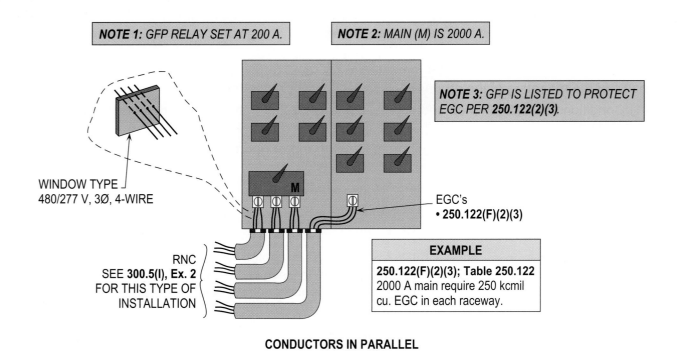

NOTE 1: *GFP RELAY SET AT 200 A.*

NOTE 2: *MAIN (M) IS 2000 A.*

NOTE 3: *GFP IS LISTED TO PROTECT EGC PER 250.122(2)(3).*

WINDOW TYPE
480/277 V, 3Ø, 4-WIRE

M

RNC
SEE **300.5(I), Ex. 2**
FOR THIS TYPE OF
INSTALLATION

EGC's
• **250.122(F)(2)(3)**

EXAMPLE
250.122(F)(2)(3); Table 250.122
2000 A main require 250 kcmil cu. EGC in each raceway.

CONDUCTORS IN PARALLEL
250.122(F)(2)(3)

Purpose of Change: To remove from the text unenforceable code requirements.

NEC Ch. 2 – Article 250
Part VI – 250.122(G)

Type of Change		Panel Action		UL	UL 508	API 500	API 505	OSHA
New Subsection		Accept in Principle		467	40.2	-	-	1910.304(a)(1)
ROP		ROC		NFPA 70E	NFPA 70B	NFPA 79	NFPA	NEMA
pg. 606	# 5-232	pg. -	# -	Article 100	29.2.2	14.2.2	-	WD 1
log: 3222	CMP: 5	log: -	Submitter: Jonathan R. Althouse			2002 NEC: -		IEC: 544.1

2005 NEC – 250.122(G) Feeder Taps

Equipment grounding conductors run with feeder taps shall not be smaller than shown in Table 250.122 based on the rating of the overcurrent device ahead of the feeder but shall not be required to be larger than the tap conductors.

Author's Substantiation. This new subsection clarifies the sizing requirements for equipment grounding conductors run with feeder taps.

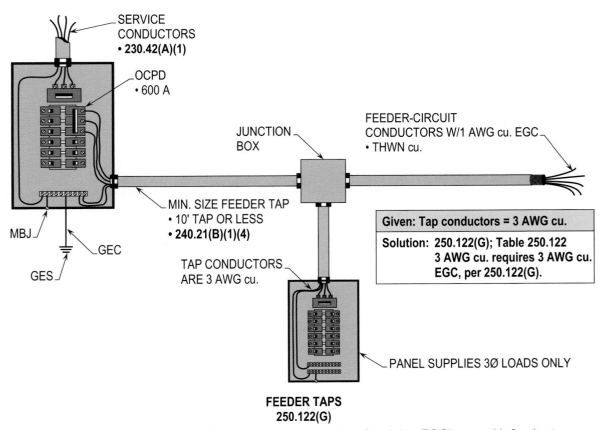

FEEDER TAPS
250.122(G)

Purpose of Change: To clarify the appropriate procedure for sizing EGC's run with feeder taps.

NEC Ch. 2 – Article 250
Part VII – 250.140 and Ex.

Type of Change		Panel Action		UL	UL 508	API 500	API 505	OSHA
Revision		Accept		467	40.2	-	-	1910.304(f)(4)(iii)
ROP		ROC		NFPA 70E	NFPA 70B	NFPA 79	NFPA	NEMA
pg. 608	# 5-235	pg. -	# -	410.10(E)(3)	29.2.2	14.2.2	-	WD 1
log: 2649	CMP: 5	log: -	Submitter: Phil Simmons			2002 NEC: 250.140		IEC: 544.1

2002 NEC – 250.140 Frames of Ranges and Clothes Dryers

~~This section shall apply to existing branch-circuit installations only. New branch-circuit installations shall comply with 250.134 and 250.138.~~ Frames of electric ranges, wall-mounted ovens, counter-mounted cooking units, clothes dryers, and outlet or junction boxes that are part of the circuit for these appliances shall be grounded in the manner specified by 250.134 or 250.138~~; or, except for mobile homes and recreational vehicles,~~ shall be permitted to be grounded to the grounded circuit conductor if all the following conditions are met.

(1) The supply circuit is 120/240-volt, single-phase, 3-wire; or 208Y/120-volt derived from a 3-phase, 4-wire, wye-connected system.

(2) The grounded conductor is not smaller than 10 AWG copper or 8 AWG aluminum.

(3) The grounded conductor is insulated, or the grounded conductor is uninsulated and part of a Type SE service-entrance cable and the branch circuit originates at the service equipment.

(4) Grounding contacts of receptacles furnished as part of the equipment are bonded to the equipment.

2005 NEC – 250.140 Frames of Ranges and Clothes Dryers

Frames of electric ranges, wall-mounted ovens, counter-mounted cooking units, clothes dryers, and outlet or junction boxes that are part of the circuit for these appliances shall be grounded in the manner specified by 250.134 or 250.138.

> **Exception:** <u>For existing branch circuit installations only where an equipment grounding conductor is not present in the outlet or junction box, the frames of electric ranges, wall-mounted ovens, counter-mounted cooking units, clothes dryers, and outlet and junction boxes that are part of the circuit for these appliances</u> shall be permitted to be grounded to the grounded circuit conductor if all the following conditions are met:
>
> **(1)** The supply circuit is 120/240-volt, single-phase, 3-wire; or 208Y/120-volt derived from a 3-phase, 4-wire, wye-connected system.
>
> **(2)** The grounded conductor is not smaller than 10 AWG copper or 8 AWG aluminum.
>
> **(3)** The grounded conductor is insulated, or the grounded conductor is uninsulated and part of a Type SE service-entrance cable and the branch circuit originates at the service equipment.
>
> **(4)** Grounding contacts of receptacles furnished as part of the equipment are bonded to the equipment.

Author's Substantiation. This revision clarifies that the frames of electric ranges, wall-mounted ovens, counter-mounted cooking units, clothes dryers, and outlet or junction boxes that are part of the circuit for these appliances shall be grounded and the new exception does not require the frames of these appliances to be grounded for existing branch circuit installations only where an equipment grounding conductor is not present in the outlet or junction box.

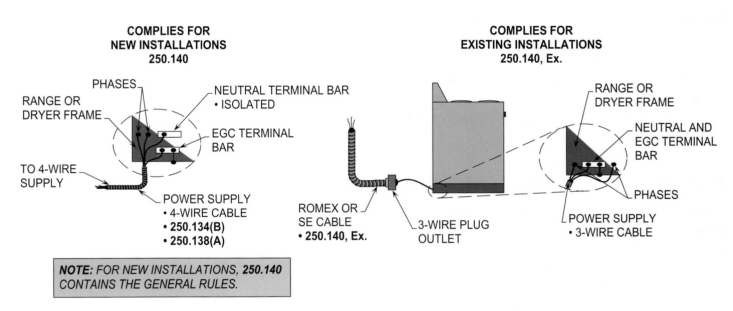

COMPLIES FOR
NEW INSTALLATIONS
250.140

PHASES

RANGE OR
DRYER FRAME

NEUTRAL TERMINAL BAR
• ISOLATED

EGC TERMINAL
BAR

TO 4-WIRE
SUPPLY

POWER SUPPLY
• 4-WIRE CABLE
• **250.134(B)**
• **250.138(A)**

*NOTE: FOR NEW INSTALLATIONS, **250.140** CONTAINS THE GENERAL RULES.*

COMPLIES FOR
EXISTING INSTALLATIONS
250.140, Ex.

RANGE OR
DRYER FRAME

NEUTRAL AND
EGC TERMINAL
BAR

PHASES

POWER SUPPLY
• 3-WIRE CABLE

ROMEX OR
SE CABLE
• **250.140, Ex.**

3-WIRE PLUG
OUTLET

FRAMES OF RANGES AND CLOTHES DRYERS
250.140 AND Ex.

Purpose of Change: To locate the general rules in **250.140** for new installations and add an exception for the rules in existing installations.

NEC Ch. 2 – Article 250
Part VII – 250.146(B)

Type of Change		Panel Action		UL	UL 508	API 500	API 505	OSHA
Revision		Accept		467	40.2	-	-	1910.305(a)(2)(iii)
ROP		ROC		NFPA 70E	NFPA 70B	NFPA 79	NFPA	NEMA
pg. 610	# 5-238a	pg. -	# -	410.2(B)(2)	27.7.3.2	14.4.5	498	WD 1
log: CP505	CMP: 5	log: -	Submitter: CMP 5			2002 NEC: 250.146(B)		IEC: 543.3

2002 NEC – 250.146(B) Contact Devices or Yokes

Contact devices or yokes designed and listed ~~for the purpose~~ shall be permitted in conjunction with the supporting screws to establish the grounding circuit between the device yoke and flush-type boxes.

2006 NEC – 250.146(B) Contact Devices or Yokes

Contact devices or yokes designed and listed <u>as self-grounding</u> shall be permitted in conjunction with the supporting screws to establish the grounding circuit between the device yoke and flush-type boxes.

Author's Substantiation. This revision clarifies any vague language and requires the contact devices or yokes be designed and listed as self-grounding.

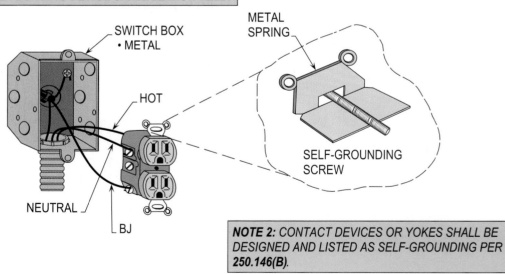

NOTE 1: IF METAL TO METAL CONTACT OF BOX AND YOKE ARE NOT MADE, A BONDING JUMPER PER **TABLE 250.122** SHALL BE USED OR A LISTED SELF-BONDING DEVICE SHALL BE INSTALLED.

METAL SPRING

SWITCH BOX
• METAL

HOT

SELF-GROUNDING SCREW

NEUTRAL

BJ

NOTE 2: CONTACT DEVICES OR YOKES SHALL BE DESIGNED AND LISTED AS SELF-GROUNDING PER 250.146(B).

CONTACT DEVICES OR YOKES
250.146(B)

Purpose of Change: To require contact devices or yokes to be listed for grounding.

NEC Ch. 2 – Article 250
Part VII – 250.147

Type of Change		Panel Action		UL	UL 508	API 500	API 505	OSHA
New Section		Accept		467	-	-	-	1910.305(b)(3)(C)
ROP		ROC		NFPA 70E	NFPA 70B	NFPA 79	NFPA	NEMA
pg. 612	# 5-244a	pg. 181	# 5-219	420.3(D)(1)	18.5	8.2.4	20	WD 1
log: CP501	CMP: 5	log: 567	Submitter: CMP 5			2002 NEC: 404.9(B)(1) and (B)(2)		IEC: 537

2005 NEC – 250.147 Bonding of General-Use Snap Switches

Snap switches, including dimmer and similar control switches, shall be connected to the equipment grounding conductor and shall provide a means to bond metal faceplates, whether or not a metal faceplate is installed. Snap switches shall be bonded by either of the following conditions.

(1) The switch is mounted with metal screws to a metal box or to a nonmetallic box with integral means for bonding the devices.

(2) An equipment grounding conductor or equipment bonding jumper is connected to the terminal for connecting the equipment grounding conductor.

> **Exception:** Where no equipment grounding conductor exists within the snap switch enclosure, a snap switch without a terminal for connecting the equipment grounding conductor shall be permitted only for replacement purposes. A snap switch wired under the provisions of this exception and located within reach of earth, grade conducting floors, or other conducting surfaces shall be provided with a faceplate of nonconducting, noncombustible material.

Author's Substantiation. This new section contains the bonding requirements of general-use snap switches. This new section was relocated from subsection 404.9(B)(1) and (B)(2) in the 2002 NEC.

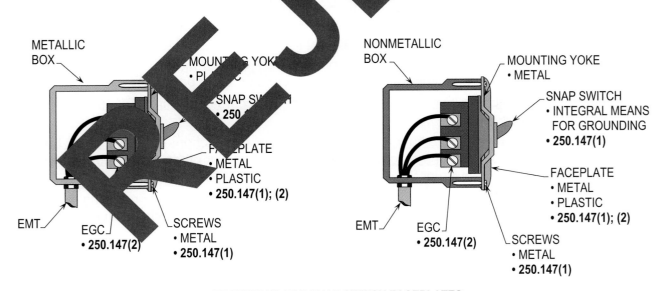

PROVISIONS FOR SNAP SWITCH FACEPLATES
250.147(1); (2)

Purpose of Change: To relocate rules from **404.9(B)(1)** and **(B)(2)** to **Article 250** and renumbered as **250.147(1)** and **(2)**.

NEC Ch. 2 – Article 250
Part VII – 250.148(A) through (E)

Type of Change		Panel Action		UL	UL 508	API 500	API 505	OSHA
Revision		Accept in Principle		467	6.6	-	-	1910.305(b)(3)(C)
ROP		ROC		NFPA 70E	NFPA 70B	NFPA 79	NFPA	NEMA
pg. 614	# 5-248	pg. -	# -	410.10(B)	-	-	514A	WD 1
log: 2650	CMP: 5	log: -	Submitter: Phil Simmons			2002 NEC: 250.148		IEC: 543.3

2002 NEC – 250.148 Continuity and Attachment of Equipment Grounding Conductors to Boxes

Where circuit conductors are spliced within a box, or terminated on equipment within or supported by a box, any ~~separate~~ equipment grounding ~~conductors~~ associated with those circuit conductors shall be spliced or joined within the box or to the box with devices suitable for the use. ~~Connections depending solely on solder shall not be used. Splices shall be made in accordance with 110.14(B) except that insulation shall not be required. The arrangement of grounding connections shall be such that the disconnection or the removal of a receptacle, luminaire (fixture), or other device fed from the box will not interfere with or interrupt the grounding continuity.~~

> **Exception:** The equipment grounding conductor permitted in 250.146(D) shall not be required to be connected to the other equipment grounding conductors or to the box.

(A) Metal Boxes. A connection shall be made between the one or more equipment grounding conductors and a metal box by means of a grounding screw that shall be used for no other purpose or a listed grounding device.

(B) Nonmetallic Boxes. One or more equipment grounding conductors brought into a nonmetallic outlet box shall be arranged ~~so~~ that a connection can be made to any fitting or device in that box requiring grounding.

2005 NEC – 250.148 Continuity and Attachment of Equipment Grounding Conductors to Boxes

Where circuit conductors are spliced within a box, or terminated on equipment within or supported by a box, any equipment grounding <u>conductor(s)</u> associated with those circuit conductors shall be spliced or joined within the box or to the box with devices suitable for the use <u>in accordance with 250.148(A) through (E)</u>.

> **Exception:** The equipment grounding conductor permitted in 250.146(D) shall not be required to be connected to the other equipment grounding conductors or to the box.

(A) <u>Connections.</u> <u>Connections and splices shall be made in accordance with 110.14(B) except that insulation shall not be required.</u>

(B) <u>Grounding Continuity.</u> <u>The arrangement of grounding connections shall be such that the disconnection or the removal of a receptacle, luminaire (fixture), or other device fed from the box does not interfere with or interrupt the grounding continuity.</u>

(C) Metal Boxes. A connection shall be made between the one or more equipment grounding conductors and a metal box by means of a grounding screw that shall be used for no other purpose or a listed grounding device.

(D) Nonmetallic Boxes. One or more equipment grounding conductors brought into a nonmetallic outlet box shall be arranged such that a connection can be made to any fitting or device in that box requiring grounding.

(E) <u>Solder.</u> <u>Connections depending solely on solder shall not be used.</u>

Author's Substantiation. This revision, with new subsections, adds clarity and user-friendliness.

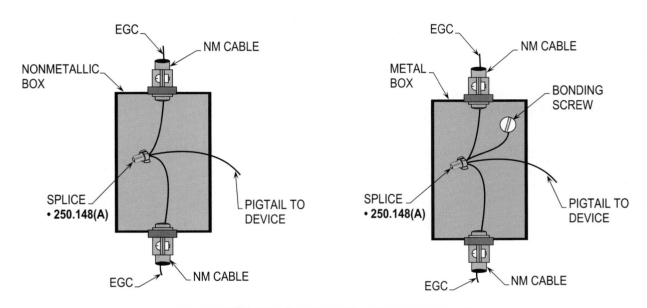

NOTE: *DEVICE CAN BE REMOVED FROM PIGTAIL WITHOUT BREAKING CONTACT WITH EGC'S PER **250.148(B)**.*

**CONTINUITY AND ATTACHMENT OF EQUIPMENT
GROUNDING CONDUCTORS TO BOXES
250.148(A) THRU (E)**

Purpose of Change: To completely revise and provide guidelines on conditions when EGC's are required to be bonded to boxes and devices.

NEC Ch. 2 – Article 250
Part VIII – 250.168

Type of Change		Panel Action		UL	UL 508	API 500	API 505	OSHA
Revision		Accept in Part		467	-	-	-	1910.304(f)(2)
ROP		ROC		NFPA 70E	NFPA 70B	NFPA 79	NFPA	NEMA
pg. 616	# 5-250	pg. -	# -	-	6.4.4.6(2)	-	1112	MG 1
log: 613	CMP: 5	log: -	Submitter: Dan Leaf			2002 NEC: 250.168		IEC: 551

2002 NEC – 250.168 Direct-Current Bonding Jumper

For dc systems, the size of the bonding jumper shall not be smaller than the system grounding conductor specified in 250.166.

2005 NEC – 250.168 Direct-Current Bonding Jumper

For dc systems, the size of the bonding jumper shall not be smaller than the system grounding <u>electrode</u> conductor specified in 250.166.

Author's Substantiation. The revision clarifies that the size of the bonding jumper shall not be smaller than the grounding electrode conductor.

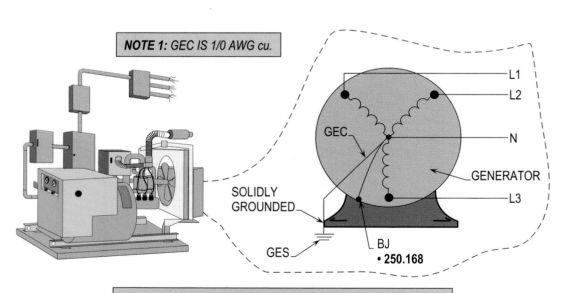

NOTE 1: *GEC IS 1/0 AWG cu.*

NOTE 2: *SECTIONS 250.168 AND 250.166(A) OR (B) REQUIRES THE BONDING JUMPER (BJ) TO BE AT LEAST 1/0 AWG cu.*

DIRECT-CURRENT BONDING JUMPER
250.168

Purpose of Change: To ensure that the bonding jumper will not be smaller than the grounding electrode conductor.

NEC Ch. 2 – Article 250
Part X – 250.184(C)

Type of Change		Panel Action		UL	UL 508	API 500	API 505	OSHA
Revision		Accept in Principle		-	-	-	-	-
ROP		ROC		NFPA 70E	NFPA 70B	NFPA 79	NFPA	NEMA
pg. 618	# 5-253	pg. 182	# 5-223	-	-	-	-	-
log: 3038	CMP: 5	log: 1086	Submitter: Elliot Rappaport			2002 NEC: 250.184(D)		IEC: -

2002 NEC – ~~250.184(D)~~ Multigrounded Neutral ~~Conductor~~

Where a multigrounded neutral system is used, the following shall apply:

(1) ~~The multigrounded neutral conductor shall be of sufficient ampacity for the load imposed on the conductor but not less than 33 1/3 percent of the ampacity of the phase conductors.~~

> ~~**Exception:** In industrial and commercial premises under engineering supervision, it shall be permissible to size the ampacity of the neutral conductor to not less than 20 percent of the ampacity of the phase conductor.~~

(2) The multigrounded neutral conductor shall be grounded at each transformer and at other additional locations by connection to a made or existing electrode.

(3) At least one grounding electrode shall be installed and connected to the multigrounded neutral circuit conductor every 400 m (1300 ft).

(4) The maximum distance between any two adjacent electrodes shall not be more than 400 m (1300 ft).

(5) It a multigrounded shielded cable system, the shielding shall be grounded at each cable joint that is exposed to personnel contact.

2005 NEC – <u>250.184(C)</u> Multigrounded Neutral <u>Systems</u>

Where a multigrounded neutral system is used, the following shall apply:

(1) <u>The neutral of a solidly grounded neutral system shall be permitted to be grounded at more than one point. Grounding shall be permitted at one or more of the following locations:</u>

<u>**a.** Transformers supplying conductors to a building or other structure</u>

<u>**b.** Underground circuits where the neutral is exposed</u>

<u>**c.** Overhead circuits installed outdoors</u>

(2) The multigrounded neutral conductor shall be grounded at each transformer and at other additional locations by connection to a made or existing electrode.

(3) At least one grounding electrode shall be installed and connected to the multigrounded neutral circuit conductor every 400 m (1300 ft).

(4) The maximum distance between any two adjacent electrodes shall not be more than 400 m (1300 ft).

(5) It a multigrounded shielded cable system, the shielding shall be grounded at each cable joint that is exposed to personnel contact.

Author's Substantiation. This revision clarifies that the neutral of a solidly grounded neutral system shall be permitted to be grounded at more than one point where transformers supplying conductors are to a building or other structure, where the neutral is exposed for underground circuits, or overhead circuits are installed outdoors.

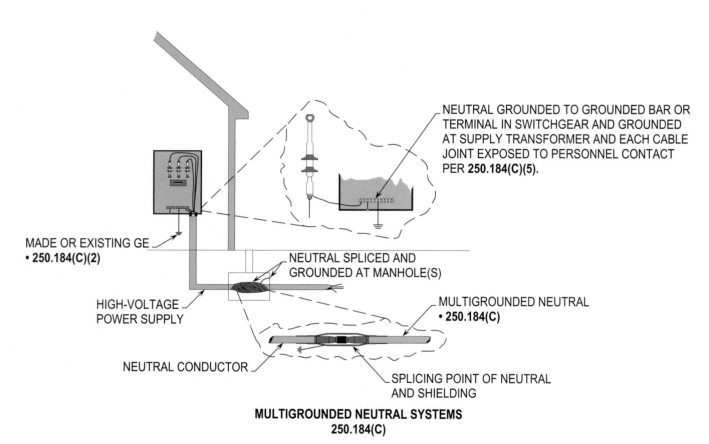

MADE OR EXISTING GE
• **250.184(C)(2)**

NEUTRAL GROUNDED TO GROUNDED BAR OR
TERMINAL IN SWITCHGEAR AND GROUNDED
AT SUPPLY TRANSFORMER AND EACH CABLE
JOINT EXPOSED TO PERSONNEL CONTACT
PER **250.184(C)(5)**.

NEUTRAL SPLICED AND
GROUNDED AT MANHOLE(S)

HIGH-VOLTAGE
POWER SUPPLY

MULTIGROUNDED NEUTRAL
• **250.184(C)**

NEUTRAL CONDUCTOR

SPLICING POINT OF NEUTRAL
AND SHIELDING

MULTIGROUNDED NEUTRAL SYSTEMS
250.184(C)

Purpose of Change: To provide requirements for grounding neutral at multiple points where each cable is exposed to contact by personnel.

NEC Ch. 2 – Article 280
Part I – 280.4(A)(3)

63

Type of Change	Panel Action	UL	UL 508	API 500	API 505	OSHA
New Item	Accept in Principle	-	-	-	-	-
ROP	ROC	NFPA 70E	NFPA 70B	NFPA 79	NFPA	NEMA
pg. 625 # 5-262	pg. 183 # 5-225	-	8.9.2	-	525	-
log: 1401 CMP: 5	log: 400 Submitter: Kenneth Brown			2002 NEC: 280.4		IEC: 543.3

2002 NEC – 280.4 Surge Arrester Selection

(A) Circuits of Less Than 1000 Volts. The rating of the surge arrester shall be equal to or greater than the maximum continuous phase-to-ground power frequency voltage available at the point of application.

Surge arresters installed on circuits of less than 1000 volts shall be listed ~~for the purpose.~~

2005 NEC – 280.4 Surge Arrester Selection

(A) Circuits of Less Than 1000 Volts. Surge arresters installed on a circuit of less than 1000 volts shall comply with all of the following:

(1) The rating of the surge arrester shall be equal to or greater than the maximum continuous phase-to-ground power frequency voltage available at the point of application.

(2) Surge arresters installed on circuits of less than 1000 volts shall be listed.

(3) Surge arresters shall be marked with a short circuit current rating and shall not be installed at a point on the system where the available fault current is in excess of that rating.

(4) Surge arresters shall not be installed on ungrounded systems, impedance grounded systems, or corner grounded delta systems unless listed specifically for use on these systems.

Author's Substantiation. This new item recognizes that surge arresters shall be marked with a short-circuit current rating and shall not be installed at a point of the system where the available fault current is in excess of that rating. It is necessary for surge arresters under 1000 volts have the short-circuit current rating marked on the arrester to ensure a safe application on the electrical system.

Stallcup's Illustrated Code Changes - 2005

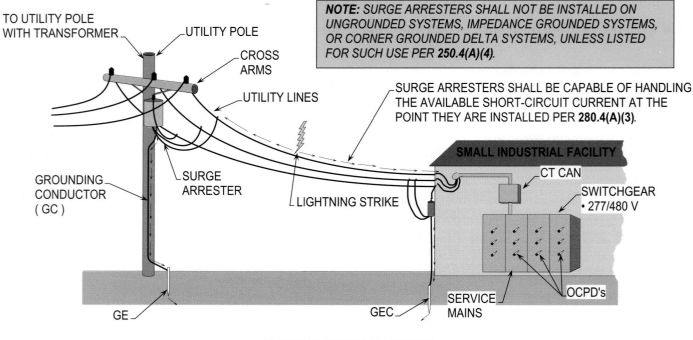

NOTE: *SURGE ARRESTERS SHALL NOT BE INSTALLED ON UNGROUNDED SYSTEMS, IMPEDANCE GROUNDED SYSTEMS, OR CORNER GROUNDED DELTA SYSTEMS, UNLESS LISTED FOR SUCH USE PER 250.4(A)(4).*

SURGE ARRESTERS SHALL BE CAPABLE OF HANDLING THE AVAILABLE SHORT-CIRCUIT CURRENT AT THE POINT THEY ARE INSTALLED PER **280.4(A)(3)**.

TO UTILITY POLE WITH TRANSFORMER

UTILITY POLE

CROSS ARMS

UTILITY LINES

GROUNDING CONDUCTOR (GC)

SURGE ARRESTER

LIGHTNING STRIKE

SMALL INDUSTRIAL FACILITY

CT CAN

SWITCHGEAR
• 277/480 V

SERVICE MAINS

OCPD's

GE

GEC

SURGE ARRESTER SELECTION
280.4(A)(3)

Purpose of Change: To verify the point where surge arresters shall be capable of handling short-circuit current without rupturing.

NEC Ch. 2 – Article 280
Part II – 280.24(A)(2)

Type of Change		Panel Action		UL	UL 508	API 500	API 505	OSHA
Revision		Accept		-	-	-	-	-
ROP		ROC		NFPA 70E	NFPA 70B	NFPA 79	NFPA	NEMA
pg. 626	# 5-265	pg. 183	# 5-228	-	-	-	-	-
log: 364	CMP: 5	log: 1217	Submitter: Jerry V. Smith			2002 NEC: 280.24(A)(2)		IEC: -

2002 NEC – 280.24(A) Metallic Interconnections

A metallic interconnection shall be made to the secondary grounded circuit conductor or the secondary circuit grounding conductor provided that, in addition to the direct grounding connection at the surge arrester, the following occurs:

(2) The grounded conductor of the secondary system is a part of a multigrounded neutral system of which the primary neutral has at least four ground connections in each mile of line in addition to a ground at each service.

2005 NEC – 280.24(A) Metallic Interconnections

A metallic interconnection shall be made to the secondary grounded circuit conductor or the secondary circuit grounding conductor provided that, in addition to the direct grounding connection at the surge arrester, the following occurs:

(2) The grounded conductor of the secondary system is a part of a multigrounded neutral system <u>or static wire</u> of which the primary neutral <u>or static wire</u> has at least four ground connections in each mile of line in addition to a ground at each service.

Author's Substantiation. This revision clarifies that the static wire shall be permitted to be used provided it has at least four ground connections in each mile of line in addition to a ground at each service.

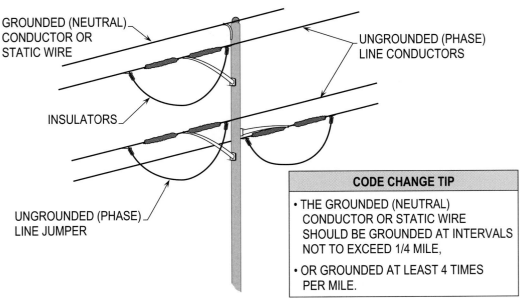

GROUNDED (NEUTRAL) CONDUCTOR OR STATIC WIRE

UNGROUNDED (PHASE) LINE CONDUCTORS

INSULATORS

UNGROUNDED (PHASE) LINE JUMPER

CODE CHANGE TIP

• THE GROUNDED (NEUTRAL) CONDUCTOR OR STATIC WIRE SHOULD BE GROUNDED AT INTERVALS NOT TO EXCEED 1/4 MILE,

• OR GROUNDED AT LEAST 4 TIMES PER MILE.

METALLIC INTERCONNECTIONS
280.24(A)(2)

Purpose of Change: To include the grounding of the static system at least 4 times per mile.

NEC Ch. 2 – Article 285
Part I – 285.3(2)

64

Type of Change		Panel Action		UL	UL 508	API 500	API 505	OSHA
Revision		Accept in Principle		551	-	-	-	-
ROP		ROC		NFPA 70E	NFPA 70B	NFPA 79	NFPA	NEMA
pg. 629	# 5-267	pg. 183	# 5-230	-	27.3	11.2.3	-	-
log: 3383	CMP: 5	log: 401	Submitter: David Shipp			2002 NEC: 285.3		IEC: 535

2002 NEC – 285.3 Uses Not Permitted

A TVSS shall not be ~~used~~ in the following:

(1) Circuits exceeding 600 volts

~~**(2)** Ungrounded electrical systems as permitted in 250.21~~

(3) Where the rating of the TVSS is less than the maximum continuous phase-to-ground power frequency voltage available at the point of application

2005 NEC – 285.3 Uses Not Permitted

A TVSS <u>device</u> shall not be <u>installed</u> in the following:

(1) Circuits exceeding 600 volts

(2) <u>On ungrounded systems, impedance grounded systems, or corner grounded delta systems unless listed specifically for use on these systems.</u>

(3) Where the rating of the TVSS is less than the maximum continuous phase-to-ground power frequency voltage available at the point of application

Author's Substantiation. This revision clarifies that transient voltage surge suppressors shall be listed specifically for use on ungrounded systems, impedance grounded systems, or corner grounded delta systems.

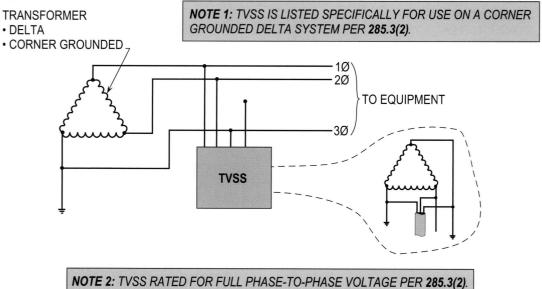

TRANSFORMER
• DELTA
• CORNER GROUNDED

NOTE 1: TVSS IS LISTED SPECIFICALLY FOR USE ON A CORNER GROUNDED DELTA SYSTEM PER 285.3(2).

TO EQUIPMENT

TVSS

NOTE 2: TVSS RATED FOR FULL PHASE-TO-PHASE VOLTAGE PER 285.3(2).

USES NOT PERMITTED
285.3(2)

Purpose of Change: To allow a listed TVSS to be installed on specific transformer systems.

NEC Ch. 2 – Article 285
Part II – 285.21(A)(1)

Type of Change		Panel Action		UL	UL 508	API 500	API 505	OSHA
Revision		Accept		551	-	-	-	-
ROP		ROC		NFPA 70E	NFPA 70B	NFPA 79	NFPA	NEMA
pg. 631	# 5-272	pg. 184	# 5-235	-	27.3	11.2.3	1449	-
log: 2533	CMP: 5	log: 1586	Submitter: Alan Manche			2002 NEC: 285.21(A)(1)		IEC: 535

2002 NEC – 285.21 Connection

Where a TVSS is installed, it shall ~~be connected as follows~~:

(A) Location.

> **(1) Service Supplied Building or Structure.** The transient voltage surge suppressor shall be connected on the load side of a service disconnect overcurrent device required in 230.91.

2005 NEC – 285.21 Connection

Where a TVSS is installed, <u>it shall comply with 285.21(A) through (C).</u>

(A) Location.

> **(1) Service Supplied Building or Structure.** The transient voltage surge suppressor shall be connected on the load side of a service disconnect overcurrent device required in 230.91, <u>unless installed in accordance with 230.82(8).</u>

Author's Substantiation. This change clarifies that the transient voltage surge suppressors shall be connected on the load side of a service disconnect overcurrent device unless connected on the supply side where installed as part of listed equipment, if suitable overcurrent protection and disconnecting means are provided.

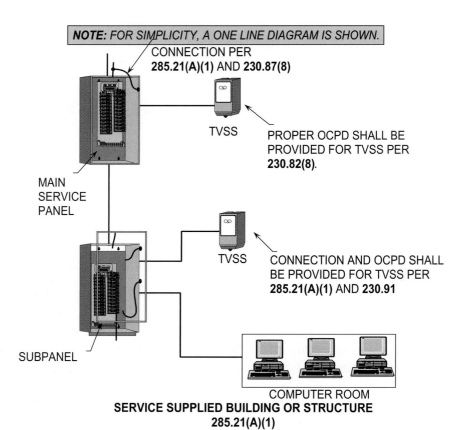

NOTE: FOR SIMPLICITY, A ONE LINE DIAGRAM IS SHOWN.

CONNECTION PER
285.21(A)(1) AND **230.87(8)**

TVSS

PROPER OCPD SHALL BE
PROVIDED FOR TVSS PER
230.82(8).

MAIN
SERVICE
PANEL

TVSS

CONNECTION AND OCPD SHALL
BE PROVIDED FOR TVSS PER
285.21(A)(1) AND **230.91**

SUBPANEL

COMPUTER ROOM
SERVICE SUPPLIED BUILDING OR STRUCTURE
285.21(A)(1)

Purpose of Change: To include the supply connection outlined in **230.82(8)** for a TVSS installation.

3

Wiring Methods and Materials

Chapter 3 of the NEC has always been referred to as the installation chapter and used by electricians to install electrical systems in a safe, dependable and reliable manner. **Chapter 3** covers the requirements needed by on-the-job electricians who are installing distribution lines and feeder-circuits and thus providing electrical power to the end of branch-circuits and on to the point of use.

Chapter 3 is also called the "rough-in" chapter by electricians. All Articles in **Chapter 3** are of the 300 series and each one contains the rules that pertain to building and installing electrical wiring methods and accessories. On-the-job working procedures of the average electrician bring him or her into almost daily contact with an Article in **Chapter 3** concerning these installation rules. For example, when installing wiring in cable trays, electricians cannot install cable tray systems and fill them with different wiring systems without knowing the requirements of **Article 392**. The same is true for electricians running rigid metal conduit. They will not know how many 90 degree bends are allowed or how often supports are required, without first studying **Article 344**.

NEC Ch. 3 – Article 300
Part I – 300.4(A)(1), Ex. 2

Type of Change	Panel Action	UL	UL 508	API 500	API 505	OSHA
New Exception	Accept in Principle in Part	-	-	-	-	-
ROP		ROC		NFPA 70E	NFPA 70B	NFPA 79	NFPA	NEMA

ROP			ROC			NFPA 70E	NFPA 70B	NFPA 79	NFPA	NEMA
pg. 643	# 3-24		pg. -	# -		-	-	-	-	-
log: 578	CMP: 3		log: -		Submitter: Vince Baclawski			2002 NEC: -		IEC: 11.2

2005 NEC – 300.4(A)(1) Bored Holes

In both exposed and concealed locations, where a cable- or raceway-type wiring method is installed through bored holes in joists, rafters, or wood members, holes shall be bored so that the edge of the hole is less than 32 mm (1 1/4 in.) from the nearest edge of the wood member. Where this distance cannot be maintained, the cable or raceway shall be protected from penetration by screws or nails by a steel plate or bushing, at least 1.6 mm (1/16 in.) thick, and of appropriate length and width installed to cover the area of the wiring.

> **Exception No. 2:** A listed and marked steel plate less than 1.6 mm (1/16 in.) thick that provides equal or better protection against nail or screw penetration shall be permitted.

Author's Substantiation. This new exception permits a listed and marked steel plate to be installed for protection against nail or screw penetration. Test have shown that heat-treated steel plates provide equivalent or better resistance to nail or screw penetration.

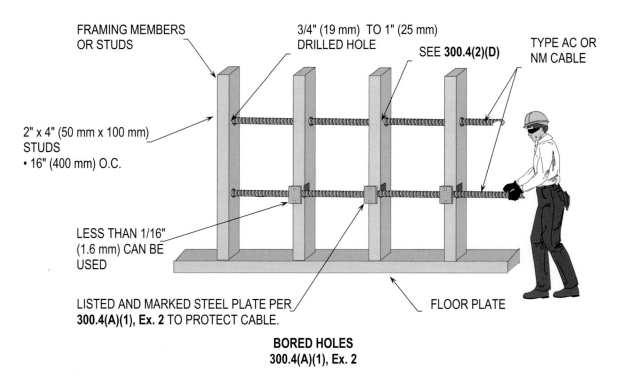

FRAMING MEMBERS OR STUDS

3/4" (19 mm) TO 1" (25 mm) DRILLED HOLE

SEE **300.4(2)(D)**

TYPE AC OR NM CABLE

2" x 4" (50 mm x 100 mm) STUDS
• 16" (400 mm) O.C.

LESS THAN 1/16" (1.6 mm) CAN BE USED

LISTED AND MARKED STEEL PLATE PER **300.4(A)(1), Ex. 2** TO PROTECT CABLE.

FLOOR PLATE

BORED HOLES
300.4(A)(1), Ex. 2

Purpose of Change: To allow a steel plate less than 1/16 inch (1.6 mm) to protect cables run through studs.

NEC Ch. 3 – Article 300
Part I – 300.4(A)(2)

Type of Change		Panel Action		UL	UL 508	API 500	API 505	OSHA
Revision		Accept in Principle in Part		-	-	-	-	-
ROP		ROC		NFPA 70E	NFPA 70B	NFPA 79	NFPA	NEMA
pg. 645	# 3-26	pg. -	# -	-	-	-	-	-
log: 2683	CMP: 3	log: -	Submitter: Phil Simmons			2002 NEC: 300.4(A)(2)		IEC: 11.2

2002 NEC – 300.4(A)(2) Notches in Wood

Where there is no objection because of weakening the building structure, in both exposed and concealed locations, cables or raceways shall be permitted to be laid in notches in wood studs, joists, rafters, or other wood members where the cable or raceway at those points is protected against nails or screws by a steel plate at least 1.6 mm (1/16 in.) thick installed before the building finish is applied.

2005 NEC – 300.4(A)(2) Notches in Wood

Where there is no objection because of weakening the building structure, in both exposed and concealed locations, cables or raceways shall be permitted to be laid in notches in wood studs, joists, rafters, or other wood members where the cable or raceway at those points is protected against nails or screws by a steel plate at least 1.6 mm (1/16 in.) thick, and of appropriate length and width, installed to cover the area of the wiring. The steel plate shall be installed before the building finish is applied.

Author's Substantiation. This revision clarifies that a steel plate at least 1.6 mm (1/16 in.) thick shall be of appropriate length and width to cover the area of the wiring.

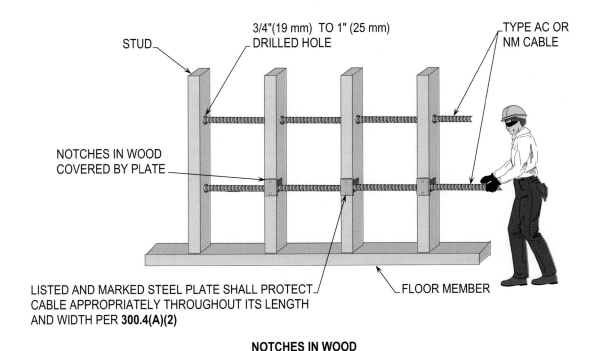

NOTCHES IN WOOD
300.4(A)(2)

Purpose of Change: To require plates to cover a cable(s) throughout its length and width.

NEC Ch. 3 – Article 300
Part I – 300.4(A)(2), Ex. 2

Type of Change	Panel Action	UL	UL 508	API 500	API 505	OSHA
New Exception	Accept in Principle in Part	-	-	-	-	-
ROP	**ROC**	**NFPA 70E**	**NFPA 70B**	**NFPA 79**	**NFPA**	**NEMA**
pg. 645	# 3-26	pg. -	# -	-	-	-	-
log: 2683	CMP: 3	log: -	Submitter: Phil Simmons		2002 NEC: -		IEC: 11.2

2005 NEC – 300.4(A)(2) Notches in Wood

Where there is no objection because of weakening the building structure, in both exposed and concealed locations, cables or raceways shall be permitted to be laid in notches in wood studs, joists, rafters, or other wood members where the cable or raceway at those points is protected against nails or screws by a steel plate at least 1.6 mm (1/16 in.) thick, and of appropriate length and width, installed to cover the area of the wiring. The steel plate shall be installed before the building finish is applied.

> **Exception No. 2:** A listed and marked steel plate less than 1.6 mm (1/16 in.) thick that provides equal or better protection against nail or screw penetration shall be permitted.

Author's Substantiation. This new exception permits a listed and marked steel plate to be installed for protection against nail or screw penetration. Test have shown that heat-treated steel plates provide equivalent or better resistance to nail or screw penetration.

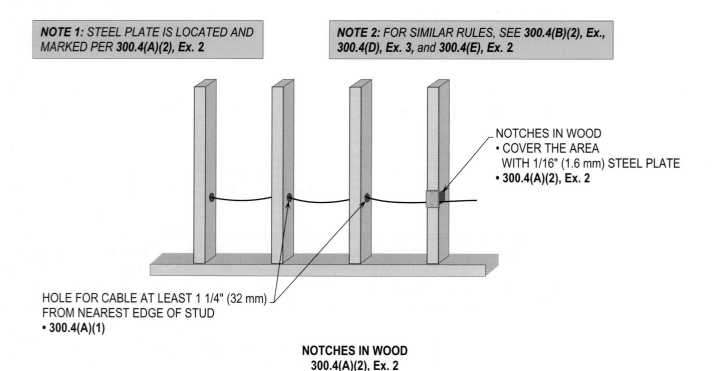

NOTE 1: STEEL PLATE IS LOCATED AND MARKED PER **300.4(A)(2), Ex. 2**

NOTE 2: FOR SIMILAR RULES, SEE **300.4(B)(2), Ex., 300.4(D), Ex. 3,** and **300.4(E), Ex. 2**

NOTCHES IN WOOD
• COVER THE AREA
  WITH 1/16" (1.6 mm) STEEL PLATE
• **300.4(A)(2), Ex. 2**

HOLE FOR CABLE AT LEAST 1 1/4" (32 mm) FROM NEAREST EDGE OF STUD
• **300.4(A)(1)**

NOTCHES IN WOOD
300.4(A)(2), Ex. 2

Purpose of Change: To verify that a steel plate less than 1/16 in. (1.6 mm) can cover cable installed in notched wood.

NEC Ch. 3 – Article 300
Part I – 300.4(D), Ex. 3

Type of Change	Panel Action	UL	UL 508	API 500	API 505	OSHA
New Exception	Accept in Principle in Part	-	-	-	-	-
ROP	**ROC**	**NFPA 70E**	**NFPA 70B**	**NFPA 79**	**NFPA**	**NEMA**
pg. 651 # 3-33	pg. - # -	-	-	-	-	-
log: 573 CMP: 3	log: -	Submitter: Vince Baclawski		2002 NEC: -		IEC: 11.2

2002 NEC – 300.4(D) Cables and Raceways Parallel to Framing Members

In both exposed and concealed locations, where a cable- or raceway-type wiring method is installed parallel to framing members, such as joists, rafters, or studs, the cable or raceway shall be installed and supported so that the nearest outside surface of the cable or raceway is not less than 32 mm (1 1/4 in.) from the nearest edge of the framing member where nails or screws are likely to penetrate. Where this distance cannot be maintained, the cable or raceway shall be protected from penetration by nails or screws by a steel plate, sleeve, or equivalent at least 1.6 mm (1/16 in.) thick.

2005 NEC – 300.4(D) Cables and Raceways Parallel to Framing Members <u>and Furring Strips</u>

In both exposed and concealed locations, where a cable- or raceway-type wiring method is installed parallel to framing members, such as joists, rafters, or studs, <u>or is installed parallel to furring strips,</u> the cable or raceway shall be installed and supported so that the nearest outside surface of the cable or raceway is not less than 32 mm (1 1/4 in.) from the nearest edge of the framing member <u>or furring strips</u> where nails or screws are likely to penetrate. Where this distance cannot be maintained, the cable or raceway shall be protected from penetration by nails or screws by a steel plate, sleeve, or equivalent at least 1.6 mm (1/16 in.) thick.

> **Exception No. 3:** <u>A listed and marked steel plate less than 1/6 mm (1/16 in.) thick that provides equal or better protection against nail or screw penetration shall be permitted.</u>

Author's Substantiation. This new exception permits a listed and marked steel plate to be installed for protection against nail or screw penetration. Test have shown that heat-treated steel plates provide equivalent or better resistance to nail or screw penetration.

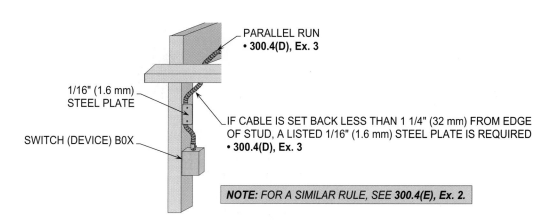

CABLES AND RACEWAYS PARALLEL TO FRAMING MEMBERS AND FURRING STRIPS
300.4(D), Ex. 3

Purpose of Change: To allow a listed and marked steel plate to protect cables run parallel with framing members.

NEC Ch. 3 – Article 300
Part I – 300.5(B)

Type of Change		Panel Action		UL	UL 508	API 500	API 505	OSHA
Relocated		Accept		6	-	-	-	1910.307(B)(3), Note
ROP		ROC		NFPA 70E	NFPA 70B	NFPA 79	NFPA	NEMA
pg. 656	# 3-39a	pg. -	# -	440.3(B)(1)	6.7.6	-	-	-
log: CP302	CMP: 3	log: -	Submitter: CMP 3			2002 NEC: 300.5(D)(5)		IEC: 11.2

2002 NEC – ~~300.5(D)(5)~~ Listing

Cables and insulated conductors installed in enclosures or raceways in underground installations shall be listed for use in wet locations.

2005 NEC – <u>300.5(B)</u> Listing

Cables and insulated conductors installed in enclosures or raceways in underground installations shall be listed for use in wet locations.

Author's Substantiation. This section has been moved and renumbered to clarify cables and insulated conductors installed in enclosures or raceways apply to all underground installations and not only direct buried installations.

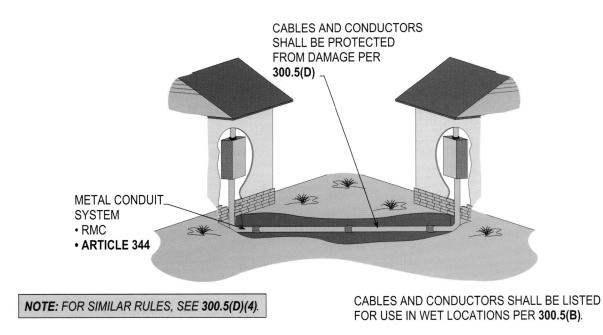

CABLES AND CONDUCTORS SHALL BE PROTECTED FROM DAMAGE PER **300.5(D)**

METAL CONDUIT SYSTEM
• RMC
• **ARTICLE 344**

NOTE: FOR SIMILAR RULES, SEE **300.5(D)(4)**.

CABLES AND CONDUCTORS SHALL BE LISTED FOR USE IN WET LOCATIONS PER **300.5(B)**.

LISTING AND PROTECTION FROM DAMAGE
300.5(B) AND (D)

Purpose of Change: For usability, existing **300.5(D)(5)** was relocated and renumbered to **300.5(B)**.

NEC Ch. 3 – Article 300
Part I – 300.5(G)

Type of Change		Panel Action		UL	UL 508	API 500	API 505	OSHA
Revision		Accept in Part		2225	-	-	-	1910.307(B)(3), Note
ROP		ROC		NFPA 70E	NFPA 70B	NFPA 79	NFPA	NEMA
pg. 659	# 3-47	pg. -	# -	440.3(B)(1)	6.7.6	-	-	-
log: 2489	CMP: 3	log: -	Submitter: Paul Dobrowsky			2002 NEC: 300.5(G)		IEC: 11.2

2002 NEC – 300.5(G) Raceway Seals

Conduits or raceways through which moisture may contact ~~energized~~ live parts shall be sealed or plugged at either or both ends.

2005 NEC – 300.5(G) Raceway Seals

Conduits or raceways through which moisture may contact live parts shall be sealed or plugged at either or both ends.

Author's Substantiation. This revision clarifies that live parts shall be sealed or plugged at either end or both ends where moisture may be present in conduits or raceways.

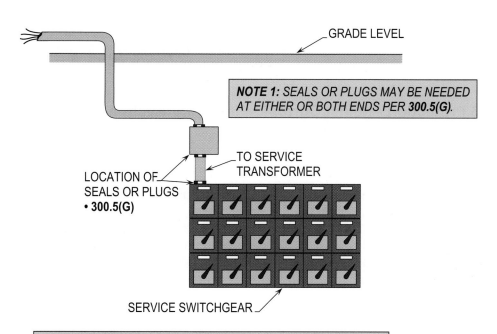

GRADE LEVEL

NOTE 1: SEALS OR PLUGS MAY BE NEEDED AT EITHER OR BOTH ENDS PER 300.5(G).

TO SERVICE TRANSFORMER

LOCATION OF SEALS OR PLUGS
• 300.5(G)

SERVICE SWITCHGEAR

NOTE 2: SEAL OR PLUG INSTALLED TO PREVENT MOISTURE FROM CONTACTING LIVE PARTS IN SWITCHGEAR PER 300.5(G).

RACEWAY SEALS
300.5(G)

Purpose of Change: To ensure that moisture will not contact live parts in any way in electrical equipment.

NEC Ch. 3 – Article 300
Part I – 300.6(A) through (C)

 67

Type of Change		Panel Action		UL	UL 508	API 500	API 505	OSHA
Revision		Accept in Principle in Part		-	-	-	-	1910.305(f)
ROP		ROC		NFPA 70E	NFPA 70B	NFPA 79	NFPA	NEMA
pg. 662	# 3-51	pg. 190	# 3-24	400.7	-	3.3.65.2	-	-
log: 1801	CMP: 3	log: 978	Submitter: William A. Wolfe			2002 NEC: 300.6		IEC: 11.2

2002 NEC – 300.6 Protection Against Corrosion

~~Metal~~ raceways, cable trays, cablebus, auxiliary gutters, cable armor, boxes, cable sheathing, cabinets, elbows, couplings, fittings, supports, and support hardware shall be of materials suitable for the environment in which they are to be installed.

(A) ~~General.~~ Ferrous raceways, cable trays, cablebus, auxiliary gutters, cable armor, boxes, cable sheathing, cabinets, metal elbows, couplings, fittings, supports, and support hardware shall be suitably protected against corrosion inside and outside (except threads at joints) by a coating of ~~approved~~ corrosion-resistant material ~~such as zinc, cadium, or enamel, they shall not be used outdoors or in wet locations as described in 300.6(C).~~ Where boxes or cabinets have an approved system or organic coatings and are marked "Raintight," "Rainproof," or "Outdoor Type," they shall be permitted outdoors. Where corrosion protection is necessary and the conduit is threaded in the field, the threads shall be coated with an approved electrically conductive, corrosion-resistant compound.

(B) **In Concrete or in Direct Contact with the Earth.** Ferrous ~~or nonferrous~~ metal raceways, cable armor, boxes, cable sheathing, cabinets, elbows, couplings, fittings, supports, and support hardware shall be permitted to be installed in concrete or in direct contact with the earth, or in areas subject to severe corrosive influences where made of material ~~judged suitable~~ for the condition, or where provided with corrosion protection approved for the condition.

2005 NEC – 300.6 Protection Against Corrosion <u>and Deterioration</u>

Raceways, cable trays, cablebus, auxiliary gutters, cable armor, boxes, cable sheathing, cabinets, elbows, couplings, fittings, supports, and support hardware shall be of materials suitable for the environment in which they are to be installed.

(A) <u>**Ferrous Metal Equipment.**</u> Ferrous <u>metal</u> raceways, cable trays, cablebus, auxiliary gutters, cable armor, boxes, cable sheathing, cabinets, metal elbows, couplings, <u>nipples,</u> fittings, supports, and support hardware shall be suitably protected against corrosion inside and outside (except threads at joints) by a coating of <u>listed</u> corrosion-resistant material. Where corrosion protection is necessary and the conduit is threaded in the field, the threads shall be coated with an approved electrically conductive, corrosion-resistant compound.

<u>**Exception:** Stainless steel shall not be required to have protective coatings.</u>

<u>(1) **Protected from Corrosion Solely by Enamel.** Where protected from corrosion solely by enamel, ferrous metal raceways, cable trays, cablebus, auxiliary gutters, cable armor, boxes, cable sheathing, cabinets, metal elbows, couplings, nipples, fittings, supports, and support hardware shall not be used outdoors or in wet locations as described in 300.6(D).</u>

<u>(2) **Organic Coatings on Boxes or Cabinets.**</u> Where boxes or cabinets have an approved system of organic coatings and <u>are</u> marked "Raintight," "Rainproof," or "Outdoor Type," they shall be permitted outdoors.

<u>(3)</u> **In Concrete or in Direct Contact with the Earth.** Ferrous metal raceways, cable armor, boxes, cable sheathing, cabinets, elbows, couplings, <u>nipples,</u> fittings, supports, and support

hardware shall be permitted to be installed in concrete or in direct contact with the earth, or in areas subject to severe corrosive influences where made of material approved for the condition, or where provided with corrosion protection <u>approved</u> for the condition.

(B) <u>Non-Ferrous Metal Equipment.</u> <u>Non-ferrous raceways, cable trays, cablebus, auxiliary gutters, cable armor, boxes, cable sheathing, cabinets, elbows, couplings, nipples, fittings, supports, and support hardware embedded or encased in concrete or in direct contact with the earth shall be provided with supplementary corrosion protection.</u>

(C) <u>Nonmetallic Equipment.</u> <u>Nonmetallic raceways, cable trays, cablebus, auxiliary gutters, boxes, cables with a nonmetallic outer jacket and internal metal armor or jacket, cable sheathing, cabinets, elbows, couplings, nipples, fittings, supports, and support hardware shall be made of material approved for the condition and shall comply with (C)(1) and (C)(2) as applicable to the specific installation.</u>

    (1) <u>Exposed to Sunlight.</u> <u>Where exposed to sunlight, the materials shall be listed as sunlight resistant or shall be identified as sunlight resistant.</u>

    (2) <u>Chemical exposure.</u> <u>Where subject to exposure to chemical solvents, vapors, splashing, or immersion, materials or coatings shall either be inherently resistant to chemicals based on its listing or be identified for the specific chemical reagent.</u>

Author's Substantiation. The section has been rewritten and divided into three groups of wiring methods for protection against corrosion and deterioration. The three groups of wiring methods are Ferrous Metal Equipment, Non-Ferrous Metal Equipment, and Nonmetallic Equipment. The words "judge suitable" have been replaced with "approved" to meet the intent of the definition of "approved" in Article 100.

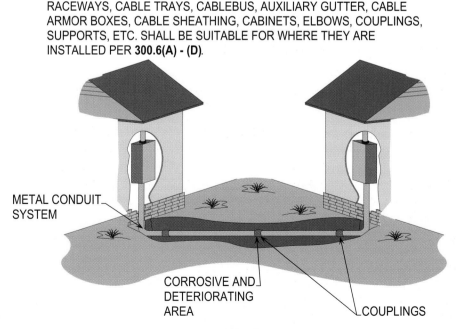

RACEWAYS, CABLE TRAYS, CABLEBUS, AUXILIARY GUTTER, CABLE ARMOR BOXES, CABLE SHEATHING, CABINETS, ELBOWS, COUPLINGS, SUPPORTS, ETC. SHALL BE SUITABLE FOR WHERE THEY ARE INSTALLED PER **300.6(A) - (D)**.

METAL CONDUIT SYSTEM

CORROSIVE AND DETERIORATING AREA

COUPLINGS

PROTECTION AGAINST CORROSION AND DETERIORATION
300.6(A) THRU (D)

Purpose of Change: To completely rewrite and revise to make clear that metallic raceways, enclosures, and associated equipment are protected from corrosion and deterioration.

NEC Ch. 3 – Article 300
Part I – 300.15(L)

68

Type of Change		Panel Action		UL	UL 508	API 500	API 505	OSHA
Revision		Accept in Principle		-	-	-	-	1910.956(b)
ROP		ROC		NFPA 70E	NFPA 70B	NFPA 79	NFPA	NEMA
pg. 678	# 3-77	pg. -	# -	-	10.2.2	-	-	-
log: 4	CMP: 3	log: -	Submitter: Don W. Jhonson			2002 NEC: 300.15(L)		IEC: 522.8.1

2002 NEC – 300.15(L) Manholes

Where accessible only to qualified persons, a box or conduit body shall not be required for conductors in manholes, except where connecting to electrical equipment. The installation shall comply with the provisions of ~~Part IV of Article 314~~.

2005 NEC – 300.15(L) Manholes <u>and Handhole Enclosures</u>

Where accessible only to qualified persons, a box or conduit body shall not be required for conductors in manholes <u>or handhole enclosures</u>, except where connecting to electrical equipment. The installation shall comply with the provisions of <u>Part V of Article 110 for manholes, and 314.30 for handhole enclosures</u>.

Author's Substantiation. This revision clarifies that a box or conduit body shall not be required for conductors in handhole enclosures where accessible only to qualified persons.

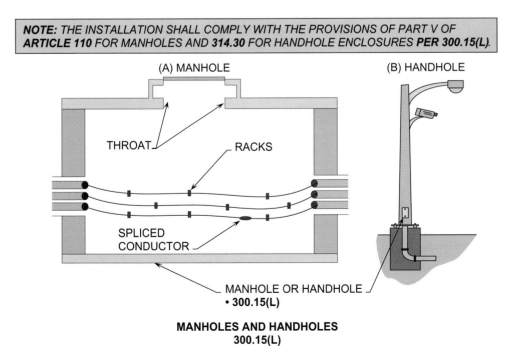

NOTE: THE INSTALLATION SHALL COMPLY WITH THE PROVISIONS OF PART V OF ARTICLE 110 FOR MANHOLES AND 314.30 FOR HANDHOLE ENCLOSURES PER 300.15(L).

MANHOLES AND HANDHOLES
300.15(L)

Purpose of Change: To add handholes as a space that is not required to have a box or conduit body for conductors.

NEC Ch. 3 – Article 300
Part I – 300.18(A), Ex.

Type of Change		Panel Action		UL	UL 508	API 500	API 505	OSHA
New Exception		Accept in Principle		183	40.2	-	-	1910.304(f)(4)
ROP		ROC		NFPA 70E	NFPA 70B	NFPA 79	NFPA	NEMA
pg. 681	# 3-81	pg. -	# -	420.1(A)(1)	29.4.4	8.2	-	-
log: 1052	CMP: 3	log: -	Submitter: Wayne H. Robinson			2002 NEC: -		IEC: 522.8.1.1

2002 NEC – 300.18(A) Complete Runs

Raceways, other than busways or exposed raceways having hinged or removable covers, shall be installed complete between outlet, junction, or splicing points prior to the installation of conductors. Where required to facilitate the installation of utilization equipment, the raceway shall be permitted to be initially installed without a terminating connection at the equipment. Prewired raceway assemblies shall be permitted only where specifically permitted in this *Code* for the applicable wiring method.

> **Exception:** Short sections of raceways used to contain conductors or cable assemblies for protection from physical damage shall not be required to be installed complete between outlet, junction, or splicing points.

Author's Substantiation. This new exception clarifies that short sections of raceways shall not be required to be installed as complete raceway sytems.

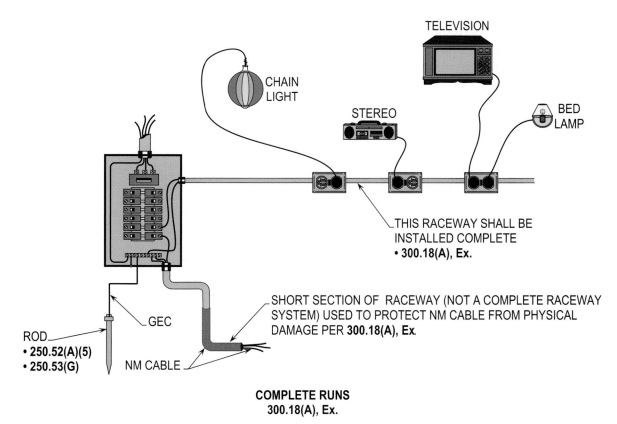

COMPLETE RUNS
300.18(A), Ex.

Purpose of Change: To clarify that short sections of raceways are not required to be installed as a complete raceway.

NEC Ch. 3 – Article 300
Part I – 300.22(B)

Type of Change		Panel Action		UL	UL 508	API 500	API 505	OSHA
Revision		Accept		1978	-	-	-	1910.305(a)(1)(ii)
ROP		ROC		NFPA 70E	NFPA 70B	NFPA 79	NFPA	NEMA
pg. 689	# 3-93	pg. -	# -	420.1(A)(3)	-	3.3.29	-	-
log: 2735	CMP: 3	log: -	Submitter: Melvin K. Sanders			2002 NEC: 300.22(B)		IEC: -

2002 NEC – 300.22(B) Ducts or Plenums Used for Environmental Air

Only wiring methods consisting of Type MI cable, Type MC cable employing a smooth or corrugated impervious metal sheath without an overall nonmetallic covering, electrical metallic tubing, flexible metallic tubing, intermediate metal conduit, or rigid metal conduit without an overall nonmetallic covering shall be installed in ducts or plenums specifically fabricated to transport environmental air. Flexible metal conduit ~~and liquidtight flexible metal conduit~~ shall be permitted, in lengths not to exceed 1.2 m (4 ft), to connect physically adjustable equipment and devices permitted to be in these ducts and plenum chambers. The connectors used with flexible metal conduit shall effectively close any openings in the connection. Equipment and devices shall be permitted within such ducts or plenum chambers only if necessary for their direct action upon, or sensing of, the contained air. Where equipment or devices are installed and illumination is necessary to facilitate maintenance and repair, enclosed gasketed-type luminaires (fixtures) shall be permitted.

2005 NEC – 300.22(B) Ducts or Plenums Used for Environmental Air

Only wiring methods consisting of Type MI cable, Type MC cable employing a smooth or corrugated impervious metal sheath without an overall nonmetallic covering, electrical metallic tubing, flexible metallic tubing, intermediate metal conduit, or rigid metal conduit without an overall nonmetallic covering shall be installed in ducts or plenums specifically fabricated to transport environmental air. Flexible metal conduit shall be permitted, in lengths not to exceed 1.2 m (4 ft), to connect physically adjustable equipment and devices permitted to be in these ducts and plenum chambers. The connectors used with flexible metal conduit shall effectively close any openings in the connection. Equipment and devices shall be permitted within such ducts or plenum chambers only if necessary for their direct action upon, or sensing of, the contained air. Where equipment or devices are installed and illumination is necessary to facilitate maintenance and repair, enclosed gasketed-type luminaires (fixtures) shall be permitted.

Author's Substantiation. This revision clarifies that liquidtight flexible metal conduit shall not be permitted for ducts or plenums used for environmental air.

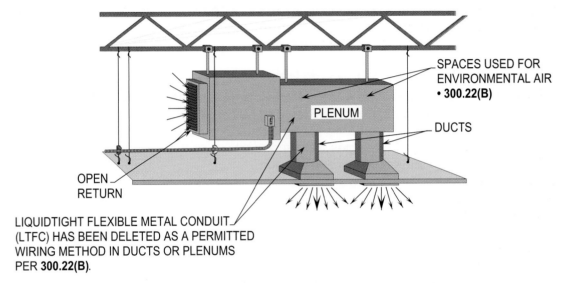

SPACES USED FOR
ENVIRONMENTAL AIR
• **300.22(B)**

PLENUM

DUCTS

OPEN
RETURN

LIQUIDTIGHT FLEXIBLE METAL CONDUIT
(LTFC) HAS BEEN DELETED AS A PERMITTED
WIRING METHOD IN DUCTS OR PLENUMS
PER **300.22(B)**.

DUCTS OR PLENUMS USED FOR ENVIRONMENTAL AIR
300.22(B)

Purpose of Change: To delete the use of LTFC in plenums or ducts used for environmental air.

NEC Ch. 3 – Article 300
Part II – 300.37

Type of Change		Panel Action		UL	UL 508	API 500	API 505	OSHA
Revision		Accept		-	-	-	-	1910.305(a)(3)(i)(B)
ROP		ROC		NFPA 70E	NFPA 70B	NFPA 79	NFPA	NEMA
pg. 693	# 3-98	pg. -	# -	420.1(C)(3)(b)	-	-	-	-
log: 98	CMP: 3	log: -	Submitter: James M. Daly			2002 NEC: 300.37		IEC: 521

2002 NEC – 300.37 Aboveground Wiring Methods

Aboveground conductors shall be installed in rigid metal conduit, in intermediate metal conduit, in electrical metallic tubing, in rigid nonmetallic conduit, in cable trays, as busways, as cablebus, in other identified raceways, or as ~~open~~ runs of metal-clad cable suitable for the use and purpose. In locations accessible to qualified persons only, ~~open~~ runs of Type MV cables, bare conductors, and bare busbars shall also be permitted. Busbars shall be permitted to be either copper or aluminum.

2005 NEC – 300.37 Aboveground Wiring Methods

Aboveground conductors shall be installed in rigid metal conduit, in intermediate metal conduit, in electrical metallic tubing, in rigid nonmetallic conduit, in cable trays, as busways, as cablebus, in other identified raceways, or as <u>exposed</u> runs of metal-clad cable suitable for the use and purpose. In locations accessible to qualified persons only, <u>exposed</u> runs of Type MV cables, bare conductors, and bare busbars shall also be permitted. Busbars shall be permitted to be either copper or aluminum.

Author's Substantiation. This revision clarifies that "exposed runs" applies to open runs and open wiring not on insulators.

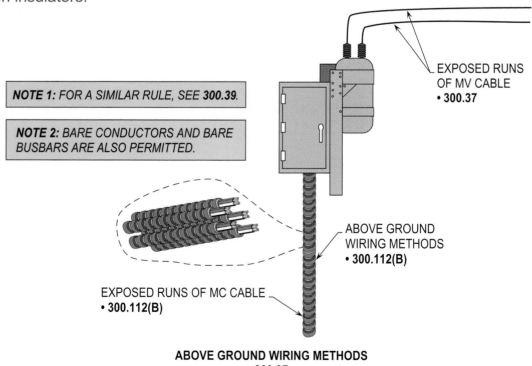

NOTE 1: FOR A SIMILAR RULE, SEE **300.39**.

NOTE 2: BARE CONDUCTORS AND BARE BUSBARS ARE ALSO PERMITTED.

EXPOSED RUNS OF MV CABLE
• 300.37

ABOVE GROUND WIRING METHODS
• 300.112(B)

EXPOSED RUNS OF MC CABLE
• 300.112(B)

ABOVE GROUND WIRING METHODS
300.37

Purpose of Change: To bring more consistency throughout the NEC, the term "open runs" was replaced with the term "exposed runs."

NEC Ch. 3 – Article 300
Part II – 300.50(D)

Type of Change		Panel Action		UL	UL 508	API 500	API 505	OSHA
Revision		Accept		-	-	-	-	-
ROP		ROC		NFPA 70E	NFPA 70B	NFPA 79	NFPA	NEMA
pg. 698	# 3-105	pg. -	# -	-	-	-	-	-
log: 1227	CMP: 3	log: -	Submitter: Melanie Roberts			2002 NEC: 300.50(D)		IEC: -

2002 NEC – 300.50(D) Backfill

Backfill containing large rocks, paving materials, cinders, large or sharply angular substances, or corrosive materials shall not be placed in an excavation where materials can damage raceways, cables, or other substructures, or prevent adequate compaction of fill, or contribute to corrosion of raceways, cables, or other substructures.

Protection in the form of granular or selected material or suitable sleeves shall be provided to prevent physical damage to the raceway or cable.

2005 NEC – 300.50(D) Backfill

Backfill containing large rocks, paving materials, cinders, large or sharply angular substances, or corrosive materials shall not be placed in an excavation where materials can damage or contribute to the corrosion of raceways, cables, or other substructures where it may prevent adequate compaction of fill.

Protection in the form of granular or selected material or suitable sleeves shall be provided to prevent physical damage to the raceway or cable.

Author's Substantiation. The subsection has been reworded for better clarity when applying the requirements for backfill.

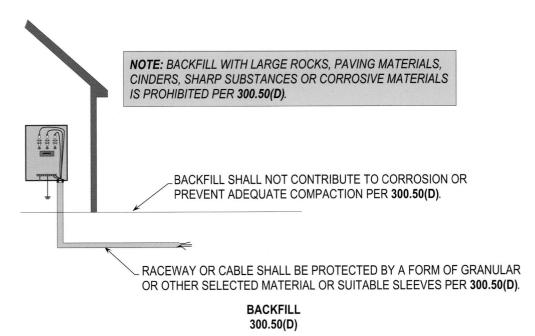

NOTE: BACKFILL WITH LARGE ROCKS, PAVING MATERIALS, CINDERS, SHARP SUBSTANCES OR CORROSIVE MATERIALS IS PROHIBITED PER **300.50(D)**.

BACKFILL SHALL NOT CONTRIBUTE TO CORROSION OR PREVENT ADEQUATE COMPACTION PER **300.50(D)**.

RACEWAY OR CABLE SHALL BE PROTECTED BY A FORM OF GRANULAR OR OTHER SELECTED MATERIAL OR SUITABLE SLEEVES PER **300.50(D)**.

**BACKFILL
300.50(D)**

Purpose of Change: To list methods for the protection of conductors and cables installed in backfill locations.

NEC Ch. 3 – Article 310
Part – 310.4

Type of Change		Panel Action		UL	UL 508	API 500	API 505	OSHA
Revision		Accept in Part		62	18.1	-	-	1910.305(f)
ROP		ROC		NFPA 70E	NFPA 70B	NFPA 79	NFPA	NEMA
pg. 700	# 6-6a	pg. 203	# 6-4	410.9(A)(1)	3.3.18	13.5	-	WC 51
log: CP600	CMP: 6	log: 764-	Submitter: -			2002 NEC: 310.4		IEC: 523.6

2002 NEC – 310.4 Conductors in Parallel

Aluminum, copper-clad aluminum, or copper conductors of size 1/0 AWG and larger, comprising each phase, neutral, or grounded circuit conductor, shall be permitted to be connected in parallel (electrically joined at both ends to form a single conductor).

2005 NEC – 310.4 Conductors in Parallel

Aluminum, copper-clad aluminum, or copper conductors of size 1/0 AWG and larger, comprising each phase, polarity, neutral, or grounded circuit conductor, shall be permitted to be connected in parallel (electrically joined at both ends).

Author's Substantiation. This revision clarifies that conductors connected in parallel shall be electrically joined at both ends. The words to form a single conductor were removed due to the confusion of applying 310.15(B)(2) when installing parallel conductors.

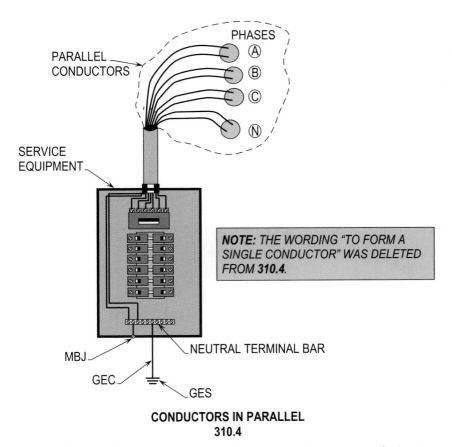

PARALLEL CONDUCTORS

SERVICE EQUIPMENT

PHASES
Ⓐ
Ⓑ
Ⓒ
Ⓝ

NOTE: THE WORDING "TO FORM A SINGLE CONDUCTOR" WAS DELETED FROM 310.4.

MBJ

GEC

NEUTRAL TERMINAL BAR

GES

CONDUCTORS IN PARALLEL
310.4

Purpose of Change: To delete a parenthetical phrase that was confusing to users.

NEC Ch. 3 – Article 310
Part – 310.4

Type of Change		Panel Action		UL	UL 508	API 500	API 505	OSHA
Revision		Accept in Part		62	18.1	-	-	1910.305(f)
ROP		ROC		NFPA 70E	NFPA 70B	NFPA 79	NFPA	NEMA
pg. 699	# 6-7	pg. -	# -	410.9(A)(1)	3.3.18	13.5	-	WC 51
log: 731	CMP: 6	log: -	Submitter: Dan Leaf			2002 NEC: 310.4		IEC: 523.6

2002 NEC – 310.4 Conductors in Parallel

The paralleled conductors in each phase, neutral, or grounded circuit conductor shall

    (1) Be of the same length

    (2) Have the same conductor material

    (3) Be the same size in circular mil area

    (4) Have the same insulation type

    (5) Be terminated in the same manner

2005 NEC – 310.4 Conductors in Parallel

The paralleled conductors in each phase, <u>polarity,</u> neutral, or grounded circuit conductor shall <u>comply with all of the following:</u>

    (1) Be of the same length

    (2) Have the same conductor material

    (3) Be the same size in circular mil area

    (4) Have the same insulation type

    (5) Be terminated in the same manner

Author's Substantiation. This revision clarifies, by adding the word "polarity," that dc circuits are included for conductors in parallel.

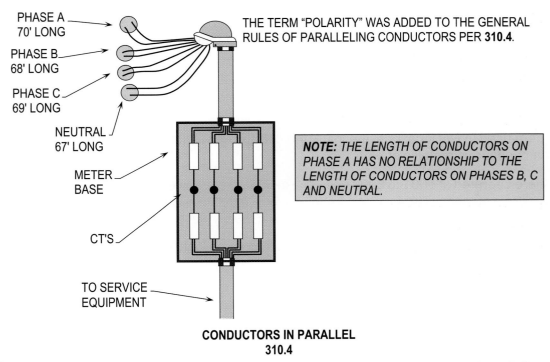

PHASE A
70' LONG

PHASE B
68' LONG

PHASE C
69' LONG

NEUTRAL
67' LONG

METER
BASE

CT'S

TO SERVICE
EQUIPMENT

THE TERM "POLARITY" WAS ADDED TO THE GENERAL
RULES OF PARALLELING CONDUCTORS PER **310.4**.

*NOTE: THE LENGTH OF CONDUCTORS ON
PHASE A HAS NO RELATIONSHIP TO THE
LENGTH OF CONDUCTORS ON PHASES B, C
AND NEUTRAL.*

**CONDUCTORS IN PARALLEL
310.4**

Purpose of Change: To clarify the term "polarity" being added to the general requirement of paralleling conductors.

NEC Ch. 3 – Article 310
Part – 310.6, Ex. (c)

70

Type of Change		Panel Action		UL	UL 508	API 500	API 505	OSHA
Revision		Accept in Part		1072	-	-	-	1910.308(a)(1)(i)
ROP		ROC		NFPA 70E	NFPA 70B	NFPA 79	NFPA	NEMA
pg. 703	# 6-12	pg. 205	# 6-15	410.1(C)(3)(c)	-	-	-	WC 74
log: 2122	CMP: 6	log: 2379	Submitter: David Brender			2002 NEC: 310.6		IEC: Ch. 43

2002 NEC – 310.6 Shielding

Solid dielectric insulated conductors operated above 2000 volts in permanent installations shall have ozone-resistant insulation and shall be shielded. All metallic insulation shields shall be grounded through an effective grounding path meeting the requirements of 250.4(A)(5) or 250.4(B)(4). Shielding shall be for the purpose of confining the voltage stresses to the insulation.

> **Exception:** Nonshielded insulated conductors listed by a qualified testing laboratory shall be permitted for use up to ~~8000~~ volts under the following conditions:
>
> **(a)** Conductors shall have insulation resistant to electric discharge and surface tracking, or the insulated conductor(s) shall be covered with a material resistant to ozone, electric discharge, and surface tracking.
>
> **(b)** Where used in wet locations, the insulated conductor(s) shall have an overall nonmetallic jacket or a continuous metallic sheath.
>
> **(c)** ~~Where operated at 5001 to 8000 volts, the insulated conductor(s) shall have a nonmetallic jacket over the insulation. The insulation shall have a specific inductive capacity not greater than 3.6, and the jacket shall have a specific inductive capacity not greater than 10 and not less than 6.~~
>
> **(d)** Insulation and jacket thickness shall be in accordance with Table 310.63.

2005 NEC – 310.6 Shielding

Solid dielectric insulated conductors operated above 2000 volts in permanent installations shall have ozone-resistant insulation and shall be shielded. All metallic insulation shields shall be grounded through an effective grounding path meeting the requirements of 250.4(A)(5) or 250.4(B)(4). Shielding shall be for the purpose of confining the voltage stresses to the insulation.

> **Exception:** Nonshielded insulated conductors listed by a qualified testing laboratory shall be permitted for use up to <u>2400</u> volts under the following conditions:
>
> **(a)** Conductors shall have insulation resistant to electric discharge and surface tracking, or the insulated conductor(s) shall be covered with a material resistant to ozone, electric discharge, and surface tracking.
>
> **(b)** Where used in wet locations, the insulated conductor(s) shall have an overall nonmetallic jacket or a continuous metallic sheath.
>
> **(c)** Insulation and jacket thickness shall be in accordance with Table 310.63.

Author's Substantiation. This exception has been revised to clarify that nonshielded insulated conductors listed by a qualified testing laboratory shall be permitted for use up to 2400 volts if certain conditions are followed.

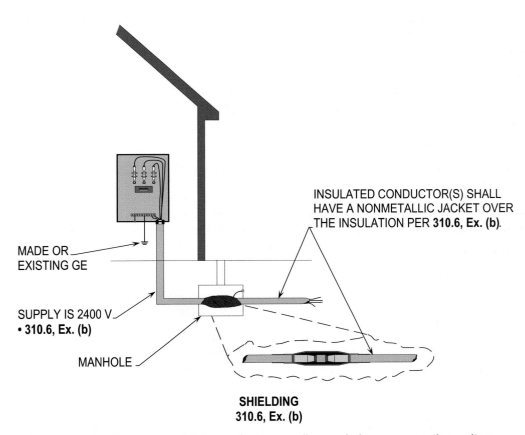

INSULATED CONDUCTOR(S) SHALL
HAVE A NONMETALLIC JACKET OVER
THE INSULATION PER **310.6, Ex. (b)**.

MADE OR
EXISTING GE

SUPPLY IS 2400 V
• **310.6, Ex. (b)**

MANHOLE

SHIELDING
310.6, Ex. (b)

Purpose of Change: To delete confusing wording and change operating voltage.

NEC Ch. 3 – Article 310
Part – 310.8(D)

71

Type of Change		Panel Action		UL	UL 508	API 500	API 505	OSHA
Revision		Accept in Principle in Part		83	21	-	-	1910.305(f)
ROP		ROC		NFPA 70E	NFPA 70B	NFPA 79	NFPA	NEMA
pg. 704	# 6-13	pg. 207	# 6-23	-	-	-	-	C 119.4
log: 2490	CMP: 6	log: 1137	Submitter: Paul Dobrowsky			2002 NEC: 310.8(D)		IEC: 521

2002 NEC – 310.8(D) Locations Exposed to Direct Sunlight

Insulated conductors and cables used where exposed to direct rays of the sun shall ~~be of a type listed for sunlight resistance or listed and marked sunlight resistant.~~

2005 NEC – 310.8(D) Locations Exposed to Direct Sunlight

Insulated conductors or cables used where exposed to direct rays of the sun shall <u>comply with one of the following:</u>

- **(1)** <u>Cables</u> listed, or listed and marked, as being sunlight resistant
- **(2)** <u>Conductors</u> listed, or listed and marked, as being sunlight resistant
- **(3)** <u>Covered with insulating material, such as tape or sleeving, that is listed, or listed and marked, as being sunlight resistant</u>

Author's Substantiation. This subsection has been placed in item form to improve readability for insulated conductors and cables exposed to direct sunlight.

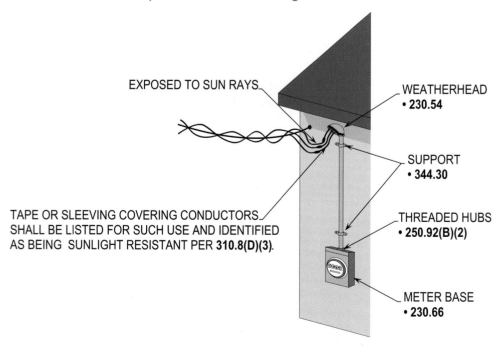

EXPOSED TO SUN RAYS

WEATHERHEAD
• **230.54**

SUPPORT
• **344.30**

THREADED HUBS
• **250.92(B)(2)**

TAPE OR SLEEVING COVERING CONDUCTORS SHALL BE LISTED FOR SUCH USE AND IDENTIFIED AS BEING SUNLIGHT RESISTANT PER **310.8(D)(3)**.

METER BASE
• **230.66**

LOCATIONS EXPOSED TO DIRECT SUNLIGHT
310.8(D)(3)

Purpose of Change: To clarify requirements for insulated conductors and cables exposed to direct rays of the sun.

NEC Ch. 3 – Article 310
Part – 310.12(C)

Type of Change		Panel Action		UL	UL 508	API 500	API 505	OSHA
Revision		Accept in Principle		83	-	-	-	1910.305(f)
ROP		ROC		NFPA 70E	NFPA 70B	NFPA 79	NFPA	NEMA
pg. 708	# 6-20	pg. -	# -	410.1	-	14.2	-	C 119.4
log: 2791	CMP: 6	log: -	Submitter: Michael I. Callanan			2002 NEC: 310.12(C)		IEC: -

2002 NEC – 310.12(C) Ungrounded Conductors

Conductors that are intended for use as ungrounded conductors, whether used as a single conductor or in multiconductor cables, shall be finished to be clearly distinguishable from grounded and grounding conductors. Distinguishing markings shall not conflict in any manner with the surface markings required by 310.11(B)(1).

Exception: Conductor identification shall be permitted in accordance with 200.7.

2005 NEC – 310.12(C) Ungrounded Conductors

Conductors that are intended for use as ungrounded conductors, whether used as a single conductor or in multiconductor cables, shall be finished to be clearly distinguishable from grounded and grounding conductors. Distinguishing markings shall not conflict in any manner with the surface markings required by 310.11(B)(1). Branch-circuit ungrounded conductors shall be identified in accordance with 210.5(C). Feeders shall be identified in accordance with 215.12.

Exception: Conductor identification shall be permitted in accordance with 200.7.

Author's Substantiation. A new sentence has been added to clarify that branch-circuit conductors and feeders shall be identified.

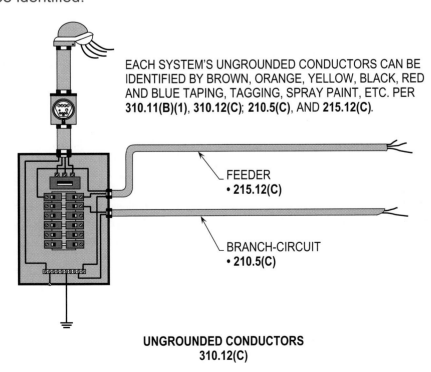

EACH SYSTEM'S UNGROUNDED CONDUCTORS CAN BE IDENTIFIED BY BROWN, ORANGE, YELLOW, BLACK, RED AND BLUE TAPING, TAGGING, SPRAY PAINT, ETC. PER **310.11(B)(1)**, **310.12(C)**; **210.5(C)**, AND **215.12(C)**.

FEEDER
• 215.12(C)

BRANCH-CIRCUIT
• 210.5(C)

UNGROUNDED CONDUCTORS
310.12(C)

Purpose of Change: To reference sections of the code that pertain to identifying (color coding) of conductors in branch and feeder circuits.

NEC Ch. 3 – Article 310
Part – 310.15(A)(1)

72

Type of Change		Panel Action		UL	UL 508	API 500	API 505	OSHA
Revision		Accept		-	-	-	-	-
ROP		ROC		NFPA 70E	NFPA 70B	NFPA 79	NFPA	NEMA
pg. 711	# 6-29	pg. -	# -	-	-	-	-	-
log: 2656	CMP: 6	log: -	Submitter: Phil Simmons			2002 NEC: 310.15(A)(1)		IEC: 523

2002 NEC – 310.15(A)(1) – Tables or Engineering Supervision

Ampacities for conductors shall be permitted to be determined by tables or under engineering supervision, as provided in 310.15(B) and (C).

2005 NEC – 310.15(A)(1) – Tables or Engineering Supervision

Ampacities for conductors shall be permitted to be determined by tables <u>as provided in 310.15(B)</u> or under engineering supervision, as provided in 310.15(C).

Author's Substantiation. This revision clarifies that Tables 310.16 through 310.19 can be used by designers, installers, and inspectors without the need of employing an electrical engineer.

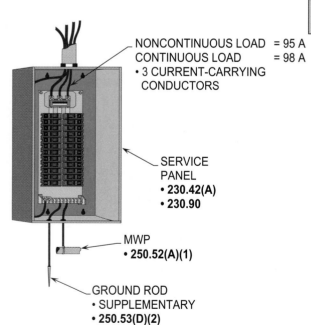

NONCONTINUOUS LOAD = 95 A
CONTINUOUS LOAD = 98 A
• 3 CURRENT-CARRYING CONDUCTORS

SERVICE PANEL
• 230.42(A)
• 230.90

MWP
• 250.52(A)(1)

GROUND ROD
• SUPPLEMENTARY
• 250.53(D)(2)

NOTE: *AMPACITIES OF CONDUCTORS CAN BE DETERMINED BY USING ENGINEERING SUPERVISION PER 310.15(C).*

What size OCPD and THWN cu. conductors are required to supply loads per 310.15(A)(1) and 310.15(B)?

Step 1: Calculating load
230.90; 230.42(A)(1)
95 A x 100% = 95 A
98 A x 125% = 122.5 A
Total load = 217.5 = 218 A

Step 2: Finding OCPD (based on load)
230.90; 240.4; 240.6(A)
218 A requires 225 A OCPD

Step 3: Finding conductors (based on load and OCPD)
310.15(A)(1); Table 310.16
225 A requires 4/0 AWG THWN

Solution: 4/0 AWG THWN cu. conductors and a 225 amp OCPD is required.

TABLES OR ENGINEERING SUPERVISION
310.15(A)(1)

Purpose of Change: To verify that **Tables 310.16** through **310.19** can be used by designers, installers, and inspectors without the need to employ an electrical engineer.

NEC Ch. 3 – Article 310
Part – 310.15(B)(2)(a)

73

Type of Change		Panel Action		UL	UL 508	API 500	API 505	OSHA
New Sentence		Accept		-	-	-	-	-
ROP		ROC		NFPA 70E	NFPA 70B	NFPA 79	NFPA	NEMA
pg. 713	# 6-33	pg. -	# -	-	-	-	-	-
log: 3097	CMP: 6	log: -	Submitter: James W. Carpenter			2002 NEC: 310.15(B)(2)(a)		IEC: Table 52-E1

2002 NEC – 310.15(B)(2) Adjustment Factors

(a) More Than Three Current-Carrying Conductors in a Raceway or Cable. Where the number of current-carrying conductors in a raceway or cable exceeds three, or where single conductors or multiconductor cables are stacked or bundled longer than 600 mm (24 in.) without maintaining spacing and are not installed in raceways, the allowable ampacity of each conductor shall be reduced as shown in Table 310.15(B)(2)(a).

2005 NEC – 310.15(B)(2) Adjustment Factors

(a) More Than Three Current-Carrying Conductors in a Raceway or Cable. Where the number of current-carrying conductors in a raceway or cable exceeds three, or where single conductors or multiconductor cables are stacked or bundled longer than 600 mm (24 in.) without maintaining spacing and are not installed in raceways, the allowable ampacity of each conductor shall be reduced as shown in Table 310.15(B)(2)(a). <u>Each current-carrying conductor of a paralleled set of conductors shall be counted as a current-carrying conductor.</u>

Author's Substantiation. A new sentence has been added to clarify that derating factors shall be applied for three or more current-carrying conductors that are paralleled.

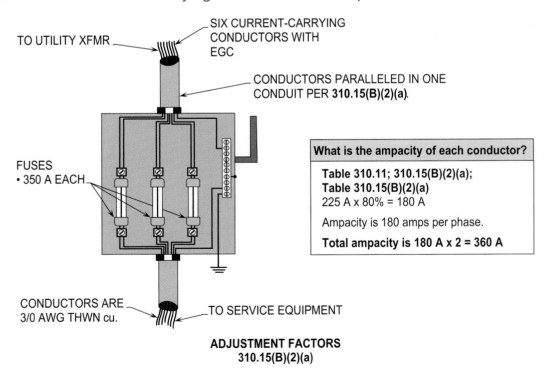

ADJUSTMENT FACTORS
310.15(B)(2)(a)

Purpose of Change: To clarify each current-carrying conductor for paralleled conductors shall be counted as a current-carrying conductor.

NEC Ch. 3 – Article 310
Part – 310.15(B)(2)(a), FPN 2

Type of Change		Panel Action		UL	UL 508	API 500	API 505	OSHA
New FPN		Accept		870	-	-	-	1910.305(f)
ROP		ROC		NFPA 70E	NFPA 70B	NFPA 79	NFPA	NEMA
pg. 712	# 6-30	pg. -	# -	410.9(A)(1)	-	13.5	-	-
log: 3350	CMP: 6	log: -	Submitter: Phil Simmons			2002 NEC: -		IEC: Table 52-E1

2005 NEC – 310.15(B)(2) Adjustment Factors

(a) More Than Three Current-Carrying Conductors in a Raceway or Cable. Where the number of current-carrying conductors in a raceway or cable exceeds three, or where single conductors or multiconductor cables are stacked or bundled longer than 600 mm (24 in.) without maintaining spacing and are not installed in raceways, the allowable ampacity of each conductor shall be reduced as shown in Table 310.15(B)(2)(a). Each current-carrying conductor of a paralleled set of conductors shall be counted as a current-carrying conductor.

> **FPN No. 2:** See 366.23(A) for correction factors for conductors in sheet metal auxiliary gutters and 376.22 for correction factors for conductors in metal wireways.

Author's Substantiation. A new FPN has been added to improve usability of the NEC when applying derating factors for conductors in sheet metal auxiliary gutter and conductors in metal wireways.

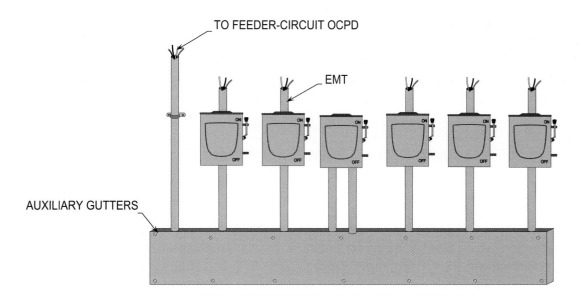

TO FEEDER-CIRCUIT OCPD

EMT

AUXILIARY GUTTERS

NOTE: *NO DERATING REQUIRED FOR 30 OR LESS CURRENT-CARRYING CONDUCTORS PER **310.15(B)(2)(a)**, **FPN 2** and **376.22**.*

ADJUSTMENT FACTORS
310.15(B)(2)(a), FPN 2

Purpose of Change: To add a new **FPN 2** that refers the user to **366.23(A)** for derating factors in sheet metal auxiliary gutters and **376.22** for derating in metal wireways.

NEC Ch. 3 – Article 310
Part – 310.15(B)(6)

Type of Change	Panel Action	UL	UL 508	API 500	API 505	OSHA
Revision	Accept in Principle in Part	870	-	-	-	1910.305(f)

ROP		ROC		NFPA 70E	NFPA 70B	NFPA 79	NFPA	NEMA
pg. 717	# 6-41	pg. -	# -	410.9(A)(1)	-	-	-	-
log: 2491	CMP: 6	log: -	Submitter: Paul Dobrowsky			2002 NEC: 310.15(B)(6)		IEC: 523

2002 NEC – 310.15(B)(6) 120/240-Volt, 3-Wire, Single-Phase Dwelling Services and Feeders

For dwelling units, conductors, as listed in Table 310.15(B)(6), shall be permitted as 120/240 volt, 3-wire, single-phase service-entrance conductors, service lateral conductors, and feeder conductors that serve as the main power feeder to a dwelling unit and are installed in raceway or cable with or without an equipment grounding conductor. For application of this section, the main power feeder shall be the feeder(s) between the main disconnect and the lighting and appliance branch-circuit panelboard(s). The feeder conductors to a dwelling unit shall not be required to be larger than their service-entrance conductors. The grounded conductor shall be permitted to be smaller than the ungrounded conductors, provided the requirements of 215.2, 220.22, and 230.42 are met.

2005 NEC – 310.15(B)(6) 120/240-Volt, 3-Wire, Single-Phase Dwelling Services and Feeders

For individual dwelling units of one family, two-family, and multifamily dwellings, conductors, as listed in Table 310.15(B)(6), shall be permitted as 120/240 volt, 3-wire, single-phase service-entrance conductors, service lateral conductors, and feeder conductors that serve as the main power feeder to each dwelling unit and are installed in raceway or cable with or without an equipment grounding conductor. For application of this section, the main power feeder shall be the feeder(s) between the main disconnect and the lighting and appliance branch-circuit panelboard(s). The feeder conductors to a dwelling unit shall not be required to have an allowable ampacity rating greater than their service-entrance conductors. The grounded conductor shall be permitted to be smaller than the ungrounded conductors, provided the requirements of 215.2, 220.61, and 230.42 are met.

Author's Substantiation. This revision improves clarity and usability of the NEC for 120/240 volt, 3-wire, single-phase dwelling services and feeders.

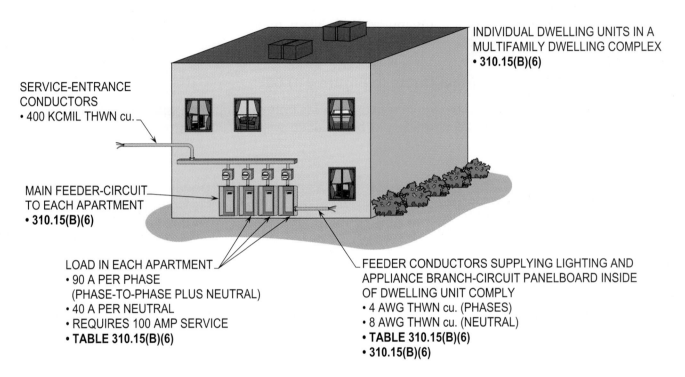

SERVICE-ENTRANCE
CONDUCTORS
• 400 KCMIL THWN cu.

MAIN FEEDER-CIRCUIT
TO EACH APARTMENT
• 310.15(B)(6)

INDIVIDUAL DWELLING UNITS IN A
MULTIFAMILY DWELLING COMPLEX
• 310.15(B)(6)

LOAD IN EACH APARTMENT
• 90 A PER PHASE
  (PHASE-TO-PHASE PLUS NEUTRAL)
• 40 A PER NEUTRAL
• REQUIRES 100 AMP SERVICE
• TABLE 310.15(B)(6)

FEEDER CONDUCTORS SUPPLYING LIGHTING AND
APPLIANCE BRANCH-CIRCUIT PANELBOARD INSIDE
OF DWELLING UNIT COMPLY
• 4 AWG THWN cu. (PHASES)
• 8 AWG THWN cu. (NEUTRAL)
• TABLE 310.15(B)(6)
• 310.15(B)(6)

120/240 VOLTS, 3-WIRE, SINGLE-PHASE
310.15(B)(6)

Purpose of Change: To revise and make this section user friendly.

NEC Ch. 3 – Article 310
Part – 310.60(C)(3)

Type of Change		Panel Action		UL	UL 508	API 500	API 505	OSHA
Deletion		Accept		1072	-	-	-	1910.308(a)
ROP		ROC		NFPA 70E	NFPA 70B	NFPA 79	NFPA	NEMA
pg. 719	# 6-47	pg. -	# -	410.9(B)(3)	-	-	-	-
log: 2010	CMP: 6	log: -	Submitter: Neil F. LaBrake, Jr.			2002 NEC: Figure 310.60		IEC: 502

2002 NEC – 310.60(C)(3) Electrical Ducts in Figure 310.60

At locations where electrical ducts enter equipment enclosures from underground, spacing between such ducts, as shown in Figure 310.60, shall be permitted to be reduced without requiring the ampacity of conductors therein to be reduced.

2005 NEC – 310.60(C)(3) Electrical Ducts in Figure 310.60

At locations where electrical ducts enter equipment enclosures from underground, spacing between such ducts, as shown in Figure 310.60, shall be permitted to be reduced without requiring the ampacity of conductors therein to be reduced.

Author's Substantiation. Detail 4 in Figure 310.60 has been deleted because the ampacity of such a duct bank is complicated to calculate since the cable in the center duct is bounded by the other cables and heat from the cable interacts with the other cables. The ampacity of this type of duct bank should be calculated only under engineering supervision as detailed in Annex B and 310.60(D).

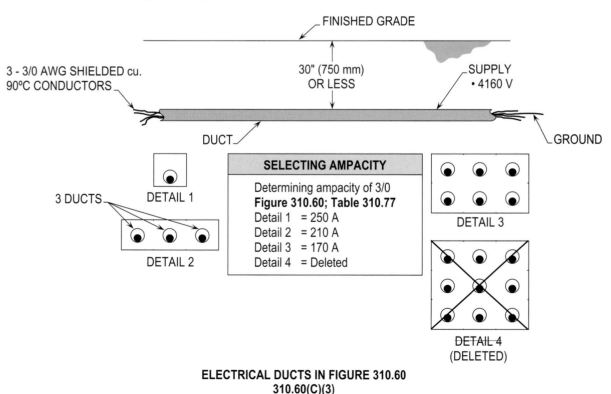

ELECTRICAL DUCTS IN FIGURE 310.60
310.60(C)(3)

Purpose of Change: To delete Detail 4 in **Figure 310.60** because it was seldom used and could only be calculated under engineering supervision.

NEC Ch. 3 – Article 310
Part – Table 310.64

Type of Change		Panel Action		UL	UL 508	API 500	API 505	OSHA
New Note		Accept		-	-	-	-	-
ROP		ROC		NFPA 70E	NFPA 70B	NFPA 79	NFPA	NEMA
pg. 721	# 6-50	pg. -	# -	-	-	-	-	-
log: 505	CMP: 6	log: -	Submitter: James M. Daly			2002 NEC: Table 310.64		IEC: 502

2002 NEC – Table 310.64 Thickness of Insulation for Shielded Solid Dielectric Insulated Conductors Rated 2001 to 35,000 Volts

Notes:

[3]173 Percent Insulation Level. Cables in this category shall be permitted to be applied under all of the following conditions:

(1) In industrial establishments where the conditions of maintenance and supervision ensure that only qualified persons service the installation

(2) Where the fault clearing time requirements of the 133 percent level category cannot be met

(3) Where an orderly shutdown is essential to protect equipment and personnel

(4) There is adequate assurance that the faulted section will be de-energized in an orderly shutdown

Also, cables with this insulation thickness shall be permitted to be used in 100 or 133 percent insulation level applications where additional insulation strength is desirable.

Author's Substantiation. A new note has been added below the table to permit the thickness of insulation for shielded solid dielectric insulated conductors rated 2001 to 35,000 volts to have an 173 percent insulation level. The last sentence of the note makes it clear that 173 percent insulation cables can be used instead of 100 or 133 percent insulation cables but the system shall still remain a 100 or 133 percent category and faults cleared as required by those two categories.

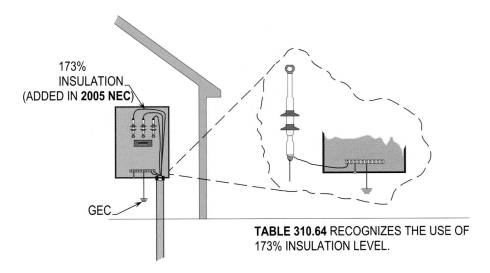

THICKNESS OF INSULATION FOR SHIELDED SOLID DIELECTRIC INSULATED CONDUCTORS RATED 2001 TO 35,000 VOLTS
TABLE 310.64

Purpose of Change: To add the use of 173 percent insulation level to **Table 310.64**.

74

Type of Change		Panel Action		UL	UL 508	API 500	API 505	OSHA
New Sentence		Accept in Principle in Part		50	7.2	-	-	1910.305(e)
ROP		ROC		NFPA 70E	NFPA 70B	NFPA 79	NFPA	NEMA
pg. 728	# 9-5	pg. 214	# 9-5	420.5	8.2.6	3.3.65.2	-	250
log: 2534	CMP: 9	log: 391	Submitter: Alan Manche			2002 NEC: 312.2(A)		IEC: 522.3

2002 NEC – 312.2(A) Damp and Wet Locations

In damp or wet locations, surface-type enclosures within the scope of this article shall be placed or equipped so as to prevent moisture or water from entering and accumulating within the cabinet or cutout box, and shall be mounted so there is at least 6 mm (1/4 in.) airspace between the enclosure and the wall or other supporting surface. Enclosures installed in wet locations shall be weatherproof.

2005 NEC – 312.2(A) Damp and Wet Locations

In damp or wet locations, surface-type enclosures within the scope of this article shall be placed or equipped so as to prevent moisture or water from entering and accumulating within the cabinet or cutout box, and shall be mounted so there is at least 6-mm (1/4-in.) airspace between the enclosure and the wall or other supporting surface. Enclosures installed in wet locations shall be weatherproof. For enclosures in wet locations, raceways or cables entering above the level of uninsulated live parts shall use fittings listed for wet locations.

Author's Substantiation. A new last sentence has been added to clarify that fittings listed for wet locations shall be used for enclosures in wet locations that have raceways or cables entering above the level of uninsulated live parts.

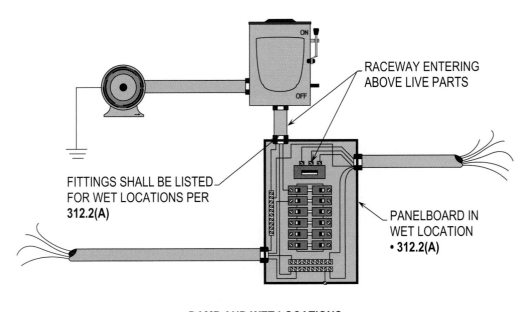

RACEWAY ENTERING ABOVE LIVE PARTS

FITTINGS SHALL BE LISTED FOR WET LOCATIONS PER 312.2(A)

PANELBOARD IN WET LOCATION
• 312.2(A)

DAMP AND WET LOCATIONS
312.2(A)

Purpose of Change: To require fittings capable of securing raceways entering enclosures above live parts.

NEC Ch. 3 – Article 312
Part I – 312.4

Type of Change		Panel Action		UL	UL 508	API 500	API 505	OSHA
Revision		Accept in Principle		-	-	-	-	1910.305(b)
ROP		ROC		NFPA 70E	NFPA 70B	NFPA 79	NFPA	NEMA
pg. 748	# 9-47	pg. 225	# 9-85	420.1(B)(2)	-	-	-	-
log: 970	CMP: 9	log: 2025	Submitter: Frederic P. Hartwell			2002 NEC: 314.21		IEC: 522.8.1

2002 NEC – ~~314.21~~ Repairing Plaster and Drywall or Plasterboard

Plaster, drywall, or plasterboard surfaces that are broken or incomplete shall be repaired so there will be no gaps or open spaces greater than 3 mm (1/8 in.) at the edge of the ~~box or fitting~~.

2005 NEC – <u>312.4</u> Repairing Plaster and Drywall or Plasterboard

Plaster, drywall, or plasterboard surfaces that are broken or incomplete shall be repaired so there will be no gaps or open spaces greater than 3 mm (1/8 in.) at the edge of the <u>cabinet or cutout box employing a flush-type cover</u>.

Author's Substantiation. This section has been moved and renumbered and revised to clarify that no gaps or open spaces greater than 3 mm (1/8 in.) at the edge of the cabinet or cutout box employing a flush-type cover are present after repairing plaster and drywall or plasterboard.

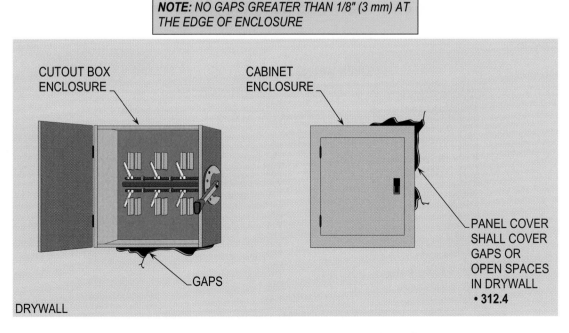

REPAIRING PLASTER AND DRYWALL OR PLASTERBOARD
312.4

Purpose of Change: To ensure that gaps or open spaces are properly covered.

NEC Ch. 3 – Article 312
Part I – 312.5(B)

Type of Change		Panel Action		UL	UL 508	API 500	API 505	OSHA
Revision		Accept		50	-	-	-	1910.305(a)(4)
ROP		ROC		NFPA 70E	NFPA 70B	NFPA 79	NFPA	NEMA
pg. 730	# 9-10	pg. -	# -	420.2(A)(2)	-	-	-	-
log: 2909	CMP: 9	log: -	Submitter: Dorothy Kellogg			2002 NEC: 312.5(B)		IEC: 522.8.1

2002 NEC – 312.5(B) Metal Cabinets, Cutout Boxes, and Meter Socket Enclosures

Where metal enclosures within the scope of this article are installed with open wiring or concealed knob-and-tube wiring, conductors shall enter through insulated bushings or, in dry locations, through flexible tubing extending from the last insulating support and firmly secured to the enclosure.

2005 NEC – 312.5(B) Metal Cabinets, Cutout Boxes, and Meter Socket Enclosures

Where metal enclosures within the scope of this article are installed with <u>messenger supported wiring,</u> open wiring <u>on insulators,</u> or concealed knob-and-tube wiring, conductors shall enter through insulated bushings or, in dry locations, through flexible tubing extending from the last insulating support and firmly secured to the enclosure.

Author's Substantiation. This revision clarifies that where metal enclosures are installed with messenger supported wiring, open wiring on insulators, or concealed knob-and-tube wiring, conductors shall enter through insulated bushings. In dry locations, conductors shall enter through flexible tubing extending from the last insulating support and firmly secured to the enclosure. The words open wiring were replaced with open wiring on insulators to clarify the proper wiring method to be used.

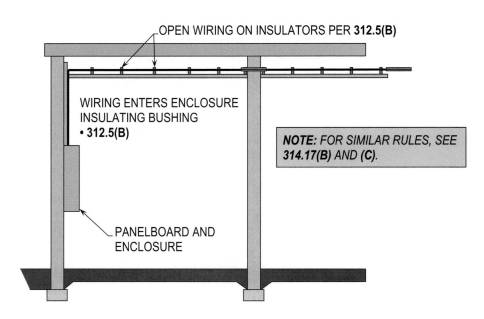

OPEN WIRING ON INSULATORS PER **312.5(B)**

WIRING ENTERS ENCLOSURE
INSULATING BUSHING
• **312.5(B)**

NOTE: FOR SIMILAR RULES, SEE 314.17(B) AND (C).

PANELBOARD AND ENCLOSURE

METAL CABINETS, CUTOUT BOXES, AND METER SOCKET ENCLOSURES
312.5(B)

Purpose of Change: To define the use of open wiring in this section.

NEC Ch. 3 – Article 314
Part II – 314.16(B)(1)

Type of Change		Panel Action		UL	UL 508	API 500	API 505	OSHA
New Sentence		Accept in Principle		514C	-	-	-	1910.305(b)
ROP		ROC		NFPA 70E	NFPA 70B	NFPA 79	NFPA	NEMA
pg. 738	# 9-27	pg. 221	# 9-65	420.2(A)(3)	-	-	-	-
log: 3315	CMP: 9	log: 2022	Submitter: Sukanta Sengupja			2002 NEC: 314.16(B)(1)		IEC: 522.8.1

2002 NEC – 314.16(B)(1) Conductor Fill

Each conductor that originates outside the box and terminates or is spliced within the box shall be counted once, and each conductor that passes through the box without splice or termination shall be counted once. The conductor fill shall be ~~computed~~ using Table 314.16(B). A conductor, no part of which leaves the box, shall not be counted.

2005 NEC – 314.16(B)(1) Conductor Fill

Each conductor that originates outside the box and terminates or is spliced within the box shall be counted once, and each conductor that passes through the box without splice or termination shall be counted once. <u>A looped, unbroken conductor not less than twice the minimum length required for free conductors in 300.14 shall be counted twice.</u> The conductor fill shall be <u>calculated</u> using Table 314.16(B). A conductor, no part of which leaves the box, shall not be counted.

Author's Substantiation. A new sentence has been added to clarify that a looped, unbroken conductor not less than twice the minimum length required for free conductors shall be counted twice. This new sentence allows more free space in the box for the installation of electronic devices to allow for its heat dissipation.

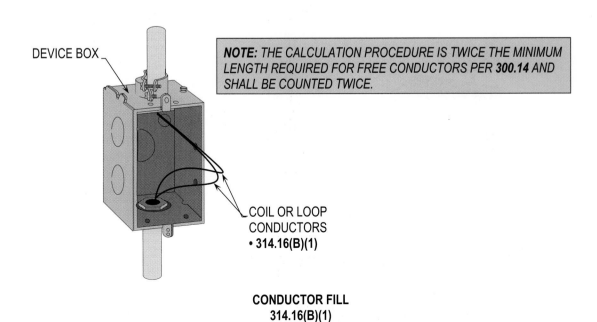

DEVICE BOX

NOTE: THE CALCULATION PROCEDURE IS TWICE THE MINIMUM LENGTH REQUIRED FOR FREE CONDUCTORS PER **300.14** AND SHALL BE COUNTED TWICE.

COIL OR LOOP CONDUCTORS
• 314.16(B)(1)

CONDUCTOR FILL
314.16(B)(1)

Purpose of Change: To address the fill count for coiled or looped conductors passing through a box.

NEC Ch. 3 – Article 314
Part II – 314.20

Type of Change		Panel Action		UL	UL 508	API 500	API 505	OSHA
Revision		Accept in Part		-	-	-	-	-
ROP		ROC		NFPA 70E	NFPA 70B	NFPA 79	NFPA	NEMA
pg. 746	# 9-43	pg. 224	# 9-82	-	-	-	-	-
log: 686	CMP: 9	log: 2905	Submitter: Dan Leaf			2002 NEC: 314.20		IEC: -

2002 NEC – 314.20 In Wall or Ceiling

In walls or ceilings with a surface of concrete, tile, gypsum, plaster, or other noncombustible material, boxes shall be installed so that the front edge of the box will not be set back of the finished surface more than 6 mm (1/4 in.).

In walls or ceilings constructed of wood or other combustible surface material, boxes shall be flush with the finished surface or project therefrom.

2005 NEC – 314.20 In Wall or Ceiling

In walls or ceilings with a surface of concrete, tile, gypsum, plaster, or other noncombustible material, boxes employing a flush-type cover or faceplate shall be installed so that the front edge of the box, plaster ring, extension ring, or listed extender will not be set back of the finished surface more than 6 mm (1/4 in.).

In walls or ceilings constructed of wood or other combustible surface material, boxes, plaster rings, extension rings, or listed extenders shall be flush with the finished surface or project therefrom.

Author's Substantiation. This revision clarifies that plaster rings, extension rings, or listed extenders shall not be set back of the finished surface more than 6 mm (1/4 in.) in walls or ceilings with a surface of concrete, tile, gypsum, plaster, or other noncombustible material.

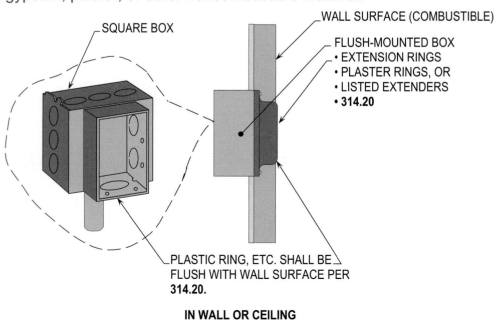

SQUARE BOX

WALL SURFACE (COMBUSTIBLE)

FLUSH-MOUNTED BOX
• EXTENSION RINGS
• PLASTER RINGS, OR
• LISTED EXTENDERS
• 314.20

PLASTIC RING, ETC. SHALL BE FLUSH WITH WALL SURFACE PER 314.20.

**IN WALL OR CEILING
314.20**

Purpose of Change: To distinguish the differences between plaster rings, extension rings, and boxes when installed in walls or ceilings.

NEC Ch. 3 – Article 314
Part II – 314.23(B)(1)

Type of Change		Panel Action		UL	UL 508	API 500	API 505	OSHA
New Sentence		Accept in Principle		-	-	-	-	-
ROP		ROC		NFPA 70E	NFPA 70B	NFPA 79	NFPA	NEMA
pg. 749	# 9-50	pg. -	# -	-	-	-	-	-
log: 596	CMP: 9	log: -	Submitter: Vince Baclawski			2002 NEC: 314.21(B)(1)		IEC: -

2002 NEC – 314.23(B)(1) Nails or Screws

Nails and screws, where used as a fastening means, shall be attached by using brackets on the outside of the enclosure, or they shall pass through the interior within 6 mm (1/4 in.) of the back or ends of the enclosure.

2005 NEC – 314.23(B)(1) Nails or Screws

Nails and screws, where used as a fastening means, shall be attached by using brackets on the outside of the enclosure, or they shall pass through the interior within 6 mm (1/4 in.) of the back or ends of the enclosure. Screws shall not be permitted to pass through the box unless exposed threads in the box are protected using approved means to avoid abrasion of conductor insulation.

Author's Substantiation. A new sentence has been added to clarify that certain types of screws used for mounting boxes, when left exposed inside a box, present a severe abrasion hazard to conductor insulation. This new sentence requires an approved means to be used to protect the conductor insulation where screws pass through the box.

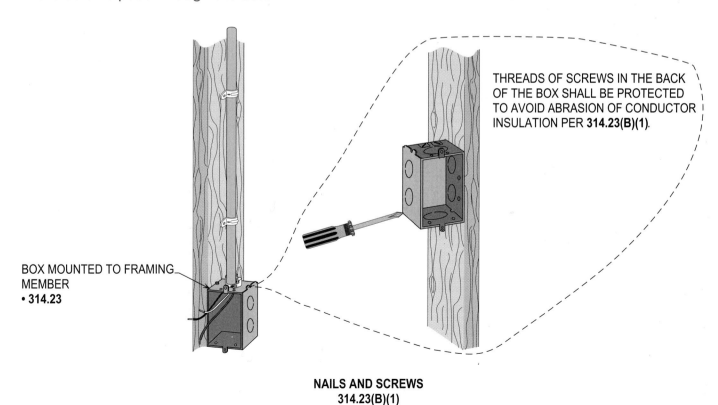

THREADS OF SCREWS IN THE BACK OF THE BOX SHALL BE PROTECTED TO AVOID ABRASION OF CONDUCTOR INSULATION PER **314.23(B)(1)**.

BOX MOUNTED TO FRAMING MEMBER
• **314.23**

NAILS AND SCREWS
314.23(B)(1)

Purpose of Change: To provide an approved means of protection from screw threads in box.

NEC Ch. 3 – Article 314
Part II – 314.23(F)

Type of Change		Panel Action		UL	UL 508	API 500	API 505	OSHA
Revision		Accept		-	-	-	-	-
ROP		ROC		NFPA 70E	NFPA 70B	NFPA 79	NFPA	NEMA
pg. 748	# 9-52	pg. -	# -	-	-	-	-	-
log: 836	CMP: 9	log: -	Submitter: Dan Leaf			2002 NEC: 314.23(F)		IEC: 521

2002 NEC – 314.23(F) Raceway Supported Enclosures, with Devices, Luminaires (Fixtures), or Lampholders

An enclosure that contains a device(s) or supports a luminaire(s) [fixture(s)], lampholder, or other equipment and is supported by entering raceways shall not exceed 1650 cm^3 (100 in.^{3}) in size. It shall have threaded entries or have hubs identified for the purpose. It shall be supported by two or more conduits threaded wrenchtight into the enclosure or hubs. Each conduit shall be secured within 450 mm (18 in.) of the enclosure.

2005 NEC – 314.23(F) Raceway Supported Enclosures, with Devices, Luminaires (Fixtures), or Lampholders

An enclosure that contains a device(s), other than splicing devices, or supports a luminaire(s) [fixture(s)], lampholder, or other equipment and is supported by entering raceways shall not exceed 1650 cm^3 (100 in.^{3}) in size. It shall have threaded entries or have hubs identified for the purpose. It shall be supported by two or more conduits threaded wrenchtight into the enclosure or hubs. Each conduit shall be secured within 450 mm (18 in.) of the enclosure.

Author's Substantiation. This revision clarifies any confusion concerning an enclosure that contains a device(s), other than splicing devices.

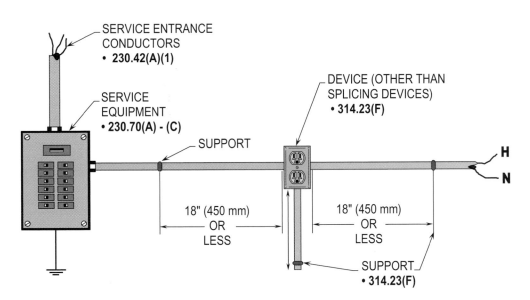

**RACEWAY SUPPORTED ENCLOSURES WITH DEVICES,
LUMINAIRES (FIXTURES), OR LAMPHOLDERS
314.23(F)**

Purpose of Change: The phrase "other than splicing devices" was added to the text because of wording that was confusing to the user.

NEC Ch. 3 – Article 314
Part II – 314.27(D)

Type of Change		Panel Action		UL	UL 508	API 500	API 505	OSHA
Revision		Accept in Principle		-	-	-	-	-
ROP		ROC		NFPA 70E	NFPA 70B	NFPA 79	NFPA	NEMA
pg. 751	# 9-57	pg. -	# -	-	-	-	-	-
log: 569	CMP: 9	log: -	Submitter: Vince Baclawski			2002 NEC: 314.27(D)		IEC: -

2002 NEC – 314.27(D) Boxes at Ceiling-Suspended (Paddle) Fan Outlets

~~Where a box is~~ used as the sole support of a ceiling-suspended (paddle) fan, ~~the box shall be listed for the application and for the weight of the fan to be supported. The installation shall comply with 422.18.~~

2005 NEC – 314.27(D) Boxes at Ceiling-Suspended (Paddle) Fan Outlets

<u>Outlet boxes or outlet box systems</u> used as the sole support of a ceiling-suspended (paddle) fan <u>shall be listed, shall be marked by their manufacturer as suitable for the purpose, and shall not support ceiling-suspended (paddle) fans that weigh more than 32 kg (70 lb). For outlet boxes or outlet box systems designed to support ceiling-suspended (paddle) fans that weigh more than 16 kg (35 lb), the required marking shall include the maximum weight to be supported.</u>

Author's Substantiation. This revision clarifies that the support requirements for boxes at ceiling-suspended (paddle) fan outlets are covered in 314.27(D) and the installation requirements are covered in 422.18.

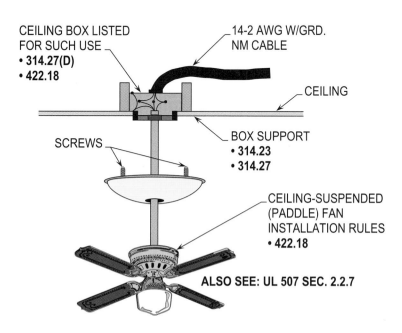

CEILING BOX LISTED
FOR SUCH USE
• 314.27(D)
• 422.18

14-2 AWG W/GRD.
NM CABLE

CEILING

BOX SUPPORT
• 314.23
• 314.27

SCREWS

CEILING-SUSPENDED
(PADDLE) FAN
INSTALLATION RULES
• 422.18

ALSO SEE: UL 507 SEC. 2.2.7

BOXES AT CEILING-SUSPENDED (PADDLE) FAN OUTLETS
314.27(D)

Purpose of Change: To include support rules in text and let installation rules remain in **422.18**.

Type of Change		Panel Action		UL	UL 508	API 500	API 505	OSHA
Revision		Accept		-	-	-	-	-
ROP		ROC		NFPA 70E	NFPA 70B	NFPA 79	NFPA	NEMA
pg. 756	# 9-68	pg. 226	# 9-92	-	-	-	-	-
log: 1406	CMP: 9	log: 466	Submitter: Ronald E. Maassen			2002 NEC: 314.29 and Ex.		IEC: 481.2

2002 NEC – 314.29 Boxes and Conduit Bodies to Be Accessible

Boxes and conduit bodies shall be installed so that the wiring contained in them can be rendered accessible without removing any part of the building or, in underground circuits, without excavating sidewalks, paving, earth, or other substance that is to be used to establish the finished grade.

> **Exception:** Listed boxes shall be permitted where covered by gravel, light aggregate, or noncohesive granulated soil if their location is effectively identified and accessible for excavation.

2005 NEC – 314.29 Boxes, Conduit Bodies, and Handhole Enclosures to Be Accessible

Boxes, conduit bodies, and handhole enclosures shall be installed so that the wiring contained in them can be rendered accessible without removing any part of the building or, in underground circuits, without excavating sidewalks, paving, earth, or other substance that is to be used to establish the finished grade.

> **Exception:** Listed boxes and handhole enclosures shall be permitted where covered by gravel, light aggregate, or noncohesive granulated soil if their location is effectively identified and accessible for excavation.

Author's Substantiation. This revision clarifies that handhole enclosures shall be accessible without removing any part of the building, or in underground circuits.

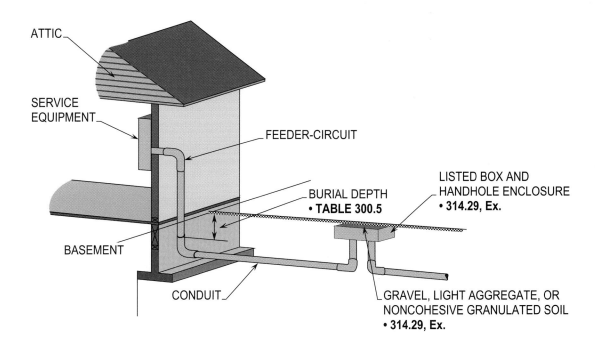

ATTIC

SERVICE
EQUIPMENT

FEEDER-CIRCUIT

LISTED BOX AND
HANDHOLE ENCLOSURE
• 314.29, Ex.

BURIAL DEPTH
• TABLE 300.5

BASEMENT

CONDUIT

GRAVEL, LIGHT AGGREGATE, OR
NONCOHESIVE GRANULATED SOIL
• 314.29, Ex.

**BOXES, CONDUIT BODIES, AND HANDHOLE
ENCLOSURES TO BE ACCESSIBLE
314.29, Ex.**

Purpose of Change: To add handhole enclosures to the main text and exception for items that shall be accessible.

NEC Ch. 3 – Article 314
Part II – 314.30(A) through (D)

79

Type of Change		Panel Action		UL	UL 508	API 500	API 505	OSHA
New Section		Accept in Principle		-	-	-	-	-
ROP		ROC		NFPA 70E	NFPA 70B	NFPA 79	NFPA	NEMA
pg. 757	# 9-6800	pg. 228	# 9-111	-	B.1.17	3.3.1	-	-
log: CP904	CMP: 9	log: 2026	Submitter: CMP 9			2002 NEC: -		IEC: -

2005 NEC – 314.30 Handhole Enclosures

Handhole enclosures shall be designed and installed to withstand all loads likely to be imposed.

> **FPN:** See ANSI/SCTE 77-2002, *Specification for Underground Enclosure Integrity*, for additional information on deliberate and nondeliberate traffic loading that can be expected to bear on underground enclosures.

(A) Size. Handhole enclosures shall be sized in accordance with 314.28(A) for conductors operating at 600 volts or below, and in accordance with 314.71 for conductors operating at over 600 volts. For handhole enclosures without bottoms where the provisions of 314.28(A)(2), Exception, or 314.71(B)(1), Exception No. 1, apply, the measurement to the removable cover shall be taken from the end of the conduit or cable assembly.

(B) Wiring Entries. Underground raceways and cable assemblies entering a handhole enclosure shall extend into the enclosure, but they shall not be required to be mechanically connected to the enclosure.

(C) Handhole Enclosures Without Bottoms. Where handhole enclosures without bottoms are installed, all enclosed conductors and any splices or terminations, if present, shall be listed as suitable for wet locations.

(D) Covers. Handhole enclosure covers shall have an identifying mark or logo that prominently identifies the function of the enclosure, such as electric. Handhole enclosure covers shall require the use of tools to open, or they shall weight over 45 kg (100 lb). Metal covers and other exposed conductive surfaces shall be bonded in accordance with 250.96(A).

Author's Substantiation. A new section has been added to address the requirements for the size of handhole enclosures, wiring entries for handhole enclosures, handhole enclosures without bottoms, and covers for handhole enclosures.

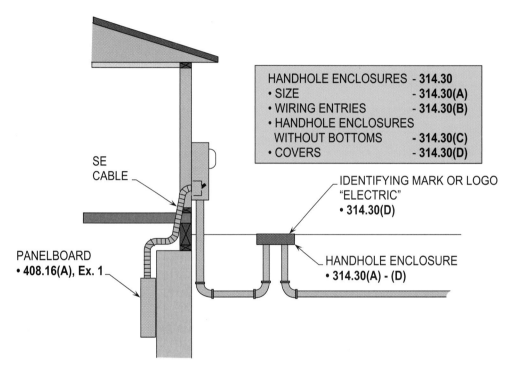

SE CABLE

PANELBOARD
• **408.16(A), Ex. 1**

HANDHOLE ENCLOSURES - **314.30**
• SIZE   - **314.30(A)**
• WIRING ENTRIES   - **314.30(B)**
• HANDHOLE ENCLOSURES
   WITHOUT BOTTOMS   - **314.30(C)**
• COVERS   - **314.30(D)**

IDENTIFYING MARK OR LOGO
"ELECTRIC"
• **314.30(D)**

HANDHOLE ENCLOSURE
• **314.30(A) - (D)**

**HANDHOLE ENCLOSURES
314.30(A) THRU (D)**

Purpose of Change: To add a section pertaining to requirements governing handhole enclosures.

NEC Ch. 3 – Article 320
Part II – 320.12

Type of Change		Panel Action		UL	UL 508	API 500	API 505	OSHA
Revision		Accept		4	-	-	-	1910.305(a)(1)
ROP		ROC		NFPA 70E	NFPA 70B	NFPA 79	NFPA	NEMA
pg. 765	# 7-13	pg. 231	# 7-19a	Article 100	-	13.7	-	RV 1
log: 2684	CMP: 7	log: CC701	Submitter: Phil Simmons			2002 NEC: 320.12		IEC: 521

2002 NEC – 320.12 Uses Not Permitted

Type AC cable shall not be used as follows:

(1) ~~In theaters and similar locations, except where permitted in 518.4~~

(2) ~~In motion picture studios~~

(3) ~~In hazardous (classified) locations except where permitted in~~

~~a. 501.4(B), Exception~~

~~b. 502.4(B), Exception No. 1~~

~~c. 504.20~~

(4) Where exposed to corrosive fumes or vapors

(5) ~~In storage battery rooms~~

(6) ~~In hoistways, or on elevators or escalators, except where permitted in 620.21~~

(7) ~~In commercial garages where prohibited in 511.4 and 511.7~~

2005 NEC – 320.12 Uses Not Permitted

Type AC cable shall not be used as follows:

(1) <u>Where subject to physical damage</u>

(2) <u>In damp or wet locations</u>

(3) <u>In air voids of masonry block or tile walls where such walls are exposed or subject to excessive moisture or dampness</u>

(4) Where exposed to corrosive fumes or vapors

(5) <u>Embedded in plaster finish on brick or other masonry in damp or wet locations</u>

Author's Substantiation. This revision clarifies the uses not permitted for Type AC cable.

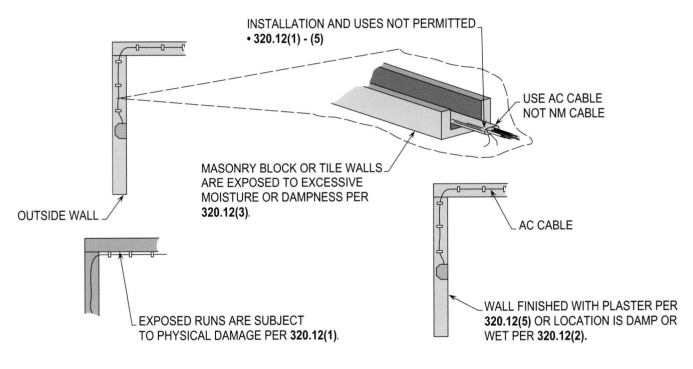

INSTALLATION AND USES NOT PERMITTED
• **320.12(1) - (5)**

USE AC CABLE
NOT NM CABLE

MASONRY BLOCK OR TILE WALLS
ARE EXPOSED TO EXCESSIVE
MOISTURE OR DAMPNESS PER
320.12(3).

OUTSIDE WALL

AC CABLE

EXPOSED RUNS ARE SUBJECT
TO PHYSICAL DAMAGE PER **320.12(1)**.

WALL FINISHED WITH PLASTER PER
320.12(5) OR LOCATION IS DAMP OR
WET PER **320.12(2)**.

USES NOT PERMITTED
320.12

Purpose of Change: To delete "uses permitted" and only reference "uses not permitted."

NEC Ch. 3 – Article 314
Part II – 320.17

Type of Change		Panel Action		UL	UL 508	API 500	API 505	OSHA
Revision		Accept		4	-	-	-	1910.305(a)(4)(iv)
ROP		ROC		NFPA 70E	NFPA 70B	NFPA 79	NFPA	NEMA
pg. 766	# 7-15	pg. -	# -	410.20(D)(4)	-	13.7	-	RV 1
log: 2117	CMP: 7	log: -	Submitter: George W. Flach			2002 NEC: 320.17		IEC: 521.1

2002 NEC – 320.17 Through or Parallel to Framing Members

Type AC cable shall be protected in accordance with 300.4 where installed through or parallel to framing members.

2005 NEC – 320.17 Through or Parallel to Framing Members

Type AC cable shall be protected in accordance with 300.4(A), (C), and (D) where installed through or parallel to framing members.

Author's Substantiation. This revision clarifies that bushings for Type AC cable shall not be required where installed in holes in metal studs.

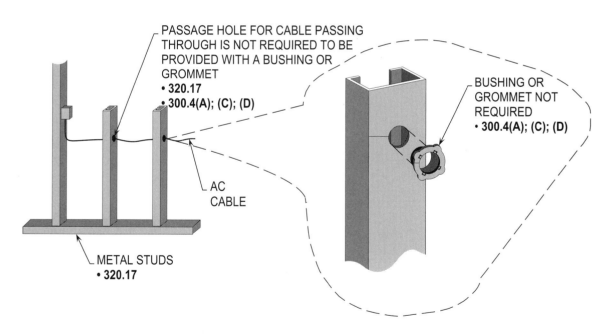

PASSAGE HOLE FOR CABLE PASSING THROUGH IS NOT REQUIRED TO BE PROVIDED WITH A BUSHING OR GROMMET
• 320.17
• 300.4(A); (C); (D)

BUSHING OR GROMMET NOT REQUIRED
• 300.4(A); (C); (D)

AC CABLE

METAL STUDS
• 320.17

THROUGH OR PARALLEL TO FRAMING MEMBERS
320.17

Purpose of Change: To clarify that **300.4(A), (C)**, and **(D)** apply when installing AC cable.

NEC Ch. 3 – Article 320
Part II – 320.30(D)(3)

Type of Change		Panel Action		UL	UL 508	API 500	API 505	OSHA
New Sentence		Accept in Principle		4	-	-	-	1910.305(i)(2)
ROP		ROC		NFPA 70E	NFPA 70B	NFPA 79	NFPA	NEMA
pg. 768	# 7-19	pg. 231	# 7-21	406.9	-	13.7	-	RV 1
log: 1718	CMP: 7	log: 2419	Submitter: Thomas E. Trainor			2002 NEC: 320.30(B)(3)		IEC: 521.1

2002 NEC – ~~320.30(B)(3)~~ Unsupported Cables

Type AC cable shall be permitted to be unsupported where the cable:

(3) Is not more than 1.8 m (6 ft) from the last point of support ~~for~~ connection~~s~~ ~~within an accessible ceiling~~ to luminaire(s) [lighting fixture(s)] or equipment.

2005 NEC – <u>320.30(D)(3)</u> Unsupported Cables

Type AC cable shall be permitted to be unsupported where the cable <u>complies with any of the following</u>:

(3) Is not more than 1.8 m (6 ft) <u>in length</u> from the last point of <u>cable</u> support <u>to the point of</u> connection to <u>a</u> luminaire(s) [lighting fixture(s)] or <u>other electrical</u> equipment <u>and the cable and point of connection are within an accessible ceiling. For the purposes of this section, Type AC cable fittings shall be permitted as a means of cable support</u>.

Author's Substantiation. A new sentence has been added to clarify that Type AC cable fittings shall be permitted as a means of cable support.

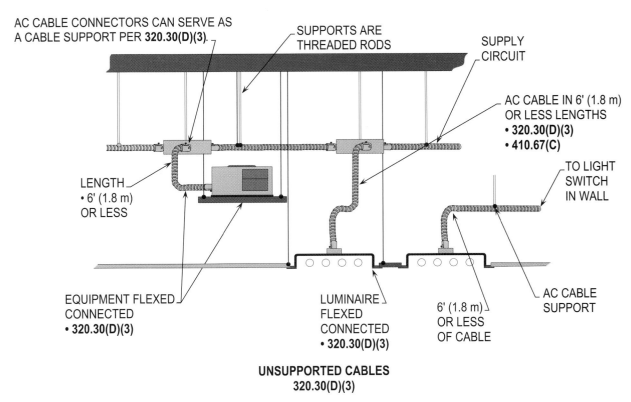

UNSUPPORTED CABLES
320.30(D)(3)

Purpose of Change: To permit AC cable fittings as a means of cable support.

NEC Ch. 3 – Article 320
Part II – 320.80(A)

81

Type of Change		Panel Action		UL	UL 508	API 500	API 505	OSHA
New Sentence		Accept		4	-	-	-	1910.304(f)
ROP		ROC		NFPA 70E	NFPA 70B	NFPA 79	NFPA	NEMA
pg. 770	# 7-22	pg. -	# -	-	-	-	-	RV 1
log: 2686	CMP: 7	log: -	Submitter: Phil Simmons			2002 NEC: 320.80(A)		IEC: 523

2002 NEC – 320.80(A) Thermal Insulation

Armored cable installed in thermal insulation shall have conductors rated at 90°C (194°F). The ampacity of cable installed in these applications shall be that of 60°C (140°F) conductors.

2005 NEC – 320.80(A) Thermal Insulation

Armored cable installed in thermal insulation shall have conductors rated at 90°C (194°F). The ampacity of cable installed in these applications shall be that of 60°C (140°F) conductors. The 90°C (194°F) rating shall be permitted to be used for ampacity derating purposes, provided the final derated ampacity does not exceed that for a 60°C (140°F) rated conductor.

Author's Substantiation. A new sentence has been added to correlate with 334.80 to permit ampacity derating purposes for Type AC cable and Type NM cable.

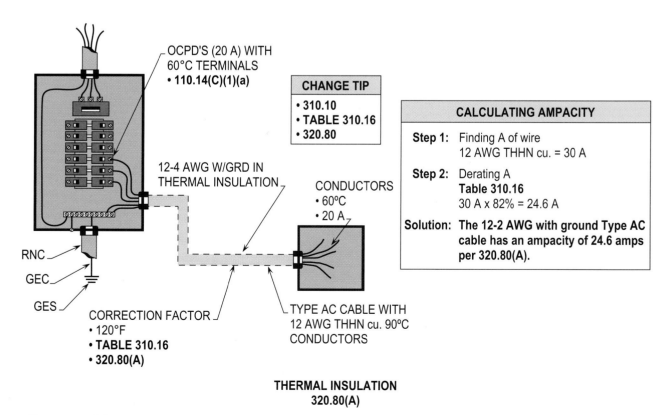

THERMAL INSULATION
320.80(A)

Purpose of Change: To correlate this section with **334.80** so that consistency is maintained for the user.

NEC Ch. 3 – Article 320
Part I – 324.6

Type of Change		Panel Action		UL	UL 508	API 500	API 505	OSHA
New Section		Accept		White Book	-	-	-	-
ROP		ROC		NFPA 70E	NFPA 70B	NFPA 79	NFPA	NEMA
pg. 777	# 7-32	pg. 235	# 7-34	-	-	-	-	-
log: 2946	CMP: 7	log: 2036	Submitter: Martin J. Brett, Jr.			2002 NEC: -		IEC: 514.2

2005 NEC – 324.6 Listing Requirements

Type FCC cable and associated fittings shall be listed.

Author's Substantiation. A new section has been added to clarify the Type FCC cable and associated fittings shall be listed.

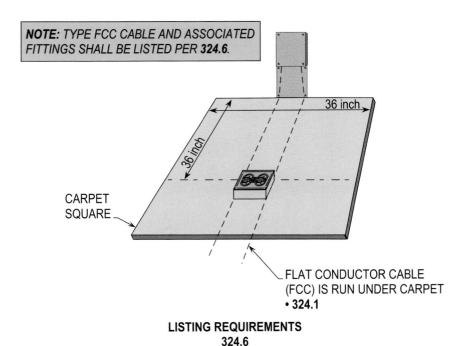

NOTE: TYPE FCC CABLE AND ASSOCIATED FITTINGS SHALL BE LISTED PER **324.6**.

36 inch

36 inch

CARPET SQUARE

FLAT CONDUCTOR CABLE (FCC) IS RUN UNDER CARPET
• **324.1**

LISTING REQUIREMENTS
324.6

Purpose of Change: To clarify that the intent is to require a listed product.

NEC Ch. 3 – Article 330
Part II – 330.30(D)(2)

Type of Change	Panel Action	UL	UL 508	API 500	API 505	OSHA
New Sentence	Accept in Principle in Part	1569	-	-	-	-
ROP		ROC		NFPA 70E	NFPA 70B	NFPA 79	NFPA	NEMA

ROP		ROC		NFPA 70E	NFPA 70B	NFPA 79	NFPA	NEMA
pg. -	# -	pg. -	# -	-	-	-	-	RV 1
log: -	CMP: -	log: -	Submitter: -			2002 NEC: -		IEC: 514.2

2002 NEC – ~~330.30(B)~~ Unsupported Cables

Type MC cable shall be permitted to be unsupported where the cable:

(2) Is not more than 1.8 m (6 ft) from the last point of support ~~for~~ connection~~s~~ ~~within an accessible ceiling~~ to luminaire~~(s)~~ [lighting fixture~~(s)~~] or equipment

2005 NEC – 330.30(D) Unsupported Cables

Type MC cable shall be permitted to be unsupported where the cable:

(2) Is not more than 1.8 m (6 ft) <u>in length</u> from the last point of <u>cable</u> support <u>to the point of</u> connection to <u>a</u> luminaire (lighting fixture<u>)</u> or <u>other piece of electrical</u> equipment <u>and the cable and point of connection are within an accessible ceiling. For the purpose of this section, Type MC cable fittings shall be permitted as a means of cable support.</u>

Author's Substantiation. A new sentence has been added to clarify that Type MC cable fittings shall be permitted as a means of cable support.

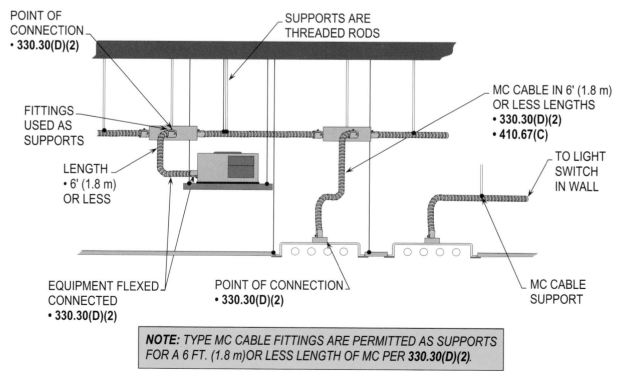

POINT OF CONNECTION
• 330.30(D)(2)

SUPPORTS ARE THREADED RODS

MC CABLE IN 6' (1.8 m) OR LESS LENGTHS
• 330.30(D)(2)
• 410.67(C)

FITTINGS USED AS SUPPORTS

TO LIGHT SWITCH IN WALL

LENGTH
• 6' (1.8 m) OR LESS

EQUIPMENT FLEXED CONNECTED
• 330.30(D)(2)

POINT OF CONNECTION
• 330.30(D)(2)

MC CABLE SUPPORT

NOTE: *TYPE MC CABLE FITTINGS ARE PERMITTED AS SUPPORTS FOR A 6 FT. (1.8 m)OR LESS LENGTH OF MC PER 330.30(D)(2).*

UNSUPPORTED CABLES
330.30(D)(2)

Purpose of Change: To revise the text so that the requirements are user friendly.

NEC Ch. 3 – Article 332
Part II – 332.30(B)

Type of Change		Panel Action		UL	UL 508	API 500	API 505	OSHA
Revision		Accept in Principle in Part		1581	-	-	-	1910.307(b)(3), Note
ROP		ROC		NFPA 70E	NFPA 70B	NFPA 79	NFPA	NEMA
pg. 807	# 7-90	pg. -	# -	440.4(A), Note	-	-	-	WC 51
log: 1715	CMP: 7	log: -	Submitter: Thomas E. Trainor			2002 NEC: 332.30(B)		IEC: 514.2

2002 NEC – 332.30(B) Unsupported Cable

Type MI cable shall be permitted to be unsupported where the cable is fished.

2005 NEC – 332.30(B) Unsupported Cable

Type MI cable shall be permitted to be unsupported where the cable is fished <u>between access points through concealed spaces in finished buildings or structures and supporting is impracticable</u>.

Author's Substantiation. This revision clarifies that Type MI cable shall be permitted to be unsupported where the cable is fished between access points through concealed spaces in finished buildings or structures.

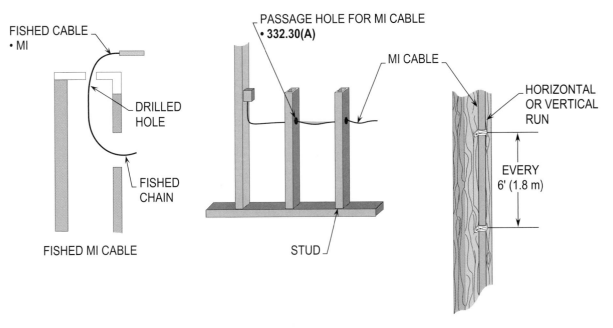

FISHED CABLE
• MI

DRILLED HOLE

FISHED CHAIN

FISHED MI CABLE

PASSAGE HOLE FOR MI CABLE
• 332.30(A)

MI CABLE

HORIZONTAL OR VERTICAL RUN

EVERY 6' (1.8 m)

STUD

SUPPORTED OR UNSUPPORTED CABLE
332.30(A) AND (B)

Purpose of Change: To clarify that MI cable can be run in unsupported lengths.

NEC Ch. 3 – Article 334
Part I – 334.2

Type of Change		Panel Action		UL	UL 508	API 500	API 505	OSHA
Revision		Accept in Principle in Part		719	-	-	-	-
ROP		ROC		NFPA 70E	NFPA 70B	NFPA 79	NFPA	NEMA
pg. 810	# 7-96	pg. -	# -	-	-	-	-	WC 5
log: 1525	CMP: 7	log: -	Submitter: James M. Daly			2002 NEC: 334.2		IEC: 514.2

2002 NEC – 334.2 Definition

Nonmetallic-Sheathed Cable. A factory assembly of two or more insulated conductors ~~having an outer sheath of~~ nonmetallic ~~material~~.

2005 NEC – 334.2 Definitions

Nonmetallic-Sheathed Cable. A factory assembly of two or more insulated conductors <u>enclosed within an overall</u> nonmetallic <u>jacket</u>.

<u>Type NM. Insulated conductors enclosed within an overall nonmetallic jacket.</u>

<u>Type NMC. Insulated conductors enclosed within an overall, corrosion resistant, nonmetallic jacket.</u>

<u>Type NMS. Insulated power or control conductors with signaling, data, and communications conductors within an overall nonmetallic jacket.</u>

Author's Substantiation. This definition has been revised by adding the three types of nonmetallic-sheathed cable to enhance the usability of the NEC.

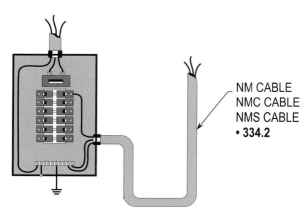

NM CABLE
NMC CABLE
NMS CABLE
• 334.2

NEW DEFINITIONS PER 334.2	
NM Cable	Insulated conductors enclosed within an overall nonmetallic jacket.
NMC Cable	Insulated conductors enclosed within an overall, corrosion resistant, nonmetallic jacket.
NMS Cable	Insulated power or control conductors with signaling, data, and communications conductors within an overall nonmetallic jacket.

DEFINITIONS
334.2

Purpose of Change: To enhance the usability of the NEC by adding the definitions of these cable types.

NEC Ch. 3 – Article 334
Part II – 334.30

Type of Change	Panel Action	UL	UL 508	API 500	API 505	OSHA
New Paragraph	Accept in Principle	719	-	-	-	-
ROP	**ROC**	**NFPA 70E**	**NFPA 70B**	**NFPA 79**	**NFPA**	**NEMA**
pg. 837 / # 7-144	pg. - / # -	-	-	-	-	WC 5
log: 1605 / CMP: 7	log: -	Submitter: David Shapiro		2002 NEC: 334.30		IEC: 514.2

2002 NEC – 334.30 Securing and Supporting

Nonmetallic-sheathed cable shall be secured by staples, cable ties, straps, hangers, or similar fittings designed and installed so as not to damage the cable at intervals not exceeding 1.4 m (4 1/2 ft) and within 300 mm (12 in.) of every cabinet, box, or fitting. Flat cables shall not be stapled on edge.

2005 NEC – 334.30 Securing and Supporting

Nonmetallic-sheathed cable shall be <u>supported and</u> secured by staples, cable ties, straps, hangers, or similar fittings designed and installed so as not to damage the cable, at intervals not exceeding 1.4 m (4 1/2 ft) and within 300 mm (12 in.) of every <u>outlet</u> box, <u>junction box,</u> cabinet, or fitting. Flat cables shall not be stapled on edge.

<u>Sections of cable protected from physical damage by raceway shall not be required to be secured within the raceway.</u>

Author's Substantiation. A new paragraph has been added to clarify that nonmetallic-sheathed cable protected from physical damage by raceway shall not be required to be secured within the raceway.

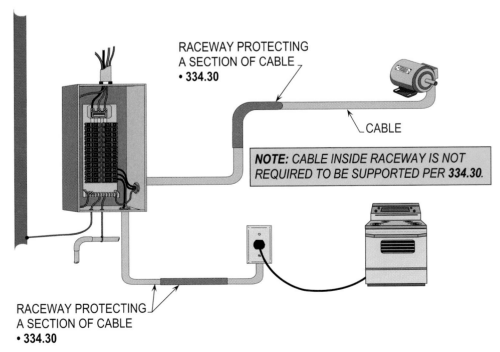

SECURING AND SUPPORTING
334.30

Purpose of Change: To allow a raceway to protect a section of cable from physical damage.

NEC Ch. 3 – Article 334
Part II – 334.80

Type of Change		Panel Action		UL	UL 508	API 500	API 505	OSHA
New Paragraph		Accept		719	-	-	-	-
ROP		ROC		NFPA 70E	NFPA 70B	NFPA 79	NFPA	NEMA
pg. 840	# 7-150a	pg. 255	# 7-120	-	-	-	-	WC 5
log: CP700	CMP: 7	log: 2384	Submitter: CMP 7			2002 NEC: 334.80		IEC: 514.2

2002 NEC – 334.80 Ampacity

The ampacity of Types NM, NMC, and NMS cable shall be determined in accordance with 310.15. The ampacity shall be in accordance with the 60°C (140°F) conductor temperature rating. The 90°C (194°F) rating shall be permitted to be used for ampacity derating purposes, provided the final derated ampacity does not exceed that for a 60°C (140°F) rated conductor. The ampacity of Types NM, NMC, and NMS cable installed in cable tray shall be determined in accordance with 392.11.

2005 NEC – 334.80 Ampacity

The ampacity of Types NM, NMC, and NMS cable shall be determined in accordance with 310.15. The ampacity shall be in accordance with the 60°C (140°F) conductor temperature rating. The 90°C (194°F) rating shall be permitted to be used for ampacity derating purposes, provided the final derated ampacity does not exceed that for a 60°C (140°F) rated conductor. The ampacity of Types NM, NMC, and NMS cable installed in cable tray shall be determined in accordance with 392.11.

Where more than two NM cables containing two or more current-carrying conductors are bundled together and pass through wood framing that is to be fire- or draft-stopped using thermal insulation or sealing foam, the allowable ampacity of each conductor shall be adjusted in accordance with Table 310.15(B)(2)(a).

Author's Substantiation. A new paragraph has been added for two or more nonmetallic-sheathed cables containing two or more current-carrying conductors, that are bundled together and pass through wood framing that is to be fire- or draft-stopped using thermal insulation or sealing foam, to have derating factors applied per Table 310.15(B)(2)(a).

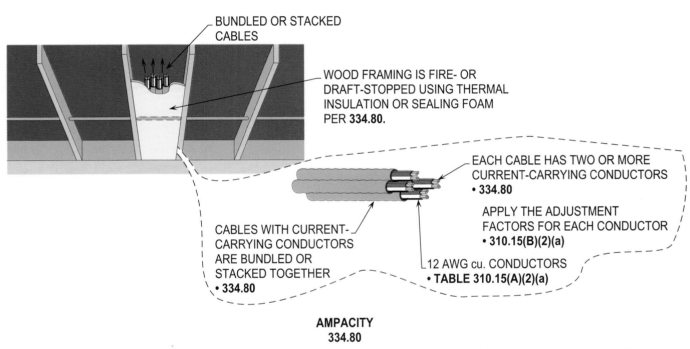

BUNDLED OR STACKED CABLES

WOOD FRAMING IS FIRE- OR DRAFT-STOPPED USING THERMAL INSULATION OR SEALING FOAM PER **334.80.**

EACH CABLE HAS TWO OR MORE CURRENT-CARRYING CONDUCTORS
• **334.80**

APPLY THE ADJUSTMENT FACTORS FOR EACH CONDUCTOR
• **310.15(B)(2)(a)**

12 AWG cu. CONDUCTORS
• **TABLE 310.15(A)(2)(a)**

CABLES WITH CURRENT-CARRYING CONDUCTORS ARE BUNDLED OR STACKED TOGETHER
• **334.80**

AMPACITY
334.80

Purpose of Change: To provide adjustment factors for bundled or stacked cables run in thermal insulation or sealing foam.

NEC Ch. 3 – Article 334
Part III – 334.108

Type of Change		Panel Action		UL	UL 508	API 500	API 505	OSHA
Revision		Accept in Part		719	-	-	-	-
ROP		ROC		NFPA 70E	NFPA 70B	NFPA 79	NFPA	NEMA
pg. 841	# 7-155	pg. -	# -	-	-	-	-	WC 5
log: 1807	CMP: 7	log: -	Submitter: William A. Wolfe		2002 NEC: 334.108		IEC: 514.2	

2002 NEC – 334.108 Equipment Grounding

In addition to the insulated conductors, the cable shall be permitted to have an insulated or bare conductor for equipment grounding purposes only. Where provided, the grounding conductor shall be sized in accordance with Article 250.

2005 NEC – 334.108 Equipment Grounding

In addition to the insulated conductors, the cable shall have an insulated or bare conductor for equipment grounding purposes only.

Author's Substantiation. This revision clarifies that nonmetallic-sheathed cable shall have an insulated or bare equipment grounding conductor for equipment grounding purposes only.

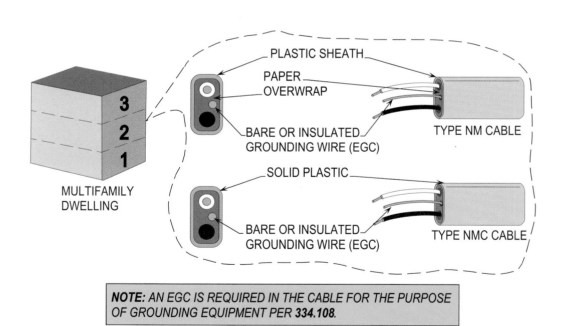

NOTE: AN EGC IS REQUIRED IN THE CABLE FOR THE PURPOSE OF GROUNDING EQUIPMENT PER **334.108**.

EQUIPMENT GROUNDING
334.108

Purpose of Change: To require an EGC for the purpose of grounding equipment.

NEC Ch. 3 – Article 340
Part III – 340.112

82

Type of Change		Panel Action		UL	UL 508	API 500	API 505	OSHA
New Sentence		Accept in Principle		719	-	-	-	-
ROP		ROC		NFPA 70E	NFPA 70B	NFPA 79	NFPA	NEMA
pg. 867	# 8-7	pg. -	# -	-	-	-	-	WC 5
log: 1891	CMP: 8	log: -	Submitter: David A. Ganiere			2002 NEC: 340.112		IEC: 514.2

2002 NEC – 340.112 Insulation

The conductors of Type UF shall be one of the moisture-resistant types listed in Table 310.13 that is suitable for branch-circuit wiring or one that is identified for such use.

2005 NEC – 340.112 Insulation

The conductors of Type UF shall be one of the moisture-resistant types listed in Table 310.13 that is suitable for branch-circuit wiring or one that is identified for such use. <u>Where installed as a substitute wiring method for NM cable, the conductor insulation shall be rated 90°C (194°F).</u>

Author's Substantiation. A new sentence has been added to clarify that Type UF conductor insulation shall be rated 90°C (194°F) where installed as a substitute wiring method for nonmetallic-sheathed cable.

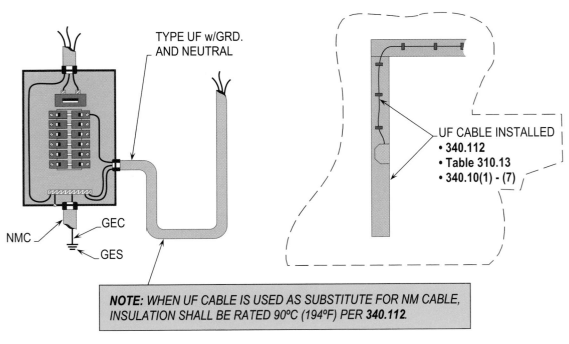

INSULATION
340.112

Purpose of Change: To require, under certain conditions of use, Type UF cable to have 90°C (194°F) insulation.

NEC Ch. 3 – Article 342
Part II – 342.22

Type of Change		Panel Action		UL	UL 508	API 500	API 505	OSHA
Revision		Accept		1242	-	-	-	1910.399
ROP		ROC		NFPA 70E	NFPA 70B	NFPA 79	NFPA	NEMA
pg. 868	# 8-8	pg. -	# -	Article 100	29.3.2	14.5.3	-	RN 2
log: 581	CMP: 8	log: -	Submitter: Vince Barclawski			2002 NEC: 342.22		IEC: 552.2

2002 NEC – 342.22 Number of Conductors

The number of conductors shall not exceed that permitted by the percentage fill specified in Table 1, Chapter 9.

Cables shall be permitted to be installed where such use is permitted by the respective cable articles. The number of cables shall not exceed the allowable percentage fill specified in Table 1, Chapter 9.

2005 NEC – 342.22 Number of Conductors

The number of conductors shall not exceed that permitted by the percentage fill specified in Table 1, Chapter 9.

Cables shall be permitted to be installed where such use is <u>not prohibited</u> by the respective cable articles. The number of cables shall not exceed the allowable percentage fill specified in Table 1, Chapter 9.

Author's Substantiation. This revision clarifies that cables shall be permitted to be pulled through intermediate metal conduit where such use is not prohibited by other cable articles.

NOTE: *SEE **344.22, 348.22, 350.22(A), 352.22, 356.22,** AND **358.22** FOR CABLES THAT ARE PERMITTED TO BE PULLED THROUGH RIGID METAL CONDUIT, FLEXIBLE METAL CONDUIT, LIQUIDTIGHT FLEXIBLE METAL CONDUIT, RIGID NONMETALLIC CONDUIT, LIQUIDTIGHT FLEXIBLE NONMETALLIC CONDUIT, AND ELECTRICAL METALLIC TUBING.*

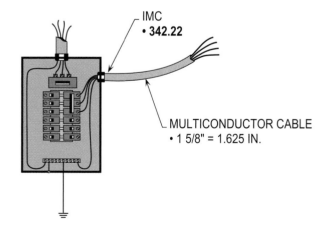

IMC
• 342.22

MULTICONDUCTOR CABLE
• 1 5/8" = 1.625 IN.

Problem: What size IMC is required to enclose a 1-5/8" multiconductor cable?

Step 1: Calculate sq. in.
342.22; Table 1, Ch. 9, Note 9
Dia.^{2} x .7854

1.625 x 1.625 x .7854 = 2.074 sq. in.

Step 2: Selecting IMC
342.22; Table 4 to Ch. 9
2.074 sq. in. = 3" IMC

Solution: A 3 in. IMC is required.

NUMBER OF CONDUCTORS
342.22

Purpose of Change: To clarify that cables can be pulled through raceways.

NEC Ch. 3 – Article 342
Part II – 342.30(B)(3)

Type of Change		Panel Action		UL	UL 508	API 500	API 505	OSHA
Revision		Accept in Principle		1242	-	-	-	1910.399
ROP		ROC		NFPA 70E	NFPA 70B	NFPA 79	NFPA	NEMA
pg. 871	# 8-16	pg. -	# -	Article 100	29.3.2	14.5.3	-	RN 2
log: 2866	CMP: 8	log: -	Submitter: Wayne A. Lilly			2002 NEC: 342.30(B)(3)		IEC: 552.2

2002 NEC – 342.30(B) Supports

IMC shall be supported in accordance with one of the following:

(3) Exposed vertical risers from industrial machinery or fixed equipment shall be permitted to be supported at intervals not exceeding 6 m (20 ft)~~,~~ if the conduit is made up with threaded couplings, the conduit is ~~firmly~~ supported at the top and bottom of the riser, and no other means of intermediate support is readily available.

2005 NEC – 342.30(B) Supports

IMC shall be supported in accordance with one of the following:

(3) Exposed vertical risers from industrial machinery or fixed equipment shall be permitted to be supported at intervals not exceeding 6 m (20 ft) if the conduit is made up with threaded couplings, the conduit is supported <u>and securely fastened</u> at the top and bottom of the riser, and no other means of intermediate support is readily available.

Author's Substantiation. This revision clarifies that intermediate metal conduit shall be supported and securely fastened at the top and bottom of the riser where installed as a vertical riser for industrial machinery or fixed equipment.

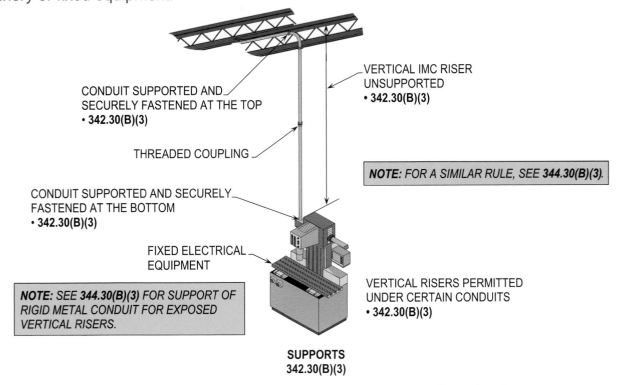

CONDUIT SUPPORTED AND SECURELY FASTENED AT THE TOP
• **342.30(B)(3)**

THREADED COUPLING

CONDUIT SUPPORTED AND SECURELY FASTENED AT THE BOTTOM
• **342.30(B)(3)**

FIXED ELECTRICAL EQUIPMENT

VERTICAL IMC RISER UNSUPPORTED
• **342.30(B)(3)**

NOTE: FOR A SIMILAR RULE, SEE **344.30(B)(3)**.

VERTICAL RISERS PERMITTED UNDER CERTAIN CONDUITS
• **342.30(B)(3)**

NOTE: SEE **344.30(B)(3)** FOR SUPPORT OF RIGID METAL CONDUIT FOR EXPOSED VERTICAL RISERS.

SUPPORTS
342.30(B)(3)

Purpose of Change: To clarify the supporting requirements for vertical risers.

NEC Ch. 3 – Article 342
Part II – 342.42(A)

Type of Change		Panel Action		UL	UL 508	API 500	API 505	OSHA
Revision		Accept		1242	-	-	-	1910.399
ROP		ROC		NFPA 70E	NFPA 70B	NFPA 79	NFPA	NEMA
pg. 872	# 8-18	pg. -	# -	Article 100	29.3.2	14.5.3	-	RN 2
log: 582	CMP: 8	log: -	Submitter: Vince Barclawski			2002 NEC: 342.42(A)		IEC: 552.2

2002 NEC – 342.42(A) Threadless

Threadless couplings and connectors used with conduit shall be made tight. Where buried in masonry or concrete, they shall be the concretetight type. Where installed in wet locations, they shall ~~be the raintight type~~. Threadless couplings and connectors shall not be used on threaded conduit ends unless listed for the purpose.

2002 NEC – 342.42(A) Threadless

Threadless couplings and connectors used with conduit shall be made tight. Where buried in masonry or concrete, they shall be the concretetight type. Where installed in wet locations, they shall <u>comply with 314.15(A)</u>. Threadless couplings and connectors shall not be used on threaded conduit ends unless listed for the purpose.

Author's Substantiation. This revision clarifies that threadless couplings and connectors for intermediate metal conduit that are installed in wet locations shall be listed for such use.

NOTE: SEE **344.42(A)** AND **358.42** FOR THREADLESS COUPLINGS AND CONNECTORS THAT ARE TO BE LISTED FOR WET LOCATIONS WHEN INSTALLING RIGID METAL CONDUIT AND ELECTRICAL METALLIC TUBING.

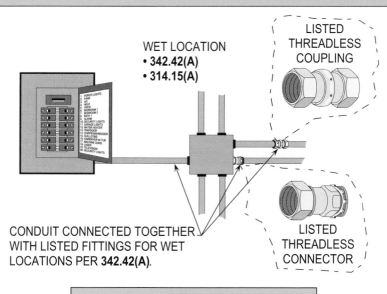

WET LOCATION
• **342.42(A)**
• **314.15(A)**

LISTED THREADLESS COUPLING

CONDUIT CONNECTED TOGETHER WITH LISTED FITTINGS FOR WET LOCATIONS PER **342.42(A)**.

LISTED THREADLESS CONNECTOR

NOTE: FOR A SIMILAR RULE, SEE **344.42(A)**.

THREADLESS
342.42(A)

Purpose of Change: To require threadless fittings used in wet locations to connect conduit to be listed for such use.

NEC Ch. 3 – Article 348
Part II – 348.30(A)

Type of Change		Panel Action		UL	UL 508	API 500	API 505	OSHA
Revision		Accept in Principle		360	-	-	-	1910.399
ROP		ROC		NFPA 70E	NFPA 70B	NFPA 79	NFPA	NEMA
pg. 884	# 8-43	pg. 270	# 8-27	Article 100	10.4	14.5.4	-	RN 2
log: 225	CMP: 8	log: 2476	Submitter: -			2002 NEC: 348.30(A)		IEC: 552.2

2002 NEC – 348.30(A) Securely Fastened

FMC shall be securely fastened in place by an approved means within 300 mm (12 in.) of each box, cabinet, conduit body, or other conduit termination and shall be supported and secured at intervals not to exceed 1.4 m (4 1/2 ft).

Exception No. 1: Where FMC is fished.

Exception No. 2: ~~Lengths not exceeding 900 mm (3 ft) at terminals where flexibility is required.~~

Exception No. 3: Lengths not exceeding 1.8 m (6 ft) from a luminaire (fixture) terminal connection for tap connections to luminaires (light fixtures) as permitted in 410.67(C).

2005 NEC – 348.30(A) Securely Fastened

FMC shall be securely fastened in place by an approved means within 300 mm (12 in.) of each box, cabinet, conduit body, or other conduit termination and shall be supported and secured at intervals not to exceed 1.4 m (4 1/2 ft).

Exception No. 1: Where FMC is fished.

Exception No. 2: <u>At terminals where flexibility is required, lengths shall not exceed the following:</u>

(1) <u>900 mm (3 ft) for metric designators 16 through 35 (trade sizes 1/2 through 1 1/4)</u>

(2) <u>1200 mm (4 ft) for metric designators 41 through 53 (trade sizes 1 1/2 through 2)</u>

(3) <u>1500 mm (5 ft) for metric designators 63 (trade size 2 1/2) and larger</u>

Exception No. 3: Lengths not exceeding 1.8 m (6 ft) from a luminaire (fixture) terminal connection for tap connections to luminaires (light fixtures) as permitted in 410.67(C).

Exception No. 4: <u>Lengths not exceeding 1.8 m (6 ft) from the last point where the raceway is securely fastened for connections within an accessible ceiling to luminaire(s) [lighting fixture(s)] or other equipment.</u>

Author's Substantiation. The revision to Exception No. 2 clarifies the supporting requirements for flexible metal conduit where flexibility is required. A new Exception No. 4 has been added to clarify the support requirements for flexible metal conduit within an accessible ceiling to luminaire(s) [lighting fixture(s)] or other equipment.

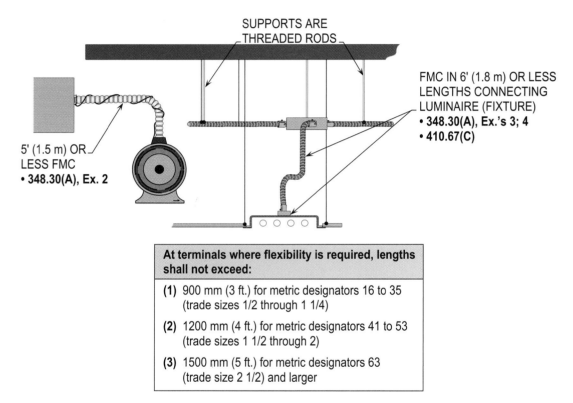

SUPPORTS ARE
THREADED RODS

FMC IN 6' (1.8 m) OR LESS
LENGTHS CONNECTING
LUMINAIRE (FIXTURE)
• **348.30(A), Ex.'s 3; 4**
• **410.67(C)**

5' (1.5 m) OR
LESS FMC
• **348.30(A), Ex. 2**

At terminals where flexibility is required, lengths shall not exceed:
(1) 900 mm (3 ft.) for metric designators 16 to 35 (trade sizes 1/2 through 1 1/4)
(2) 1200 mm (4 ft.) for metric designators 41 to 53 (trade sizes 1 1/2 through 2)
(3) 1500 mm (5 ft.) for metric designators 63 (trade size 2 1/2) and larger

SECURELY FASTENED
348.30(A)

Purpose of Change: To clarify the supporting requirements for flexible metal conduit.

NEC Ch. 3 – Article 350
Part II – 350.30(A), Ex. 4

 84

Type of Change		Panel Action		UL	UL 508	API 500	API 505	OSHA
New Exception		Accept in Principle		360	-	-	-	1910.399
ROP		ROC		NFPA 70E	NFPA 70B	NFPA 79	NFPA	NEMA
pg. 895	# 8-64	pg. -	# -	Article 100	10.4	14.5.4	-	RN 2
log: 19	CMP: 8	log: -	Submitter: Joseph A. Ross			2002 NEC: 350.30(A)		IEC: -

2002 NEC – 350.30(A) Securely Fastened

LFMC shall be securely fastened in place by an approved means within 300 mm (12 in.) of each box, cabinet, conduit body, or other conduit termination and shall be supported and secured at intervals not to exceed 1.4 m (4 1/2 ft).

> **Exception No. 1:** Where LFMC is fished.
>
> **Exception No. 2:** Lengths not exceeding 900 mm (3 ft) at terminals where flexibility is necessary.
>
> **Exception No. 3:** Lengths not exceeding 1.8 m (6 ft) from a luminaire (fixture) terminal connection for tap conductors to luminaires (light fixtures), as permitted in 410.67(C).

2005 NEC – 350.30(A) Securely Fastened

LFMC shall be securely fastened in place by an approved means within 300 mm (12 in.) of each box, cabinet, conduit body, or other conduit termination and shall be supported and secured at intervals not to exceed 1.4 m (4 1/2 ft).

> **Exception No. 1:** Where LFMC is fished.
>
> **Exception No. 2:** Lengths not exceeding 900 mm (3 ft) at terminals where flexibility is necessary.
>
> **Exception No. 3:** Lengths not exceeding 1.8 m (6 ft) from a luminaire (fixture) terminal connection for tap conductors to luminaires (light fixtures), as permitted in 410.67(C).
>
> **Exception No. 4:** Lengths not exceeding 1.8 m (6 ft) from the last point where the raceway is securely fastened for connections within an accessible ceiling to luminaire(s) [lighting fixture(s)] or other equipment.

Author's Substantiation. A new Exception No. 4 has been added to clarify the support requirements for liquidtight flexible metal conduit within an accessible ceiling to luminaire(s) [lighting fixture(s)] or other equipment.

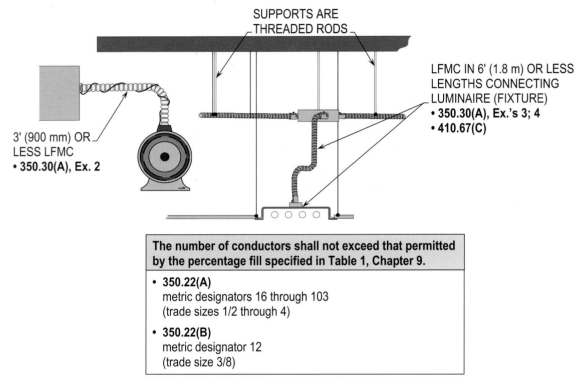

SUPPORTS ARE
THREADED RODS

LFMC IN 6' (1.8 m) OR LESS
LENGTHS CONNECTING
LUMINAIRE (FIXTURE)
• **350.30(A), Ex.'s 3; 4**
• **410.67(C)**

3' (900 mm) OR
LESS LFMC
• **350.30(A), Ex. 2**

**The number of conductors shall not exceed that permitted
by the percentage fill specified in Table 1, Chapter 9.**

• **350.22(A)**
  metric designators 16 through 103
  (trade sizes 1/2 through 4)

• **350.22(B)**
  metric designator 12
  (trade size 3/8)

SECURELY FASTENED
350.30(A), Ex. 4

Purpose of Change: To clarify the supporting requirements for liquidtight flexible metal conduit.

NEC Ch. 3 – Article 353
Part – Article 353

Type of Change		Panel Action		UL	UL 508	API 500	API 505	OSHA
New Article		Accept in Principle in Part		360	-	-	-	1910.399
ROP		**ROC**		**NFPA 70E**	**NFPA 70B**	**NFPA 79**	**NFPA**	**NEMA**
pg. 911	# 8-96	pg. 275	# 8-51	Article 100	10.4	14.5.4	-	RN 2
log: 1252	CMP: 8	log: 2057	Submitter: David H. Kendall			2002 NEC: -		IEC: 522.2

Author's Substantiation. A new article has been added for High Density Polyethylene Conduit — Type HDPE conduit that is a listed product and is restricted in its uses.

THIS NEW ARTICLE IS DIVIDED INTO THREE PARTS:

PART I	GENERAL
PART II	INSTALLATION
PART III	CONSTRUCTION SPECIFICATIONS

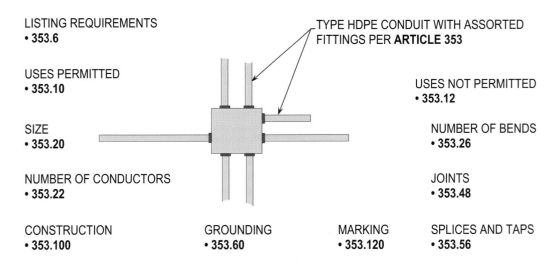

LISTING REQUIREMENTS
• **353.6**

TYPE HDPE CONDUIT WITH ASSORTED FITTINGS PER **ARTICLE 353**

USES PERMITTED
• **353.10**

USES NOT PERMITTED
• **353.12**

SIZE
• **353.20**

NUMBER OF BENDS
• **353.26**

NUMBER OF CONDUCTORS
• **353.22**

JOINTS
• **353.48**

CONSTRUCTION
• **353.100**

GROUNDING
• **353.60**

MARKING
• **353.120**

SPLICES AND TAPS
• **353.56**

**HIGH DENSITY POLYETHYLENE CONDUIT (TYPE HDPE)
ARTICLE 353**

Purpose of Change: To introduce a new **Article** pertaining to a new wiring method.

NEC Ch. 3 – Article 356
Part II – 356.30

Type of Change	Panel Action	UL	UL 508	API 500	API 505	OSHA
New Item	Accept in Part	1660	-	-	-	1910.399
ROP		**ROC**	**NFPA 70E**	**NFPA 70B**	**NFPA 79**	**NFPA**	**NEMA**
pg. 922	# 8-114	pg. 279	# 8-67	Article 100	10.4	14.5.4	-	RN 1
log: 20	CMP: 8	log: 2453	Submitter: Joseph A. Ross			2002 NEC: -		IEC: -

2002 NEC – 356.30 Securing and Supporting

Type LFNC-B shall be securely fastened and supported in accordance with one of the following:

(1) The conduit shall be securely fastened at intervals not exceeding 900 mm (3 ft) and within 300 mm (12 in.) on each side of every outlet box, junction box, cabinet, or fitting.

(2) Securing or supporting of the conduit shall not be required where it is fished, installed in lengths not exceeding 900 mm (3 ft) at terminals where flexibility is required, or installed in lengths not exceeding 1.8 m (6 ft) from a luminaire (fixture) terminal connection for tap conductors to luminaires (lighting fixtures) permitted in 410.67(C).

(3) Horizontal runs of LFNC supported by openings through framing members at intervals not exceeding 900 mm (3 ft) and securely fastened within 300 mm (12 in.) of termination points shall be permitted.

2005 NEC – 356.30 Securing and Supporting

Type LFNC-B shall be securely fastened and supported in accordance with one of the following:

(1) Where installed in lengths exceeding 1.8 m (6 ft), the conduit shall be securely fastened at intervals not exceeding 900 mm (3 ft) and within 300 mm (12 in.) on each side of every outlet box, junction box, cabinet, or fitting.

(2) Securing or supporting of the conduit shall not be required where it is fished, installed in lengths not exceeding 900 mm (3 ft) at terminals where flexibility is required, or installed in lengths not exceeding 1.8 m (6 ft) from a luminaire (fixture) terminal connection for tap conductors to luminaires (lighting fixtures) permitted in 410.67(C).

(3) Horizontal runs of LFNC supported by openings through framing members at intervals not exceeding 900 mm (3 ft) and securely fastened within 300 mm (12 in.) of termination points shall be permitted.

(4) Securing or supporting of LFNC-B shall not be required where installed in lengths not exceeding 1.8 m (6 ft) from the last point where the raceway is securely fastened for connections within an accessible ceiling to luminaire(s) [lighting fixture(s)] or other equipment.

Author's Substantiation. A new item (4) has been added to clarify the support requirements for liquidtight flexible nonmetallic conduit within an accessible ceiling to luminaire(s) [lighting fixture(s)] or other equipment.

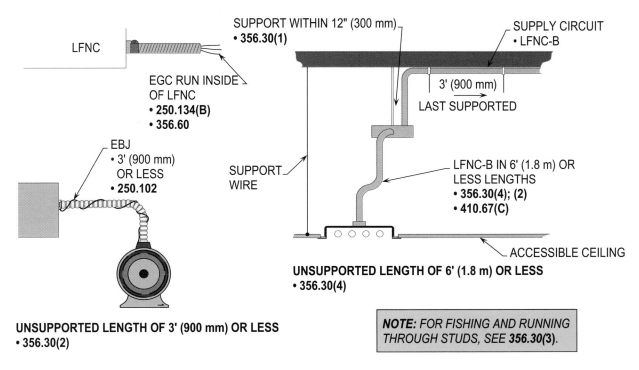

LFNC

SUPPORT WITHIN 12" (300 mm)
• **356.30(1)**

SUPPLY CIRCUIT
• LFNC-B

EGC RUN INSIDE
OF LFNC
• **250.134(B)**
• **356.60**

3' (900 mm)
LAST SUPPORTED

EBJ
• 3' (900 mm)
OR LESS
• **250.102**

SUPPORT
WIRE

LFNC-B IN 6' (1.8 m) OR
LESS LENGTHS
• **356.30(4); (2)**
• **410.67(C)**

ACCESSIBLE CEILING

UNSUPPORTED LENGTH OF 6' (1.8 m) OR LESS
• **356.30(4)**

UNSUPPORTED LENGTH OF 3' (900 mm) OR LESS
• **356.30(2)**

NOTE: *FOR FISHING AND RUNNING THROUGH STUDS, SEE **356.30(3)**.*

SECURING AND SUPPORTING
356.30(4)

Purpose of Change: To clarify the supporting requirements for LFNC-B.

NEC Ch. 3 – Article 362
Part II – 362.10(2), Ex.

Type of Change		Panel Action		UL	UL 508	API 500	API 505	OSHA
Revision		Accept		3	-	-	-	-
ROP		ROC		NFPA 70E	NFPA 70B	NFPA 79	NFPA	NEMA
pg. 936	# 8-147	pg. -	# -	-	-	-	-	TC 13
log: 284	CMP: 8	log: -	Submitter: James M. Daly			2002 NEC: 362.10(2), Ex.		IEC: 552.2

2002 NEC – 362.10(2), Ex.

Where a fire sprinkler system(s) is installed in accordance with NFPA 13-~~1999~~, *Standard for the Installation of Sprinkler Systems*, on all floors, ENT ~~is~~ permitted to be used within walls, floors, and ceilings, exposed or concealed, in buildings exceeding three floors above grade.

2005 NEC – 362.10(2), Ex.

Where a fire sprinkler system(s) is installed in accordance with NFPA 13-2002, *Standard for the Installation of Sprinkler Systems*, on all floors, ENT <u>shall be</u> permitted to be used within walls, floors, and ceilings, exposed or concealed, in buildings exceeding three floors above grade.

Author's Substantiation. This revision clarifies that electrical nonmetallic tubing shall be permitted to be installed in walls, ceilings, and floors, exposed or concealed, in buildings exceeding three floors above grade where a fire sprinkler system is installed in accordance with NFPA 13-2002. See Ex. to 362.10(5) for a similar rule pertaining to electrical nonmetallic tubing installed above suspended ceilings in buildings exceeding three floors above grade.

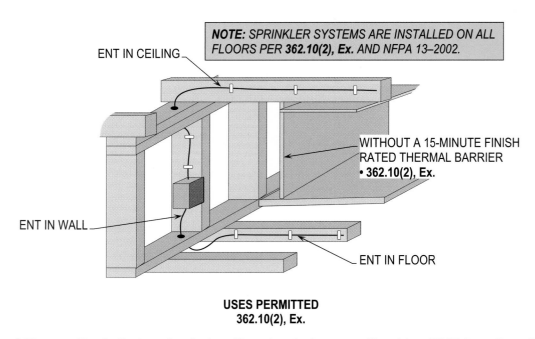

USES PERMITTED
362.10(2), Ex.

Purpose of Change: To clarify the rules for installing electrical nonmetallic tubing (ENT) in walls, ceilings, and floors based upon a fire sprinkler system(s).

NEC Ch. 3 – Article 374
Part I – 374.11

Type of Change	Panel Action	UL	UL 508	API 500	API 505	OSHA
New Sentence	Accept in Principle	884	-	-	-	-
ROP	**ROC**	**NFPA 70E**	**NFPA 70B**	**NFPA 79**	**NFPA**	**NEMA**
pg. 763 # 8-206	pg. - # -	-	-	-	-	-
log: 2703 CMP: 8	log: -	Submitter: George Straniero		2002 NEC: 374.11		IEC: 552.2

2002 NEC – 374.11 Connection to Cabinets and Extensions from Cells

Connections between raceways and distribution centers and wall outlets shall be made by means of flexible metal conduit where not installed in concrete, rigid metal conduit, intermediate metal conduit, electrical metallic tubing, or approved fittings. Where there are provisions for the termination of an equipment grounding conductor, nonmetallic conduit, electrical nonmetallic tubing, or liquidtight flexible nonmetallic conduit ~~where not installed in concrete~~ shall be permitted.

2005 NEC – 374.11 Connection to Cabinets and Extensions from Cells

Connections between raceways and distribution centers and wall outlets shall be made by means of liquidtight flexible metal conduit, flexible metal conduit where not installed in concrete, rigid metal conduit, intermediate metal conduit, electrical metallic tubing, or approved fittings. Where there are provisions for the termination of an equipment grounding conductor, nonmetallic conduit, electrical nonmetallic tubing, or liquidtight flexible nonmetallic conduit shall be permitted. <u>Where installed in concrete, liquidtight flexible nonmetallic conduit shall be listed and marked for direct burial.</u>

> **FPN:** <u>Liquidtight flexible metal conduit and liquidtight flexible nonmetallic conduit that is suitable for installation in concrete is listed and marked for direct burial.</u>

Author's Substantiation. This revision clarifies that liquidtight flexible nonmetallic conduit is a UL listed product as suitable for use in poured concrete when marked for direct burial.

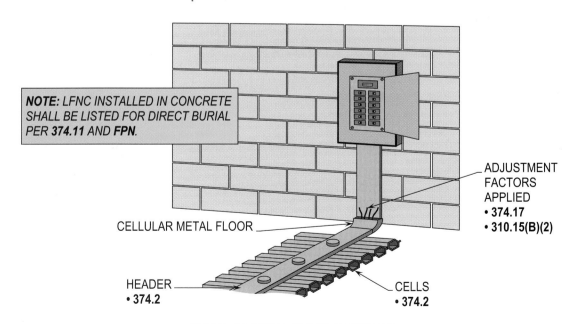

NOTE: LFNC INSTALLED IN CONCRETE SHALL BE LISTED FOR DIRECT BURIAL PER 374.11 AND FPN.

CELLULAR METAL FLOOR

ADJUSTMENT FACTORS APPLIED
• 374.17
• 310.15(B)(2)

HEADER
• 374.2

CELLS
• 374.2

CELLULAR METAL FLOOR RACEWAYS
ARTICLE 374

Purpose of Change: To clarify that listed liquidtight flexible nonmetallic conduit is suitable for direct burial.

NEC Ch. 3 – Article 376
Part II – 376.23(A)

Type of Change		Panel Action		UL	UL 508	API 500	API 505	OSHA
Revision		Accept		870	-	-	-	1910.399
ROP		ROC		NFPA 70E	NFPA 70B	NFPA 79	NFPA	NEMA
pg. 968	# 8-214	pg. 292	# 8-125	-	-	3.3.108	-	RN 1
log: 2857	CMP: 8	log: 1031	Submitter: Jim Pauley			2002 NEC: 376.23(A)		IEC: 552.2

2002 NEC – 376.23(A) Deflected Insulated Conductors

Where insulated conductors are deflected within a metallic wireway, either at the ends or where conduits, fittings, or other raceways or cables enter or leave the metallic wireway, or where the direction of the metallic wireway is deflected greater than 30 degrees, dimensions corresponding to ~~312.6(A)~~ shall apply.

2005 NEC – 376.23(A) Deflected Insulated Conductors

Where insulated conductors are deflected within a metallic wireway, either at the ends or where conduits, fittings, or other raceways or cables enter or leave the metallic wireway, or where the direction of the metallic wireway is deflected greater than 30 degrees, dimensions corresponding to <u>one wire per terminal in Table 312.6(A)</u> shall apply.

Author's Substantiation. This revision clarifies that one wire per terminal in Table 312.6(A) shall apply to deflected insulated conductors when sizing metal wireways.

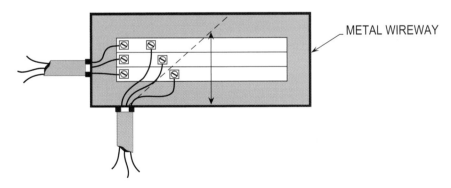

INSULATED CONDUCTORS ARE DEFLECTED WITH A METAL WIREWAY EITHER

- AT THE ENDS, OR
- WHERE CONDUITS, FITTINGS, OR OTHER RACEWAYS OR CABLES ENTER OR LEAVE THE WIREWAY, OR
- WHERE THE DIRECTION OF THE WIREWAY IS DEFLECTED GREATER THAN 30°.

THE DIMENSIONS CORRESPONDING TO ONE WIRE PER TERMINAL IN **TABLE 312.6(A)** SHALL APPLY.

DEFLECTED INSULATED CONDUCTORS
376.23(A)

Purpose of Change: To reference **Table 312.6(A)** instead of **312.6(A)** and to outline the required dimensions.

NEC Ch. 3 – Article 376
Part II – 376.23(B)

Type of Change		Panel Action		UL	UL 508	API 500	API 505	OSHA
New Sentence		Accept in Part		870	-	-	-	1910.399
ROP		ROC		NFPA 70E	NFPA 70B	NFPA 79	NFPA	NEMA
pg. 969	# 8-216	pg. -	# -	-	-	3.3.108	-	RN 1
log: 3057	CMP: 8	log: -	Submitter: Truman C. Surbrook			2002 NEC: 376.23(B)		IEC: 552.2

2002 NEC – 376.23(B) Metallic Wireways Used as Pull Boxes

Where insulated conductors 4 AWG or larger are pulled through a wireway, the distance between raceway and cable entries enclosing the same conductor shall not be less than that required in 314.28(A)(1) for straight pulls and 314.28(A)(2) for angle pulls.

2005 NEC – 376.23(B) Metallic Wireways Used as Pull Boxes

Where insulated conductors 4 AWG or larger are pulled through a wireway, the distance between raceway and cable entries enclosing the same conductor shall not be less than that required in 314.28(A)(1) for straight pulls and 314.28(A)(2) for angle pulls. <u>When transposing cable size into raceway size, the minimum metric designator (trade size) raceway required for the number and size of conductors in the cable shall be used.</u>

Author's Substantiation. A new sentence has been added to clarify that the minimum metric designator (trade size) raceways required for the number and size of conductors in the cable shall be used where transposing cable size into raceway for sizing metallic wireways used as pull boxes.

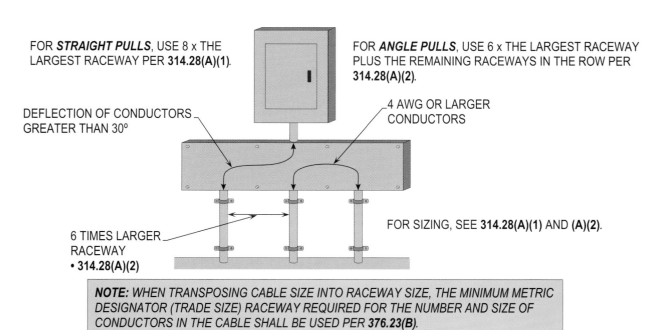

METALLIC WIREWAYS USED AS PULL BOXES
376.23(B)

Purpose of Change: To clarify cable size into raceway size using the metric designator (trade size) based on conductors in the cable.

NEC Ch. 3 – Article 378
Part II – 378.23(A)

Type of Change		Panel Action		UL	UL 508	API 500	API 505	OSHA
Revision		Accept		870	-	-	-	1910.399
ROP		ROC		NFPA 70E	NFPA 70B	NFPA 79	NFPA	NEMA
pg. 973	# 8-221	pg. 293	# 8-133	-	3.3.108	-	-	RN 1
log: 2861	CMP: 8	log: 1033	Submitter: Jim Pauley			2002 NEC: 378.23(A)		IEC: 552.2

2002 NEC - 2002 NEC – 378.23(A) Deflected Insulated Conductors

Where insulated conductors are deflected within a nonmetallic wireway, either at the ends or where conduits, fittings, or other raceways or cables enter or leave the nonmetallic wireway, or where the direction of the nonmetallic wireway is deflected greater than 30 degrees, dimensions corresponding to 312.6(A) shall apply.

2005 NEC – 378.23(A) Deflected Insulated Conductors

Where insulated conductors are deflected within a nonmetallic wireway, either at the ends or where conduits, fittings, or other raceways or cables enter or leave the nonmetallic wireway, or where the direction of the nonmetallic wireway is deflected greater than 30 degrees, dimensions corresponding to <u>one wire per terminal in Table 312.6(A)</u> shall apply.

Author's Substantiation. This revision clarifies that one wire per terminal in Table 312.6(A) shall apply to deflected insulated conductors when sizing nonmetallic wireways.

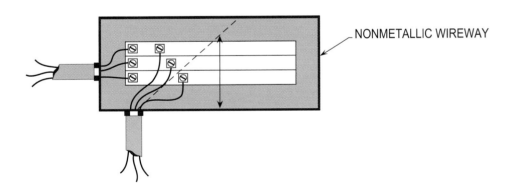

NONMETALLIC WIREWAY

INSULATED CONDUCTORS ARE DEFLECTED WITH A NONMETALLIC WIREWAY EITHER

• AT THE ENDS, OR

• WHERE CONDUITS, FITTINGS, OR OTHER RACEWAYS OR CABLES ENTER OR LEAVE THE WIREWAY, OR

• WHERE THE DIRECTION OF THE WIREWAY IS DEFLECTED GREATER THAN 30°.

THE DIMENSIONS CORRESPONDING TO ONE WIRE PER TERMINAL IN **TABLE 312.6(A)** SHALL APPLY.

DEFLECTED INSULATED CONDUCTORS
378.23(A)

Purpose of Change: To reference **Table 312.6(A)** instead of **312.6(A)** and to outline the required dimensions.

NEC Ch. 3 – Article 378
Part II – 378.23(B)

Type of Change		Panel Action		UL	UL 508	API 500	API 505	OSHA
New Sentence		Accept in Part		870	-	-	-	1910.399
ROP		ROC		NFPA 70E	NFPA 70B	NFPA 79	NFPA	NEMA
pg. 974	# 8-223	pg. -	# -	-	-	3.3.108	-	RN 1
log: 3056	CMP: 8	log: -	-	Submitter: Truman C. Surbrook		2002 NEC: 378.23(B)		IEC: 552.2

2002 NEC – 378.23(B) Nonmetallic Wireways Used as Pull Boxes

Where insulated conductors 4 AWG or larger are pulled through a wireway, the distance between raceway and cable entries enclosing the same conductor shall not be less than that required in 314.28(A)(1) for straight pulls and 314.28(A)(2) for angle pulls.

2005 NEC – 378.23(B) Nonmetallic Wireways Used as Pull Boxes

Where insulated conductors 4 AWG or larger are pulled through a wireway, the distance between raceway and cable entries enclosing the same conductor shall not be less than that required in 314.28(A)(1) for straight pulls and 314.28(A)(2) for angle pulls. <u>When transposing cable size into raceway size, the minimum metric designator (trade size) raceway required for the number and size of conductors in the cable shall be used.</u>

Author's Substantiation. A new sentence has been added to clarify that the minimum metric designator (trade size) raceways required for the number and size of conductors in the cable shall be used where transposing cable size into raceway for sizing nonmetallic wireways used as pull boxes.

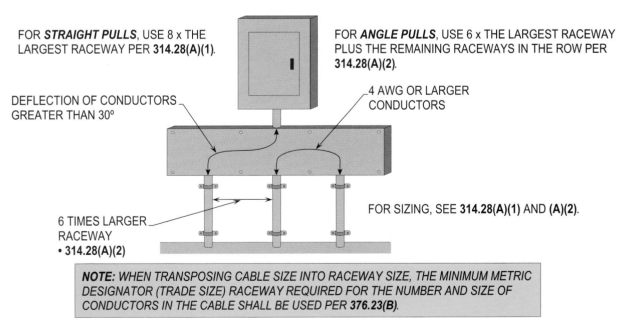

FOR *STRAIGHT PULLS*, USE 8 x THE LARGEST RACEWAY PER **314.28(A)(1)**.

FOR *ANGLE PULLS*, USE 6 x THE LARGEST RACEWAY PLUS THE REMAINING RACEWAYS IN THE ROW PER **314.28(A)(2)**.

DEFLECTION OF CONDUCTORS GREATER THAN 30º

4 AWG OR LARGER CONDUCTORS

6 TIMES LARGER RACEWAY
• **314.28(A)(2)**

FOR SIZING, SEE **314.28(A)(1)** AND **(A)(2)**.

NOTE: WHEN TRANSPOSING CABLE SIZE INTO RACEWAY SIZE, THE MINIMUM METRIC DESIGNATOR (TRADE SIZE) RACEWAY REQUIRED FOR THE NUMBER AND SIZE OF CONDUCTORS IN THE CABLE SHALL BE USED PER **376.23(B)**.

METALLIC WIREWAYS USED AS PULL BOXES
376.23(B)

Purpose of Change: To clarify cable size into raceway size using the metric designator (trade size) based on conductors in the cable.

NEC Ch. 3 – Article 390
Part – 390.17

 87

Type of Change	Panel Action	UL	UL 508	API 500	API 505	OSHA
New Section	Accept in Principle	884	-	-	-	-
ROP	ROC	NFPA 70E	NFPA 70B	NFPA 79	NFPA	NEMA
pg. 994 # 8-258	pg. - # -	-	-	-	-	-
log: 952 CMP: 8	log: -	Submitter: Roger D. Wilson		2002 NEC: -		IEC: 552.2

2005 NEC – 390.17 Ampacity of Conductors

The ampacity adjustment factors, in 310.15(B)(2), shall apply to conductors installed in underfloor raceways.

Author's Substantiation. A new section has been added to clarify that ampacity adjustment factors shall be applied to conductors installed in underfloor raceways.

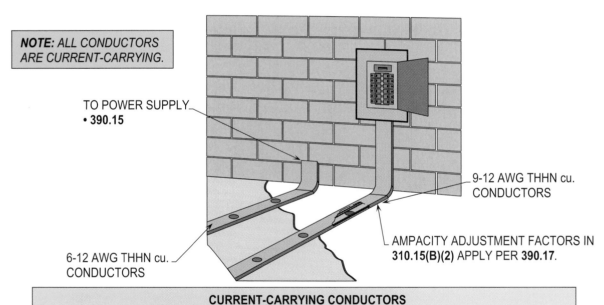

NOTE: ALL CONDUCTORS ARE CURRENT-CARRYING.

TO POWER SUPPLY
• 390.15

9-12 AWG THHN cu. CONDUCTORS

6-12 AWG THHN cu. CONDUCTORS

AMPACITY ADJUSTMENT FACTORS IN 310.15(B)(2) APPLY PER 390.17.

CURRENT-CARRYING CONDUCTORS	
NUMBER	NUMBER
6 conductors require 80% adjustment factor **Table 310.1(B)(2)(a); 310.16** 30 A x 80% = 24 A Ampacity is 24 A after adjustment factor is applied.	9 conductors require 70% adjustment factor **Table 310.1(B)(2)(a); 310.16** 30 A x 70% = 21 A Ampacity is 21 A after adjustment factor is applied.

AMPACITY OF CONDUCTORS
390.17

Purpose of Change: To reference **310.15(B)(2)** for application of adjustment factors for four or more current-carrying conductors.

NEC Ch. 3 – Article 392
Part – 392.3

Type of Change		Panel Action		UL	UL 508	API 500	API 505	OSHA
Revision		Accept in Principle		1277	-	-	-	1910.305(c)
ROP		ROC		NFPA 70E	NFPA 70B	NFPA 79	NFPA	NEMA
pg. 995	# 8-262	pg. -	# -	420.1(C)(2)	Ch. 24	14.5.10	-	VE 1 and 2
log: 705	CMP: 8	log: -	Submitter: Dan Leaf			2002 NEC: 392.3		IEC: 552.2

2002 NEC – 392.3 Uses Permitted

Cable tray shall be permitted to be used as a support system for ~~services~~, feeders, branch circuits, communications circuits, control circuits, and signaling circuits. Cable tray installations shall not be limited to industrial establishments. Where exposed to direct rays of the sun, insulated conductors and jacketed cables shall be identified as being sunlight resistant. Cable trays and their associated fittings shall be identified for the intended use.

2005 NEC – 392.3 Uses Permitted

Cable tray shall be permitted to be used as a support system for <u>service conductors,</u> feeders, branch circuits, communications circuits, control circuits, and signaling circuits. Cable tray installations shall not be limited to industrial establishments. Where exposed to direct rays of the sun, insulated conductors and jacketed cables shall be identified as being sunlight resistant. Cable trays and their associated fittings shall be identified for the intended use.

Author's Substantiation. This revision clarifies that service conductors apply to the conductors from the service point to the service disconnecting means.

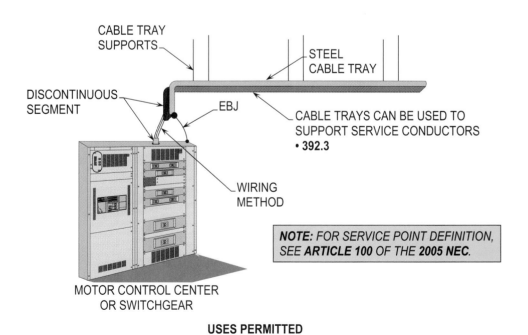

CABLE TRAY SUPPORTS

STEEL CABLE TRAY

DISCONTINUOUS SEGMENT

EBJ

CABLE TRAYS CAN BE USED TO SUPPORT SERVICE CONDUCTORS
• 392.3

WIRING METHOD

NOTE: FOR SERVICE POINT DEFINITION, SEE ARTICLE 100 OF THE 2005 NEC.

MOTOR CONTROL CENTER OR SWITCHGEAR

USES PERMITTED
392.3

Purpose of Change: To replace the word "services" with "service conductors" to correlate with **392.2** of the **2005 NEC**.

NEC Ch. 3 – Article 392
Part – 392.3(B)(1)(a)

Type of Change		Panel Action		UL	UL 508	API 500	API 505	OSHA
Revision		Accept		-	-	-	-	-
ROP		ROC		NFPA 70E	NFPA 70B	NFPA 79	NFPA	NEMA
pg. 997	# 8-265	pg. -	# -	-	-	-	-	-
log: 203	CMP: 8	log: -	Submitter: Nate Periard			2002 NEC: 392.3(B)(1)(a)		IEC: -

2002 NEC – 392.3(B)(1) Single Conductors

Single-conductor cables shall be permitted to be installed in accordance with the following:

(a) Single-conductor cable shall be 1/0 AWG or larger and shall be of a type listed and marked on the surface for use in cable trays. Where 1/0 AWG through 4/0 AWG single-conductor cables are installed in ladder cable tray, the maximum allowable rung spacing for the ladder cable tray shall be ~~230 mm~~ (9 in.).

2005 NEC – 392.3(B)(1) Single Conductors

Single-conductor cables shall be permitted to be installed in accordance with (B)(1)(a) through (B)(1)(c).

(a) Single-conductor cable shall be 1/0 AWG or larger and shall be of a type listed and marked on the surface for use in cable trays. Where 1/0 AWG through 4/0 AWG single-conductor cables are installed in ladder cable tray, the maximum allowable rung spacing for the ladder cable tray shall be <u>225 mm</u> (9 in.).

Author's Substantiation. This revision is consistent with other references in this article such as Table 392.9 and Table 392.10(A).

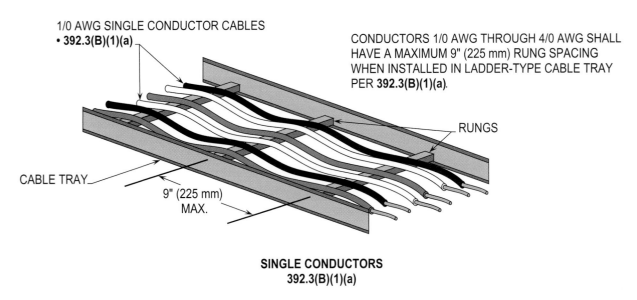

1/0 AWG SINGLE CONDUCTOR CABLES
• 392.3(B)(1)(a)

CONDUCTORS 1/0 AWG THROUGH 4/0 AWG SHALL HAVE A MAXIMUM 9" (225 mm) RUNG SPACING WHEN INSTALLED IN LADDER-TYPE CABLE TRAY PER 392.3(B)(1)(a).

RUNGS

CABLE TRAY

9" (225 mm) MAX.

SINGLE CONDUCTORS
392.3(B)(1)(a)

Purpose of Change: To change 230 mm to 225 mm to be consistent with **Tables 392.9** and **392.10(A)** of the **2005 NEC**.

NEC Ch. 3 – Article 392
Part – 392.4

Type of Change		Panel Action		UL	UL 508	API 500	API 505	OSHA
Revision		Accept		-	-	-	-	-
ROP		ROC		NFPA 70E	NFPA 70B	NFPA 79	NFPA	NEMA
pg. 999	# 8-271	pg. -	# -	-	-	-	-	-
log: 1740	CMP: 8	log: -	Submitter: Marcelo M. Hirschler			2002 NEC: 392.4		IEC: -

2002 NEC – 392.4 Uses Not Permitted

Cable tray systems shall not be used in hoistways or where subject to severe physical damage. Cable tray systems shall not be used in ~~environmental airspaces~~, except as permitted in 300.22, to support wiring methods recognized for use in such spaces.

2005 NEC – 392.4 Uses Not Permitted

Cable tray systems shall not be used in hoistways or where subject to severe physical damage. Cable tray systems shall not be used in <u>ducts, plenums, and other air-handling spaces,</u> except as permitted in 300.22, to support wiring methods recognized for use in such spaces.

Author's Substantiation. This revision correlates with the language in NFPA 90A for uses not permitted when installing cable tray.

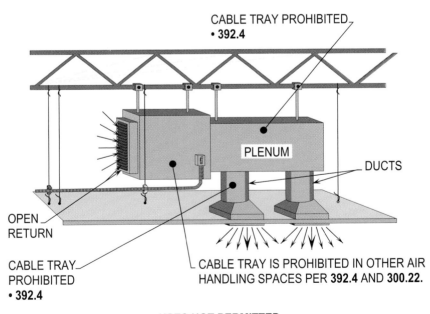

CABLE TRAY PROHIBITED
• 392.4

PLENUM

DUCTS

OPEN RETURN

CABLE TRAY PROHIBITED
• 392.4

CABLE TRAY IS PROHIBITED IN OTHER AIR HANDLING SPACES PER **392.4** AND **300.22.**

USES NOT PERMITTED
392.4

Purpose of Change: To revise the language in the NEC to correlate with the language in NFPA 90A.

NEC Ch. 3 – Article 392
Part – 392.7(A)

Type of Change		Panel Action		UL	UL 508	API 500	API 505	OSHA
Revision		Accept		1277	-	-	-	1910.305(c)
ROP		ROC		NFPA 70E	NFPA 70B	NFPA 79	NFPA	NEMA
pg. 1001	# 8-274	pg. 307	# 8-189	420.1(C)(2)	Ch. 24	14.5.10		VE 1 and 2
log: 1272	CMP: 8	log: 3449	Submitter: David H. Kendall			2002 NEC: 392.7(A)		IEC: 552.2

2002 NEC – 392.7(A) Metallic Cable Trays

Metallic cable trays that support electrical conductors shall be grounded as required for conductor enclosures in ~~Article 250~~.

2005 NEC – 392.7(A) Metallic Cable Trays

Metallic cable trays that support electrical conductors shall be grounded as required for conductor enclosures in <u>accordance with 250.96</u>.

Author's Substantiation. This revision clarifies that the user of the NEC shall reference 250.96 instead of Article 250 for grounding of metallic cable trays.

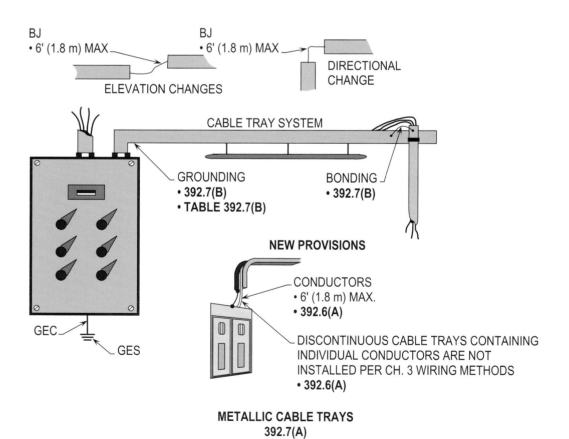

METALLIC CABLE TRAYS
392.7(A)

Purpose of Change: To reference only **250.96** instead of the entire Article.

NEC Ch. 3 – Article 392
Part – 392.10(A)(1)

Type of Change		Panel Action		UL	UL 508	API 500	API 505	OSHA
Revision		Accept in Principle		1277	-	-	-	1910.399
ROP		ROC		NFPA 70E	NFPA 70B	NFPA 79	NFPA	NEMA
pg. 1008	# 8-287	pg. 309	# 8-199	420.1(C)(2)	Ch. 24	14.5.10	-	VE 1 and 2
log: 3218	CMP: 8	log: 1386	Submitter: Brenda A. Carter			2002 NEC: 392.10(A))(1)		IEC: 552.2

2002 NEC – 392.10(A) Ladder or Ventilated Trough Cable Trays

Where ladder or ventilated trough cable trays contain single-conductor cables, the maximum number of single conductors shall conform to the following:

(1) Where all of the cables are 1000 kcmil or larger, the sum of the diameters of all single-conductor cables shall not exceed the cable tray width.

2005 NEC – 392.10(A) Ladder or Ventilated Trough Cable Trays

Where ladder or ventilated trough cable trays contain single-conductor cables, the maximum number of single conductors shall conform to the following:

(1) Where all of the cables are 1000 kcmil or larger, the sum of the diameters of all single-conductor cables shall not exceed the cable tray width, and the cables shall be installed in a single layer. Conductors that are bound together to comprise each circuit group shall be permitted to be installed in other than a single layer.

Author's Substantiation. This revision clarifies that cables larger than 1000 kcmil shall be installed in a single layer unless conductors are bound together to comprise each circuit group.

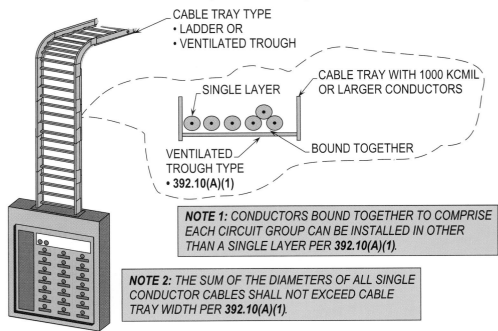

LADDER OR VENTILATED TROUGH CABLE TRAYS
392.10(A)(1)

Purpose of Change: To outline the requirements for installing single conductor and multiconductor cables in a single layer in a cable tray system.

NEC Ch. 3 – Article 392
Part – 392.11(B)(3), Ex.

Type of Change		Panel Action		UL	UL 508	API 500	API 505	OSHA
New Exception		Accept		1277	-	-	-	1910.305(c)
ROP		ROC		NFPA 70E	NFPA 70B	NFPA 79	NFPA	NEMA
pg. 1010	# 8-291	pg. -	# -	420.1(C)(2)	Ch. 24	14.5.10	-	VE 1 and 2
log: 1653	CMP: 8	log: -	Submitter: Richard J. Buschart			2002 NEC: -		IEC: 552.2

2002 NEC – 392.11(B)(3) Single-Conductor Cables

The allowable ampacity of single-conductor cables shall be as permitted by 310.15(A)(2). The derating factors of 310.15(B)(2)(a) shall not apply to the ampacity of cables in cable trays. The ampacity of single-conductor cables, or single conductors cabled together (triplexed, quadruplexed, etc.), nominally rated 2000 volts or less shall comply with the following:

(3) Where single conductors are installed in a single layer in uncovered cable trays, with a maintained space of not less than one cable diameter between individual conductors, the ampacity of 1/0 AWG and larger cables shall not exceed the allowable ampacities in Tables 310.17 and 310.19.

2005 NEC – 392.11(B)(3) Single-Conductor Cables

The allowable ampacity of single-conductor cables shall be as permitted by 310.15(A)(2). The derating factors of 310.15(B)(2)(a) shall not apply to the ampacity of cables in cable trays. The ampacity of single-conductor cables, or single conductors cabled together (triplexed, quadruplexed, etc.), nominally rated 2000 volts or less, shall comply with the following:

(3) Where single conductors are installed in a single layer in uncovered cable trays, with a maintained space of not less than one cable diameter between individual conductors, the ampacity of 1/0 AWG and larger cables shall not exceed the allowable ampacities in Table 310.17 and Table 310.19.

Exception to (B)(3): For solid bottom cable trays the ampacity of single conductor cables shall be determined by 310.15(C).

Author's Substantiation. A new exception has been added to clarify that the ampacity of single conductor cables shall be determined under engineering supervision per 310.15(C), due to solid bottom cable trays restricting air flow to single conductor cables.

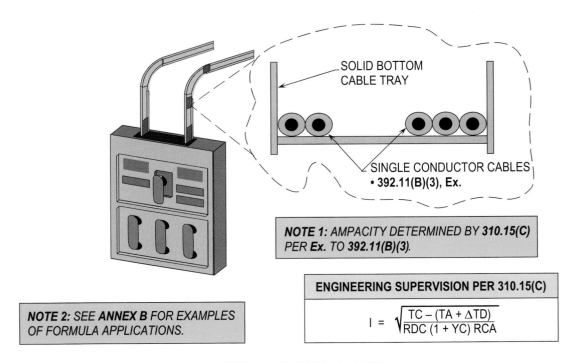

SOLID BOTTOM
CABLE TRAY

SINGLE CONDUCTOR CABLES
• **392.11(B)(3), Ex.**

NOTE 1: *AMPACITY DETERMINED BY **310.15(C)** PER **Ex.** TO **392.11(B)(3)**.*

NOTE 2: *SEE **ANNEX B** FOR EXAMPLES OF FORMULA APPLICATIONS.*

ENGINEERING SUPERVISION PER 310.15(C)

$$I = \sqrt{\dfrac{TC - (TA + \Delta TD)}{RDC\,(1 + YC)\,RCA}}$$

SINGLE-CONDUCTOR CABLES
392.11(B)(3), Ex.

Purpose of Change: To provide an alternate method for determining ampacity of the conductor in the cable.

NEC Ch. 3 – Article 398
Part II – 398.15(A)

Type of Change		Panel Action		UL	UL 508	API 500	API 505	OSHA
Revision		Accept		511	-	-	-	1910.305(a)(4)
ROP		ROC		NFPA 70E	NFPA 70B	NFPA 79	NFPA	NEMA
pg. 1013	# 7-211	pg. -	# -	-	-	-	-	HV 2
log: 947	CMP: 7	log: -	Submitter: James M. Daly			2002 NEC: 398.15(A)		IEC: -

2002 NEC – 398.15(A) Dry Locations

In dry locations, where not exposed to ~~severe~~ physical damage, conductors shall be permitted to be separately enclosed in flexible nonmetallic tubing. The tubing shall be in continuous lengths not exceeding 4.5 m (15 ft) and secured to the surface by straps at intervals not exceeding 1.4 m (4 1/2 ft).

2005 NEC – 398.15(A) Dry Locations

In dry locations, where not exposed to physical damage, conductors shall be permitted to be separately enclosed in flexible nonmetallic tubing. The tubing shall be in continuous lengths not exceeding 4.5 m (15 ft) and secured to the surface by straps at intervals not exceeding 1.4 m (4 1/2 ft).

Author's Substantiation. This revision clarifies that open wiring insulators shall not be exposed to any type of physical damage.

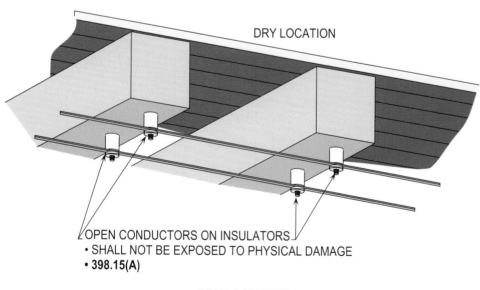

DRY LOCATION

OPEN CONDUCTORS ON INSULATORS
• SHALL NOT BE EXPOSED TO PHYSICAL DAMAGE
• 398.15(A)

DRY LOCATIONS
398.15(A)

Purpose of Change: To change the wording from "severe physical damage" to "physical damage" because the term "severe" was found to be too vague.

Equipment for General Use

Chapter 4 has always been utilized by users and maintainers, who have the responsibility of servicing, repairing, and maintaining electrical equipment.

If an electrician or user is replacing a ballast in a fluorescent luminaire (lighting fixture), **410.73** of the NEC must be reviewed before attempting to replace such ballast to ensure that the correct ballast and installation procedure is used. Consider a U frame motor that has been replaced with a T frame motor. In order to verify that the overloads are the proper size, **430.32(A)(1)** must be complied with. When maintenance must be performed on electrical equipment, the maintenance electrician or user must reference **Chapter 4** and use one of the Articles and Sections that cover the particular piece of equipment being repaired or replaced.

Electricians must not get confused when servicing or repairing specialized equipment that is not included in **Chapter 4** of the NEC. For example, to work on a crane, electricians must reference **Article 610 of Chapter 6**, which covers special equipment. To service an elevator, **Article 620** must be referenced and the proper Section selected that deals with the problem causing the trouble; note that other standards are sometimes needed to completely pull maintenance on various types of equipment.

Note that the rules of the NEC may apply when replacing equipment and components in existing equipment and apparatus.

The key Articles and Sections in **Chapter 4** are of the 400 series and are related to installing, replacing, servicing, repairing or using electrical equipment.

NEC Ch. 4 – Article 400
Part I – 400.5

Type of Change		Panel Action		UL	UL 508	API 500	API 505	OSHA
New Sentence		Accept in Part		62	-	-	-	1910.305(g)
ROP		ROC		NFPA 70E	NFPA 70B	NFPA 79	NFPA	NEMA
pg. 1016	# 6-78	pg. 311	# 6-62	410.6 and 7	19.5	Ch. 13	-	WC 5
log: 1521	CMP: 6	log: 3101	Submitter: Noel Williams			2002 NEC: 400.5		IEC: 522.7

2002 NEC – 400.5 Ampacities for Flexible Cords and Cables

Table 400.5(A) provides the allowable ampacities, and Table 400.5(B) provides the ampacities for flexible cords and cables with not more than three current-carrying conductors. These tables shall be used in conjunction with applicable end-use product standards to ensure selection of the proper size and type. If the number of current-carrying conductors exceeds three, the allowable ampacity or the ampacity of each conductor shall be reduced from the 3-conductor rating as shown in Table 400.5.

2005 NEC – 400.5 Ampacities for Flexible Cords and Cables

(A) Ampacity Tables. Table 400.5(A) provides the allowable ampacities, and Table 400.5(B) provides the ampacities for flexible cords and cables with not more than three current-carrying conductors. These tables shall be used in conjunction with applicable end-use product standards to ensure selection of the proper size and type. Where cords are used in ambient temperatures exceeding 30°C (86°F), the temperature correction factors from Table 310.16 that correspond to the temperature rating of the cord shall be applied to the ampacity for Table 400.5(B). Where the number of current-carrying conductors exceeds three, the allowable ampacity or the ampacity of each conductor shall be reduced from the 3-conductor rating as shown in Table 400.5.

Author's Substantiation. A new sentence has been added to clarify that the temperature correction factors from Table 310.16 that correspond to the temperature rating of the cord shall be applied to the ampacity for Table 400.5 where such cords are used in ambient temperatures exceeding 30°C (86°F).

TEMPORARY INSTALLATIONS PER **ARTICLE 590**

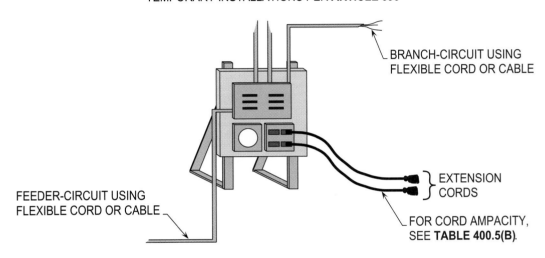

BRANCH-CIRCUIT USING
FLEXIBLE CORD OR CABLE

EXTENSION
CORDS

FEEDER-CIRCUIT USING
FLEXIBLE CORD OR CABLE

FOR CORD AMPACITY,
SEE **TABLE 400.5(B)**.

NOTE: *WHERE CORDS ARE RUN IN AMBIENT TEMPERATURES EXCEEDING 86°F (30°C), CORRECTION FACTORS PER* **TABLE 310.16** *SHALL BE APPLIED, BASED ON SUCH TEMPERATURES PER* **400.5**.

AMPACITIES FOR FLEXIBLE CORDS AND CABLES
400.5

Purpose of Change: To refer the user to Tables for selecting ampacities and correction factors where there are excessive temperatures surrounding the cords and cables.

NEC Ch. 4 – Article 400
Part I – 400.7(A)(10)

Type of Change		Panel Action		UL	UL 508	API 500	API 505	OSHA
Revision		Accept in Part		62	-	-	-	1910.305(g)
ROP		ROC		NFPA 70E	NFPA 70B	NFPA 79	NFPA	NEMA
pg. 1018	# 6-85	pg. 312	# 6-64	410.6 and 7	19.5	Ch. 13	-	WC 5
log: 717	CMP: 6	log: 101	Submitter: Dan Leaf			2002 NEC: 400.7(A)		IEC: 522.7

2002 NEC – 400.7 Uses Permitted

(A) Uses. Flexible cords and cables shall be used only for the following:

(1) Pendants

(2) Wiring of luminaires (fixtures)

(3) Connection of portable lamps, portable and mobile signs, or appliances

(4) Elevator cables

(5) Wiring of cranes and hoists

(6) Connection of utilization equipment to facilitate frequent interchange

(7) Prevention of the transmission of noise or vibration

(8) Appliances where the fastening means and mechanical connections are specifically designed to permit ready removal for maintenance and repair, and the appliance is intended or identified for flexible cord connection

(9) ~~Data processing cables as permitted by 645.5~~

(10) Connection of moving parts

(11) ~~Temporary wiring as permitted in 527.4(B) and 527.4(C)~~

2005 NEC – 400.7 Uses Permitted

(A) Uses. Flexible cords and cables shall be used only for the following:

(1) Pendants

(2) Wiring of luminaires (fixtures)

(3) Connection of portable lamps, portable and mobile signs, or appliances

(4) Elevator cables

(5) Wiring of cranes and hoists

(6) Connection of utilization equipment to facilitate frequent interchange

(7) Prevention of the transmission of noise or vibration

(8) Appliances where the fastening means and mechanical connections are specifically designed to permit ready removal for maintenance and repair, and the appliance is intended or identified for flexible cord connection

(9) Connection of moving parts

(10) <u>Where specifically permitted elsewhere in this *Code*</u>

Author's Substantiation. This revision clarifies that Chapters 1 through 4 apply except as amended by Chapters 5, 6, and 7 for the particular conditions.

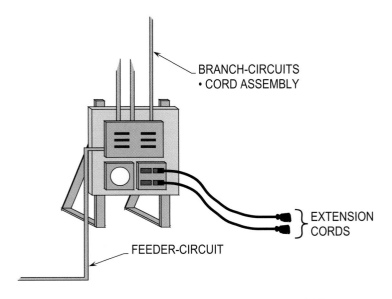

BRANCH-CIRCUITS
• CORD ASSEMBLY

EXTENSION CORDS

FEEDER-CIRCUIT

(B) REFERENCE TO TEMPORARY INSTALLATIONS IN **400.7(A)(11)** WAS DELETED IN THE **2005 NEC**.

(A) REFERENCE TO DATA PROCESSING CABLES IN **400.7(A)(9)** WAS DELETED IN THE **2005 NEC**.

USES PERMITTED
400.7(A)(10)

Purpose of Change: The words "Where specifically permitted elsewhere in this *Code*" were added to clarify that Chapter 1 through 4 apply, except as amended by Chapters 5, 6, and 7 for the particular conditions.

NEC Ch. 4 – Article 400
Part I – 400.14

Type of Change	Panel Action	UL	UL 508	API 500	API 505	OSHA
New Paragraph	Accept	62	-	-	-	1910.305(g)
ROP		ROC	NFPA 70E	NFPA 70B	NFPA 79	NFPA	NEMA

ROP		ROC		NFPA 70E	NFPA 70B	NFPA 79	NFPA	NEMA
pg. 1024	# 6-95	pg. 313	# 6-77	410.6 and 7	19.5	Ch. 13	-	WC 5
log: 911	CMP: 6	log: 575	Submitter: Samuel B. Friedman			2002 NEC: 400.14		IEC: 522.7

2002 NEC – 400.14 Protection from Damage

Flexible cords and cables shall be protected by bushings or fittings where passing through holes in covers, outlet boxes, or similar enclosures.

2005 NEC – 400.14 Protection from Damage

Flexible cords and cables shall be protected by bushings or fittings where passing through holes in covers, outlet boxes, or similar enclosures.

<u>In industrial establishments where the conditions of maintenance and supervision ensure that only qualified persons service the installation, flexible cords and cables shall be permitted to be installed in aboveground raceways that are not longer than 15 m (50 ft) to protect the flexible cord or cable from physical damage. Where more than three current-carrying conductors are installed within the raceway, the allowable ampacity shall be reduced in accordance with Table 400.5.</u>

Author's Substantiation. A new paragraph has been added to clarify that flexible cords and cables shall be permitted to be installed in aboveground raceways that are not longer than 15 m (50 ft) to protect the flexible cord or cable from physical damage. This type of installation shall be in industrial establishments only where the conditions of maintenance and supervision ensure that only qualified persons service the installation. The allowable ampacity shall be reduced in accordance with Table 400.5 where more than three current-carrying conductors are installed within the raceway.

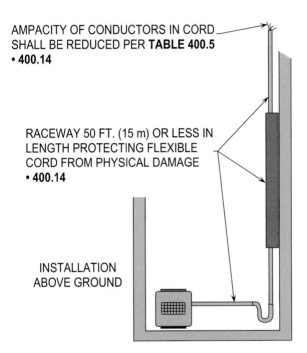

AMPACITY OF CONDUCTORS IN CORD SHALL BE REDUCED PER **TABLE 400.5**
• **400.14**

RACEWAY 50 FT. (15 m) OR LESS IN LENGTH PROTECTING FLEXIBLE CORD FROM PHYSICAL DAMAGE
• **400.14**

INSTALLATION ABOVE GROUND

PROTECTION FROM DAMAGE
400.14

Purpose of Change: To recognize the use of a raceway to protect a flexible cord from physical damage.

NEC Ch. 4 – Article 404
Part I – 404.7, Ex. 2

91

Type of Change		Panel Action		UL	UL 508	API 500	API 505	OSHA
New Exception		Accept		20	15	-	-	1910.305(c)(2)
ROP		ROC		NFPA 70E	NFPA 70B	NFPA 79	NFPA	NEMA
pg. 1032	# 9-80a	pg. -	# -	420.3	11.5	3.3.100	-	WD 1
log: CP903	CMP: 9	log: -	Submitter: CMP 9			2002 NEC: -		IEC: 537

2005 NEC – 404.7 Indicating

General-use and motor-circuit switches, circuit breakers, and molded case switches, where mounted in an enclosure as described in 404.3, shall clearly indicate whether they are in the open (off) or closed (on) position.

Where these switch or circuit breaker handles are operated vertically rather than rotationally or horizontally, the up position of the handle shall be the (on) position.

> **Exception No. 2.** On busway installations, tap switches employing a center-pivoting handle shall be permitted to be open or closed with either end of the handle in the up or down position. The switch position shall be clearly indicating and shall be visible from the floor or from the usual point of operation.

Author's Substantiation. A new Exception No. 2 has been added to address busway switch designs that have one end of the handle in the up position at all times, enabling a downward pull to change the switch position at all times.

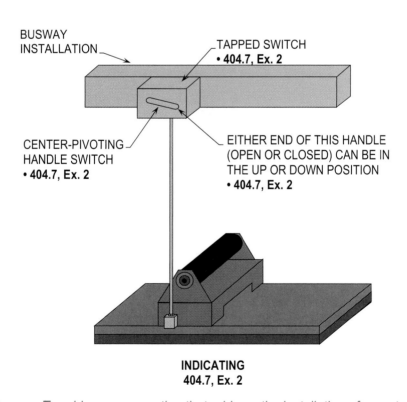

BUSWAY INSTALLATION

TAPPED SWITCH
• **404.7, Ex. 2**

CENTER-PIVOTING HANDLE SWITCH
• **404.7, Ex. 2**

EITHER END OF THIS HANDLE (OPEN OR CLOSED) CAN BE IN THE UP OR DOWN POSITION
• **404.7, Ex. 2**

INDICATING
404.7, Ex. 2

Purpose of Change: To add a new exception that address the installation of a center-pivoting handle switch on a busway.

NEC Ch. 4 – Article 404
Part I – 404.8(B)

Type of Change		Panel Action		UL	UL 508	API 500	API 505	OSHA
Revision		Accept in Principle		20	15	-	-	1910.305(c)(2)
ROP		ROC		NFPA 70E	NFPA 70B	NFPA 79	NFPA	NEMA
pg. 1035	# 9-86	pg. 317	# 9-117	420.3	16.5	3.3.100	-	WD 1
log: 1274	CMP: 9	log: 2029	Submitter: David H. Kendall			2002 NEC: 404.8(B)		IEC: 537

2002 NEC – 404.8(B) Voltage Between Adjacent Devices

A snap switch shall not be grouped or ganged in enclosures with other snap switches, receptacles, or similar devices, unless they are arranged so that the voltage between adjacent devices does not exceed 300 volts, or unless they are installed in enclosures equipped with ~~permanently~~ installed barriers between adjacent devices.

2005 NEC – 404.8(B) Voltage Between Adjacent Devices

A snap switch shall not be grouped or ganged in enclosures with other snap switches, receptacles, or similar devices, unless they are arranged so that the voltage between adjacent devices does not exceed 300 volts, or unless they are installed in enclosures equipped with <u>identified, securely</u> installed barriers between adjacent devices.

Author's Substantiation. This revision clarifies that a snap switch shall not be grouped or ganged in enclosures with other snap switches, receptacles, or similar devices, unless they are installed in enclosures equipped with identified, securely installed barriers between adjacent devices.

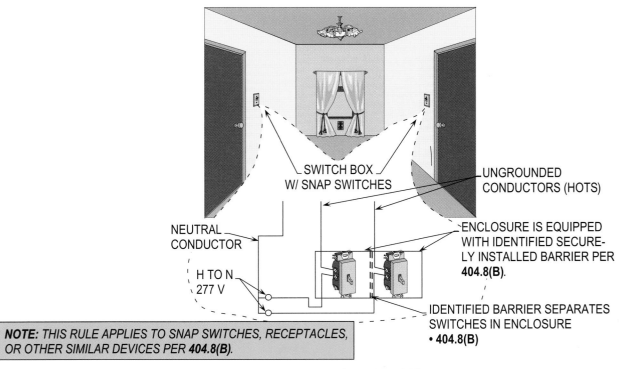

NEUTRAL CONDUCTOR

SWITCH BOX W/ SNAP SWITCHES

UNGROUNDED CONDUCTORS (HOTS)

ENCLOSURE IS EQUIPPED WITH IDENTIFIED SECURE-LY INSTALLED BARRIER PER 404.8(B).

H TO N 277 V

IDENTIFIED BARRIER SEPARATES SWITCHES IN ENCLOSURE • 404.8(B)

*NOTE: THIS RULE APPLIES TO SNAP SWITCHES, RECEPTACLES, OR OTHER SIMILAR DEVICES PER **404.8(B)**.*

VOLTAGE BETWEEN ADJACENT DEVICES
404.8(B)

Purpose of Change: To provide requirements for multiple fed general-use snap switches.

NEC Ch. 4 – Article 406
Part – 406.4(B)

Type of Change		Panel Action		UL	UL 508	API 500	API 505	OSHA
Revision		Accept in Principle in Part		498	15	-	-	1910.305(g)
ROP		ROC		NFPA 70E	NFPA 70B	NFPA 79	NFPA	NEMA
pg. 1046	# 18-20	pg. -	# -	420.3(D)	11.5	3.3.100	-	PR 4
log: 748	CMP: 18	log: -	Submitter: Dan Leaf			2002 NEC: 406.4(B)		IEC: 413.1.12

2002 NEC – 406.4(B) Boxes That Are Flush

Receptacles mounted in boxes that are flush with the ~~wall~~ surface or project therefrom shall be installed ~~so~~ that the mounting yoke or strap of the receptacle is held rigidly against the box or ~~raised~~ box cover.

2005 NEC – 406.4(B) Boxes That Are Flush

Receptacles mounted in boxes that are flush with the <u>finished</u> surface or project therefrom shall be installed <u>such</u> that the mounting yoke or strap of the receptacle is held rigidly against the box or box cover.

Author's Substantiation. This revision clarifies that receptacles mounted in boxes shall be set flush with the finished surface.

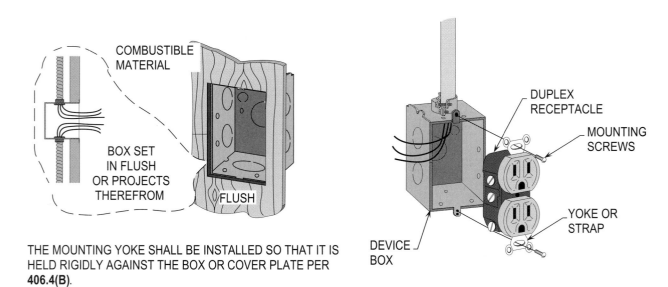

COMBUSTIBLE MATERIAL

BOX SET IN FLUSH OR PROJECTS THEREFROM

FLUSH

THE MOUNTING YOKE SHALL BE INSTALLED SO THAT IT IS HELD RIGIDLY AGAINST THE BOX OR COVER PLATE PER **406.4(B)**.

DUPLEX RECEPTACLE

MOUNTING SCREWS

YOKE OR STRAP

DEVICE BOX

BOXES THAT ARE FLUSH
406.4(B)

Purpose of Change: To provide requirements for mounting of receptacles to boxes set flush with the finished surface.

NEC Ch. 4 – Article 406
Part – 406.4(D), Ex.

92

Type of Change		Panel Action		UL	UL 508	API 500	API 505	OSHA
New Exception		Accept in Part		498	15	-	-	1910.305(c)(2)
ROP		ROC		NFPA 70E	NFPA 70B	NFPA 79	NFPA	NEMA
pg. 1047	# 18-21	pg. 320	# 8-12a	420.3(D)	11.5	3.3.100	-	PR 4
log: 3209	CMP: 18	log: CC1800	Submitter: Stewart Rappaport			2002 NEC: -		IEC: 537

2002 NEC – 406.4(D) Position of Receptacle Faces

After installation, receptacle faces shall be flush with or project from faceplates of insulating material and shall project a minimum of 0.4 mm (0.015 in.) from metal faceplates.

2005 NEC – 406.4(D) Position of Receptacle Faces

After installation, receptacle faces shall be flush with or project from faceplates of insulating material and shall project a minimum of 0.4 mm (0.015 in.) from metal faceplates.

> **Exception No. 1:** Listed kits or assemblies encompassing receptacles and nonmetallic faceplates that cover the receptacle face, where the plate cannot be installed on any other receptacle, shall be permitted.

Author's Substantiation. A new exception has been added to clarify that listed kits or assemblies encompassing receptacles and nonmetallic faceplates that cover the receptacle face shall not be required to be flush with or project from faceplates of insulating material where the plate cannot be installed on any other receptacle.

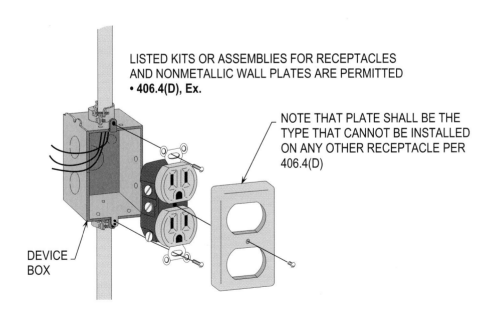

LISTED KITS OR ASSEMBLIES FOR RECEPTACLES AND NONMETALLIC WALL PLATES ARE PERMITTED
• 406.4(D), Ex.

NOTE THAT PLATE SHALL BE THE TYPE THAT CANNOT BE INSTALLED ON ANY OTHER RECEPTACLE PER 406.4(D)

DEVICE BOX

POSITION OF RECEPTACLE FACES
406.4(D), Ex.

Purpose of Change: To allow the use of listed kits or assemblies making up types of cover plates.

NEC Ch. 4 – Article 406
Part – 406.6

93

Type of Change	Panel Action	UL	UL 508	API 500	API 505	OSHA
Revision	Acceot in Principle in Part	498	15	-	-	1910.305(g)
ROP		ROC		NFPA 70E	NFPA 70B	NFPA 79	NFPA	NEMA
pg. 1049	# 18-27	pg. -	# -	420.3(D)	11.5	3.3.100	-	PR 4
log: 742	CMP: 18	log: -	Submitter: Dan Leaf			2002 NEC: 406.6		IEC: 413.1.12

2002 NEC – 406.6 Attachment Plugs

All attachment plugs and cord connectors shall be listed ~~for the purpose~~ and marked with the manufacturer s name or identification and voltage and ampere ratings.

2005 NEC – 406.6 Attachment Plugs, Cord Connectors, and Flanged Surface Devices

All attachment plugs, cord connectors, and flanged surface devices (inlets and outlets) shall be listed and marked with the manufacturer s name or identification and voltage and ampere ratings.

Author's Substantiation. This revision recognizes that flanged surface devices (inlets and outlets) shall be listed and marked with the manufacturer's name or identification and voltage and ampere rating. A flanged surface inlet shall be installed such that the prongs, blades, or pins are not energized unless an energized cord connector is inserted into it.

NOTE 1: *ATTACHMENT PLUGS, CORD CONNECTORS, AND FLANGED SURFACE DEVICES (INLETS AND OUTLETS) SHALL BE LISTED AND MARKED PER 406.6.*

NOTE 2: *FOR INSTALLATION REQUIREMENTS, SEE 406.6(A) THRU (D) OF THE 2005 NEC.*

**ATTACHMENT PLUGS, CORD CONNECTORS, AND FLANGED SURFACE DEVICES
406.6**

Purpose of Change: To recognize that cord connectors and flanged surface devices shall be listed and marked.

NEC Ch. 4 – Article 406
Part – 406.9(B)(4)

Type of Change		Panel Action		UL	UL 508	API 500	API 505	OSHA
Revision		Accept		498	40.1.5	-	-	1910.305(c)(2)
ROP		ROC		NFPA 70E	NFPA 70B	NFPA 79	NFPA	NEMA
pg. 1060	# 18-48	pg. 323	# 18-25	420.3(D)	18.2.2	5.2	-	PR 4
log: 605	CMP: 18	log: 526	Submitter: Vince Baclawski			2002 NEC: 406.9(B)(4)		IEC: 413.1.12

2002 NEC – 406.9(B) Grounding-Pole Identification

Grounding-type receptacles, adapters, cord connections, and attachment plugs shall have a means for connection of a grounding conductor to the grounding pole.

A terminal for connection to the grounding pole shall be designated by one of the following:

(4) If the terminal for the ~~equipment~~ grounding conductor is not visible, the conductor entrance hole shall be marked with the word *green* or *ground*, the letters *G* or *GR*, ~~or the~~ grounding symbol, ~~as shown in Figure 406.9(B)(4),~~ or otherwise identified by a distinctive green color. If the terminal for the equipment grounding conductor is readily removable, the area adjacent to the terminal shall be similarly marked.

Figure 406.9(B)(4) ~~Grounding Symbol~~

2005 NEC – 406.9(B) Grounding-Pole Identification

Grounding-type receptacles, adapters, cord connections, and attachment plugs shall have a means for connection of a grounding conductor to the grounding pole.

A terminal for connection to the grounding pole shall be designated by one of the following:

(4) If the terminal for the grounding conductor is not visible, the conductor entrance hole shall be marked with the word *green* or *ground*, the letters *G* or *GR*, a grounding symbol, or otherwise identified by a distinctive green color. If the terminal for the equipment grounding conductor is readily removable, the area adjacent to the terminal shall be similarly marked.

FPN: See FPN Figure 406.9(B)(4).

FPN Figure 406.9(B)(4): One Example of a Symbol Used to Identify the Termination Point for an Equipment Grounding Conductor.

Author's Substantiation. A new FPN has been added to clarify that more than one symbol is used to identify the termination point for an equipment grounding conductor.

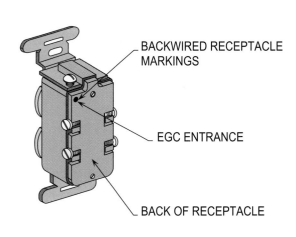

BACKWIRED RECEPTACLE
MARKINGS

EGC ENTRANCE

BACK OF RECEPTACLE

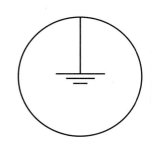

THIS SYMBOL IS ONE EXAMPLE TO
IDENTIFY THE TERMINATION POINT
FOR AN EQUIPMENT GROUNDING
CONDUCTOR.

CHOICE OF MARKING
WITH THE WORDS:
• GREEN
• GROUND
WITH THE LETTERS:
• "G"
• "GR"
OR WITH THE GROUNDING SYMBOLS
SHOWN
• **FIGURE 406.9(B)(4)**

GROUNDING–POLE IDENTIFICATION
406.9(B)(4)

Purpose of Change: To add another acceptable grounding symbol without the circle around it.

NEC Ch. 4 – Article 408
Part I – 408.3(C)

Type of Change		Panel Action		UL	UL 508	API 500	API 505	OSHA
Revision		Accept in Principle		16 and 1558	-	-	-	1910.305(d)
ROP		ROC		NFPA 70E	NFPA 70B	NFPA 79	NFPA	NEMA
pg. 1064	# 9-103	pg. -	# -	420.4	27.2.2.6	-	-	PB 1
log: 1927	CMP: 9	log: -	Submitter: Donald A. Ganiere			2002 NEC: 408.3(C)		IEC: 53

2002 NEC – 408.3(C) Used as Service Equipment

Each switchboard or panelboard, if used as service equipment, shall be provided with a main bonding jumper sized in accordance with 250.28(D) or the equivalent placed within the panelboard or one of the sections of the switchboard for connecting the grounded service conductor on its supply side to the switchboard or panelboard frame. All sections of a switchboard shall be bonded together using an equipment ~~grounding~~ conductor sized in accordance with Table 250.122.

2005 NEC – 408.3(C) Used as Service Equipment

Each switchboard or panelboard, if used as service equipment, shall be provided with a main bonding jumper sized in accordance with 250.28(D) or the equivalent placed within the panelboard or one of the sections of the switchboard for connecting the grounded service conductor on its supply side to the switchboard or panelboard frame. All sections of a switchboard shall be bonded together using an equipment <u>bonding</u> conductor sized in accordance with Table 250.122 <u>or Table 250.66 as appropriate</u>.

Author's Substantiation. This revision clarifies the procedures for sizing the main bonding jumper and equipment bonding jumper.

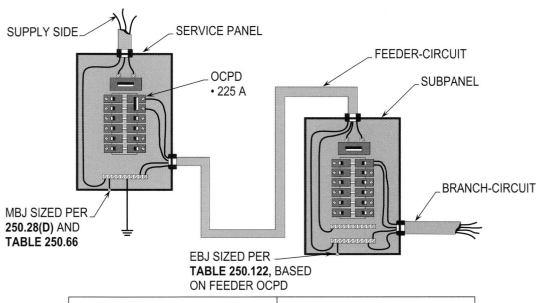

SUPPLY SIDE

SERVICE PANEL

FEEDER-CIRCUIT

OCPD
• 225 A

SUBPANEL

BRANCH-CIRCUIT

MBJ SIZED PER
250.28(D) AND
TABLE 250.66

EBJ SIZED PER
TABLE 250.122, BASED
ON FEEDER OCPD

Example:	Supply 400 kcmil cu.	**Example:**	Feeder OCPD 225 A
Solution:	**A 400 kcmil cu. requires a 1/0 AWG MBJ.**	**Solution:**	**A 225 A CB requires a 4 AWG EBJ.**

USED AS SERVICE EQUIPMENT
408.3(C)

Purpose of Change: To clarify the procedure for sizing the main bonding jumper and the equipment bonding jumper on the supply and load side of the electrical system.

NEC Ch. 4 – Article 408
Part I – 408.5

Type of Change	Panel Action	UL	UL 508	API 500	API 505	OSHA
Relocated	Accept in Principle	67 and 1558	-	-	-	1910.305(d)
ROP		ROC		NFPA 70E	NFPA 70B	NFPA 79	NFPA	NEMA
pg. 1066	# 9-107	pg. -	# -	400.24(F)	24.4	-	-	PB 1
log: 2862	CMP: 9	log: -	Submitter: Alan Manche			2002 NEC: 408.10		IEC: 53

2002 NEC – ~~408.10~~ Clearance for Conductors Entering Bus Enclosures

Where conduits or other raceways enter a switchboard, floor-standing panelboard, or similar enclosure at the bottom, sufficient space shall be provided to permit installation of conductors in the enclosure. The wiring space shall not be less than shown in ~~Table 408.10~~ where the conduit or raceways enter or leave the enclosure below the busbars, their supports, or other obstructions. The conduit or raceways, including their end fittings, shall not rise more than 75 mm (3 in.) above the bottom of the enclosure.

2005 NEC – 408.5 Clearance for Conductor Entering Bus Enclosures

Where conduits or other raceways enter a switchboard, floor-standing panelboard, or similar enclosure at the bottom, sufficient space shall be provided to permit installation of conductors in the enclosure. The wiring space shall not be less than shown in Table 408.5 where the conduit or raceways enter or leave the enclosure below the busbars, their supports, or other obstructions. The conduit or raceways, including their end fittings, shall not rise more than 75 mm (3 in.) above the bottom of the enclosure.

Author's Substantiation. This section has been relocated and renumbered to clarify that this is a general requirement for conduits or other raceways that enter a switchboard, floor-standing panelboard, or similar enclosure at the bottom where sufficient space is to be provided.

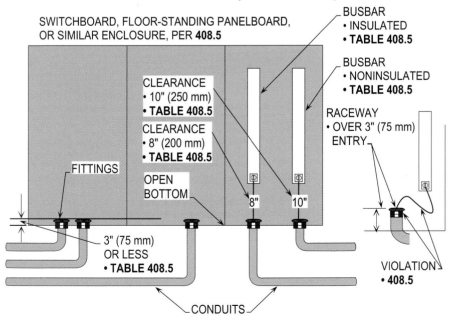

CLEARANCE FOR CONDUCTOR ENTERING BUS ENCLOSURES
408.5

Purpose of Change: To relocate this section as a general requirement for conduits or other raceways that enter a switchboard, floor-standing panelboard, or similar enclosure at the bottom where sufficient space is to be provided.

NEC Ch. 4 – Article 408
Part III – 408.30

Type of Change		Panel Action		UL	UL 508	API 500	API 505	OSHA
Revision		Accept		67 and 1558	-	-	-	1910.305(d)
ROP		ROC		NFPA 70E	NFPA 70B	NFPA 79	NFPA	NEMA
pg. 1066	# 9-108	pg. -	# -	420.4	27.2.2.6	16.2	-	PB 1
log: 1320	CMP: 9	log: -	Submitter: TCC			2002 NEC: 408.13		IEC: 53

2002 NEC – ~~408.13~~ General

All panelboards shall have a rating not less than the minimum feeder capacity required for the load ~~computed~~ in accordance with Article 220. Panelboards shall be durably marked by the manufacturer with the voltage and the current rating and the number of phases for which they are designed and with the manufacturer s name or trademark in such a manner so as to be visible after installation, without disturbing the interior parts or wiring.

2002 NEC – <u>408.30</u> General

All panelboards shall have a rating not less than the minimum feeder capacity required for the load <u>calculated</u> in accordance with Article 220. Panelboards shall be durably marked by the manufacturer with the voltage and the current rating and the number of phases for which they are designed and with the manufacturer s name or trademark in such a manner so as to be visible after installation, without disturbing the interior parts or wiring.

Author's Substantiation. This change clarifies that the term "calculate" will be standardized language throughout the NEC for determining load values.

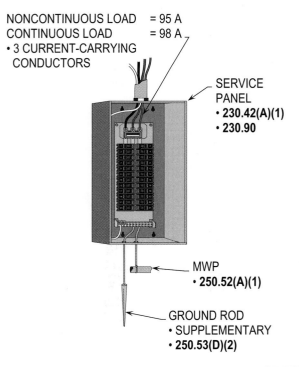

NONCONTINUOUS LOAD = 95 A
CONTINUOUS LOAD = 98 A
• 3 CURRENT-CARRYING CONDUCTORS

SERVICE PANEL
• 230.42(A)(1)
• 230.90

MWP
• 250.52(A)(1)

GROUND ROD
• SUPPLEMENTARY
• 250.53(D)(2)

What size OCPD and THWN cu. conductors are required to supply loads?

Step 1: Calculating load
230.90; 230.42(A)(1)
95 A x 100% = 95 A
98 A x 125% = 122.5 A
Total load = 217.5 = 218 A

Step 2: Finding OCPD (based on load)
230.90; 240.4; 240.6(A)
218 A requires 225 A OCPD

Step 3: Finding conductors (based on load)
230.42(A)(1); 230.90; Table 310.16
225 A requires 4/0 AWG THWN

Solution: 4/0 AWG THWN cu. conductors and a 225 amp OCPD is required.

GENERAL
408.30

Purpose of Change: To replace the term "compute" with the term "calculate" so that consistency can be achieved throughout the NEC.

NEC Ch. 4 – Article 408
Part III – 408.36(B)

Type of Change		Panel Action		UL	UL 508	API 500	API 505	OSHA
Revision		Accept in Principle		67 and 1558	-	-	-	1910.305(d)
ROP		ROC		NFPA 70E	NFPA 70B	NFPA 79	NFPA	NEMA
pg. 1067	# 9-112	pg. -	# -	420.4	27.2.2.6	16.2	-	PB 1
log: 1400	CMP: 9	log: -	Submitter: Raj Awesti			2002 NEC: 408.16(B)		IEC: 53

2002 NEC – ~~408.16(B)~~ Power Panelboard Protection

In addition to the requirements of ~~408.13~~, a power panelboard with supply conductors that include a neutral and having more than 10 percent of its overcurrent devices protecting branch circuits rated 30 amperes or less shall be protected by an overcurrent protective device having a rating not greater than that of the panelboard. ~~The~~ overcurrent protective device shall be located within or at any point on the supply side of the panelboard.

2005 NEC – 408.36(B) Power Panelboard Protection

In addition to the requirements of 408.30, a power panelboard with supply conductors that include a neutral, and having more than 10 percent of its overcurrent devices protecting branch circuits rated 30 amperes or less, shall be protected by an overcurrent protective device having a rating not greater than that of the panelboard. This overcurrent protective device shall be located within or at any point on the supply side of the panelboard.

Author's Substantiation. This revision correlates with the exception for describing the individual protection that shall not be required for a power panelboard used as service equipment with multiple disconnecting means.

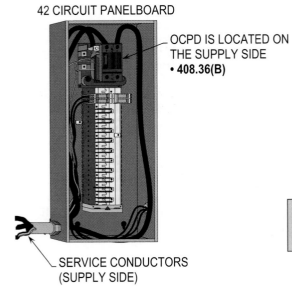

42 CIRCUIT PANELBOARD

OCPD IS LOCATED ON THE SUPPLY SIDE
• 408.36(B)

SERVICE CONDUCTORS (SUPPLY SIDE)

Is this a power panelboard?
Step 1: Applying 10% rule **408.36(B)** 42 circuits x 10% = 4.2
Step 2: Type of panel **408.36(B)** 6 -SP's exceed 4.2 SP's
Solution: Yes, this panel is a power panelboard.

*NOTE: IN THIS PANELBOARD, THERE ARE 6 CIRCUITS WITH NEUTRAL CONNECTIONS RATED AT 30 A OR LESS PER **408.36(B)**.*

POWER PANELBOARD PROTECTION
408.36(B)

Purpose of Change: To revise the text to correlate with the wording of the exception.

NEC Ch. 4 – Article 409
Part I – 409.1

 94.1

Type of Change		Panel Action		UL	UL 508	API 500	API 505	OSHA
New Article		Accept		508	18	-	-	1910.305(j)(4)
ROP		ROC		NFPA 70E	NFPA 70B	NFPA 79	NFPA	NEMA
pg. 1076	# 11-5	pg. -	# -	400.11	21.4.5	Ch. 9, 10, and 12	75	ICS 1, 2, and 5
log: 2536	CMP: 11	log: CC1101	Submitter: Alan Manche			2002 NEC: -		IEC: 60204-1

2005 NEC – 409.1 Scope

This article covers industrial control panels intended for general use and operating at 600 volts or less.

FPN: UL 508A is a safety standard for industrial control panels.

Author's Substantiation. A new article has been added to address industrial control panels intended for general use and operating at 600 volts or less.

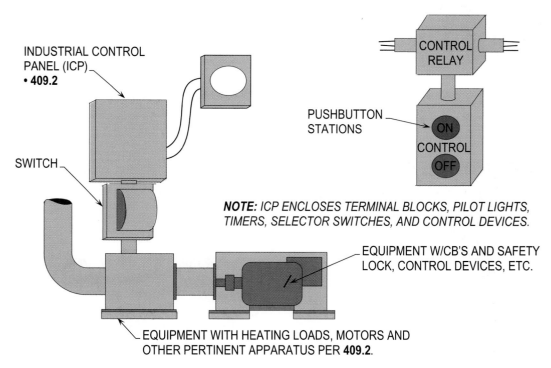

INDUSTRIAL CONTROL
PANEL (ICP)
• **409.2**

SWITCH

CONTROL RELAY

PUSHBUTTON
STATIONS

ON
CONTROL
OFF

NOTE: *ICP ENCLOSES TERMINAL BLOCKS, PILOT LIGHTS, TIMERS, SELECTOR SWITCHES, AND CONTROL DEVICES.*

EQUIPMENT W/CB'S AND SAFETY
LOCK, CONTROL DEVICES, ETC.

EQUIPMENT WITH HEATING LOADS, MOTORS AND
OTHER PERTINENT APPARATUS PER **409.2**.

**INDUSTRIAL CONTROL PANELS
409.1 AND 409.2**

Purpose of Change: To provide a new Article to address the design and installation of industrial control panels (ICP's) and their associated apparatus.

NEC Ch. 4 – Article 409
Part I – 409.2

 94.2

Type of Change		Panel Action		UL	UL 508	API 500	API 505	OSHA
New Article		Accept		508	18	-	-	1910.305(j)(4)
ROP		ROC		NFPA 70E	NFPA 70B	NFPA 79	NFPA	NEMA
pg. 1076	# 11-5	pg. 325	# 11-3a	400.11	21.4.5	Ch. 9, 10, and 12	75	ICS 1, 2, and 5
log: 2536	CMP: 11	log: CC1101	Submitter: Alan Manche			2002 NEC: -		IEC: 60204-1

2005 NEC – 409.2 Definitions

Industrial Control Panel. An assembly of a systematic and standard arrangement of two or more components such as motor controllers, overload relays, fused disconnect switches, and circuit breakers and related control devices such as pushbutton stations, selector switches, timers, switches, control relays, and the like with associated wiring, terminal blocks, pilot lights, and similar components. The industrial control panel does not include the controlled equipment.

Author's Substantiation. A new article has been added to address industrial control panels intended for general use and operating at 600 volts or less. A new definition has been added to describe an industrial control panel.

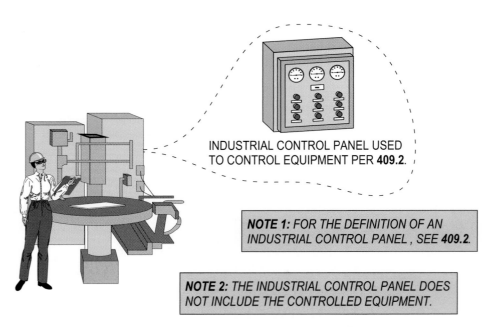

INDUSTRIAL CONTROL PANEL USED TO CONTROL EQUIPMENT PER **409.2**.

NOTE 1: *FOR THE DEFINITION OF AN INDUSTRIAL CONTROL PANEL, SEE **409.2**.*

NOTE 2: *THE INDUSTRIAL CONTROL PANEL DOES NOT INCLUDE THE CONTROLLED EQUIPMENT.*

INDUSTRIAL CONTROL PANEL
409.2

Purpose of Change: To define an industrial control panel in new **Article 409**.

NEC Ch. 4 – Article 409
Part II – 409.20

94.3

Type of Change		Panel Action		UL	UL 508	API 500	API 505	OSHA
New Article		Accept		508	21	-	-	1910.305(j)(4)
ROP		ROC		NFPA 70E	NFPA 70B	NFPA 79	NFPA	NEMA
pg. 1076	# 11-5	pg. -	# -	400.11	3.3.18	7.2.13	75	ICS 1, 2, and 5
log: 2536	CMP: 11	log: -	Submitter: Alan Manche			2002 NEC: -		IEC: 60204-1

2005 NEC – 409.20 Conductor – Minimum Size and Ampacity

The size of the industrial control panel supply conductor shall have an ampacity not less than 125 percent of the full-load current rating of all resistance heating loads plus 125 percent of the full-load current rating of the highest rated motor plus the sum of the full-load current ratings of all other connected motors and apparatus based on their duty cycle that may be in operation at the same time.

Author's Substantiation. A new article has been added to address industrial control panels intended for general use and operating at 600 volts or less. This section addresses the procedures for sizing and selecting conductors supplying an industrial control panel.

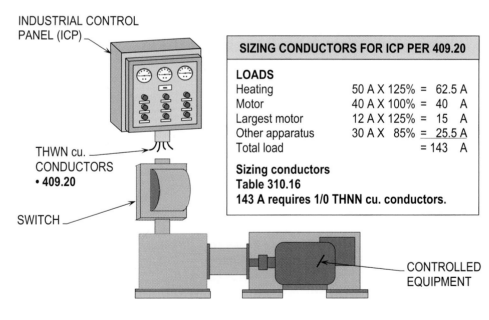

INDUSTRIAL CONTROL PANEL (ICP)

THWN cu. CONDUCTORS
• 409.20

SWITCH

CONTROLLED EQUIPMENT

SIZING CONDUCTORS FOR ICP PER 409.20

LOADS
Heating	50 A X 125% = 62.5 A
Motor	40 A X 100% = 40 A
Largest motor	12 A X 125% = 15 A
Other apparatus	30 A X 85% = 25.5 A
Total load	= 143 A

Sizing conductors
Table 310.16
143 A requires 1/0 THNN cu. conductors.

CONDUCTOR – MINIMUM SIZE AND AMPACITY
409.20

Purpose of Change: To provide a new Article addressing the procedure for sizing and selecting conductors supplying an industrial control panel (ICP).

NEC Ch. 4 – Article 409
Part II – 409.21(A) and (B)

 94.4

Type of Change		Panel Action		UL	UL 508	API 500	API 505	OSHA
New Article		Accept		508	18	-	-	1910.305(j)(4)
ROP		ROC		NFPA 70E	NFPA 70B	NFPA 79	NFPA	NEMA
pg. 1076	# 11-5	pg. 325	# 11-3a	400.11	21.4.5	Ch. 9, 10, and 11	75	ICS 1, 2, and 5
log: 2536	CMP: 11	log: CC1101	Submitter: Alan Manche			2002 NEC: -		IEC: 60204-1

2005 NEC – 409.21 Overcurrent Protection

(A) General. Industrial control panels shall be provided with overcurrent protection in accordance with Parts I, II, and IX of Article 240.

(B) Location. This protection shall be provided by either of the following:

(1) An overcurrent protective device located ahead of the industrial control panel.

(2) A single main overcurrent protective device located within the industrial control panel. Where overcurrent protection is provided as part of the industrial control panel, the supply conductors shall be considered as either feeder or taps as covered by 240.21.

Author's Substantiation. A new article has been added to address industrial control panels intended for general use and operating at 600 volts or less. This section addresses where the overcurrent protection devices protecting the electrical system shall be located.

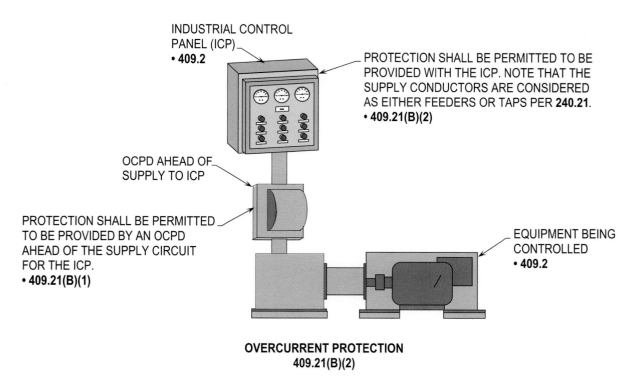

INDUSTRIAL CONTROL
PANEL (ICP)
• **409.2**

PROTECTION SHALL BE PERMITTED TO BE
PROVIDED WITH THE ICP. NOTE THAT THE
SUPPLY CONDUCTORS ARE CONSIDERED
AS EITHER FEEDERS OR TAPS PER **240.21**.
• **409.21(B)(2)**

OCPD AHEAD OF
SUPPLY TO ICP

PROTECTION SHALL BE PERMITTED
TO BE PROVIDED BY AN OCPD
AHEAD OF THE SUPPLY CIRCUIT
FOR THE ICP.
• **409.21(B)(1)**

EQUIPMENT BEING
CONTROLLED
• **409.2**

OVERCURRENT PROTECTION
409.21(B)(2)

Purpose of Change: To clarify where the OCPD protecting the electrical system shall be located.

NEC Ch. 4 – Article 409
Part II – 409.21(C)

94.5

Type of Change		Panel Action		UL	UL 508	API 500	API 505	OSHA
New Article		Accept		508	18	-	-	1910.305(j)(4)
ROP		ROC		NFPA 70E	NFPA 70B	NFPA 79	NFPA	NEMA
pg. 1076	# 11-5	pg. 325	# 11-3a	400.11	21.4.5	Ch. 9, 10, and 11	75	ICS 1, 2, and 5
log: 2536	CMP: 11	log: CC1101	Submitter: Alan Manche			2002 NEC: -		IEC: 60204-1

2005 NEC – <u>409.21 Overcurrent Protection</u>

(C) <u>**Rating.** The rating or setting of the overcurrent protective device for the circuit supplying the industrial control panel shall not be greater than the sum of the largest rating or setting of the branch-circuit short-circuit and ground-fault protective device provided with the industrial control panel, plus 125 percent of the full-load current rating of all resistance heating loads, plus the sum of the full-load currents of all other motors and apparatus that could be in operation at the same time.</u>

<u>**Exception:** Where one or more instantaneous trip circuit breakers or motor short-circuit protectors are used for motor branch-circuit short-circuit and ground-fault protection as permitted in 430.52(C), the procedure specified above for determining the maximum rating of the protective device for the circuit supplying the industrial control panel shall apply with the following provisions: For the purpose of the calculation, each instantaneous trip circuit breaker or motor short-circuit protector shall be assumed to have a rating not exceeding the maximum percentage of motor full-load current permitted by Table 430.52 for the type of control panel supply circuit protective device employed.</u>

<u>Where no branch-circuit short-circuit and ground-fault protective device is provided with the industrial control panel for motor or combination of motor and non-motor loads, the rating or setting of the overcurrent protective device shall be based on 430.52 and 430.53, as applicable.</u>

Author's Substantiation. A new article has been added to address industrial control panels intended for general use and operating at 600 volts or less. The section addresses the requirements for sizing the overcurrent protection devices ahead of the circuit supplying the industrial control panel.

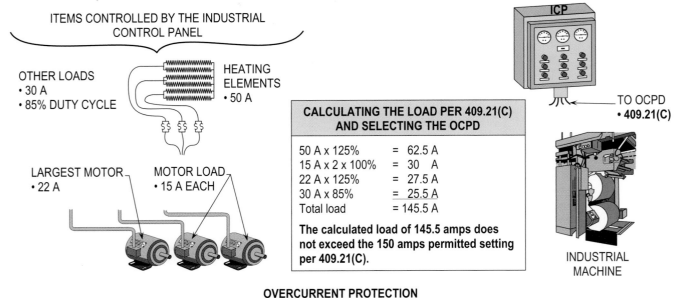

OVERCURRENT PROTECTION
409.21(C)

Purpose of Change: To provide requirements for sizing the OCPD ahead of the circuit supplying an industrial control panel (ICP).

NEC Ch. 4 – Article 409
Part III – 409.100 through 409.110

94.6

Type of Change		Panel Action		UL	UL 508	API 500	API 505	OSHA
New Article		Accept		508	12	-	-	1910.305(j)(4)
ROP		ROC		NFPA 70E	NFPA 70B	NFPA 79	NFPA	NEMA
pg. 1076	# 11-5	pg. 325	# 11-3a	400.11	21.4.5	Ch. 9, 10, and 11	75	ICS 1, 2, and 5
log: 2536	CMP: 11	log: CC1101	Submitter: Alan Manche			2002 NEC: -		IEC: 60204-1

Author's Substantiation. A new article has been added to address industrial control panels intended for general use and operating at 600 volts or less. Sections 409.100 through 409.110 address the construction specifications for installing industrial control panels operating on electrical systems rated 600 volts or less.

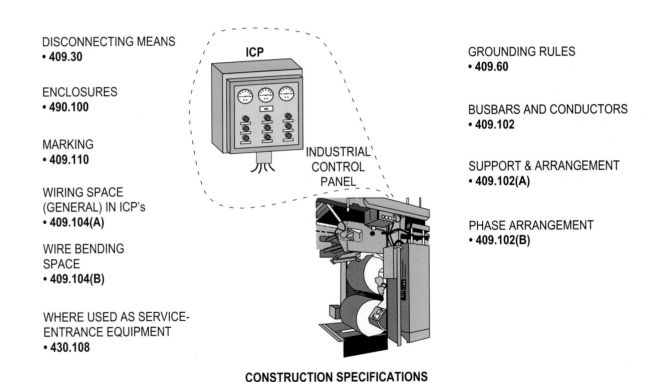

DISCONNECTING MEANS
• **409.30**

ENCLOSURES
• **490.100**

MARKING
• **409.110**

WIRING SPACE
(GENERAL) IN ICP's
• **409.104(A)**

WIRE BENDING
SPACE
• **409.104(B)**

WHERE USED AS SERVICE-
ENTRANCE EQUIPMENT
• **430.108**

ICP

INDUSTRIAL
CONTROL
PANEL

GROUNDING RULES
• **409.60**

BUSBARS AND CONDUCTORS
• **409.102**

SUPPORT & ARRANGEMENT
• **409.102(A)**

PHASE ARRANGEMENT
• **409.102(B)**

CONSTRUCTION SPECIFICATIONS
409.100 THRU 409.110

Purpose of Change: To outline the construction specifications for installing industrial control panels operating on electrical systems rated 600 volts or less.

NEC Ch. 4 – Article 410
Part II – 410.4(D)

Type of Change		Panel Action		UL	UL 508	API 500	API 505	OSHA
Revision		Accept in Principle		1570 and 1571	-	-	-	1910.305(d)
ROP		ROC		NFPA 70E	NFPA 70B	NFPA 79	NFPA	NEMA
pg. 1084	# 18-55	pg. 331	# 18-31	420.10	B.1.11	16.2.4	73	OS 1
log: 1572	CMP: 18	log: 1856	Submitter: Michael J. Johnston			2002 NEC: 410.4(D)		IEC: 701

2002 NEC – 410.4(D) Bathtub and Shower Areas

No parts of cord-connected ~~hanging~~ luminaires (fixtures), lighting track, pendants, or ceiling-suspended (paddle) fans shall be located within a zone measured 900 mm (3 ft) horizontally and 2.5 m (8 ft) vertically from the top of the bathtub rim or shower stall threshold. This zone is all encompassing and includes the zone directly over the tub or shower stall.

2005 NEC – 410.4(D) Bathtub and Shower Areas

No parts of cord-connected luminaires (fixtures), <u>chain-, cable-, or cord-suspended-luminaires (fixtures),</u> lighting track, pendants, or ceiling-suspended (paddle) fans shall be located within a zone measured 900 mm (3 ft) horizontally and 2.5 m (8 ft) vertically from the top of the bathtub rim or shower stall threshold. This zone is all encompassing and includes the zone directly over the tub or shower stall. <u>Luminaires (lighting fixtures) located in this zone shall be listed for damp locations, or listed for wet locations where subject to shower spray.</u>

Author's Substantiation. This revision clarifies the types of luminaires (fixtures) and ceiling-suspended (paddle) fans not permitted within a zone measured 900 mm (3 ft) horizontally and 2.5 m (8 ft) vertically from the top of the bathtub rim or shower stall threshold. A new sentence has been added to clarify that luminaires (lighting fixtures) installed in this zone shall be listed for damp locations or wet locations where subject to shower spray.

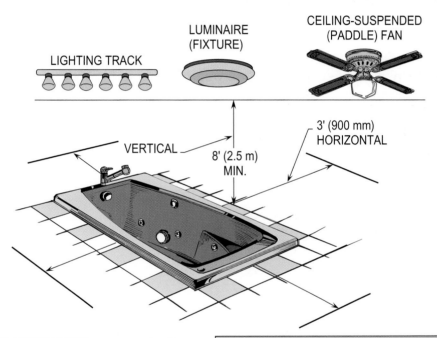

LIGHTING TRACK

LUMINAIRE
(FIXTURE)

CEILING-SUSPENDED
(PADDLE) FAN

VERTICAL

8' (2.5 m)
MIN.

3' (900 mm)
HORIZONTAL

NOTE 1: *NO CORD, CHAIN, OR CABLE TYPE LUMINAIRES (FIXTURES) SHALL BE INSTALLED PER 410.4(D).*

NOTE 2: *LUMINAIRES (LIGHTING FIXTURES) LOCATED IN THIS ZONE SHALL BE LISTED FOR DAMP OR WET LOCATIONS WHERE SUBJECT TO SHOWER SPRAY.*

BATH AND SHOWER AREAS
410.4(D)

Purpose of Change: To clarify the installation procedures for luminaires and paddle fans above and around bathtubs and shower areas.

NEC Ch. 4 – Article 410
Part II – 410.4(E)

95

Type of Change		Panel Action		UL	UL 508	API 500	API 505	OSHA
New Subsection		Accept in Principle		1572	-	-	-	1910.305(d)
ROP		ROC		NFPA 70E	NFPA 70B	NFPA 79	NFPA	NEMA
pg. 1086	# 18-57	pg. 331	# 18-35	420.10	B.1.11	16.2.4	-	OS 1
log: 2331	CMP: 18	log: 2598	Submitter: Vince Baclawski			2002 NEC: -		IEC: 714

2005 NEC – 410.4(E) Luminaires (Fixtures) in Indoor Sports, Mixed-Use, and All-Purpose Facilities

Luminaires (fixtures) subject to physical damage, using a mercury vapor or metal halide lamp, installed in playing and spectator seating areas of indoor sports, mixed-use, or all-purpose facilities shall be of the type that protects the lamp with a glass or plastic lens. Such luminaires (fixtures) shall be permitted to have an additional guard.

Author's Substantiation. A new subsection has been added to clarify that luminaires (fixtures) that are installed in playing and spectator seating areas of indoor sports, mixed-use, or all-purpose facilities and that are subject to physical damage shall be of the type that protects the lamp with a glass or plastic lens.

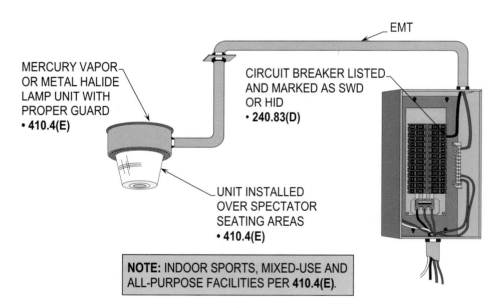

EMT

MERCURY VAPOR OR METAL HALIDE LAMP UNIT WITH PROPER GUARD
• 410.4(E)

CIRCUIT BREAKER LISTED AND MARKED AS SWD OR HID
• 240.83(D)

UNIT INSTALLED OVER SPECTATOR SEATING AREAS
• 410.4(E)

NOTE: INDOOR SPORTS, MIXED-USE AND ALL-PURPOSE FACILITIES PER 410.4(E).

LUMINAIRES (FIXTURES) IN INDOOR SPORTS, MIXED-USE, AND ALL-PURPOSE FACILITIES
410.4(E)

Purpose of Change: To provide requirements for lighting units installed indoors over spectator seating areas.

NEC Ch. 4 – Article 410
Part III – 410.14(B)

Type of Change		Panel Action		UL	UL 508	API 500	API 505	OSHA
Revision		Accept in Principle		1570	-	-	-	1910.305(d)
ROP		ROC		NFPA 70E	NFPA 70B	NFPA 79	NFPA	NEMA
pg. 1087	# 18-60	pg. 332	# 18-36	420.10	B.1.11	16.2.4	-	OS 1
log: 757	CMP: 18	log: 1035	Submitter: Dan Leaf			2002 NEC: 410.14(B)		IEC: 714

2002 NEC – 410.14(B) Access to Boxes

Electric-discharge luminaires (~~lighting~~ fixtures) surface mounted over concealed outlet, pull, or junction boxes shall be ~~installed~~ with suitable openings in back of the fixture to provide access to the ~~boxes~~.

2005 NEC – 410.14(B) Access to Boxes

Electric-discharge luminaires (fixtures) surface mounted over concealed outlet, pull, or junction boxes <u>and designed not to be supported solely by the outlet box</u> shall be <u>provided</u> with suitable openings in back of the <u>luminaire</u> (fixture) to provide access to the <u>wiring in the box</u>.

Author's Substantiation. This revision clarifies that access shall be provided to the wiring in the box if a luminaire (fixture) is installed over a box.

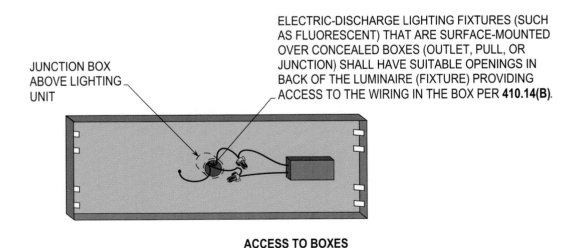

JUNCTION BOX ABOVE LIGHTING UNIT

ELECTRIC-DISCHARGE LIGHTING FIXTURES (SUCH AS FLUORESCENT) THAT ARE SURFACE-MOUNTED OVER CONCEALED BOXES (OUTLET, PULL, OR JUNCTION) SHALL HAVE SUITABLE OPENINGS IN BACK OF THE LUMINAIRE (FIXTURE) PROVIDING ACCESS TO THE WIRING IN THE BOX PER **410.14(B)**.

ACCESS TO BOXES
410.14(B)

Purpose of Change: To ensure that there is access to box and wiring when a luminaire (lighting fixture) is installed over a box.

NEC Ch. 4 – Article 410
Part IV – 410.15(B)

96

Type of Change		Panel Action		UL	UL 508	API 500	API 505	OSHA
Revision		Accept		-	-	-	-	-
ROP		ROC		NFPA 70E	NFPA 70B	NFPA 79	NFPA	NEMA
pg. 1088	# 18-61	pg. -	# -	-	-	-	-	OS 1
log: 2873	CMP: 18	log: -	Submitter: Ray Carter			2002 NEC: -		IEC: 714

2002 NEC – 410.15(B) Metal Poles Supporting Luminaires (Lighting Fixtures)

Metal poles shall be permitted to be used to support luminaires (lighting fixtures) and as a raceway to enclose supply conductors, provided the following conditions are met:

2005 NEC – 410.15(B) Metal <u>or Nonmetallic</u> Poles Supporting Luminaires (Lighting Fixtures)

Metal <u>or nonmetallic</u> poles shall be permitted to be used to support luminaires (lighting fixtures) and as a raceway to enclose supply conductors, provided the following conditions are met:

Author's Substantiation. This revision clarifies that nonmetallic poles shall be permitted to support luminaires (lighting fixtures) and as a raceway to enclose supply conductors.

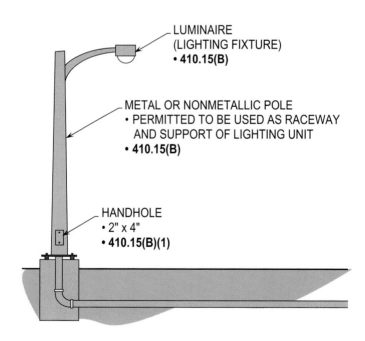

LUMINAIRE
(LIGHTING FIXTURE)
• **410.15(B)**

METAL OR NONMETALLIC POLE
• PERMITTED TO BE USED AS RACEWAY
  AND SUPPORT OF LIGHTING UNIT
• **410.15(B)**

HANDHOLE
• 2" x 4"
• **410.15(B)(1)**

METAL OR NONMETALLIC POLES SUPPORTING LUMINAIRES (LIGHTING FIXTURES)
410.15(B)

Purpose of Change: To include a nonmetallic pole to support a luminare (fixture) and enclose the wiring.

NEC Ch. 4 – Article 410
Part IV – 410.16(A)

Type of Change		Panel Action		UL	UL 508	API 500	API 505	OSHA
Revision		Accept in Principle in Part		514 and 514C	-	-	-	1910.305(j)
ROP		ROC		NFPA 70E	NFPA 70B	NFPA 79	NFPA	NEMA
pg. 1090	# 18-65	pg. -	# -	420.2	Ch. 17	16.2.4	-	OS 1
log: 604	CMP: 18	log: -	Submitter: Vince Baclawski			2002 NEC: 410.16(A)		IEC: 714

2002 NEC – 410.16(A) Outlet Boxes

Outlet boxes or fittings installed as required by 314.23 shall be permitted to support luminaires (fixtures).

2005 NEC – 410.16(A) Outlet Boxes

Outlet boxes or fittings installed as required by 314.23 <u>and complying with the provisions of 314.27(A) and 314.27(B)</u> shall be permitted to support luminaires (fixtures).

Author's Substantiation. This revision adds references for specific requirements when boxes are used to support luminaires (fixtures).

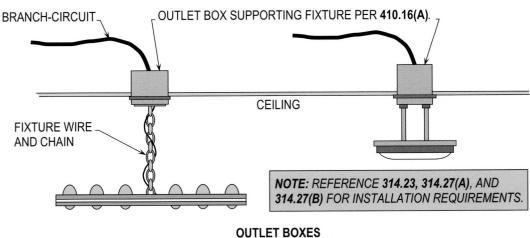

BRANCH-CIRCUIT

OUTLET BOX SUPPORTING FIXTURE PER **410.16(A)**.

CEILING

FIXTURE WIRE AND CHAIN

NOTE: REFERENCE **314.23, 314.27(A)**, AND **314.27(B)** FOR INSTALLATION REQUIREMENTS.

OUTLET BOXES
410.16(A)

Purpose of Change: For clearer installation procedures, sections were added to aid the user.

NEC Ch. 4 – Article 410
Part V – 410.18(B), Ex. 2

Type of Change	Panel Action	UL	UL 508	API 500	API 505	OSHA
New Exception	Accept in Principle	1053	-	-	-	1926.404(b)(1)(ii)

ROP		ROC		NFPA 70E	NFPA 70B	NFPA 79	NFPA	NEMA
pg. 1091	# 18-69	pg. 332	# 18-38	410.2(B)(4)(b)	14.3.3	16.2.1.2	-	-
log: 1083	CMP: 18	log: 1857	Submitter: Russell LeBlanc			2002 NEC: -		IEC: 714

2005 NEC – 410.18(B) Made of Insulating Material

Luminaires (fixtures) directly wired or attached to outlets supplied by a wiring method does not provide a ready means for grounding shall be of insulating material and shall have no exposed conductive parts.

> **Exception No. 2:** Where no equipment grounding conductor exists at the outlet, replacement luminaires (fixtures) that are GFCI protected shall not be required to be connected to an equipment grounding conductor.

Author's Substantiation. A new Exception No. 2 has been added to clarify that replacement luminaires (fixtures) that are GFCI protected shall not be required to be connected to an equipment grounding conductor, where no equipment grounding conductor exists at the outlet.

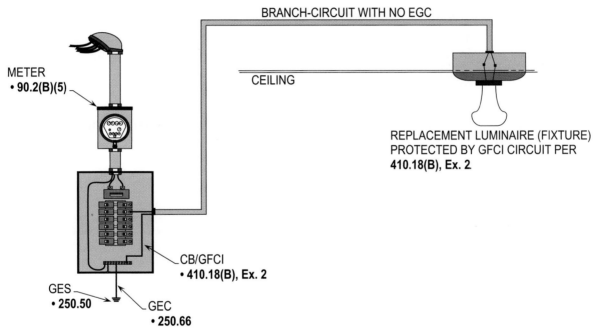

BRANCH-CIRCUIT WITH NO EGC

CEILING

METER
• **90.2(B)(5)**

REPLACEMENT LUMINAIRE (FIXTURE)
PROTECTED BY GFCI CIRCUIT PER
410.18(B), Ex. 2

CB/GFCI
• **410.18(B), Ex. 2**

GES
• **250.50**

GEC
• **250.66**

EXPOSED LUMINAIRE (FIXTURE) PARTS
410.18(B), Ex. 2

Purpose of Change: To allow a metallic fixture to be installed with a GFCI protected circuit without an EGC.

NEC Ch. 4 – Article 410
Part VI – 410.30(C)(1)(2)

Type of Change		Panel Action		UL	UL 508	API 500	API 505	OSHA
Revision		Accept		62	-	-	-	1910.305(g)
ROP		ROC		NFPA 70E	NFPA 70B	NFPA 79	NFPA	NEMA
pg. 1096	# 18-79	pg. -	# -	410.6 and 420.7	19.5	13.6.2	-	WC 5
log: 2332	CMP: 18	log: -	Submitter: Vince Baclawski			2002 NEC: 410.30(C)(1)(2)		IEC: 522.8.1.8

2002 NEC – 410.30(C)(1)

A listed luminaire (fixture) or a listed assembly shall be permitted to be cord connected if the following conditions apply:

(2) The flexible cord meets all the following:

a. Is visible for its entire length outside the luminaire (fixture)

b. Is not subject to strain or physical damage

c. Is terminated in a grounding-type attachment plug cap or busway plug or has a luminaire (fixture) assembly with a strain relief and canopy

2005 NEC – 410.30(C)(1) <u>Cord Connected Installation</u>

A listed luminaire (fixture) or a listed assembly shall be permitted to be cord connected if the following conditions apply:

(2) The flexible cord meets all the following:

a. Is visible for its entire length outside the luminaire (fixture)

b. Is not subject to strain or physical damage

c. Is terminated in a grounding-type attachment plug cap or busway <u>plug, or is a part of a listed assembly incorporating a manufactured wiring system connector in accordance with 604.6(C),</u> or has a luminaire (fixture) assembly with a strain relief and canopy

Author's Substantiation. The revision recognizes that a listed luminaire (fixture) shall be permitted to be connected with flexible cord from a listed assembly incorporating a manufactured wiring system connector in accordance with 604.6(C).

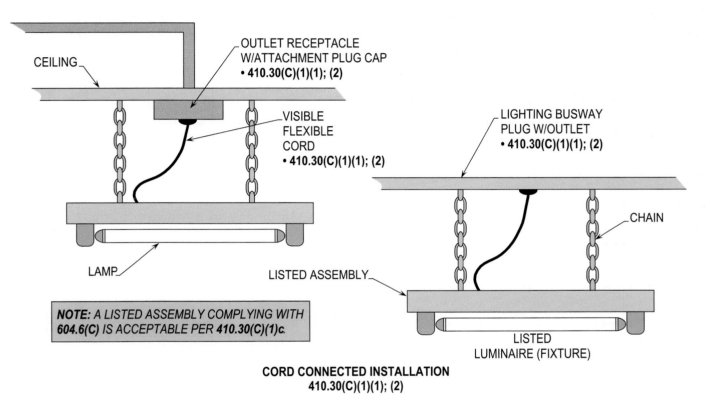

NOTE: A LISTED ASSEMBLY COMPLYING WITH 604.6(C) IS ACCEPTABLE PER 410.30(C)(1)c.

CORD CONNECTED INSTALLATION
410.30(C)(1)(1); (2)

Purpose of Change: To recognize cord connected luminaires (fixtures) as manufactured wiring systems.

NEC Ch. 4 – Article 410
Part VII – 410.36

Type of Change	Panel Action	UL	UL 508	API 500	API 505	OSHA
Revision	Accept	-	-	-	-	-
ROP	ROC	NFPA 70E	NFPA 70B	NFPA 79	NFPA	NEMA
pg. 1097	# 18-81a	pg. -	# -	-	-	-	-
log: CP1800	CMP: 18	log: -	Submitter: CMP 18	2002 NEC: 410.36		IEC: 514

2002 NEC – 410.36 Design and Material

Luminaires (fixtures) shall be constructed of metal, wood, or other material suitable for the application and shall be designed and assembled so as to secure requisite mechanical strength and rigidity. Wiring compartments, including their entrances, shall be ~~such that~~ conductors ~~may~~ be drawn in and withdrawn without physical damage.

2005 NEC – 410.36 Design and Material

Luminaires (fixtures) shall be constructed of metal, wood, or other material suitable for the application and shall be designed and assembled so as to secure requisite mechanical strength and rigidity. Wiring compartments, including their entrances, shall be <u>designed and constructed to permit</u> conductors <u>to</u> be drawn in and withdrawn without physical damage.

Author's Substantiation. The revision removes any vague language pertaining to the design and construction of wiring compartments, including their entrances.

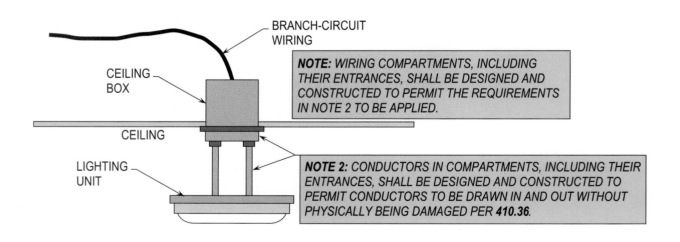

DESIGN AND MATERIAL
410.36

Purpose of Change: To provide guidelines for pulling and removing conductors from luminaires (fixtures) and associated apparatus.

NEC Ch. 4 – Article 410
Part XIII – 410.73(E)(4)

Type of Change	Panel Action	UL	UL 508	API 500	API 505	OSHA
Revision	Accept in Principle	935	-	-	-	-
ROP	ROC	NFPA 70E	NFPA 70B	NFPA 79	NFPA	NEMA
pg. 1102 # 18-90	pg. - # -	-	17.4 and B.1.19			C 82.13
log: 737 CMP: 18	log: -	Submitter: Dan Leaf		2002 NEC: 410.73(E)(4)		IEC: 422.2

2002 NEC – 410.73(E)(4) ~~Emergency~~ Egress Luminaires (Fixtures)

A ballast in a fluorescent luminaire (fixture) that is used for egress lighting and energized only during ~~an emergency~~ shall not have thermal protection.

2005 NEC – 410.73(E)(4) Egress Luminaires (Fixtures)

A ballast in a fluorescent luminaire (fixture) that is used for egress lighting and energized only during <u>a failure of the normal supply</u> shall not have thermal protection.

Author's Substantiation. The term "emergency" is not defined in the NEC and therefore has been deleted. This revision clarifies that egress luminaires (fixtures) shall not be equipped with thermal protection.

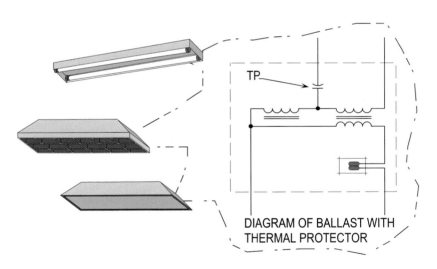

TP

DIAGRAM OF BALLAST WITH
THERMAL PROTECTOR

EMERGENCY EGRESS LUMINAIRES SHALL NOT BE EQUIPPED
WITH THERMAL PROTECTION PER **410.73(E)(4)**

EGRESS LUMINAIRES (FIXTURES)
410.73(E)(4)

Purpose of Change: To eliminate or make clear that luminaires (fixtures) are not to be equipped with thermal protection.

NEC Ch. 4 – Article 410
Part XIII – 410.73(F)(5)

Type of Change		Panel Action		UL	UL 508	API 500	API 505	OSHA
New Subdivision		Accept in Principle		1571	-	-	-	1910.305(d)
ROP		ROC		NFPA 70E	NFPA 70B	NFPA 79	NFPA	NEMA
pg. 1103	# 18-91	pg. 334	# 18-48	420.10	-	-	-	OS 1
log: 3006	CMP: 18	log: 2600	Submitter: Fred Carpenter			2002 NEC: -		IEC: 422.2

2005 NEC – 410.73(F)(5) Metal Halide Lamp Containment

Luminaires (fixtures) that use a metal halide lamp other than a thick-glass parabolic reflector lamp (PAR) shall be provided with a containment barrier that encloses the lamp, or shall be provided with a physical means that only allows the use of a lamp that is Type O.

> **FPN:** See ANSI Standard C78.387, *American National Standard for Electric Lamps – Metal Halide Lamps, Methods of Measuring Characteristics.*

Author's Substantiation. A new subsection has been added to recognize the protection that shall be provided for metal halide lamps.

A METAL HALIDE LAMP OTHER THAN A THICK-GLASS PARABOLIC
REFLECTOR LAMP SHALL HAVE A CONTAINMENT BARRIER
ENCLOSING THE LAMP PER **410.73(F)(5)**.

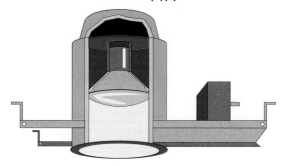

A PHYSICAL MEANS THAT ALLOWS THE USE OF A
TYPE O LAMP PER **410.73(F)(5)**.

METAL HALIDE LAMP CONTAINMENT
410.73(F)(5)

Purpose of Change: To provide, when necessary, a containment barrier for metal halide lamps in HID lighting units.

NEC Ch. 4 – Article 410
Part XIII – 410.73(G)

98

Type of Change	Panel Action	UL	UL 508	API 500	API 505	OSHA
New Subsection	Accept in Principle in Part	-	-	-	-	-

ROP		ROC		NFPA 70E	NFPA 70B	NFPA 79	NFPA	NEMA
pg. 1106	# 18-93	pg. 335	# 18-52	-	-	-	-	-
log: 3421	CMP: 18	log: 20	Submitter: Danny Liggett			2002 NEC: -		IEC: -

2005 NEC – 410.73(G) Disconnecting Means

In indoor locations, other than dwellings and associated accessory structures, fluorescent luminaires (fixtures) that utilize double-ended lamps and contain ballast(s) that can be serviced in place or ballasted luminaires that are supplied from multiwire branch circuits and contain ballast(s) that can be serviced in place shall have a disconnecting means either internal or external to each luminaire (fixture), to disconnect simultaneously from the source of supply all conductors of the ballast, including the grounded conductor if any. The line side terminals of the disconnecting means shall be guarded. The disconnecting means shall be located so as to be accessible to qualified persons before servicing or maintaining the ballast. This requirement shall become effective January 1, 2008.

> **Exception No. 1:** A disconnecting means shall not be required for luminaires (fixtures) installed in hazardous (classified) location(s).

> **Exception No. 2:** A disconnecting means shall not be required for emergency illumination required in 700.16.

> **Exception No. 3:** For cord-and-plug-connected luminaires, an accessible separable connector or an accessible plug and receptacle shall be permitted to serve as the disconnecting means.

> **Exception No. 4:** A disconnecting means shall not be required in industrial establishments with restricted public access where conditions of maintenance and supervision ensure that only qualified persons service the installation by written procedures.

> **Exception No. 5:** Where more than one luminaire is installed and supplied by other than a multiwire branch circuit, a disconnecting means shall not be required for every luminaire when the design of the installation includes locally accessible disconnects, such that the illuminated space cannot be left in total darkness.

Author's Substantiation. A new subsection has been added to address the requirements for providing a disconnecting means when replacing or troubleshooting a ballast in a luminaire (fixture).

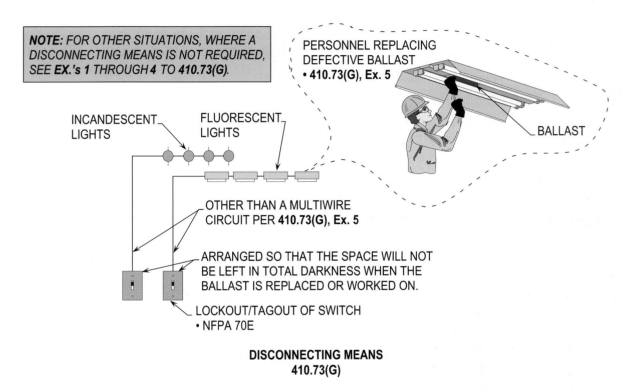

NOTE: *FOR OTHER SITUATIONS, WHERE A DISCONNECTING MEANS IS NOT REQUIRED, SEE EX.'s 1 THROUGH 4 TO 410.73(G).*

PERSONNEL REPLACING DEFECTIVE BALLAST
• **410.73(G), Ex. 5**

BALLAST

INCANDESCENT LIGHTS

FLUORESCENT LIGHTS

OTHER THAN A MULTIWIRE CIRCUIT PER **410.73(G), Ex. 5**

ARRANGED SO THAT THE SPACE WILL NOT BE LEFT IN TOTAL DARKNESS WHEN THE BALLAST IS REPLACED OR WORKED ON.

LOCKOUT/TAGOUT OF SWITCH
• NFPA 70E

DISCONNECTING MEANS
410.73(G)

Purpose of Change: To provide requirements when replacing or troubleshooting a ballast in a luminaire (fixture).

NEC Ch. 4 – Article 411
Part – 411.5(C)

Type of Change	Panel Action	UL	UL 508	API 500	API 505	OSHA
Revision	Accept in Principle	1570	-	-	-	-
ROP		ROC		NFPA 70E	NFPA 70B	NFPA 79	NFPA	NEMA

ROP		ROC		NFPA 70E	NFPA 70B	NFPA 79	NFPA	NEMA
pg. 1112	# 18-102	pg. 339	# 18-65	-	-	-	-	270
log: 715	CMP: 18	log: 1038	Submitter: Dan Leaf			2002 NEC: 411.5(C)		IEC: -

2002 NEC – 411.5(C) Bare Conductors

Exposed bare conductors and current-carrying parts shall be permitted. Bare conductors shall not be installed less than 2.1 m (7 ft) above the finished floor, unless specifically listed for a lower installation height.

2005 NEC – 411.5(C) Bare Conductors

Exposed bare conductors and current-carrying parts shall be permitted <u>for indoor installations only</u>. Bare conductors shall not be installed less than 2.1 m (7 ft) above the finished floor, unless specifically listed for a lower installation height.

Author's Substantiation. This revision clarifies that exposed bare conductors and current-carrying parts are restricted to indoor installations only.

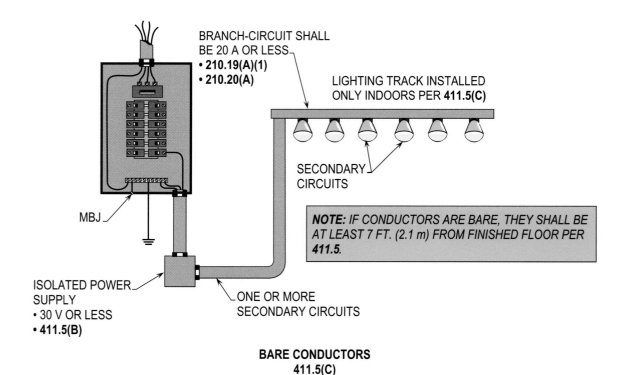

BRANCH-CIRCUIT SHALL BE 20 A OR LESS
• **210.19(A)(1)**
• **210.20(A)**

LIGHTING TRACK INSTALLED ONLY INDOORS PER **411.5(C)**

SECONDARY CIRCUITS

MBJ

*NOTE: IF CONDUCTORS ARE BARE, THEY SHALL BE AT LEAST 7 FT. (2.1 m) FROM FINISHED FLOOR PER **411.5**.*

ISOLATED POWER SUPPLY
• 30 V OR LESS
• **411.5(B)**

ONE OR MORE SECONDARY CIRCUITS

BARE CONDUCTORS
411.5(C)

Purpose of Change: To restrict bare conductors to areas located indoors.

NEC Ch. 4 – Article 422
Part II – 422.12, Ex. 2

Type of Change	Panel Action	UL	UL 508	API 500	API 505	OSHA
New Exception	Accept in Principle	1453	-	-	-	1910.305(j)(3)(i)
ROP		ROC		NFPA 70E	NFPA 70B	NFPA 79	NFPA	NEMA

ROP		ROC		NFPA 70E	NFPA 70B	NFPA 79	NFPA	NEMA
pg. 1119	# 17-12	pg. 350	# 17-42	420.10	-	-	-	DC 3
log: 1603	CMP: 17	log: 1865	Submitter: David Shapiro			2002 NEC: -		IEC: 424.2

2005 NEC – 422.12 Central Heating Equipment

Central heating equipment other than fixed electric space-heating equipment shall be supplied by an individual branch circuit.

> **Exception No. 2:** Permanently connected air-conditioning equipment shall be permitted to be connected to the same branch circuit.

Author's Substantiation. A new exception has been added to permit permanently connected air-conditioning equipment to be connected to the same branch circuit as the central heating equipment.

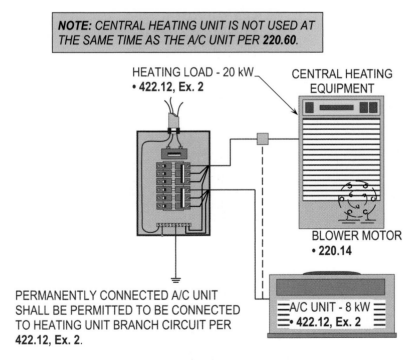

CENTRAL HEATING EQUIPMENT
422.12, Ex. 2

Purpose of Change: To allow an A/C unit to be connected to the same branch-circuit as the central heating equipment because they will not be used simultaneously.

NEC Ch. 4 – Article 422
Part II – 422.13

Type of Change		Panel Action	UL	UL 508	API 500	API 505	OSHA
Revision		Accept	1453	-	-	-	1910.305(j)(3)(i)
ROP		ROC	NFPA 70E	NFPA 70B	NFPA 79	NFPA	NEMA
pg. 1120	# 17-15	pg. -	# -	420.10	-	-	-	DC 3
log: 709	CMP: 17	log: -	Submitter: Dan Leaf		2002 NEC: 422.13		IEC: 424.2

2002 NEC – 422.13 Storage-Type Water Heaters

A ~~branch circuit supplying a~~ fixed storage-type water heater that has a capacity of 450 L (120 gal) or less shall ~~have a rating not less than 125 percent of the nameplate rating of the water heater~~.

2005 NEC – 422.13 Storage-Type Water Heaters

A fixed storage-type water heater that has a capacity of 450 L (120 gal) or less shall <u>be considered a continuous load</u>.

Author's Substantiation. This revision clarifies that storage-type water heaters shall be considered as continuous loads.

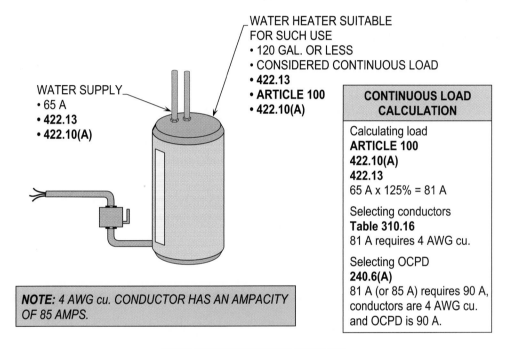

WATER HEATER SUITABLE
FOR SUCH USE
• 120 GAL. OR LESS
• CONSIDERED CONTINUOUS LOAD
• **422.13**
• **ARTICLE 100**
• **422.10(A)**

WATER SUPPLY
• 65 A
• **422.13**
• **422.10(A)**

CONTINUOUS LOAD CALCULATION
Calculating load **ARTICLE 100** **422.10(A)** **422.13** 65 A x 125% = 81 A
Selecting conductors **Table 310.16** 81 A requires 4 AWG cu.
Selecting OCPD **240.6(A)** 81 A (or 85 A) requires 90 A, conductors are 4 AWG cu. and OCPD is 90 A.

NOTE: 4 AWG cu. CONDUCTOR HAS AN AMPACITY OF 85 AMPS.

STORAGE-TYPE WATER HEATERS
422.13

Purpose of Change: To clarify that storage-type water heaters shall be considered as continuous loads.

NEC Ch. 4 – Article 422
Part II – 422.14

Type of Change	Panel Action	UL	UL 508	API 500	API 505	OSHA
Revision	Accept	73	-	-	-	-
ROP	**ROC**	**NFPA 70E**	**NFPA 70B**	**NFPA 79**	**NFPA**	**NEMA**
pg. 1121　# 17-16a	pg. -　# -	-	-	-	-	-
log: CP1702　CMP: 17	log: -	Submitter: CMP 17		2002 NEC: 422.14		IEC: -

2002 NEC – 422.14 Infrared Lamp Industrial Heating Appliances

Infrared ~~industrial~~ heating appliance lampholders shall be permitted to be ~~connected to any of the branch circuits in Article 210 and, in industrial occupancies, shall be permitted to be~~ operated in series on circuits of over 150 volts to ground, provided the voltage rating of the lampholders is not less than the circuit voltage.

Each section, panel, or strip carrying a number of infrared lampholders (including the internal wiring of such section, panel, or strip) shall be considered an appliance. The terminal connection block of each such assembly shall be considered an individual outlet.

2005 NEC – 422.14 Infrared Lamp Industrial Heating Appliances

<u>In industrial occupancies,</u> infrared heating appliance lampholders shall be permitted to be operated in series on circuits of over 150 volts to ground, provided the voltage rating of the lampholders is not less than the circuit voltage.

Each section, panel, or strip carrying a number of infrared lampholders (including the internal wiring of such section, panel, or strip) shall be considered an appliance. The terminal connection block of each such assembly shall be considered an individual outlet.

Author's Substantiation. This revision clarifies that requirements of Article 210 apply generally and such reference is not needed for connection of infrared lamp industrial heating appliances to branch circuits.

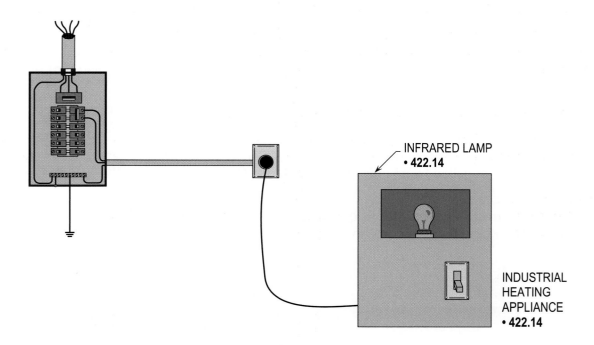

INFRARED LAMP INDUSTRIAL HEATING APPLIANCES
422.14

Purpose of Change: To revise the text and address industrial occupancies when applying the requirements of this section.

NEC Ch. 4 – Article 422
Part II – 422.16(B)(4)

100

Type of Change	Panel Action	UL	UL 508	API 500	API 505	OSHA
New Subdivision	Accept in Principle in Part	197	-	-	-	1910.304(f)(4)(ii)
ROP		**ROC**		**NFPA 70E**	**NFPA 70B**	**NFPA 79**	**NFPA**	**NEMA**

ROP		ROC		NFPA 70E	NFPA 70B	NFPA 79	NFPA	NEMA
pg. 1122	# 17-21	pg. -	# -	410.10(E)(3)	-	-	-	-
log: 219	CMP: 17	log: -	Submitter: Don A. Hursey			2002 NEC: -		IEC: 532

2005 NEC – 422.16(B)(4) Range Hoods

Range hoods shall be permitted to be cord-and-plug connected with a flexible cord identified as suitable for use on range hoods in the installation instructions of the appliance manufacturer, where all of the following conditions are met:

(1) The flexible cord is terminated with a grounding-type attachment plug.

> **Exception:** A listed range hood distinctly marked to identify it as protected by a system of double insulation, or its equivalent, shall not be required to be terminated with a grounding-type attachment plug.

(2) The length of the cord is not less than 450 mm (18 in.) and not over 900 mm (36 in.).

(3) Receptacles are located to avoid physical damage to the flexible cord.

(4) The receptacle is accessible.

(5) The receptacle is supplied by an individual branch circuit.

Author's Substantiation. A new subdivision has been added to recognize the requirements for cord-and-plug connections using flexible cord and accessories to connect range hoods to the supply circuit.

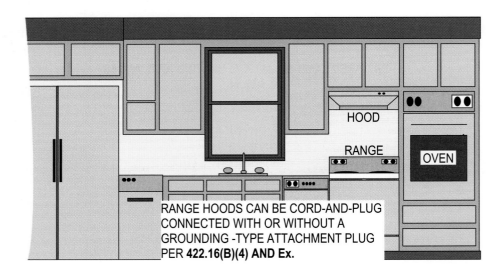

RANGE HOODS CAN BE CORD-AND-PLUG CONNECTED WITH OR WITHOUT A GROUNDING -TYPE ATTACHMENT PLUG PER **422.16(B)(4) AND Ex.**

RANGE HOODS
422.16(B)(4) AND Ex.

Purpose of Change: To provide requirements for cord-and-plug connections using flexible cord and accessories to connect range hoods to the supply circuit.

NEC Ch. 4 – Article 422
Part III – 422.31(B)

101

Type of Change	Panel Action	UL	UL 508	API 500	API 505	OSHA
New Sentence	Accept in Principle	94	-	-	-	1910.305(j)(3)
ROP	**ROC**	**NFPA 70E**	**NFPA 70B**	**NFPA 79**	**NFPA**	**NEMA**
pg. 1124 # 17-24	pg. 359 # 17-83	420.10(D)(2)	11.5	5.3	-	DC 3
log: 1690 CMP: 17	log: 1591 Submitter: James T. Dollard			2002 NEC: 422.31(B)		IEC: Ch. 46

2002 NEC – 422.31(B) Appliances Rated Over 300 Volt-Amperes or 1/8 Horsepower

For permanently connected appliances rated over 300 volt-amperes or 1/8 hp, the branch-circuit switch or circuit breaker shall be permitted to serve as the disconnecting means where the switch or circuit breaker is within sight from the appliance or is capable of being locked in the open position.

2005 NEC – 422.31(B) Appliances Rated Over 300 Volt-Amperes or 1/8 Horsepower

For permanently connected appliances rated over 300 volt-amperes or 1/8 hp, the branch-circuit switch or circuit breaker shall be permitted to serve as the disconnecting means where the switch or circuit breaker is within sight from the appliance or is capable of being locked in the open position. The provision for locking or adding a lock to the disconnecting means shall be installed on or at the switch or circuit breaker used as the disconnecting means and shall remain in place with or without the lock installed.

Author's Substantiation. A new sentence has been added to clarify that appliances rated over 300 volt-amperes or 1/8 horsepower shall have provisions for locking or adding a lock to the disconnecting means at the switch or circuit breaker used as the disconnecting means and shall remain in place with or without the lock installed.

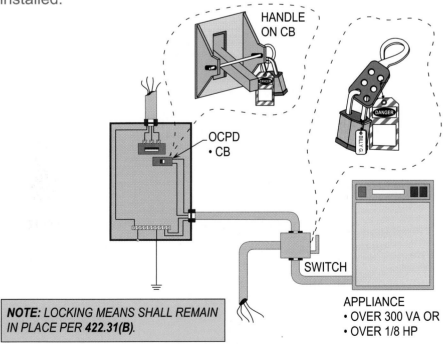

APPLIANCES RATED OVER 300 VOLT-AMPERES OR 1/8 HORSEPOWER
422.31(B)

Purpose of Change: New wording was added, and as a result, a provision for the locking means to remain in place is required.

NEC Ch. 4 – Article 422
Part IV – 422.51

102

Type of Change		Panel Action		UL	UL 508	API 500	API 505	OSHA
New Section		Accept in Principle in Part		541	-	-	-	-
ROP		ROC		NFPA 70E	NFPA 70B	NFPA 79	NFPA	NEMA
pg. 1114	# 17-6	pg. 339	# 17-3	-	-	-	90 A	-
log: 2451	CMP: 17	log: 1864	Submitter: William H. King, Jr.			2002 NEC: 422.51		IEC: Ch. 46

2005 NEC – 422.51 Cord-and-Plug-Connected Vending Machines

Cord-and-plug-connected vending machines manufactured or re-manufactured on or after January 1, 2005, shall include a ground-fault circuit-interrupter as an integral part of the attachment plug or located in the power supply cord within 300 mm (12 in.) of the attachment plug. Cord-and-plug-connected vending machines not incorporating integral GFCI protection shall be connected to a GFCI protected outlet.

Author's Substantiation. A new section has been added to provide requirements for cord-and-plug-connected vending machines. Cord-and-plug-connected vending machines shall include a ground-fault circuit-interrupter as an integral part of the attachment plug or located in the power supply cord within 300 (12 in.) of the attachment plug. If integral GFCI protection is not incorporated into the attachment plug or power supply cord, the cord-and-plug-connected vending machine shall be connected to a GFCI protected outlet.

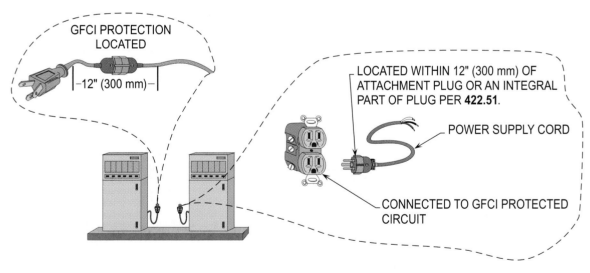

GFCI PROTECTION LOCATED

–12" (300 mm)–

LOCATED WITHIN 12" (300 mm) OF ATTACHMENT PLUG OR AN INTEGRAL PART OF PLUG PER **422.51**.

POWER SUPPLY CORD

CONNECTED TO GFCI PROTECTED CIRCUIT

VENDING MACHINES MANUFACTURED OR RE-MANUFACTURED ON OR AFTER JAN. 1, 2005 SHALL INCLUDE GFCI PROTECTION.

CORD-AND-PLUG CONNECTED VENDING MACHINES
422.51

Purpose of Change: To provide GFCI protection requirements for cord-and-plug-connected vending machines.

NEC Ch. 4 – Article 424
Part I – 424.3(B)

Type of Change		Panel Action		UL	UL 508	API 500	API 505	OSHA
Revision		Accept in Principle		1042	-	-	-	-
ROP		ROC		NFPA 70E	NFPA 70B	NFPA 79	NFPA	NEMA
pg. 1127	# 17-30	pg. -	# -	-	J.2.1.1	-	-	DC 3
log: 779	CMP: 17	log: -	Submitter: Dan Leaf			2002 NEC: 424.3(B)		IEC: 463

2002 NEC – 424.3(B) Branch-Circuit Sizing

~~The ampacity of the branch-circuit conductors and the rating or setting of overcurrent protective devices supplying fixed electric space-heating equipment consisting of resistance elements with or without a motor shall not be less than 125 percent of the total load of the motors and the heaters. The rating or setting of overcurrent protective devices shall be permitted in accordance with 240.4(B). A contactor, thermostat, relay, or similar device, listed for continuous operation at 100 percent of its rating, shall be permitted to supply its full-rated load as provided in 210.19(A), Exception.~~

~~The size of the branch-circuit conductors and overcurrent protective devices supplying fixed electric space-heating equipment, including a hermetic refrigerant motor-compressor with or without resistance units, shall be computed in accordance with 440.34 and 440.35. The provisions of this section shall not apply to conductors that form an integral part of approved fixed electric space-heating equipment.~~

2005 NEC – 424.3(B) Branch-Circuit Sizing

Fixed electric space heating equipment shall be considered continuous load.

Author's Substantiation. This revision clarifies that fixed electric space-heating equipment shall be considered a continuous load.

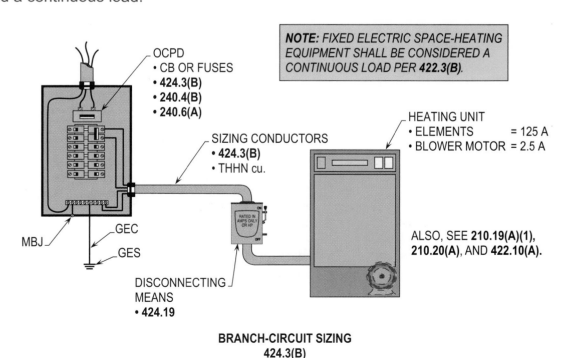

NOTE: FIXED ELECTRIC SPACE-HEATING EQUIPMENT SHALL BE CONSIDERED A CONTINUOUS LOAD PER **422.3(B)**.

OCPD
• CB OR FUSES
• **424.3(B)**
• **240.4(B)**
• **240.6(A)**

SIZING CONDUCTORS
• **424.3(B)**
• THHN cu.

HEATING UNIT
• ELEMENTS   = 125 A
• BLOWER MOTOR = 2.5 A

MBJ

GEC

GES

DISCONNECTING MEANS
• **424.19**

ALSO, SEE **210.19(A)(1)**, **210.20(A)**, AND **422.10(A)**.

BRANCH-CIRCUIT SIZING
424.3(B)

Purpose of Change: To clarify that fixed electric space-heating is considered a continuous load.

NEC Ch. 4 – Article 424
Part V – 424.44(G)

Type of Change		Panel Action		UL	UL 508	API 500	API 505	OSHA
Revision		Accept in Part		943	-	-	-	-
ROP		ROC		NFPA 70E	NFPA 70B	NFPA 79	NFPA	NEMA
pg. 1130	# 17-38	pg. -	# -	-	-	-	-	280
log: 2664	CMP: 17	log: -	Submitter: Phil Simmons			2002 NEC: 424.44(G)		IEC: 43

2002 NEC – 424.44(G) Ground-Fault Circuit-Interrupter Protection ~~for Heated Floors of Bathrooms, and in Hydromassage Bathtub, Spa, and Hot Tub Locations~~

Ground-fault circuit-interrupter protection for personnel shall be provided for electrically heated floors ~~in~~ bathrooms, and in hydromassage bathtub~~, spa, and hot tub~~ locations.

2005 NEC – 424.44(G) Ground-Fault Circuit-Interrupter Protection

Ground-fault circuit-interrupter protection for personnel shall be provided for <u>cables installed in</u> electrically heated floors <u>of</u> bathrooms and in hydromassage bathtub locations.

Author's Substantiation. This revision clarifies that ground-fault circuit-interrupter protection shall be provided for the cables.

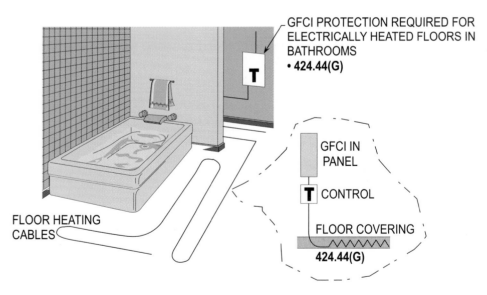

GFCI PROTECTION REQUIRED FOR ELECTRICALLY HEATED FLOORS IN BATHROOMS
• 424.44(G)

GFCI IN PANEL

T CONTROL

FLOOR COVERING
424.44(G)

FLOOR HEATING CABLES

GROUND-FAULT CIRCUIT-INTERRUPTER PROTECTION
424.44(G)

Purpose of Change: To verify that cables shall be GFCI-protected when installed in electrically heated floors.

NEC Ch. 4 – Article 426
Part I – 426.4

Type of Change		Panel Action		UL	UL 508	API 500	API 505	OSHA
Revision		Accept		943	-	-	-	-
ROP		ROC		NFPA 70E	NFPA 70B	NFPA 79	NFPA	NEMA
pg. 1132	# 17-42	pg. -	# -	-	-	-	-	280
log: 766	CMP: 17	log: -	Submitter: Dan Leaf			2002 NEC: 426.4		IEC: 43

2002 NEC – 426.4 ~~Branch-Circuit Sizing~~

~~The ampacity of branch-circuit conductors and the rating or setting of overcurrent protective devices supplying fixed outdoor electric deicing and snow-melting equipment shall not be less than 125 percent of the total load of the heaters. The rating or setting of overcurrent protective devices shall be permitted in accordance with 240.4(B).~~

2005 NEC – 426.4 <u>Continuous Load</u>

<u>Fixed outdoor electric deicing and snow-melting equipment shall be considered as a continuous load.</u>

Author's Substantiation. This revision clarifies that fixed outdoor electric deicing and snow-melting equipment shall be considered as a continuous load.

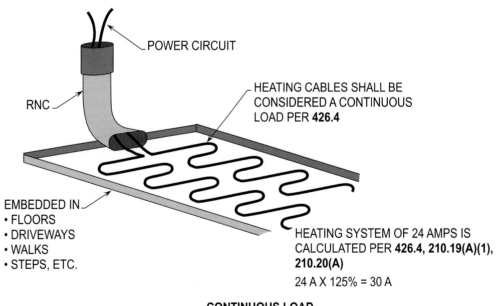

POWER CIRCUIT

RNC

HEATING CABLES SHALL BE CONSIDERED A CONTINUOUS LOAD PER **426.4**

EMBEDDED IN
• FLOORS
• DRIVEWAYS
• WALKS
• STEPS, ETC.

HEATING SYSTEM OF 24 AMPS IS CALCULATED PER **426.4, 210.19(A)(1), 210.20(A)**
24 A X 125% = 30 A

CONTINUOUS LOAD
426.4

Purpose of Change: To recognize electrically energized heating systems as a continuous load.

NEC Ch. 4 – Article 426
Part VI – 426.50(A)

Type of Change		Panel Action		UL	UL 508	API 500	API 505	OSHA
Revision		Accept in Principle		943	-	-	-	-
ROP		ROC		NFPA 70E	NFPA 70B	NFPA 79	NFPA	NEMA
pg. 1133	# 17-39	pg. 362	# 17-98	-	-	-	-	280
log: 772	CMP: 17	log: 3199	Submitter: Dan Leaf			2002 NEC: 426.50(A)		IEC: 43

2002 NEC – 426.50(A) Disconnection

All fixed outdoor deicing and snow-melting equipment shall be provided with a means for disconnection from all ungrounded conductors. Where readily accessible to the user of the equipment, the branch-circuit switch or circuit breaker shall be permitted to serve as the disconnecting means. ~~Switches used as~~ the disconnecting means shall be of the indicating type.

2005 NEC – 426.50(A) Disconnection

All fixed outdoor deicing and snow-melting equipment shall be provided with a means for disconnection from all ungrounded conductors. Where readily accessible to the user of the equipment, the branch-circuit switch or circuit breaker shall be permitted to serve as the disconnecting means. The disconnecting means shall be of the indicating type and be provided with a positive lockout in the "off" position.

Author's Substantiation. This revision clarifies that a disconnecting means shall be provided for disconnection from all ungrounded conductors with a positive lockout in the "off" position.

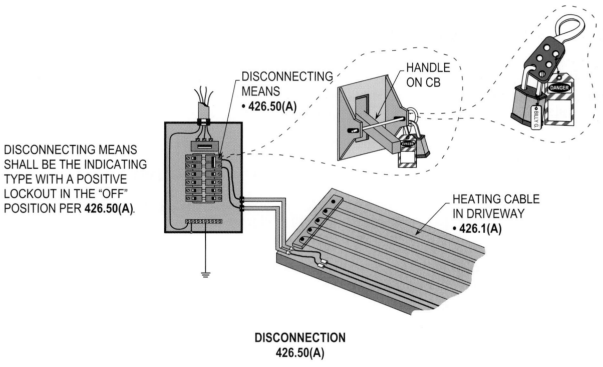

DISCONNECTING
MEANS
• **426.50(A)**

HANDLE
ON CB

DISCONNECTING MEANS
SHALL BE THE INDICATING
TYPE WITH A POSITIVE
LOCKOUT IN THE "OFF"
POSITION PER **426.50(A)**.

HEATING CABLE
IN DRIVEWAY
• **426.1(A)**

DISCONNECTION
426.50(A)

Purpose of Change: To outline the characteristics of the device used for lockout of the heating cables.

NEC Ch. 4 – Article 427
Part IV – 427.27, Ex.

103

Type of Change		Panel Action		UL	UL 508	API 500	API 505	OSHA
New Exception		-		943	-	-	-	-
ROP		ROC		NFPA 70E	NFPA 70B	NFPA 79	NFPA	NEMA
pg. 1137	# 17-53	pg. 362	# 17-99	-	-	-	-	280
log: 228	CMP: 17	log: 1870	Submitter: C. J. Erickson			2002 NEC: -		IEC: 43

2005 NEC – 427.27 Voltage Limitations

Unless protected by ground-fault circuit-interrupter protection for personnel, the secondary winding of the isolation transformer connected to the pipeline or vessel being heated shall not have an output voltage greater than 30 volts ac.

Where ground-fault circuit-interrupter protection for personnel is provided, the voltage shall be permitted to be greater than 30 but not more than 80 volts.

> **Exception:** In industrial establishments, the isolation transformer connected to the pipeline or vessel being heated shall be permitted to have an output voltage not greater than 132 volts ac to ground where all of the following conditions apply:
>
> **(1)** Conditions of maintenance and supervision ensure that only qualified persons service the installed systems.
>
> **(2)** Ground fault protection of equipment is provided.
>
> **(3)** The pipeline or vessel being heated is completely enclosed in a grounded metal enclosure.
>
> **(4)** The transformer secondary connections to the pipeline or vessel being heated are completely enclosed in a grounded metal mesh or metal enclosure.

Author's Substantiation. A new exception has been added to recognize that in industrial establishments, where certain conditions of use are followed, an isolation transformer connected to a pipeline or vessel being heated shall be permitted to have an output voltage not greater than 132 volts to ground.

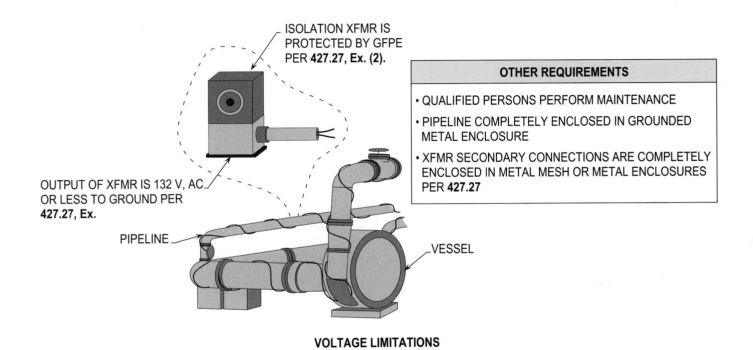

ISOLATION XFMR IS
PROTECTED BY GFPE
PER **427.27, Ex. (2).**

OTHER REQUIREMENTS

- QUALIFIED PERSONS PERFORM MAINTENANCE
- PIPELINE COMPLETELY ENCLOSED IN GROUNDED METAL ENCLOSURE
- XFMR SECONDARY CONNECTIONS ARE COMPLETELY ENCLOSED IN METAL MESH OR METAL ENCLOSURES PER **427.27**

OUTPUT OF XFMR IS 132 V, AC
OR LESS TO GROUND PER
427.27, Ex.

PIPELINE

VESSEL

VOLTAGE LIMITATIONS
427.27, Ex.'s (1) THRU (4)

Purpose of Change: To allow a voltage of 132 volts or less to ground to be used under certain conditions.

NEC Ch. 4 – Article 427
Part VII – 427.57

Type of Change		Panel Action		UL	UL 508	API 500	API 505	OSHA
Revision		Accept		1663	-	-	-	-
ROP		ROC		NFPA 70E	NFPA 70B	NFPA 79	NFPA	NEMA
pg. 1137	# 17-54	pg. -	# -	-	-	-	-	WD 6
log: 763	CMP: 17	log: -	Submitter: Dan Leaf			2002 NEC: 427.57		IEC: 43

2002 NEC – 427.57 Overcurrent Protection

Heating equipment shall be considered as protected against overcurrent where supplied by a branch circuit as specified in ~~427.4~~.

2005 NEC – 427.57 Overcurrent Protection

Heating equipment shall be considered as protected against overcurrent where supplied by a branch circuit as specified in 210.3 and 210.23.

Author's Substantiation. This revision clarifies the specific references for selecting a branch circuit rating to protect cables supplying pipelines and vessels.

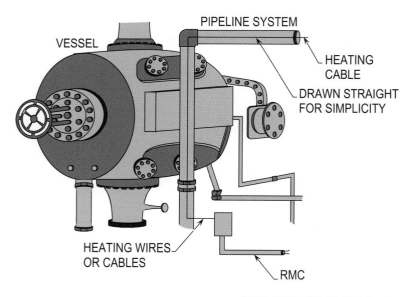

BRANCH-CIRCUIT SIZES PER 427.57	
210.3	210.23
15 A	15 A & 20 A
20 A	30 A
30 A	40 A & 50 A
40 A	50 A OR LESS
50 A	

OVERCURRENT PROTECTION
427.57

Purpose of Change: To reference **210.3** and **210.23** for selecting a branch-circuit rating to protect cables supplying pipelines and vessels.

NEC Ch. 4 – Article 430
Part I – 430.2

Type of Change	Panel Action	UL	UL 508	API 500	API 505	OSHA
New Definition	Accept	508C	-	-	-	-

ROP		ROC		NFPA 70E	NFPA 70B	NFPA 79	NFPA	NEMA
pg. 1140	# 11-10	pg. -	# -	-	20.22	7.3.1.2	-	ICS 7
log: 2537	CMP: 11	log: -	Submitter: Alan Manche			2002 NEC: -		IEC: 146-2

2005 NEC – <u>430.2 Definitions</u>

<u>**Adjustable Speed Drive.**</u> <u>A combination of the power converter, motor, and motor mounted auxiliary devices such as encoders, tachometers, thermal switches and detectors, air blowers, heaters, and vibration sensors.</u>

Author's Substantiation. A new definition has been added to define an adjustable speed drive and its accessories.

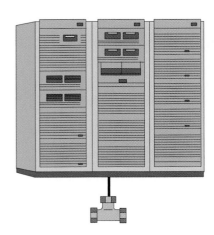

ADJUSTABLE SPEED DRIVE PER 430.2 IS A COMBINATION OF:
- THE POWER CONVERTER
- MOTOR
- AUXILIARY DEVICES (MOTOR)
  - ENCODERS
  - TACHOMETERS
  - THERMAL SWITCHES AND DETECTORS
  - AIR BLOWERS
  - HEATERS
  - VIBRATION SENSORS

ADJUSTABLE SPEED DRIVE
430.2

Purpose of Change: To define an adjustable speed drive and its accessories.

Type of Change		Panel Action		UL	UL 508	API 500	API 505	OSHA
New Definition		Accept		508C	-	-	-	-
ROP		ROC		NFPA 70E	NFPA 70B	NFPA 79	NFPA	NEMA
pg. 1140	# 11-10	pg. -	# -	-	70.22	7.3.1.2	-	ICS 7
log: 2537	CMP: 11	log: -	Submitter: Alan Manche			2002 NEC: 430.2		IEC: 146-2

2002 NEC – 430.2 Adjustable-Speed Drive Systems

~~The incoming branch circuit or feeder to power conversion equipment included as a part of an adjustable-speed drive system shall be based on the rated input to the power conversion equipment. Where the power conversion equipment is marked to indicate that overload protection is included, additional overload protection shall not be required.~~

~~The disconnecting means shall be permitted to be in the incoming line to the conversion equipment and shall have a rating not less than 115 percent of the rated input current of the conversion unit.~~

2005 NEC – 430.2 Definitions

Adjustable-Speed Drive System. An interconnected combination of equipment that provides a means of adjusting the speed of a mechanical load coupled to a motor. A drive system typically consists of an adjustable speed drive and auxiliary electrical apparatus.

Author's Substantiation. A new definition has been added to define an adjustable-speed drive system and its ability to speed control a mechanical load.

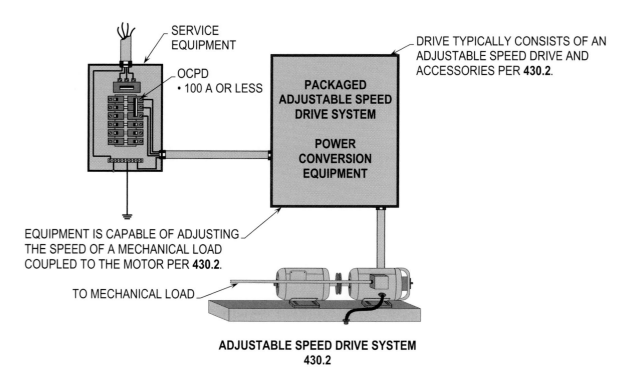

ADJUSTABLE SPEED DRIVE SYSTEM
430.2

Purpose of Change: To define an adjustable speed drive system and its ability to speed control a mechanical load.

NEC Ch. 4 – Article 430
Part I – 430.2

Type of Change		Panel Action		UL	UL 508	API 500	API 505	OSHA
Relocated		Accept in Principle in Part		1004 and 845	-	-	-	-
ROP		ROC		NFPA 70E	NFPA 70B	NFPA 79	NFPA	NEMA
pg. 1139	# 11-8	pg. 363	# 11-12	-	Ch. 11	Ch. 15	-	MG 1
log: 1573	CMP: 11	log: 404	Submitter: Michael J. Johnston			2002 NEC: 430.81(A)		IEC: 432 and 433

2005 NEC – <u>430.2 Definitions</u>

Controller. For the purpose of this article, a controller is any switch or device that is normally used to start and stop a motor by making and breaking the motor circuit current.

Author's Substantiation. A new definition has been added to define a controller and its associated components.

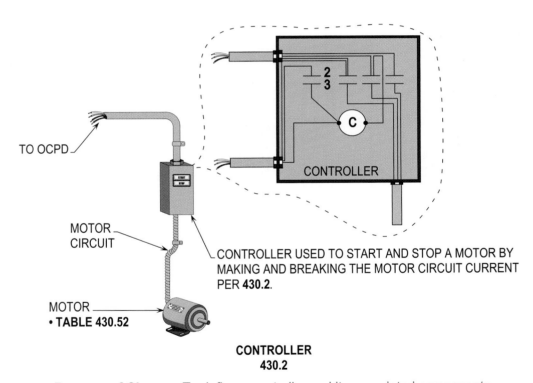

CONTROLLER
430.2

Purpose of Change: To define a controller and its associated components.

NEC Ch. 4 – Article 430
Part I – 430.2

Type of Change		Panel Action		UL	UL 508	API 500	API 505	OSHA
New Definition		Accept		1004 and 845	-	-	-	1910.304(f)(4)
ROP		ROC		NFPA 70E	NFPA 70B	NFPA 79	NFPA	NEMA
pg. 1140	# 11-9	pg. -	# -	420.10(E)	Ch. 11	Ch. 15	-	MG 1
log: 1574	CMP: 11	log: -	Submitter: Michael J. Johnston			2002 NEC: 430.72		IEC: 432 and 433

2005 NEC – 430.2 Definitions

Motor Control Circuit. The circuit of a control apparatus or system that carries the electric signals directing the performance of the controller but does not carry the main power current.

Author's Substantiation. A new definition has been added to define a motor control circuit and its task for receiving and directing signals to energize and deenergize the coil of a controller.

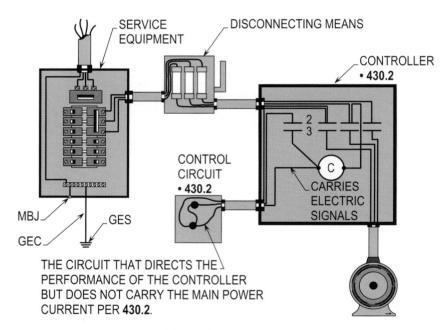

MOTOR CONTROL CIRCUIT
430.2

Purpose of Change: To define a motor control circuit and its task in receiving and directing signals to energize and deenergize the coil of a controller.

NEC Ch. 4 – Article 430
Part I – 430.2

106

Type of Change		Panel Action		UL	UL 508	API 500	API 505	OSHA
New Definition		Accept		1004 and 1047	-	-	-	1910.305(f)(4)
ROP		ROC		NFPA 70E	NFPA 70B	NFPA 79	NFPA	NEMA
pg. 1176	# 11-74	pg. 370	# 11-45a	420.10(E)	Ch. 11	Ch. 15	-	MG 1
log: 1029	CMP: 11	log: CC1102	Submitter: William E. Anderson			2002 NEC: -		IEC: Ch.. 46

2005 NEC – <u>430.2 Definitions</u>

<u>**System Isolation Equipment.** A redundantly monitored, remotely operated contactor-isolating system, packaged to provide the disconnection/isolation function, capable of verifiable operation from multiple remote locations by means of lockout switches, each having the capability of being padlocked in the off (open) position.</u>

Author's Substantiation. A new definition has been added to define system isolation equipment and how it provides service in a motor system.

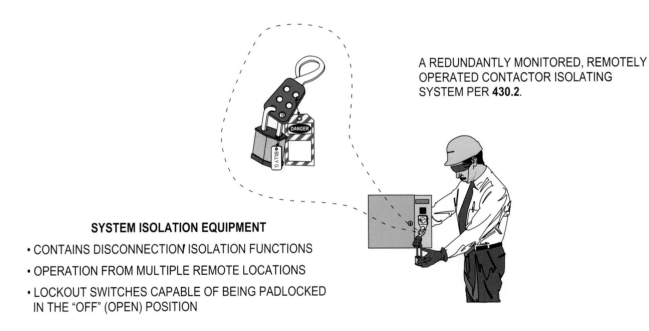

A REDUNDANTLY MONITORED, REMOTELY OPERATED CONTACTOR ISOLATING SYSTEM PER **430.2**.

SYSTEM ISOLATION EQUIPMENT

• CONTAINS DISCONNECTION ISOLATION FUNCTIONS

• OPERATION FROM MULTIPLE REMOTE LOCATIONS

• LOCKOUT SWITCHES CAPABLE OF BEING PADLOCKED IN THE "OFF" (OPEN) POSITION

SYSTEM ISOLATION EQUIPMENT
430.2

Purpose of Change: To define what "system isolation equipment" is and how it can provide a service in a motor system.

Type of Change	Panel Action	UL	UL 508	API 500	API 505	OSHA
Revision	Accept in Principle	1004	-	-	-	1910.305(j)(4)

ROP		ROC		NFPA 70E	NFPA 70B	NFPA 79	NFPA	NEMA
pg. 1143	# 11-15	pg. -	# -	420.10(E)	Ch. 11	Ch. 15	-	MG 1
log: 22	CMP: 11	log: -	Submitter: TCC			2002 NEC: 430.6(A)(1)		IEC: 432 and 433

2002 NEC – 430.6(A)(1) Table Values

The values given in Tables ~~430.147, 430.148, 430.149, and 430.150, including notes,~~ shall be used to determine the ampacity of conductors or ampere ratings of switches, branch-circuit short-circuit and ground-fault protection, instead of the actual current rating marked on the motor nameplate. Where a motor is marked in amperes, but not horsepower, the horsepower rating shall be assumed to be that corresponding to the value given in ~~Tables 430.147, 430.148, 430.149, and 430.150~~, interpolated if necessary.

2005 NEC – 430.6(A)(1) Table Values

Other than for motors built for low speeds (less than 1200 RPM) or high torques, and for multispeed motors, the values given in Table 430.247, Table 430.248, Table 430.249, and Table 430.250 shall be used to determine the ampacity of conductors or ampere ratings of switches, branch-circuit short-circuit and ground-fault protection, instead of the actual current rating marked on the motor nameplate. Where a motor is marked in amperes, but not horsepower, the horsepower rating shall be assumed to be that corresponding to the value given in Table 430.247, Table 430.248, Table 430.249, and Table 430.250, interpolated if necessary. Motors built for low speeds (less than 1200 RPM) or high torques may have higher full-load currents, and multispeed motors will have full-load current varying with speed, in which case the nameplate current ratings shall be used.

Author's Substantiation. Tables 430.147 through 430.150 have been renumbered as Tables 430.247 through 430.250.

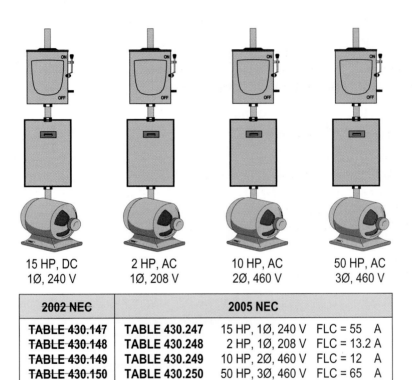

| 15 HP, DC | 2 HP, AC | 10 HP, AC | 50 HP, AC |
| 1Ø, 240 V | 1Ø, 208 V | 2Ø, 460 V | 3Ø, 460 V |

2002 NEC	2005 NEC		
TABLE 430.147	TABLE 430.247	15 HP, 1Ø, 240 V	FLC = 55 A
TABLE 430.148	TABLE 430.248	2 HP, 1Ø, 208 V	FLC = 13.2 A
TABLE 430.149	TABLE 430.249	10 HP, 2Ø, 460 V	FLC = 12 A
TABLE 430.150	TABLE 430.250	50 HP, 3Ø, 460 V	FLC = 65 A

TABLE VALUES
430.6(A)(1)

Purpose of Change: To add the new number of **Tables 430.247** through **430.250** to **430.6(A)** of the **2005 NEC**.

NEC Ch. 4 – Article 430
Part I – 430.7(A)(9)

108

Type of Change		Panel Action		UL	UL 508	API 500	API 505	OSHA
Revision		Accept		1004	-	-	-	1910.305(j)(4)
ROP		ROC		NFPA 70E	NFPA 70B	NFPA 79	NFPA	NEMA
pg. 1143	# 11-16	pg. -	# -	420.10(E)	Ch. 11	Ch. 15	-	MG 1
log: 2308	CMP: 11	log: -	Submitter: Vince Baclawski			2002 NEC: 430.7(A)(8)		IEC: 432 and 433

2002 NEC – 430.7(A) Usual Motor Applications

A motor shall be marked with the following information.

(9) Design letter for design B, C, D, ~~or E~~ motors

2005 NEC – 430.7(A) Usual Motor Applications

A motor shall be marked with the following information.

(9) Design letter for design B, C, or D motors.

Author's Substantiation. This revision clarifies that Design E motors are no longer manufactured and referenced in the NEMA Standards Publication MG 1-1998, *Motors and Generators*.

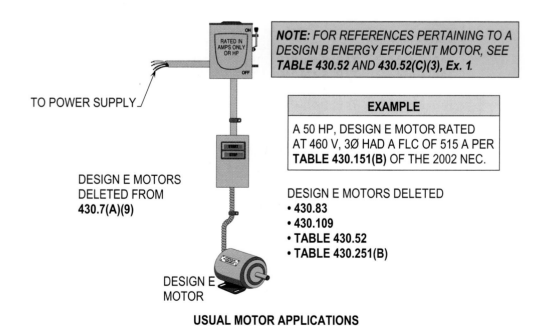

TO POWER SUPPLY

RATED IN AMPS ONLY OR HP

ON

OFF

START STOP

DESIGN E MOTORS DELETED FROM 430.7(A)(9)

DESIGN E MOTOR

NOTE: *FOR REFERENCES PERTAINING TO A DESIGN B ENERGY EFFICIENT MOTOR, SEE TABLE 430.52 AND 430.52(C)(3), Ex. 1.*

EXAMPLE

A 50 HP, DESIGN E MOTOR RATED AT 460 V, 3Ø HAD A FLC OF 515 A PER **TABLE 430.151(B)** OF THE 2002 NEC.

DESIGN E MOTORS DELETED
- **430.83**
- **430.109**
- **TABLE 430.52**
- **TABLE 430.251(B)**

USUAL MOTOR APPLICATIONS
430.7(A)(9)

Purpose of Change: To delete Design E motors because they are no longer manufactured and referenced in NEMA MG 1.

NEC Ch. 4 – Article 430
Part I – 430.7(A)(15)

Type of Change		Panel Action		UL	UL 508	API 500	API 505	OSHA
New Item		Accept		1004	-	-	-	1910.305(j)(4)
ROP		ROC		NFPA 70E	NFPA 70B	NFPA 79	NFPA	NEMA
pg. 1143	# 11-17	pg. -	# -	420.10(E)	Ch. 11	Ch. 15	-	MG 1
log: 1339	CMP: 11	log: -	Submitter: Thomas F. Mueller			2002 NEC: -		IEC: 432 and 433

2005 NEC – 430.7(A) Usual Motor Applications

A motor shall be marked with the following information.

(15) Motors equipped with electrically powered condensation prevention heaters shall be marked with the rated heater voltage, number of phases, and the rated power in watts.

Author's Substantiation. A new item has been added to recognize electrically powered prevention heaters to reduce condensation in motors.

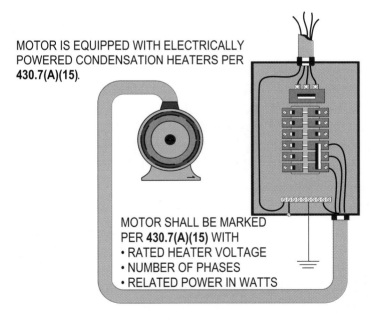

MOTOR IS EQUIPPED WITH ELECTRICALLY POWERED CONDENSATION HEATERS PER **430.7(A)(15)**.

MOTOR SHALL BE MARKED PER **430.7(A)(15)** WITH
• RATED HEATER VOLTAGE
• NUMBER OF PHASES
• RELATED POWER IN WATTS

**USUAL MOTOR APPLICATIONS
430.7(A)(15)**

Purpose of Change: To provide guidelines for electrically powered prevention heaters used to reduce condensation in motors.

NEC Ch. 4 – Article 430
Part I – 430.8, Ex.'s 1 through 4

Type of Change		Panel Action		UL	UL 508	API 500	API 505	OSHA
Revision		Accept in Principle		1004 and 845	-	-	-	1910.305(f)(4)
ROP		ROC		NFPA 70E	NFPA 70B	NFPA 79	NFPA	NEMA
pg. 1145	# 11-19	pg. 364	# 11-16	420.10(E)	Ch. 11	Ch. 15	-	MG 1
log: 3348	CMP: 11	log: 3454	Submitter: Brannon Wiltse			2002 NEC: 430.8		IEC: 432 and 433

2002 NEC – 430.8 Marking on Controllers

A controller shall be marked with the manufacturer s name or identification, the voltage, the current or horsepower rating, and such other necessary data to properly indicate the ~~motors~~ for which it is suitable. A controller that includes motor overload protection suitable for group motor application shall be marked with the motor overload protection and the maximum branch-circuit short-circuit and ground-fault protection for such applications.

2005 NEC – 430.8 Marking on Controllers

A controller shall be marked with the manufacturer s name or identification, the voltage, the current or horsepower rating, the short-circuit current rating, and such other necessary data to properly indicate the applications for which it is suitable.

> **Exception No. 1:** The short-circuit current rating is not required for controllers applied in accordance with 430.81(A) or (B).
>
> **Exception No. 2:** The short-circuit rating is not required to be marked on the controller when the short-circuit current rating of the controller is marked elsewhere on the assembly.
>
> **Exception No. 3:** The short-circuit rating is not required to be marked on the controller when the assembly into which it is installed has a marked short-circuit current rating.
>
> **Exception No. 4:** Short-circuit ratings are not required for controllers rated less than 2 hp at 300 V or less and listed exclusively for general-purpose branch circuits.

A controller that includes motor overload protection suitable for group motor application shall be marked with the motor overload protection and the maximum branch-circuit short-circuit and ground-fault protection for such applications.

Author's Substantiation. The short-circuit rating is not required to be marked on controllers where in accordance with 430.81(A) or (C), where the short-circuit current rating is marked elsewhere on the assembly, when the assembly into which it is installed has a marked short-circuit current rating, and where controllers are rated less than 2 hp at 300 V or less and listed exclusively for general-purpose branch circuits.

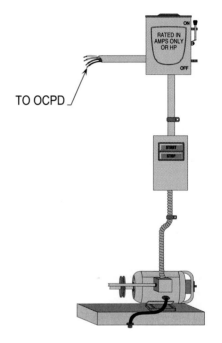

TO OCPD

**CONDITIONS WHERE SHORT-CIRCUIT
CURRENT RATINGS (SCCR) ARE NOT REQUIRED**

Ex. 1: SCCR IS NOT REQUIRED FOR CONTROLLERS PER **430.81(A)** OR **(B)**.

Ex. 2: SCCR IS NOT REQUIRED ON THE CONTROLLER, IF MARKED ELSE-WHERE ON THE ASSEMBLY.

Ex. 3: SCCR IS NOT REQUIRED ON CONTROLLER WHEN THE ASSEMBLY INTO WHICH IT IS INSTALLED HAS A SCCR ON IT.

Ex. 4: SCCR IS NOT REQUIRED ON CONTROLLER RATED LESS THAN 2 HP AT 300 V OR LESS AND LISTED FOR GENERAL-PURPOSE BRANCH-CIRCUIT.

**MARKING ON CONTROLLERS
430.8, Ex.'s 1 THRU 4**

Purpose of Change: To include a list of conditions where short-circuit current ratings are not required on controllers.

NEC Ch. 4 – Article 430
Part II – 430.22(A)

Type of Change		Panel Action		UL	UL 508	API 500	API 505	OSHA
Revision		Accept		1004	-	-	-	1910.305(f)(4)
ROP		ROC		NFPA 70E	NFPA 70B	NFPA 79	NFPA	NEMA
pg. 1147	# 11-23	pg. -	# -	420.10(E)	Ch. 11	Ch. 15	-	MG 1
log: 780	CMP: 11	log: -	Submitter: Dan Leaf			2002 NEC: 430.22(A)		IEC: 432 and 433

2002 NEC – 430.22(A) General

~~Branch-circuit~~ conductors that supply a single motor used in a continuous duty application shall have an ampacity of not less than 125 percent of the motor s full-load current rating as determined by 430.6(A)(1).

2005 NEC – 430.22(A) General

Conductors that supply a single motor used in a continuous duty application shall have an ampacity of not less than 125 percent of the motor s full-load current rating as determined by 430.6(A)(1).

Author's Substantiation. This revision clarifies that a single motor used in a continuous duty application applies to service, feeder, and branch circuit conductors.

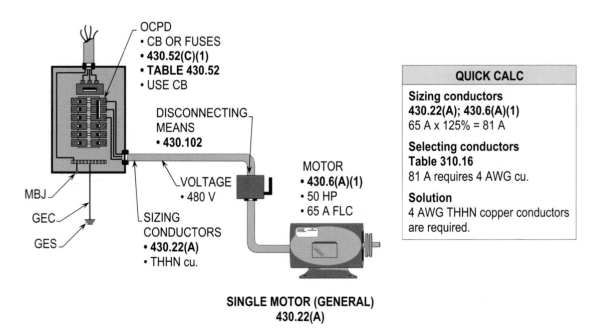

SINGLE MOTOR (GENERAL)
430.22(A)

Purpose of Change: To delete "branch-circuit" before "conductors" in an effort to make this section easier to read and understand.

NEC Ch. 4 – Article 430
Part III – 430.32(A)

Type of Change		Panel Action		UL	UL 508	API 500	API 505	OSHA
Revision		Accept		1004	-	-	-	1910.305(f)(4)
ROP		ROC		NFPA 70E	NFPA 70B	NFPA 79	NFPA	NEMA
pg. 1152	# 11-30	pg. -	# -	420.10(E)	Ch. 11	Ch. 15	-	MG 1
log: 195	CMP: 11	log: -	Submitter: Michael Owen			2002 NEC: 430.32(A)		IEC: 432 and 433

2002 NEC – 430.32(A) More Than 1 Horsepower

Each ~~continuous-duty~~ motor rated more than 1 hp shall be protected against overload by one of the means in 430.32(A)(1) through (A)(4).

2005 NEC – 430.32(A) More Than 1 Horsepower

Each motor <u>used in a continuous-duty application and</u> rated more than 1 hp shall be protected against overload by one of the means in 430.32(A)(1) through (A)(4).

Author's Substantiation. This revision clarifies that the duty cycle of motor is rated by the motor s application.

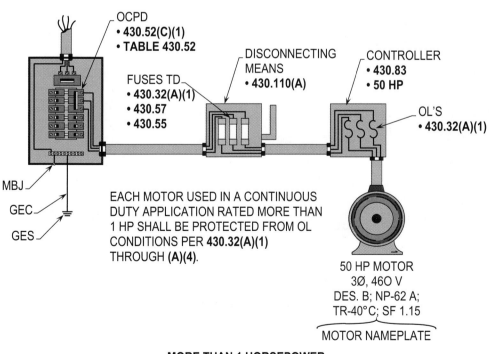

EACH MOTOR USED IN A CONTINUOUS DUTY APPLICATION RATED MORE THAN 1 HP SHALL BE PROTECTED FROM OL CONDITIONS PER **430.32(A)(1)** THROUGH **(A)(4)**.

MORE THAN 1 HORSEPOWER
430.32(A)

Purpose of Change: To clear up the confusion that motors are rated for continuous-duty use only.

NEC Ch. 4 – Article 430
Part IV – 430.52(C)(1)

Type of Change	Panel Action	UL	UL 508	API 500	API 505	OSHA
Revision	Accept	1004	-	-	-	1910.305(f)(4)

ROP		ROC		NFPA 70E	NFPA 70B	NFPA 79	NFPA	NEMA
pg. 1155	# 11-36	pg. -	# -	420.10(E)	Ch.. 11	Ch. 15	-	MG 1
log: 2309	CMP: 11	log: -	Submitter: Vince Baclawski			2002 NEC: Table 430.52		IEC: 432 and 433

2002 NEC – 430.52(C)(1) In Accordance with Table 430.52

A protective device that has a rating or setting not exceeding the value calculated according to the values given in Table 430.52 shall be used.

2005 NEC – 430.52(C)(1) In Accordance with Table 430.52

A protective device that has a rating or setting not exceeding the value calculated according to the values given in Table 430.52 shall be used.

Author's Substantiation. This revision clarifies that Design E motors are no longer manufactured and referenced in the NEMA Standards Publication MG 1-1998, *Motors and Generators*. The reference to Design E motors has been deleted from Table 430.52.

TABLE 430.52 (Sample)				
Percentage of Full-Load Current				
Type of Motor	Nontime Delay Fuse[1]	Dual Element (Time-Delay) Fuse[1]	Instantaneous Trip Breaker	Inverse Time Breaker[2]
Squirrel cage – other than ~~Design E or~~ Design B energy-efficient	300	175	800	250
~~Design E or~~ Design B energy-efficient	300	175	1100	250

For Example: What is the percentage listed to allow a Design B energy efficient motor to start and run using an INST CB?

Step 1: Table 430.52
1100%

Solution: The allowable percentage is 1100% per **Table 430.52.**

Note that all of the table percentages are not shown.

NOTE: DESIGN E MOTOR DESIGNATION HAS BEEN REMOVED FROM TABLE 430.52 AS WELL AS OTHER SECTIONS IN ARTICLE 430.

IN ACCORDANCE WITH TABLE 430.52
430.52(C)(1)

Purpose of Change: To delete Design E motors from **Table 430.52** because NEMA no longer references this motor in NEMA MG 1.

NEC Ch. 4 – Article 430
Part IV – 430.52(C)(6), FPN

109

Type of Change	Panel Action	UL	UL 508	API 500	API 505	OSHA
New FPN	Accept in Principle	1004 and 845	-	-	-	1910.305(f)(4)
ROP		**ROC**		**NFPA 70E**	**NFPA 70B**	**NFPA 79**	**NFPA**	**NEMA**

pg. 1155	# 11-37	pg. -	# -	420.10(E)	Ch. 11	Ch. 15	-	MG 1
log: 3019	CMP: 11	log: -	Submitter: Brannon Wiltse			2002 NEC: -		IEC: 432 and 433

2002 NEC – 430.52(C)(6) Self-Protected Combination Controller

A listed self-protected combination controller shall be permitted in lieu of the devices specified in Table 430.52. Adjustable instantaneous-trip settings shall not exceed 1300 percent of full-load motor current for other than ~~Design E motors or~~ Design B energy efficient motors and not more than 1700 percent of full-load motor current for ~~Design E motors or~~ Design B energy efficient motors.

2005 NEC – 430.52(C)(6) Self-Protected Combination Controller

A listed self-protected combination controller shall be permitted in lieu of the devices specified in Table 430.52. Adjustable instantaneous-trip settings shall not exceed 1300 percent of full-load motor current for other than Design B energy-efficient motors and not more than 1700 percent of full-load motor current for Design B energy-efficient motors.

> **FPN:** Proper application of self-protected combination controllers on 3-phase systems, other than solidly grounded wye, particularly on corner grounded delta systems, considers the self-protected combination controllers individual pole-interrupting capability.

Author's Substantiation. A new FPN has been added to recognize the proper application of self-protected combination controllers for systems other than solidly grounded wye.

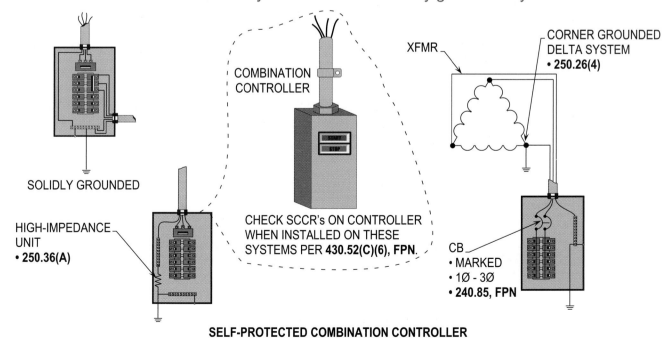

SELF-PROTECTED COMBINATION CONTROLLER
430.52(C)(6), FPN

Purpose of Change: A new FPN has been added to clarify the proper application of self-protected combination controllers on 3-phase systems other than solidly grounded wye.

NEC Ch. 4 – Article 430
Part IV – 430.53(C)(3) and (C)(6)

Type of Change		Panel Action		UL	UL 508	API 500	API 505	OSHA
New Item		Accept		1004	-	-	-	1910.305(f)(4)
ROP		ROC		NFPA 70E	NFPA 70B	NFPA 79	NFPA	NEMA
pg. 1158	# 11-44	pg. 366	# 11-28	420.10(E)	Ch. 11	Ch. 15	-	MG 1
log: 2329	CMP: 11	log: 405	Submitter: Vince Baclawski			2002 NEC: 430.53(C)(3)		IEC: 432 and 433

2002 NEC – 430.53(C) Other Group Installations

Two or more motors of any rating or one or more motors and other load(s), with each motor having individual overload protection, shall be permitted to be connected to one branch circuit where the motor controller(s) and overload device(s) are (1) installed as a listed factory assembly and the motor branch-circuit short-circuit and ground-fault protective device either is provided as part of the assembly or is specified by a marking on the assembly, or (2) the motor branch-circuit short-circuit and ground-fault protective device, the motor controller(s), and overload device(s) are field-installed as separate assemblies listed for such use and provided with manufacturers' instructions for use with each other, and (3) all of the following conditions are complied with:

(3) Each circuit breaker is ~~one~~ of the inverse time type ~~and listed for group installation~~.

2005 NEC – 430.53(C) Other Group Installations

Two or more motors of any rating or one or more motors and other load(s), with each motor having individual overload protection, shall be permitted to be connected to one branch circuit where the motor controller(s) and overload device(s) are (1) installed as a listed factory assembly and the motor branch-circuit short-circuit and ground-fault protective device either is provided as part of the assembly or is specified by a marking on the assembly, or (2) the motor branch-circuit short-circuit and ground-fault protective device, the motor controller(s), and overload device(s) are field-installed as separate assemblies listed for such use and provided with manufacturers' instructions for use with each other, and (3) all of the following conditions are complied with:

(3) Each circuit breaker is <u>listed and is</u> of the inverse time type.

(6) <u>Overcurrent protection for loads other than motor loads shall be in accordance with Parts I through VII of Article 240.</u>

Author's Substantiation. A new item has been added to recognize that overcurrent protection shall be in accordance with Article 240 for loads other than motor loads.

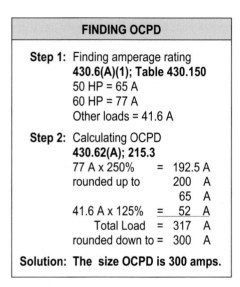

FINDING OCPD

Step 1: Finding amperage rating
430.6(A)(1); Table 430.150
50 HP = 65 A
60 HP = 77 A
Other loads = 41.6 A

Step 2: Calculating OCPD
430.62(A); 215.3

77 A x 250%	=	192.5 A
rounded up to		200 A
		65 A
41.6 A x 125%	=	52 A
Total Load	=	317 A
rounded down to =		300 A

Solution: The size OCPD is 300 amps.

NOTE: OCPD FOR OTHER LOADS IS CALCULATED PER 430.53(C)(6) AND ARTICLE 240.

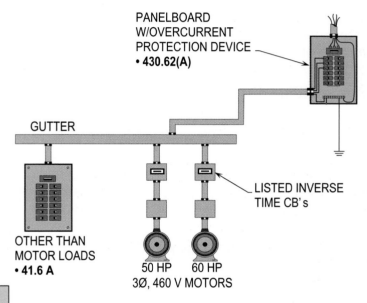

PANELBOARD
W/OVERCURRENT
PROTECTION DEVICE
• **430.62(A)**

GUTTER

LISTED INVERSE
TIME CB's

OTHER THAN
MOTOR LOADS
• **41.6 A**

50 HP   60 HP
3Ø, 460 V MOTORS

OTHER GROUP INSTALLATIONS
430.53(C)(3) AND (C)(6)

Purpose of Change: To refer to **Article 240** for protection of loads other than motors in a group installation.

NEC Ch. 4 – Article 430
Part VII – 430.83(E)

Type of Change		Panel Action		UL	UL 508	API 500	API 505	OSHA
Revision		Accept in Principle		1004 and 845	-	-	-	1910.305(f)(4)
ROP		ROC		NFPA 70E	NFPA 70B	NFPA 79	NFPA	NEMA
pg. 1164	# 11-58a	pg. -	# -	420.10(E)	Ch. 11	Ch. 15	-	MG 1
log: 3050	CMP: 11	log: -	Submitter: Todd F. Lottmann			2002 NEC: 430.83(E)		IEC: 432 and 433

2002 NEC – 430.83(E) Voltage Rating

A controller with a straight voltage rating, for example, 240 volts or 480 volts, shall be permitted to be applied in a circuit in which the nominal voltage between any two conductors does not exceed the controller's voltage rating. A controller with a slash rating, for example, 120/240 volts or 480/277 volts, shall only be applied in a circuit in which the nominal voltage to ground from any conductor does not exceed the lower of the two values of the controller s voltage rating and the nominal voltage between any two conductors does not exceed the higher value of the controller s voltage rating.

2005 NEC – 430.83(E) Voltage Rating

A controller with a straight voltage rating, for example, 240 volts or 480 volts, shall be permitted to be applied in a circuit in which the nominal voltage between any two conductors does not exceed the controller's voltage rating. A controller with a slash rating, for example, 240/120 volts or 480Y/277 volts, shall only be applied in a solidly grounded circuit in which the nominal voltage to ground from any conductor does not exceed the lower of the two values of the controller s voltage rating and the nominal voltage between any two conductors does not exceed the higher value of the controller s voltage rating.

Author's Substantiation. This revision clarifies that a controller with a slash rating shall only be used in a solidly grounded system.

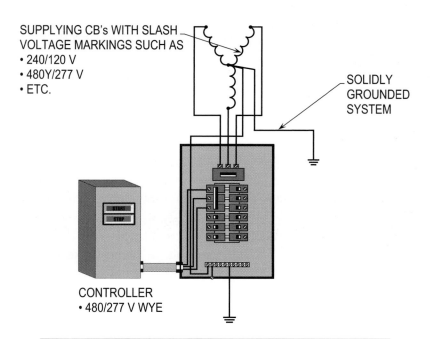

SUPPLYING CB's WITH SLASH
VOLTAGE MARKINGS SUCH AS
• 240/120 V
• 480Y/277 V
• ETC.

SOLIDLY
GROUNDED
SYSTEM

CONTROLLER
• 480/277 V WYE

NOTE: *CONTROLLERS WITH A SLASH RATING SHALL BE USED*
ONLY ON A SOLIDLY GROUNDED SYSTEM PER ***430.83(E).***

VOLTAGE RATING
430.83(E)

Purpose of Change: To make it clear that controllers with a slash rating shall be used in a solidly grounded system only.

NEC Ch. 4 – Article 430
Part IX – 430.102(B), Ex.

 110

Type of Change		Panel Action		UL	UL 508	API 500	API 505	OSHA
Revision		Accept		1004	-	-	-	1910.305(f)(4)
ROP		ROC		NFPA 70E	NFPA 70B	NFPA 79	NFPA	NEMA
pg. 1171	# 11-67	pg. 369	# 11-42	420.10(E)	Ch. 11	Ch. 15	-	MG 1
log: 3387	CMP: 11	log: 2574	Submitter: John W. Young			2002 NEC: 430.102(B), Ex.		IEC: 432 and 433

2002 NEC – 430.102(B) Motor

A disconnecting means shall be located in sight from the motor location and the driven machinery location. The disconnecting means required in accordance with 430.102(A) shall be permitted to serve as the disconnecting means for the motor if it is located in sight from the motor location and the driven machinery location.

> **Exception:** The disconnecting means shall not be required to be in sight from the motor and the driven machinery location under either condition (a) or (b), provided the disconnecting means required in accordance with 430.102(A) is individually capable of being locked in the open position. The provision for locking or adding a lock to the disconnecting means shall be ~~permanently~~ installed on or at the switch or circuit breaker used as the disconnecting means.

2005 NEC – 430.102(B) Motor

A disconnecting means shall be located in sight from the motor location and the driven machinery location.

> **Exception:** The disconnecting means shall not be required to be in sight from the motor and the driven machinery location under either condition (a) or (b), provided the disconnecting means required in accordance with 430.102(A) is individually capable of being locked in the open position. The provision for locking or adding a lock to the disconnecting means shall be installed on or at the switch or circuit breaker used as the disconnecting means <u>and shall remain in place with or without the lock installed</u>.
>
> The disconnecting means required in accordance with 430.102(A) shall be permitted to serve as the disconnecting means for the motor if it is located in sight from the motor location and the driven machinery location.

Author's Substantiation. This revision clarifies any vague language pertaining to the provisions for locking or adding a lock on or at the switch or circuit breaker. The locking device shall remain in place with or without the lock installed.

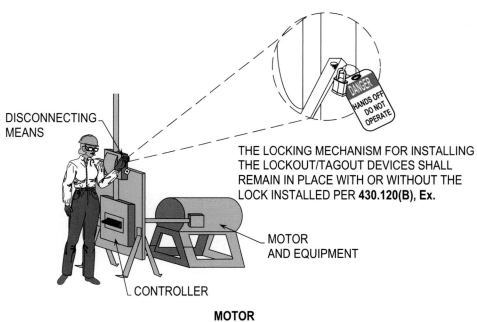

DISCONNECTING MEANS

THE LOCKING MECHANISM FOR INSTALLING THE LOCKOUT/TAGOUT DEVICES SHALL REMAIN IN PLACE WITH OR WITHOUT THE LOCK INSTALLED PER **430.120(B), Ex.**

MOTOR AND EQUIPMENT

CONTROLLER

MOTOR
430.102(B), Ex.

Purpose of Change: To revise the wording in an effort to make this section easier to read and understand.

NEC Ch. 4 – Article 430
Part IX – 430.108

Type of Change		Panel Action		UL	UL 508	API 500	API 505	OSHA
Revision		Accept in Part		1004 and 1087	-	-	-	1910.305(f)(4)
ROP		ROC		NFPA 70E	NFPA 70B	NFPA 79	NFPA	NEMA
pg. 1174	# 11-72	pg. -	# -	420.10(E)	Ch. 11	Ch. 15	-	MG 1
log: 739	CMP: 11	log: -	Submitter: Dan Leaf			2002 NEC: 430.108		IEC: 432 and 433

2002 NEC – 430.108 Every ~~Switch~~

Every disconnecting means in the motor circuit between the point of attachment to the feeder and the point of connection to the motor shall comply with the requirements of 430.109 and 430.110.

2005 NEC – 430.108 Every **Disconnecting Means**

Every disconnecting means in the motor circuit between the point of attachment to the feeder and the point of connection to the motor shall comply with the requirements of 430.109 and 430.110.

Author's Substantiation. This revision clarifies that a switch and a disconnecting means are not necessarily the same device.

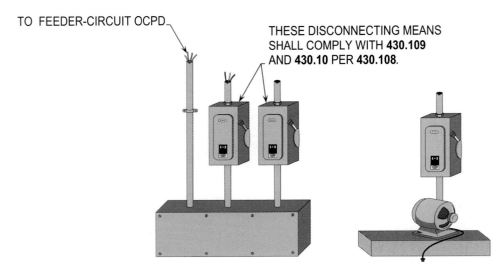

TO FEEDER-CIRCUIT OCPD

THESE DISCONNECTING MEANS
SHALL COMPLY WITH **430.109**
AND **430.10** PER **430.108**.

EVERY DISCONNECTING MEANS
430.108

Purpose of Change: To more accurately address the disconnecting means, instead of a switch, in the title head.

NEC Ch. 4 – Article 430
Part IX – 430.109(A)(6)

 111

Type of Change		Panel Action		UL	UL 508	API 500	API 505	OSHA
New Sentences		Accept		1004 and 845	-	-	-	1910.305(f)(4)
ROP		ROC		NFPA 70E	NFPA 70B	NFPA 79	NFPA	NEMA
pg. 1175	# 11-73	pg. 370	# 11-45	420.10(E)	Ch. 11	Ch. 15	-	MG 1
log: 900	CMP: 11	log: 731	Submitter: Gordon Davis			2002 NEC: 430.109(A)(6)		IEC: 432 and 433

2002 NEC – 430.109(A)(6) Manual Motor Controller

Listed manual motor controllers additionally marked Suitable as Motor Disconnect shall be permitted as a disconnecting means where installed between the final motor branch-circuit short-circuit ~~and ground-fault~~ protective device and the motor.

2005 NEC – 430.109(A)(6) Manual Motor Controller

Listed manual motor controllers additionally marked Suitable as Motor Disconnect shall be permitted as a disconnecting means where installed between the final motor branch-circuit short-circuit protective device and the motor. <u>Listed manual motor controllers additionally marked Suitable as Motor Disconnect shall be permitted as disconnecting means on the line side of the fuses permitted in 430.52(C)(5). In this case, the fuses permitted in 430.52(C)(5) shall be considered supplementary fuses, and suitable branch-circuit short-circuit and ground-fault protective devices shall be installed on the line side of the manual motor controller additionally marked Suitable as Motor Disconnect.</u>

Author's Substantiation. New sentences have been added to clarify the requirements for a listed manual motor controller used as a disconnecting means on the line side of fuses.

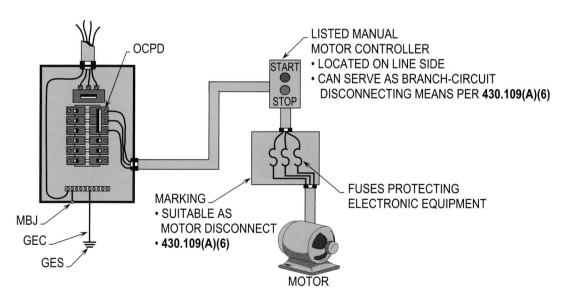

LISTED MANUAL MOTOR CONTROLLER
• LOCATED ON LINE SIDE
• CAN SERVE AS BRANCH-CIRCUIT DISCONNECTING MEANS PER **430.109(A)(6)**

FUSES PROTECTING ELECTRONIC EQUIPMENT

MARKING
• SUITABLE AS MOTOR DISCONNECT
• **430.109(A)(6)**

MANUAL MOTOR CONTROLLER
430.109(A)(6)

Purpose of Change: To provide requirements when a listed manual motor controller is used as a disconnecting means on the line side of fuses.

Type of Change		Panel Action		UL	UL 508	API 500	API 505	OSHA
New Subdivision		Accept		1004 and 1047	-	-	-	1910.305(f)(4)
ROP		ROC		NFPA 70E	NFPA 70B	NFPA 79	NFPA	NEMA
pg. 1176	# 11-74	pg. 370	# 11-45a	420.10(E)(2)	Ch. 11	Ch. 15	-	MG 1
log: 1029	CMP: 11	log: CC1102	Submitter: William E. Anderson			2002 NEC: -		IEC: Ch. 46

2005 NEC – 430.109(A)(7) System Isolation Equipment

System isolation equipment shall be listed for disconnection purposes. System isolation equipment shall be installed on the load side of the overcurrent protection and its disconnecting means. The disconnecting means shall be one of the types permitted by 430.109(A)(1) through (A)(3).

Author's Substantiation. A new subdivision has been added to recognize the use of system isolation equipment as a disconnecting means.

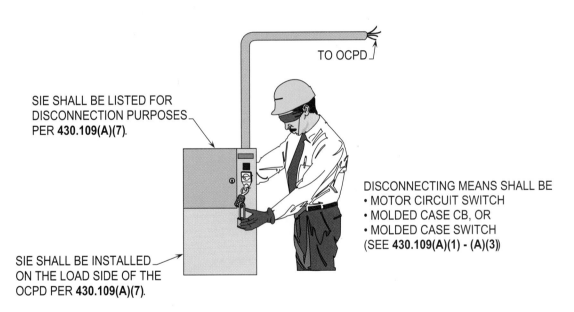

SIE SHALL BE LISTED FOR DISCONNECTION PURPOSES PER **430.109(A)(7)**.

TO OCPD

DISCONNECTING MEANS SHALL BE
• MOTOR CIRCUIT SWITCH
• MOLDED CASE CB, OR
• MOLDED CASE SWITCH
(SEE **430.109(A)(1) - (A)(3)**)

SIE SHALL BE INSTALLED ON THE LOAD SIDE OF THE OCPD PER **430.109(A)(7)**.

SYSTEM ISOLATION EQUIPMENT
430.109(A)(7)

Purpose of Change: To recognize the use of system isolation equipment (SIE) as a disconnecting means.

NEC Ch. 4 – Article 430
Part IX – 430.110(A) and Ex.

Type of Change		Panel Action		UL	UL 508	API 500	API 505	OSHA
Revision		Accept in Part		1004 and 1087	-	-	-	1910.305(f)(4)
ROP		ROC		NFPA 70E	NFPA 70B	NFPA 79	NFPA	NEMA
pg. 1178	# 11-79	pg. -	# -	420.10(E)(2)	Ch. 11	Ch. 15	-	MG 1
log: 746	CMP: 11	log: -	Submitter: Dan Leaf			2002 NEC: 430.110(A) and Ex.		IEC: Ch. 46

2002 NEC – 430.110(A) General

The disconnecting means for motor circuits rated 600 volts, nominal, or less shall have an ampere rating ~~of at least~~ 115 percent of the full-load current rating of the motor.

> **Exception:** A listed nonfused motor-circuit switch having a horsepower rating ~~equal to or greater~~ than the motor horsepower shall be permitted to have an ampere rating less than 115 percent of the full-load current rating of the motor.

2005 NEC – 430.110(A) General

The disconnecting means for motor circuits rated 600 volts, nominal, or less shall have an ampere rating <u>not less than</u> 115 percent of the full-load current rating of the motor.

> **Exception:** A listed nonfused motor-circuit switch having a horsepower rating <u>not less</u> than the motor horsepower shall be permitted to have an ampere rating less than 115 percent of the full-load current rating of the motor.

Author's Substantiation. This revision clarifies that the disconnecting means shall not be less than 115 percent of the full-load current rating of the motor.

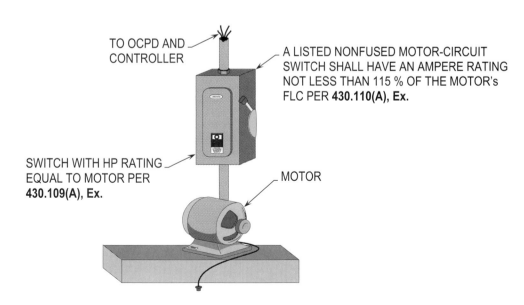

AMPERE RATING AND INTERRUPTING CAPACITY (GENERAL)
430.110(A) AND Ex.

Purpose of Change: To require the disconnecting means to be rated not less than 115 percent of the motor's FLC in amps.

NEC Ch. 4 – Article 430
Part X – 430.122(A)

 113.1

Type of Change		Panel Action		UL	UL 508	API 500	API 505	OSHA
Relocated		Accept		508C	-	-	-	-
ROP		ROC		NFPA 70E	NFPA 70B	NFPA 79	NFPA	NEMA
pg. 1146	# 11-21	pg. -	# -	-	70.22	7.3.1.2	-	ICS 7
log: 2538	CMP: 11	log: -		Submitter: Alan Manche		2002 NEC: 430.22(A), Ex. 2		IEC: 146-2

2005 NEC – <u>430.122(A) Branch/Feeder Circuit Conductors</u>

<u>Circuit conductors supplying power conversion equipment included as part of an adjustable-speed drive system shall have an ampacity not less than 125 percent of the rated input to the power conversion equipment.</u>

Author's Substantiation. A new Part X to Article 430 has been added to recognize the installation of adjustable speed drive systems. Subsection 430.22(A), Ex. 2 has been relocated and renumbered as 430.122(A) in Part X of Article 430.

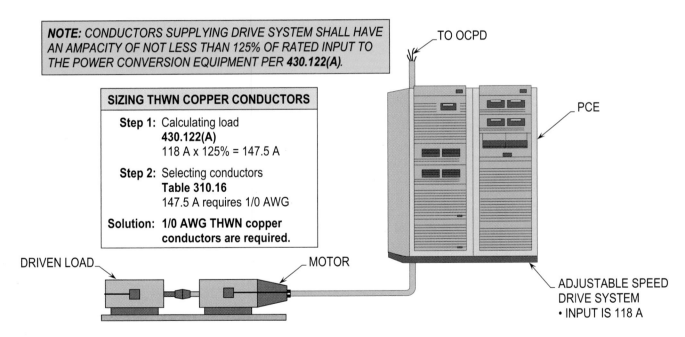

NOTE: *CONDUCTORS SUPPLYING DRIVE SYSTEM SHALL HAVE AN AMPACITY OF NOT LESS THAN 125% OF RATED INPUT TO THE POWER CONVERSION EQUIPMENT PER 430.122(A).*

TO OCPD

PCE

SIZING THWN COPPER CONDUCTORS

Step 1: Calculating load
430.122(A)
118 A x 125% = 147.5 A

Step 2: Selecting conductors
Table 310.16
147.5 A requires 1/0 AWG

Solution: 1/0 AWG THWN copper conductors are required.

DRIVEN LOAD

MOTOR

ADJUSTABLE SPEED DRIVE SYSTEM
• INPUT IS 118 A

BRANCH/FEEDER-CIRCUIT CONDUCTORS
430.122(A)

Purpose of Change: To relocate and renumber the existing **430.22(A), Ex. 2** to **430.122(A)** in the new **Part X** of the **2005 NEC**.

Type of Change		Panel Action		UL	UL 508	API 500	API 505	OSHA
New Part		Accept		508C	-	-	-	-
ROP		ROC		NFPA 70E	NFPA 70B	NFPA 79	NFPA	NEMA
pg. 1138	# 11-6	pg. -	# -	-	70.22	7.3.1.2	-	ICS 7
log: 2540	CMP: 11	log: -	Submitter: Alan Manche			2002 NEC: -		IEC: 146.2

2005 NEC – 430.122(B) Bypass Device

For an adjustable speed drive system that utilizes a bypass device, the conductor ampacity shall not be less than required by 430.6. The ampacity of circuit conductors supplying power conversion equipment included as part of an adjustable-speed drive system that utilizes a bypass device shall be the larger of either of the following:

(1) 125 percent of the rated input to the power conversion equipment

(2) 125 percent of the motor full-load current rating as determined by 430.6

Author's Substantiation. A new Part X to Article 430 has been added to recognize the installation of adjustable speed drive systems. This new subsection provides the requirements for sizing conductors based on the drive input or the motor s full-load current in amps.

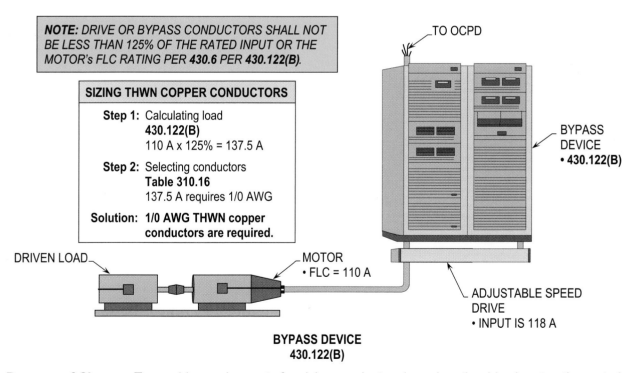

Purpose of Change: To provide requirements for sizing conductors based on the drive input or the motor's FLC in amps.

NEC Ch. 4 – Article 430
Part X – 430.124(A) through (C)

113.3

Type of Change		Panel Action		UL	UL 508	API 500	API 505	OSHA
New Part		Accept		508C	-	-	-	-
ROP		ROC		NFPA 70E	NFPA 70B	NFPA 79	NFPA	NEMA
pg. 1138	# 11-6	pg. -	# -	-	70.22	7.3.1.2	-	ICS 7
log: 2540	CMP: 11	log: -	Submitter: Alan Manche			2002 NEC: -		IEC: 146-2

2005 NEC – 430.124 Overload Protection

Overload protection of the motor shall be provided.

(A) Included in Power Conversion Equipment. Where the power conversion equipment is marked to indicate that motor overload protection is included, additional overload protection shall not be required.

(B) Bypass Circuits. For adjustable speed drive systems that utilize a bypass device to allow motor operation at rated full load speed, motor overload protection as described in Article 430, Part III, shall be provided in the bypass circuit.

(C) Multiple Motor Applications. For multiple motor application, individual motor overload protection shall be provided in accordance with Article 430, Part III.

Author's Substantiation. A new Part X to Article 430 has been added to recognize the installation of adjustable speed drive systems. This subsection provides the requirements for overload protection for motors and drive systems.

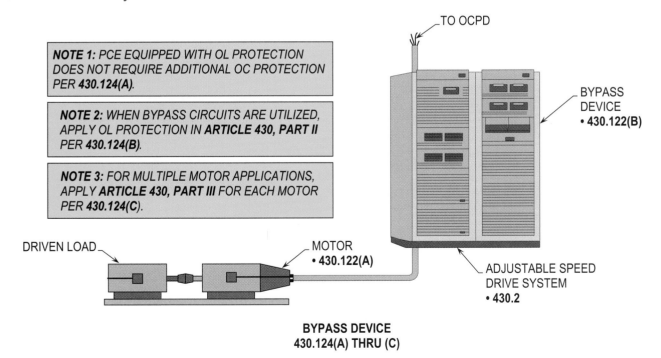

NOTE 1: PCE EQUIPPED WITH OL PROTECTION DOES NOT REQUIRE ADDITIONAL OC PROTECTION PER **430.124(A)**.

NOTE 2: WHEN BYPASS CIRCUITS ARE UTILIZED, APPLY OL PROTECTION IN **ARTICLE 430, PART II** PER **430.124(B)**.

NOTE 3: FOR MULTIPLE MOTOR APPLICATIONS, APPLY **ARTICLE 430, PART III** FOR EACH MOTOR PER **430.124(C)**.

TO OCPD

BYPASS DEVICE
• 430.122(B)

DRIVEN LOAD

MOTOR
• 430.122(A)

ADJUSTABLE SPEED
DRIVE SYSTEM
• 430.2

BYPASS DEVICE
430.124(A) THRU (C)

Purpose of Change: To include overload protection for motors and drive systems under certain conditions of use when using the systems.

NEC Ch. 4 – Article 430
Part X – 430.126(A)(1) through (A)(4)

113.4

Type of Change		Panel Action		UL	UL 508	API 500	API 505	OSHA
New Part		Accept		508C	-	-	-	-
ROP		ROC		NFPA 70E	NFPA 70B	NFPA 79	NFPA	NEMA
pg. 1138	# 11-6	pg. 371	# 11-50	-	70.22	7.3.1.2	-	ICS 7
log: 2540	CMP: 11	log: 2571	Submitter: Alan Manche			2002 NEC: -		IEC: 146-2

2005 NEC – 430.126(A) General

Adjustable speed drive systems shall protect against motor overtemperature conditions. Overtemperature protection is in addition to the conductor protection required by 430.32. Protection shall be provided by one of the following means.

(1) Motor thermal protector in accordance with 430.32

(2) Adjustable speed drive controller with load and speed-sensitive overload protection and thermal memory retention upon shutdown or power loss

(3) Overtemperature protection relay utilizing thermal sensors embedded in the motor and meeting the requirements of 430.32(A)(2) or (B)(2)

(4) Thermal sensors embedded in the motor that is received and acted upon by an adjustable speed drive

Author's Substantiation. A new Part X to Article 430 has been added to recognize the installation of adjustable speed drive systems. This subsection provides requirements for motor overtemperature protection to protect the drive systems from excessive temperatures.

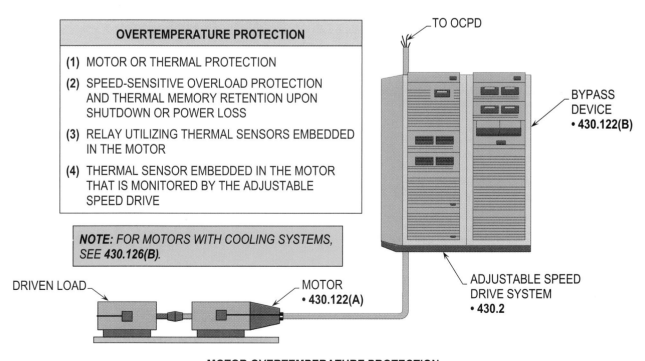

MOTOR OVERTEMPERATURE PROTECTION
430.126(A)(1) THRU (A)(4)

Purpose of Change: To include motor overtemperature protection for drive systems with excessive temperatures.

NEC Ch. 4 – Article 430
Part X – 430.126(D)

113.5

Type of Change		Panel Action		UL	UL 508	API 500	API 505	OSHA
New Part		Accept		508C	-	-	-	-
ROP		ROC		NFPA 70E	NFPA 70B	NFPA 79	NFPA	NEMA
pg. 1138	# 11-6	pg. -	# -	-	70.22	7.3.1.2	-	ICS 7
log: 2540	CMP: 11	log: -	Submitter: Alan Manche			2002 NEC: -		IEC: 146-2

2005 NEC – 430.126(D) Automatic Restarting and Orderly Shutdown

The provisions of 430.43 and 430.44 shall apply to the motor overtemperature protection means.

Author's Substantiation. A new Part X to Article 430 has been added to recognize the installation of adjustable speed drive systems. This subsection provides requirements for multiple motor applications, including automatic restarting and orderly shutdown.

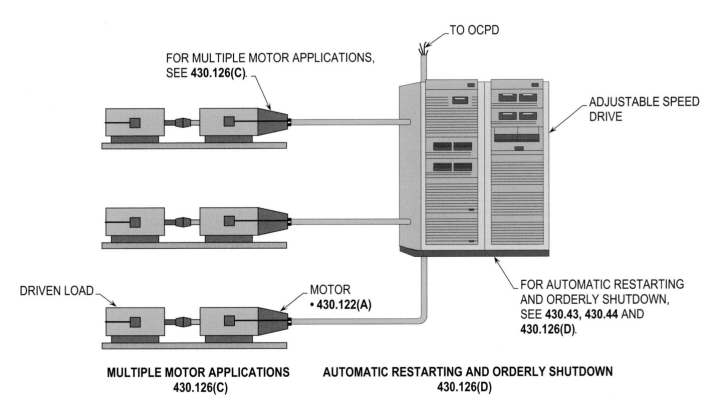

TO OCPD

FOR MULTIPLE MOTOR APPLICATIONS, SEE **430.126(C)**.

ADJUSTABLE SPEED DRIVE

DRIVEN LOAD

MOTOR
• **430.122(A)**

FOR AUTOMATIC RESTARTING AND ORDERLY SHUTDOWN, SEE **430.43**, **430.44** AND **430.126(D)**.

MULTIPLE MOTOR APPLICATIONS
430.126(C)

AUTOMATIC RESTARTING AND ORDERLY SHUTDOWN
430.126(D)

Purpose of Change: To include requirements that address multiple motor applications, including automatic restarting and orderly shutdown rules.

NEC Ch. 4 – Article 430
Part X – 430.128

Type of Change		Panel Action		UL	UL 508	API 500	API 505	OSHA
New Part		Accept		508C	-	-	-	-
ROP		ROC		NFPA 70E	NFPA 70B	NFPA 79	NFPA	NEMA
pg. 1138	# 11-6	pg. -	# -	-	70.22	7.3.1.2	-	ICS 7
log: 2540	CMP: 11	log: -		Submitter: Alan Manche		2002 NEC: -		IEC: 146-2

2005 NEC – 430.128 Disconnecting Means

The disconnecting means shall be permitted to be in the incoming line to the conversion equipment and shall have a rating not less than 115 percent of the rated input current of the conversion unit.

Author's Substantiation. A new Part X to Article 430 has been added to recognize the installation of adjustable speed drive systems. This section provides requirements for sizing, selecting, and locating the disconnecting means.

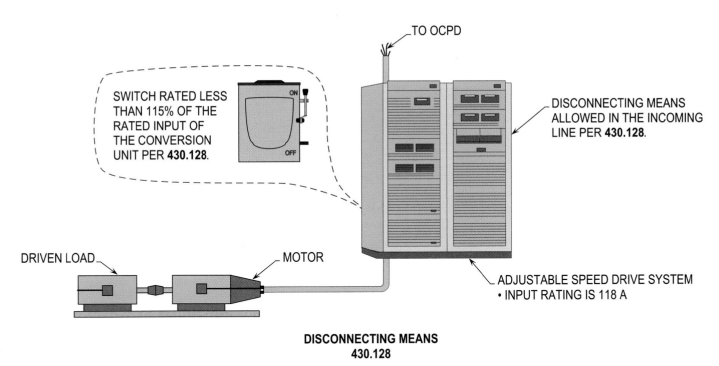

TO OCPD

SWITCH RATED LESS THAN 115% OF THE RATED INPUT OF THE CONVERSION UNIT PER **430.128**.

DISCONNECTING MEANS ALLOWED IN THE INCOMING LINE PER **430.128**.

DRIVEN LOAD

MOTOR

ADJUSTABLE SPEED DRIVE SYSTEM
• INPUT RATING IS 118 A

DISCONNECTING MEANS
430.128

Purpose of Change: To include requirements for sizing, selecting, and locating the disconnecting means.

NEC Ch. 4 – Article 430
Part XIV – Table 430.247 through 251

Type of Change		Panel Action		UL	UL 508	API 500	API 505	OSHA
Renumbered		Accept		1004	-	-	-	1910.305(j)(4)
ROP		**ROC**		**NFPA 70E**	**NFPA 70B**	**NFPA 79**	**NFPA**	**NEMA**
pg. 1138	# 11-6	pg. -	# -	420.10(E)	Ch. 11	Ch. 15	-	MG 1
log: 2540	CMP: 11	log: -	Submitter: Alan Manche			2002 NEC: Tables 430.147 thru 430.251		IEC: 432 and 433

Author's Substantiation. Tables 430.147 through 430.151 have been renumbered as Tables 430.247 through 430.251.

MOTOR AND
CONTROLLER

MOTOR	2005 NEC	2002 NEC
DC MOTORS	**TABLE 430.247**	~~TABLE 430.147~~
1Ø MOTORS	**TABLE 430.248**	~~TABLE 430.148~~
2Ø MOTORS	**TABLE 430.249**	~~TABLE 430.149~~
3Ø MOTORS	**TABLE 430.250**	~~TABLE 420.150~~
1Ø LOCKED ROTOR CURRENT	**TABLE 430.251(A)**	~~TABLE 420.151(A)~~
3Ø LOCKED ROTOR CURRENT	**TABLE 430.251(B)**	~~TABLE 420.151(B)~~

EXAMPLE: A DESIGN B, 50 HP, 3Ø, 460 V MOTOR HAS A FLC OF 65 A PER **TABLE 430.250** AND A LOCKED ROTOR CURRENT OF 363 A PER **TABLE 430.251(B)** OF THE **2005 NEC**.

MOTOR FLC AND LRC VALUES
TABLES 430.247 THRU 430.251(B)

Purpose of Change: To include a 200 series of numbers for over 600 volt motors and Tables for selecting FLC and LRC ratings in amps.

NEC Ch. 4 – Article 440
Part – 440.4(B), Ex. 3

Type of Change		Panel Action		UL	UL 508	API 500	API 505	OSHA
New Exception		Accept in Principle		884	-	-	-	1910.305(j)(3)(ii)
ROP		ROC		NFPA 70E	NFPA 70B	NFPA 79	NFPA	NEMA
pg. 1187	# 11-95	pg. 373	# 11-59	420.10(D)(3)	-	-	-	-
log: 3320	CMP: 11	log: 2865	Submitter: Brannon Wiltse			2002 NEC: -		IEC: Ch. 43

2005 NEC – 440.4(B) Multimotor and Combination-Load Equipment

Multimotor and combination-load equipment shall be provided with a visible nameplate marked with the maker s name, the rating in volts, frequency and number of phases, minimum supply circuit conductor ampacity, the maximum rating of the branch-circuit short-circuit and ground-fault protective device, and the short-circuit current rating of the motor controllers or industrial control panel. The ampacity shall be calculated by using Part IV and counting all the motors and other loads that will be operated at the same time. The branch-circuit short-circuit and ground-fault protective device rating shall not exceed the value calculated by using Part III. Multimotor or combination-load equipment for use on two or more circuits shall be marked with the above information for each circuit.

> **Exception No. 3:** Multimotor and combination-load equipment used in one- and two-family dwellings, cord-and-attachment-plug-connected equipment, or equipment supplied from a branch circuit protected at 60 A or less shall not be required to be marked with a short-circuit current rating.

Author's Substantiation. A new exception has been added to include requirements for multimotor and combination-load equipment installed in one- and two-family dwellings.

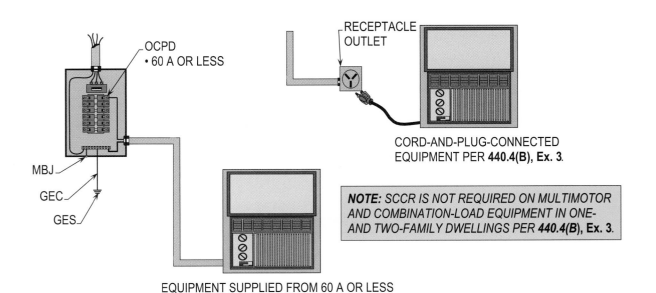

MULTIMOTOR AND COMBINATION-LOAD EQUIPMENT
440.4(B), Ex. 3

Purpose of Change: To include requirements for multimotor and combination-load equipment installed in one- and two-family dwellings.

NEC Ch. 4 – Article 440
Part II – 440.14, Ex. 1

114

Type of Change	Panel Action	UL	UL 508	API 500	API 505	OSHA
New Sentence	Accept in Part	884	-	-	-	1910.305(j)(3)(ii)
ROP		ROC	NFPA 70E	NFPA 70B	NFPA 79	NFPA	NEMA
pg. 1190	# 11-100	pg. -	# -	420.10(D)(2)	-	-	-	-
log: 2816	CMP: 11	log: -	Submitter: Michael I. Callanan		2002 NEC: 440.14, Ex. 1		IEC: Ch. 43

2002 NEC – 440.14 Location

Disconnecting means shall be located within sight from and readily accessible from the air-conditioning or refrigerating equipment. The disconnecting means shall be permitted to be installed on or within the air-conditioning or refrigerating equipment.

The disconnecting means shall not be located on panels that are designed to allow access to the air-conditioning or refrigerating equipment.

> **Exception No. 1:** Where the disconnecting means provided in accordance with 430.102(A) is capable of being locked in the open position, and the refrigerating or air-conditioning equipment is essential to an industrial process in a facility where the conditions of maintenance and ~~the~~ supervision ensure that only qualified persons service the equipment, a disconnecting means within sight from the equipment shall not be required.

2005 NEC – 440.14 Location

Disconnecting means shall be located within sight from and readily accessible from the air-conditioning or refrigerating equipment. The disconnecting means shall be permitted to be installed on or within the air-conditioning or refrigerating equipment.

The disconnecting means shall not be located on panels that are designed to allow access to the air-conditioning or refrigerating equipment.

> **Exception No. 1:** Where the disconnecting means provided in accordance with 430.102(A) is capable of being locked in the open position, and the refrigerating or air-conditioning equipment is essential to an industrial process in a facility <u>with written safety procedures, and</u> where the conditions of maintenance and supervision ensure that only qualified persons service the equipment, a disconnecting means within sight from the equipment shall not be required. <u>The provision for locking or adding a lock to the disconnecting means shall be permanently installed on or at the switch or circuit breaker used as the disconnecting means.</u>

Author's Substantiation. A new sentence has been added to clarify that permanent provisions shall be made for locking or adding a lock to the disconnecting means.

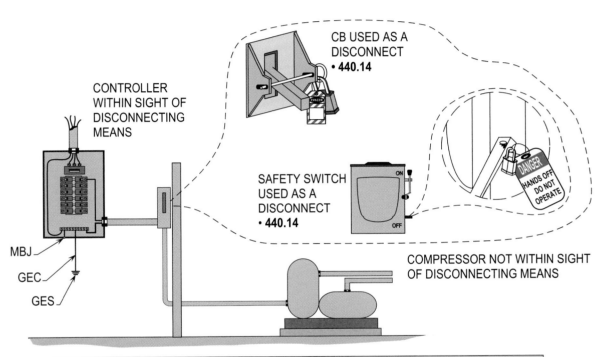

CB USED AS A
DISCONNECT
• **440.14**

CONTROLLER
WITHIN SIGHT OF
DISCONNECTING
MEANS

SAFETY SWITCH
USED AS A
DISCONNECT
• **440.14**

ON

OFF

DANGER
HANDS OFF
DO NOT
OPERATE

MBJ

GEC

GES

COMPRESSOR NOT WITHIN SIGHT
OF DISCONNECTING MEANS

NOTE: *PROVISIONS FOR LOCKING OR ADDING A LOCK TO THE DISCONNECTING MEANS SHALL BE PERMANENTLY INSTALLED ON THE SWITCH OR CB PER **440.14, Ex. 1**.*

LOCATION
440.14, Ex. 1

Purpose of Change: To include a permanent means for installing a lock to lock open the circuit supplying the compressor.

NEC Ch. 4 – Article 440
Part IV – 440.32

Type of Change	Panel Action	UL	UL 508	API 500	API 505	OSHA
Revision	Accept in Part	884	-	-	-	1910.305(j)(3)
ROP		ROC		NFPA 70E	NFPA 70B	NFPA 79	NFPA	NEMA
pg. 1192	# 11-105	pg. 375	# 11-63	420.10(D)	-	-	-	-
log: 1418	CMP: 11	log: 1990	Submitter: Jamie McNamara			2002 NEC: 440.32		IEC: Ch. 13

2002 NEC – 440.32 Single Motor-Compressor

Branch-circuit conductors supplying a single motor-compressor shall have an ampacity not less than 125 percent of either the motor-compressor rated-load current or the branch-circuit selection current, whichever is greater.

For a wye-start, delta-run connected motor-compressor, the selection of the branch-circuit conductors between the controller and the motor-compressor shall be permitted to be based on 58 percent of either the motor-compressor rated-load current or the branch-circuit selection current, whichever is greater.

2005 NEC – 440.32 Single Motor-Compressor

Branch-circuit conductors supplying a single motor-compressor shall have an ampacity not less than 125 percent of either the motor-compressor rated-load current or the branch-circuit selection current, whichever is greater.

For a wye-start, delta-run connected motor-compressor, the selection of the branch-circuit conductors between the controller and the motor-compressor shall be permitted to be based on 72 percent of either the motor-compressor rated-load current or the branch-circuit selection current, whichever is greater.

> **FPN:** The individual motor circuit conductors of wye-start, delta-run connected motor-compressors carry 58 percent of the rated load current. The multiplier of 72 percent is obtained by multiplying 58 percent by 1.25.

Author's Substantiation. This revision and new FPN clarify the procedures for selecting the conductors between the controller and compressor for a wye-start, delta-run unit.

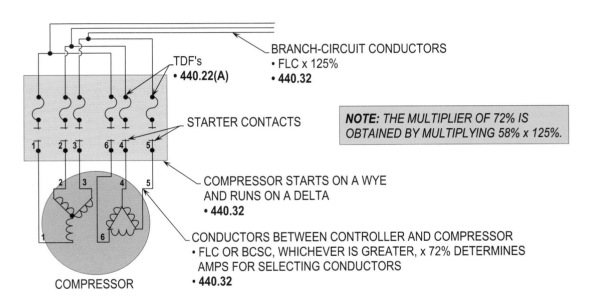

TDF's
• **440.22(A)**

BRANCH-CIRCUIT CONDUCTORS
• FLC x 125%
• **440.32**

STARTER CONTACTS

NOTE: *THE MULTIPLIER OF 72% IS OBTAINED BY MULTIPLYING 58% x 125%.*

COMPRESSOR STARTS ON A WYE AND RUNS ON A DELTA
• **440.32**

CONDUCTORS BETWEEN CONTROLLER AND COMPRESSOR
• FLC OR BCSC, WHICHEVER IS GREATER, x 72% DETERMINES AMPS FOR SELECTING CONDUCTORS
• **440.32**

COMPRESSOR

SINGLE MOTOR-COMPRESSOR
440.32

Purpose of Change: To include the procedure for selecting the conductors between the controller and compressor for a wye-start and delta-run unit.

NEC Ch. 4 – Article 445
Part – 445.11

Type of Change		Panel Action		UL	UL 508	API 500	API 505	OSHA
Revision		Accept		1004	-	-	-	-
ROP		ROC		NFPA 70E	NFPA 70B	NFPA 79	NFPA	NEMA
pg. 1197	# 13-5	pg. -	# -	-	6.4.4.6(2)	-	-	MG 1
log: 2923	CMP: 13	log: -	Submitter: Dorothy Kellogg			2002 NEC: 445.11		IEC: Ch. 43

2002 NEC – 445.11 Marking

Each generator shall be provided with a nameplate giving the manufacturer's name, the rated frequency, power factor, number of phases if of alternating current, the rating in kilowatts or kilovolt amperes, the normal volts and amperes corresponding to the rating, rated revolutions per minute, insulation system class and rated ambient temperature or rated temperature rise, and time rating.

2005 NEC – 445.11 Marking

Each generator shall be provided with a nameplate giving the manufacturer's name, the rated frequency, power factor, number of phases if of alternating current, <u>the subtransient and transient impedances,</u> the rating in kilowatts or kilovolt amperes, the normal volts and amperes corresponding to the rating, rated revolutions per minute, insulation system class and rated ambient temperature or rated temperature rise, and time rating.

Author's Substantiation. This revision clarifies that the impedance of the generator shall be provided in order to calculate the available short-circuit current.

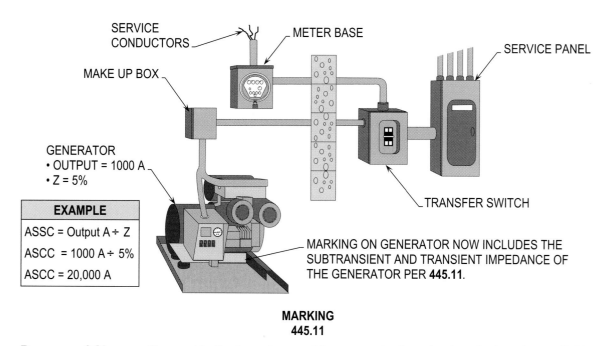

SERVICE CONDUCTORS

METER BASE

MAKE UP BOX

SERVICE PANEL

GENERATOR
• OUTPUT = 1000 A
• Z = 5%

EXAMPLE
ASSC = Output A ÷ Z
ASCC = 1000 A ÷ 5%
ASCC = 20,000 A

TRANSFER SWITCH

MARKING ON GENERATOR NOW INCLUDES THE SUBTRANSIENT AND TRANSIENT IMPEDANCE OF THE GENERATOR PER **445.11**.

**MARKING
445.11**

Purpose of Change: To provide the impedance of the generator in order to calculate the available short-circuit current (ASCC).

NEC Ch. 4 – Article 445
Part – 445.18

Type of Change		Panel Action		UL	UL 508	API 500	API 505	OSHA
Revision		Accept in Principle		1004	-	-	-	-
ROP		ROC		NFPA 70E	NFPA 70B	NFPA 79	NFPA	NEMA
pg. 1198	# 13-7	pg. 380	# 13-6	-	6.4.4.6(2)	-	-	MG 1
log: 3004	CMP: 13	log: 3109	Submitter: James R. Harvey			2002 NEC: 445.18		IEC: Ch. 43

2002 NEC – 445.18 Disconnecting Means Required for Generators

Generators shall be equipped with a disconnect by means of which the generator and all protective devices and control apparatus are able to be disconnected entirely from the circuits supplied by the generator except where:

(1) The driving means for the generator can be readily shut down; and

(2) The generator is not arranged to operate in parallel with another generator or other source of voltage.

2005 NEC – 445.18 Disconnecting Means Required for Generators

Generators shall be equipped with disconnect(s) by means of which the generator and all protective devices and control apparatus are able to be disconnected entirely from the circuits supplied by the generator except where both of the following conditions apply:

(1) The driving means for the generator can be readily shut down.

(2) The generator is not arranged to operate in parallel with another generator or other source of voltage.

Author's Substantiation. This revision clarifies that more than one disconnect at the generator shall be permitted.

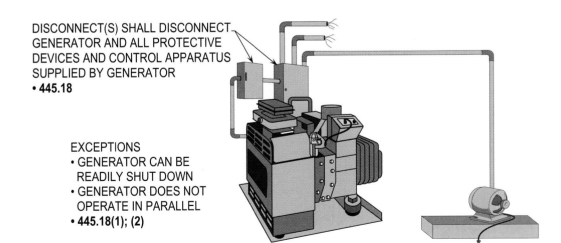

DISCONNECT(S) SHALL DISCONNECT GENERATOR AND ALL PROTECTIVE DEVICES AND CONTROL APPARATUS SUPPLIED BY GENERATOR
• **445.18**

EXCEPTIONS
• GENERATOR CAN BE READILY SHUT DOWN
• GENERATOR DOES NOT OPERATE IN PARALLEL
• **445.18(1); (2)**

DISCONNECTING MEANS REQUIRED FOR GENERATORS
445.18

Purpose of Change: To allow one or more disconnect(s) at generator for disconnecting downstream apparatus.

NEC Ch. 4 – Article 450
Part – 450.5

Type of Change		Panel Action		UL	UL 508	API 500	API 505	OSHA
New Sentence		Accept in Principle		1561	42	-	-	1910.304(f)(1)(iv)
ROP		ROC		NFPA 70E	NFPA 70B	NFPA 79	NFPA	NEMA
pg. 1201	# 9-132	pg. 381	# 9-133	410.10(C)(4)	9.1.3	7.2.7.1	-	ST 20
log: 2542	CMP: 9	log: 1595	Submitter: Alan Manche			2002 NEC: 450.5		IEC: Ch. 43

2002 NEC – 450.5 Grounding Autotransformers

Grounding autotransformers covered in this section are zigzag or T-connected transformers connected to 3-phase, 3-wire ungrounded systems for the purpose of creating a 3-phase, 4-wire distribution system or providing a neutral reference for grounding purposes. Such transformers shall have a continuous per-phase current rating and a continuous neutral current rating.

2005 NEC – 450.5 Grounding Autotransformers

Grounding autotransformers covered in this section are zigzag or T-connected transformers connected to 3-phase, 3-wire ungrounded systems for the purpose of creating a 3-phase, 4-wire distribution system or providing a neutral reference for grounding purposes. Such transformers shall have a continuous per-phase current rating and a continuous neutral current rating. <u>Zig-zag connected transformers shall not be installed on the load side of any system grounding connection, including those made in accordance with 250.24(B), 250.30(A)(1), or 250.32(B)(2).</u>

Author's Substantiation. A new sentence has been added to clarify that zigzag transformers shall not be installed on the load side of any system grounding scheme.

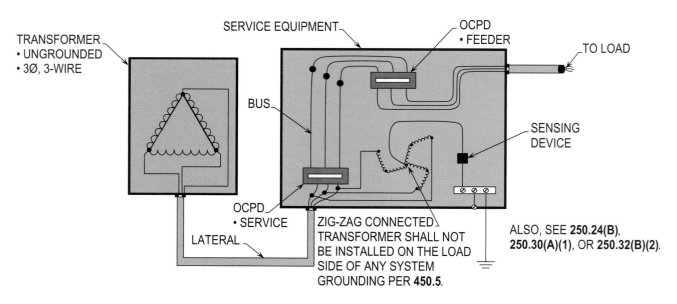

GROUNDING AUTOTRANSFORMERS
450.5

Purpose of Change: To add requirements for grounding a zig-zag transformer to avoid grounding it on the load side of any system grounding scheme.

NEC Ch. 4 – Article 450
Part – 450.6(C)

Type of Change	Panel Action	UL	UL 508	API 500	API 505	OSHA
New Subsection	Accept in Principle in Part	1561	42	-	-	1910.304(f)(1)(iv)
ROP		**ROC**		**NFPA 70E**	**NFPA 70B**	**NFPA 79**	**NFPA**	**NEMA**

ROP		ROC		NFPA 70E	NFPA 70B	NFPA 79	NFPA	NEMA
pg. 1202	# 9-133	pg. -	# -	410.10(C)(4)	9.1.3	-	-	ST 20
log: 758	CMP: 9	log: -	Submitter: Dan Leaf			2002 NEC: -		IEC: Ch. 43

2005 NEC – <u>450.6(C) Grounding</u>

<u>Where the secondary tie system is grounded, each transformer secondary supplying the tie system shall be grounded in accordance with the requirements of 250.30 for separately derived systems.</u>

Author's Substantiation. A new subsection has been added to recognize that under certain conditions of use, the transformer secondaries shall be grounded based on a separately derived system.

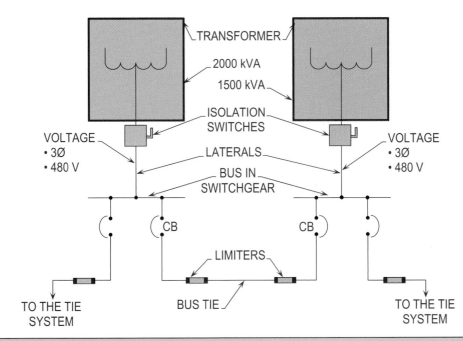

NOTE: *WHERE THE SECONDARY TIE SYSTEM IS GROUNDED, EACH XFMR SECONDARY SUPPLYING THE TIE SHALL BE GROUNDED PER* **250.30** *AS REQUIRED BY* **450.6(C)**.

GROUNDING
450.6(C)

Purpose of Change: To clarify that under certain conditions of use, the XFMR secondary shall be grounded based on a separately derived system.

NEC Ch. 4 – Article 490
Part III – 490.46

Type of Change		Panel Action		UL	UL 508	API 500	API 505	OSHA
New Section		Accept in Principle in Part		1558	-	-	-	1910.308(a)(3)
ROP		ROC		NFPA 70E	NFPA 70B	NFPA 79	NFPA	NEMA
pg. 1214	# 9-141	pg. 381	# 9-136	400.21(F)	8.2 and H.4	-	-	SG 5
log: 3269	CMP: 9	log: 1278	Submitter: Barry M. Hornberger			2002 NEC: -		IEC: Ch. 43

2005 NEC – 490.46 Metal Enclosed and Metal Clad Service Equipment

Metal enclosed and metal clad switchgear installed as high-voltage service equipment shall include a ground bus for the connection of service cable shields and to facilitate the attachment of safety grounds for personnel protection. This bus shall be extended into the compartment where the service conductors are terminated.

Author's Substantiation. A new section has been added to recognize the concept of providing a means for grounding service conductors in the compartment where they are terminated.

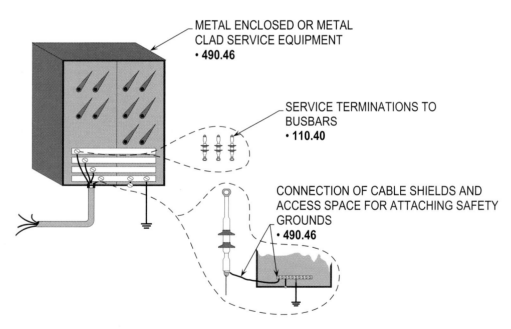

METAL ENCLOSED AND METAL CLAD SERVICE EQUPMENT
490.46

Purpose of Change: To provide requirements pertaining to service equipment rated over 600 volts.

Special Occupancies

Chapter 5 in the NEC has always been known as the special chapter, that covers a specific subject that has a particular function or purpose. **Chapter 5** deals with places or locations where people work, and such occupancies can have built-in electrical hazards.

The main function of this chapter is to protect personnel and equipment from the electrical hazards that could occur in a particular work area, based upon the type of occupancy. For example, designers, installers and inspectors use the requirements of **Chapter 5** to design, install and inspect the electrical wiring methods and equipment that are located in hazardous areas where persons are assembled, confined, living, playing or being entertained.

Chapter 5 is special because it is based upon the type of occupancy that prohibits the greater portion of its design and installation techniques from falling under the requirements of **Chapters 1 through 4** in the NEC.

NEC Ch. 5 – Article 500
Part – 500.2

Type of Change		Panel Action		UL	UL 508	API 500	API 505	OSHA
Revision		Accept in Part		2062	-	3.2.17	3.2.19	1910.307(b)(3), Note
ROP		ROC		NFPA 70E	NFPA 70B	NFPA 79	NFPA 496	NEMA
pg. 1218	# 14-8	pg. 384	# 14-6	440.4(A), FPN	22.1	3.3.53 and A.4.1	3.3.8	-
log: 1678	CMP: 14	log: 650	Submitter: David N. Bishop			2002 NEC: 500.2		IEC: 60079-20

2002 NEC – 500.2 Definitions

For purposes of Articles 500 through 504 and Articles 510 through 516, the following definitions apply.

Purged and Pressurized. The process of supplying an enclosure with a protective gas at a sufficient flow and positive pressure to reduce the concentration of any flammable gas or vapor initially present to an acceptable level.

> **FPN:** For further information, see ANSI/NFPA 496-~~1998~~, *Purged and Pressurized Enclosures for Electrical Equipment.*

2005 NEC – 500.2 Definitions

For purposes of Articles 500 through 504 and Articles 510 through 516, the following definitions apply.

Purged and Pressurized. The process of <u>(1) purging,</u> supplying an enclosure with a protective gas at a sufficient flow and positive pressure to reduce the concentration of any flammable gas or vapor initially present to an acceptable level<u>; and (2) pressurization, supplying an enclosure with a protective gas with or without continuous flow at sufficient pressure to prevent the entrance of a flammable gas or vapor, a combustible dust, or an ignitible fiber</u>.

> **FPN:** For further information, see ANSI/NFPA 496-<u>2003</u>, *Purged and Pressurized Enclosures for Electrical Equipment.*

Author's Substantiation. This revision clarifies the requirements for pressurization shall be included to provide safety.

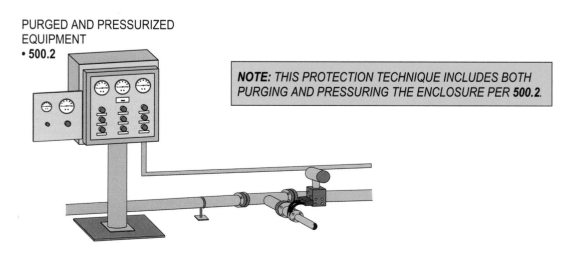

PURGED AND PRESSURIZED
EQUIPMENT
• 500.2

NOTE: THIS PROTECTION TECHNIQUE INCLUDES BOTH PURGING AND PRESSURING THE ENCLOSURE PER 500.2.

DEFINITIONS – PURGED AND PRESSURIZED
500.2

Purpose of Change: To clarify that for this protection technique to provide safety, pressurization shall be included.

NEC Ch. 5 – Article 500
Part – 500.5(C)(2)(1)

Type of Change		Panel Action		UL	UL 508	API 500	API 505	OSHA
Revision		Accept in Principle		2062	-	-	-	1910.307(b)(3), Note
ROP		ROC		NFPA 70E	NFPA 70B	NFPA 79	NFPA 499	NEMA
pg. 1219	# 14-11	pg. 385	# 14-8	440.4(A), FPN	22.1	3.3.53 and A.4.1	4.1.3.2	-
log: 1799	CMP: 14	log: 651	Submitter: David Wechsler			2002 NEC: 500.5(C)(2)(1)		IEC: 61241-10

2002 NEC – 500.5(C)(2) Class II, Division 2

A Class II, Division 2 location is a location

(1) ~~Where~~ combustible dust ~~is not normally~~ in the air in quantities sufficient to produce explosive or ignitible mixtures~~, and dust accumulations are normally insufficient to interfere with the normal operation of electrical equipment or other apparatus, but combustible dust may be in suspension in the air as a result of infrequent malfunctioning of handling or processing equipment and~~

2002 NEC – 500.5(C)(2) Class II, Division 2

A Class II, Division 2 location is a location

(1) <u>In which</u> combustible dust <u>due to abnormal operations may be present</u> in the air in quantities sufficient to produce explosive or ignitible mixtures<u>; or</u>

Author's Substantiation. This revision clarifies the conditions for determining a Class II, Division 2 location.

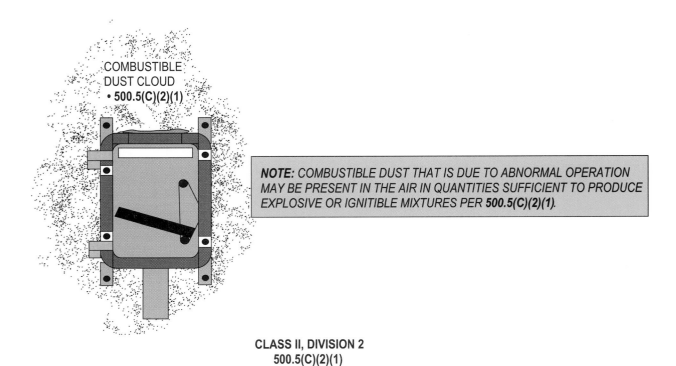

COMBUSTIBLE
DUST CLOUD
• 500.5(C)(2)(1)

NOTE: COMBUSTIBLE DUST THAT IS DUE TO ABNORMAL OPERATION MAY BE PRESENT IN THE AIR IN QUANTITIES SUFFICIENT TO PRODUCE EXPLOSIVE OR IGNITIBLE MIXTURES PER 500.5(C)(2)(1).

CLASS II, DIVISION 2
500.5(C)(2)(1)

Purpose of Change: To revise the text in an effort to make it easier to read and understand.

NEC Ch. 5 – Article 500
Part – 500.5(C)(2)(2)

Type of Change		Panel Action		UL	UL 508	API 500	API 505	OSHA
Revision		Accept in Principle		2062	-	-	-	1910.399; 1910.307(b)(3)
ROP		ROC		NFPA 70E	NFPA 70B	NFPA 79	NFPA 499	NEMA
pg. 1219	# 14-11	pg. 385	# 14-8	440.4(A), FPN	22.1	3.3.53 and A.4.1	5.3	-
log: 1799	CMP: 14	log: 651	Submitter: David Wechsler			2002 NEC: 500.5(C)(2)(1)		IEC: 61241-10

2002 NEC – 500.5(C)(2) Class II, Division 2

A Class II, Division 2 location is a location

(1) Where combustible ~~dust is not normally in the air in quantities sufficient to produce explosive or ignitible mixtures, and~~ dust accumulations are normally insufficient to interfere with the normal operation of electrical equipment or other apparatus, but ~~combustible dust may be in suspension in the air~~ as a result of infrequent malfunctioning of handling or processing equipment ~~and~~

2005 NEC – 500.5(C)(2) Class II, Division 2

A Class II, Division 2 location is a location

(2) Where combustible dust accumulations are <u>present but are</u> normally insufficient to interfere with the normal operation of electrical equipment or other apparatus, but <u>could</u> as a result of infrequent malfunctioning of handling or processing equipment <u>become suspended in the air; or</u>

Author's Substantiation. This revision clarifies the conditions for determining a Class II, Division 2 location.

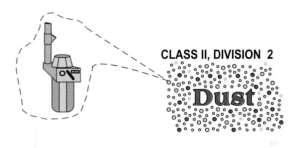

NOTE 1: *DUST IS USUALLY INSUFFICIENT TO INTERFERE WITH NORMAL OPERATION OF EQUIPMENT PER 500.5(C)(2).*

CLASS II, DIVISION 2

NOTE 2: *DUST IS A RESULT OF INFREQUENT MALFUNCTIONING OF EQUIPMENT PER 500.5(C)(2).*

STEPS TO APPLY IN ACCEPTANCE OF EQUIPMENT	
APPROVAL OF EQUIPMENT	110.2
DEFINITION OF APPROVED	ARTICLE 100
DEFINITION OF IDENTIFICATION	ARTICLE 100
LISTED AND LABELED	ARTICLE 100
EXAMINATION OF EQUIPMENT FOR SAFETY	90.7
EXAMINATION	110.3(A)
INSTALLATION AND USE	110.3(B)

CLASS II, DIVISION 2
500.5(C)(2)(2)

Purpose of Change: To revise the text in an effort to make it easier to read and understand.

NEC Ch. 5 – Article 500
Part – 500.5(C)(2)(3)

Type of Change		Panel Action		UL	UL 508	API 500	API 505	OSHA
Revision		Accept in Principle		2062	-	-	-	1910.399; 1910.307(b)(3)
ROP		ROC		NFPA 70E	NFPA 70B	NFPA 79	NFPA 499	NEMA
pg. 1219	# 14-11	pg. 385	# 14-8	440.4(A), FPN	22.1	3.3.53 and A.4.1	5.3	-
log: 1799	CMP: 14	log: 651	Submitter: David Wechsler			2002 NEC: 500.5(C)(2)(2)		IEC: 61241-10

2002 NEC – 500.5(C)(2) Class II, Division 2

A Class II, Division 2 location is a location

(2) ~~Where~~ combustible dust accumulations on, in, or in the vicinity of the electrical equipment ~~may~~ be sufficient to interfere with the safe dissipation of heat from electrical equipment or ~~may~~ be ignitible by abnormal operation or failure of electrical equipment.

2005 NEC – 500.5(C)(2) Class II, Division 2

A Class II, Division 2 location is a location

(3) In which combustible dust accumulations on, in, or in the vicinity of the electrical equipment could be sufficient to interfere with the safe dissipation of heat from electrical equipment, or could be ignitible by abnormal operation or failure of electrical equipment.

Author's Substantiation. This revision clarifies the conditions for determining a Class II, Division 2 location.

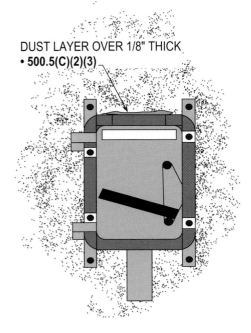

DUST LAYER OVER 1/8" THICK
• 500.5(C)(2)(3)

NOTE: DUST ACCUMULATIONS ON, IN, OR IN THE VICINITY OF THE ELECTRICAL EQUIPMENT COULD INTERFERE WITH SAFE DISSIPATION OF OPERATING HEAT PER 500.5(C)(2)(3).

CLASS II, DIVISION 2
500.5(C)(2)(3)

Purpose of Change: To revise the text in an effort to make it easier to read and understand.

NEC Ch. 5 – Article 500
Part – 500.7(K)

116

Type of Change		Panel Action		UL	UL 508	API 500	API 505	OSHA
New Sentence		Accept in Principle		2062	-	6.5	6.8	1910.307(b)(3), Note
ROP		ROC		NFPA 70E	NFPA 70B	NFPA 79	NFPA	NEMA
pg. 1220	# 14-13	pg. -	# -	440.4(A), FPN	22.1	3.3.53 and A.4.1	-	-
log: 2975	CMP: 14	log: -	Submitter: Mark Goodman			2002 NEC: 500.7(K)		IEC: 60079-20

2002 NEC – 500.7(K) Combustible Gas Detection System

A combustible gas detection system shall be permitted as a means of protection in industrial establishments with restricted public access and where the conditions of maintenance and supervision ensure that only qualified persons service the installation. Gas detection equipment shall be listed for detection of the specific gas or vapor to be encountered. Where such a system is installed, equipment specified in 500.7(K)(1), (2), or (3) shall be permitted.

(1) Inadequate Ventilation. In a Class I, Division 1 location that is so classified due to inadequate ventilation, electrical equipment suitable for Class I, Division 2 locations shall be permitted.

(2) Interior of a Building. In a building located in, or with an opening into, a Class I, Division 2 location where the interior does not contain a source of flammable gas or vapor, electrical equipment for unclassified locations shall be permitted.

(3) Interior of a Control Panel. In the interior of a control panel containing instrumentation utilizing or measuring flammable liquids, gases, or vapors, electrical equipment suitable for Class I, Division 2 locations shall be permitted.

FPN No. 1: For further information, see ANSI/ISA-12.13.01, *Performance Requirements, Combustible Gas Detectors*.

FPN No. 2: For further information, see ANSI/API RP 500, *Recommended Practice for Classification of Locations for Electrical Installations at Petroleum Facilities Classified as Class I, Division 1 or Division 2*.

FPN No. 3: For further information, see ISA-RP12.13.02, *Installation, Operation, and Maintenance of Combustible Gas Detection Instruments*.

2005 NEC – 500.7(K) Combustible Gas Detection System

A combustible gas detection system shall be permitted as a means of protection in industrial establishments with restricted public access and where the conditions of maintenance and supervision ensure that only qualified persons service the installation. Gas detection equipment shall be listed for detection of the specific gas or vapor to be encountered. Where such a system is installed, equipment specified in 500.7(K)(1), (K)(2), or (K)(3) shall be permitted.

The type of detection equipment, its listing, installation location(s), alarm and shutdown criteria, and calibration frequency shall be documented when combustible gas detectors are used as a protection technique.

FPN No. 1: For further information, see ANSI/ISA-12.13.01, *Performance Requirements, Combustible Gas Detectors*.

FPN No. 2: For further information, see ANSI/API RP 500, *Recommended Practice for Classification of Locations for Electrical Installations at Petroleum Facilities Classified as Class I, Division 1 or Division 2*.

FPN No. 3: For further information, see ISA-RP12.13.02, *Installation, Operation, and Maintenance of Combustible Gas Detection Instruments*.

(1) Inadequate Ventilation. In a Class I, Division 1 location that is so classified due to inadequate ventilation, electrical equipment suitable for Class I, Division 2 locations shall be permitted.

(2) Interior of a Building. In a building located in, or with an opening into, a Class I, Division 2 location where the interior does not contain a source of flammable gas or vapor, electrical equipment for unclassified locations shall be permitted.

(3) Interior of a Control Panel. In the interior of a control panel containing instrumentation utilizing or measuring flammable liquids, gases, or vapors, electrical equipment suitable for Class I, Division 2 locations shall be permitted.

Author's Substantiation. A new sentence has been added to clarify that documentation shall be provided where combustible gas detectors are used as a protection technique. The FPN s have been relocated under the main text so as not to appear to only applying to subsection (K)(3).

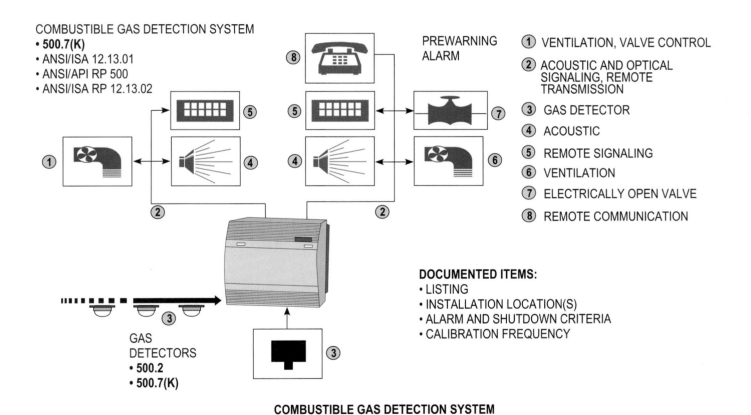

COMBUSTIBLE GAS DETECTION SYSTEM
500.7(K)

Purpose of Change: To provide guidelines to document proof of installation techniques before installing this protective scheme.

NEC Ch. 5 – Article 500
Part – 500.8(A)(4)

 117

Type of Change		Panel Action		UL	UL 508	API 500	API 505	OSHA
Revision		Accept		135.6	-	1.1.2	1.1.2	1910.307(b)(3), Note
ROP		ROC		NFPA 70E	NFPA 70B	NFPA 79	NFPA 497	NEMA
pg. 1222	# 14-16	pg. 386	# 14-14	440.4(A), FPN	22.1	3.3.53 and A.4.1	1.1.2	-
log: 2925	CMP: 14	log: 3457	Submitter: Dorothy Kellogg			2002 NEC: 500.8(A)		IEC: 60079-20

2002 NEC – 500.8(A) Approval for Class and Properties

(4) Equipment, ~~regardless of the classification of the location in which it is installed,~~ that depends on a single compression seal, diaphragm, or tube to prevent flammable or combustible fluids from entering the equipment shall be identified for a Class I, Division 2 location. Equipment installed in a Class I, Division 1 location shall be identified for the Class I, Division 1 location.

2005 NEC – 500.8(A) Approval for Class and Properties

(4) Equipment that depends on a single compression seal, diaphragm, or tube to prevent flammable or combustible fluids from entering the equipment shall be identified for a Class I, Division 2 location <u>even if installed in an unclassified location</u>. Equipment installed in a Class I, Division 1 location shall be identified for the Class I, Division 1 location.

Author's Substantiation. This revision removes any ambiguities from the existing text and permits a uniform enforcement of the requirement.

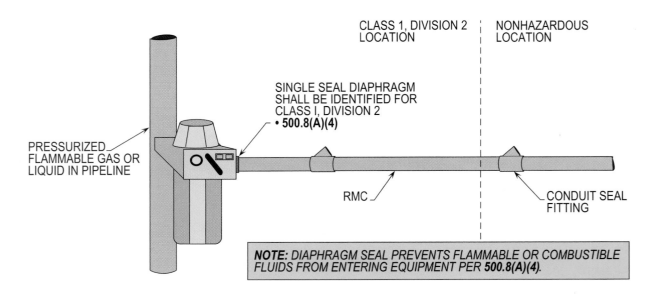

APPROVAL FOR CLASS AND PROPERTIES
500.8(A)(4)

Purpose of Change: To define the use of equipment and seals in Class I and Division 1 and 2 locations.

NEC Ch. 5 – Article 500
Part – 500.8(B)(1) through (B)(6)

Type of Change		Panel Action		UL	UL 508	API 500	API 505	OSHA
Revision		Accept in Principle		2062	1.1.2	1.1.2	-	1910.307(b)(2)(ii)
ROP		ROC		NFPA 70E	NFPA 70B	NFPA 79	NFPA 497	NEMA
pg. 1224	# 14-17a	pg. 386	# 14-16	440.3(D)	22.1	3.3.53 and A.4.1	1.1.2	-
log: CP1402	CMP: 14	log: 1926	Submitter: CMP 14			2002 NEC: 500.8(B)		IEC: 60079-20

2002 NEC – 500.8(B) Marking

Equipment shall be marked to show ~~the class, group, and operating temperature or temperature class referenced to a 40°C ambient.~~

Exception No. 1: Equipment of the non-heat-producing type, such as junction boxes, conduit, and fittings, and equipment of the heat-producing type having a maximum temperature not more than 100°C ~~(212°F)~~ shall not be required to have a marked operating temperature or temperature class.

Exception No. 2: Fixed luminaires (lighting fixtures) marked for use in Class I, Division 2 or Class II, Division 2 locations ~~only~~ shall not be required to be marked to indicate the group.

Exception No. 3: Fixed general-purpose equipment in Class I locations, other than fixed luminaires (lighting fixtures), that is acceptable for use in Class I, Division 2 locations shall not be required to be marked with the class, group, division, ~~or operating temperature.~~

Exception No. 4: Fixed dusttight equipment other than fixed luminaires (lighting fixtures) that is acceptable for use in Class II, Division 2 and Class III locations shall not be required to be marked with the class, group, division, ~~or operating temperature.~~

Exception No. 5: ~~Electric equipment suitable for ambient temperatures exceeding 40°C (104°F) shall be marked with both the maximum ambient temperature and the operating temperature or temperature class at that ambient temperature.~~

FPN: Equipment not marked to indicate a division, or marked Division 1 or Div. 1, is suitable for both Division 1 and 2 locations. Equipment marked Division 2 or Div. 2 is suitable for Division 2 locations only.

The temperature class, if provided, shall be indicated using the temperature class (T Codes) shown in Table 500.8(B). ~~The temperature class (T Code) marked on equipment nameplates shall be in accordance with Table 500.8(B).~~ Equipment for Class I and Class II shall be marked with the maximum safe operating temperature, as determined by simultaneous exposure to the combinations of Class I and Class II conditions.

FPN: ~~Since there is no consistent relationship between explosion properties and ignition temperature, the two are independent requirements.~~

2005 NEC – 500.8(B) Marking

Equipment shall be marked to show <u>the environment for which it has been evaluated. Unless otherwise specified or allowed in (B)(6), the marking shall include the information specified in (B)(1) through (B)(5).</u>

(1) Class. <u>The making shall specify the class(es) for which the equipment is suitable.</u>

(2) Division. <u>The making shall specify the division if the equipment is suitable for Division 2 only. Equipment suitable for Division 1 shall be permitted to omit the division marking.</u>

FPN: Equipment not marked to indicate a division, or marked Division 1 or Div. 1, is suitable for both Division 1 and 2 locations<u>; see 500.8(A)(2).</u> Equipment marked Division 2 or Div. 2 is suitable for Division 2 locations only.

(3) Material Classification Group. <u>The marking shall specify the applicable material classification group(s) in accordance with 500.6.</u>

Exception: Fixed luminaires (lighting fixtures) marked for use only in Class I, Division 2 or Class II, Division 2 locations shall not be required to indicate the group.

(4) Equipment Temperature. The marking shall specify the temperature class or operating temperature at a 40°C ambient termperature, or at the higher ambient temperature if the equipment is rated and marked for an ambient temperature of greater than 40°C. The temperature class, if provided, shall be indicated using the temperature class (T Codes) shown in Table 500.8(B). Equipment for Class I and Class II shall be marked with the maximum safe operating temperature, as determined by simultaneous exposure to the combinations of Class I and Class II conditions.

Exception: Equipment of the non-heat-producing type, such as junction boxes, conduit, and fittings, and equipment of the heat-producing type having a maximum temperature not more than 100°C shall not be required to have a marked operating temperature or temperature class.

FPN: More than one marked temperature class or operating temperature, for gases and vapors, dusts, and different ambient temperatures, may appear.

(5) Ambient Temperature Range. For equipment rated for a temperature range other than -25°C to +40°C, the marking shall specify the special range of ambient temperatures. The marking shall include either the symbol "Ta" or "Tamb."

FPN: As an example, such a marking might be "-30°C $\leq$ Ta $\leq$ +40°C."

(6) Allowances

    (a) General Purpose Equipment. Fixed general-purpose equipment in Class I locations, other than fixed luminaires (lighting fixtures), that is acceptable for use in Class I, Division 2 locations shall not be required to be marked with the class, division, group, temperature class, or ambient temperature range.

    (b) Dusttight Equipment. Fixed dusttight equipment, other than fixed luminaires (lighting fixtures), that is acceptable for use in Class II, Division 2 and Class III locations shall not be required to be marked with the class, division, group, temperature class, or ambient temperature range.

    (c) Associated Apparatus. Associated intrinsically safe apparatus and associated nonincendive field wiring apparatus that are not protected by an alternative type of protection shall not be marked with the class, division, group, or temperature class. Associated intrinsically safe apparatus and associated nonincendive field wiring apparatus shall be marked with the class, division, and group of the apparatus to which it is to be connected.

    (d) Simple Apparatus. "Simple apparatus" as defined by Article 504, shall not be required to be marked with class, division, group, temperature class, or ambient temperature range.

Author's Substantiation. This section has been completely rewritten with title heads and subsections in an effort to make these requirements easier to read and understand.

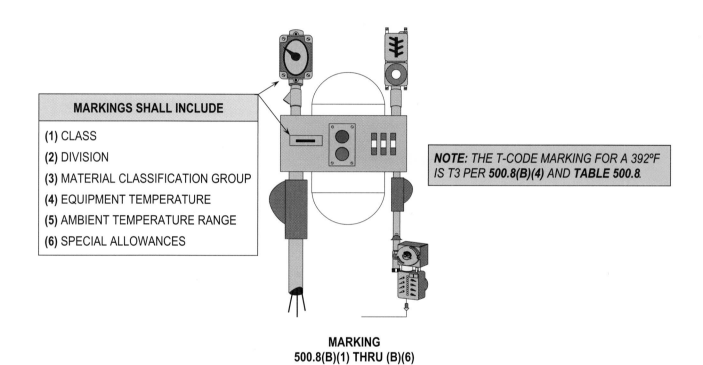

MARKINGS SHALL INCLUDE

(1) CLASS

(2) DIVISION

(3) MATERIAL CLASSIFICATION GROUP

(4) EQUIPMENT TEMPERATURE

(5) AMBIENT TEMPERATURE RANGE

(6) SPECIAL ALLOWANCES

NOTE: *THE T-CODE MARKING FOR A 392°F IS T3 PER* **500.8(B)(4)** *AND* **TABLE 500.8.**

MARKING
500.8(B)(1) THRU (B)(6)

Purpose of Change: To completely rewrite this section and assign title heads to subsections in an attempt to make these requirements easier to read and understand.

NEC Ch. 5 – Article 501
Parts I through III – 501.1 through 501.150

 118

Type of Change		Panel Action		UL	UL 508	API 500	API 505	OSHA
Renumbered		Accept		2062	-	1.1.2	1.1.2	1910.307(b)(3), Note
ROP		ROC		NFPA 70E	NFPA 70B	NFPA 79	NFPA 497	NEMA
pg. 1226	# 14-19a	pg. 388	# 14-18	440.4(A), FPN	22.1	3.3.53 and A.4.1	1.1.2	-
log: CP1406	CMP: 14	log: 654	Submitter: CMP 14			2002 NEC: Article 501		IEC: 61241-10

Author's Substantiation. Article 501 has been reorganized and renumbered under the guidance of the NFPA Task Group on Usability. See cross reference table for a better understanding of these new renumbered sections of Article 501.

CROSS REFERENCE		
2005 NEC	**2002 NEC**	**SECTION HEADING**
PART I – GENERAL		
501.1	——	Scope
501.5	501.1	General
PART II – WIRING		
501.10	501.4	Wiring Methods
501.15	501.5	Sealing and Drainage
501.20	501.13	Conductor Insulation, Class I, Divisions 1 and 2
501.25	501.15	Uninsulated Exposed Parts, Class I, Divisions 1 and 2
501.30	501.16	Grounding and Bonding, Class I, Divisions 1 and 2
501.35	501.17	Surge Protection
501.40	501.18	Multiwire Branch-Circuits
PART III – EQUIPMENT		
501.100	501.2	Transformers and Capacitors
501.105	501.3	Meters, Instruments, and Relays
501.115	501.6	Switches, Circuit Breakers, Motor Controllers, and Fuses
501.120	501.7	Control Transformers and Resistors
501.125	501.8	Motors and Generators
501.130	501.9	Luminaires (Lighting Fixtures)
501.135	501.10	Utilization Equipment
501.140	501.11	Flexible Cords, Class I, Divisions 1 and 2
501.145	501.12	Receptacles and Attachment Plugs, Class I, Divisions 1 and 2
501.150	501.14	Signaling, Alarm, Remote-Controlled, and Communications Systems

NEC Ch. 5 – Article 501
Part II – 501.15(B)(2)

Type of Change	Panel Action	UL	UL 508	API 500	API 505	OSHA
Revision	Accept in Principle in Part	2062	-	6.2.1.1(b)	-	1910.399
ROP		ROC		NFPA 70E	NFPA 70B	NFPA 79	NFPA 497	NEMA

ROP			ROC			NFPA 70E	NFPA 70B	NFPA 79	NFPA 497	NEMA
pg. 1238	# 14-34		pg. 394	# 14-43		Article 100	22.1	3.3.53 and A.4.1	4.1.3.2	-
log: 2937	CMP: 14		log: 1932	Submitter: Dorothy Kellogg				2002 NEC: 501.5(B)(2)		IEC: 60079-20

2002 NEC – 501.5(B)(2) Class I, Division 2 Boundary

In each conduit run passing from a Class I, Division 2 location into an unclassified location. The sealing fitting shall be permitted on either side of the boundary of such location within 3.05 m (10 ft) of the boundary and shall be designed and installed so as to minimize the amount of gas or vapor within the Division portion of the conduit from being communicated to the conduit beyond the seal. Rigid metal conduit or threaded steel intermediate metal conduit shall be used between the sealing fitting and the point at which the conduit leaves the Division 2 location, and a threaded connection shall be used at the sealing fitting. Except for listed explosionproof reducers at the conduit seal, there shall be no union, coupling, box, or fitting between the conduit seal and the point at which the conduit leaves the Division 2 location.

2005 NEC – 501.15(B)(2) Class I, Division 2 Boundary

In each conduit run passing from a Class I, Division 2 location into an unclassified location. The sealing fitting shall be permitted on either side of the boundary of such location within 3.05 m (10 ft) of the boundary. Rigid metal conduit or threaded steel intermediate metal conduit shall be used between the sealing fitting and the point at which the conduit leaves the Division 2 location, and a threaded connection shall be used at the sealing fitting. Except for listed reducers at the conduit seal, there shall be no union, coupling, box, or fitting between the conduit seal and the point at which the conduit leaves the Division 2 location. Conduit shall be sealed to minimize the amount of gas or vapor within the Division 2 portion of the conduit from being communicated to the conduit beyond the seal. Such seals shall not be required to be explosionproof but shall be identified for the purpose of minimizing passage of gases under normal operation conditions and shall be accessible.

Author's Substantiation. This revision clarifies that explosionproof seals shall not be required as boundary seals between Division 2 and unclassified locations.

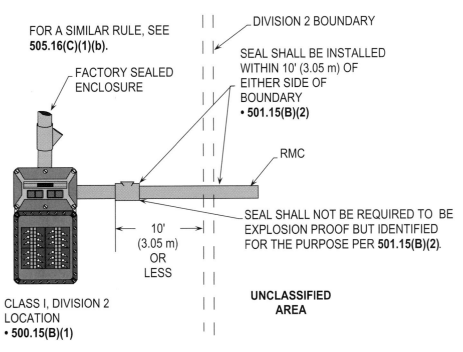

FOR A SIMILAR RULE, SEE
505.16(C)(1)(b).

FACTORY SEALED
ENCLOSURE

DIVISION 2 BOUNDARY

SEAL SHALL BE INSTALLED
WITHIN 10' (3.05 m) OF
EITHER SIDE OF
BOUNDARY
• **501.15(B)(2)**

RMC

SEAL SHALL NOT BE REQUIRED TO BE
EXPLOSION PROOF BUT IDENTIFIED
FOR THE PURPOSE PER **501.15(B)(2)**.

10'
(3.05 m)
OR
LESS

CLASS I, DIVISION 2
LOCATION
• **500.15(B)(1)**

**UNCLASSIFIED
AREA**

**CLASS I, DIVISIONS 1 AND 2
501.15(B)(2)**

Purpose of Change: To recognize a seal that is not explosionproof to be used at the Class I, Division 2 boundary.

NEC Ch. 5 – Article 501
Part II – 501.15(C)

Type of Change		Panel Action		UL	UL 508	API 500	API 505	OSHA
Revision		Accept in Principle		2062	-	1.1.2	1.1.2	1910.307(b)(3), Note
ROP		ROC		NFPA 70E	NFPA 70B	NFPA 79	NFPA 497	NEMA
pg. 1239	# 14-36	pg. 394	# 14-44	440.4(A), FPN	22.1	3.3.53 and A.4.1	1.1.2	-
log: 2936	CMP: 14	log: 983	Submitter: Dorothy Kellogg			2002 NEC: 501.5(C)		IEC: 60079-20

2002 NEC – ~~501.5(C)~~ Class I, Division 1 and 2

~~Where required,~~ seals in Class I, Division 1 and 2 locations shall comply with ~~501.5~~(C)(1) through (C)(6).

2005 NEC – 501.15(C) Class I, Division 1 and 2

Seals installed in Class I, Division 1 and Division 2 locations shall comply with 501.15(C)(1) through (C)(6).

Author's Substantiation. This revision clarifies that seals shall be required in all Class I, Division 1 and Division 2 locations.

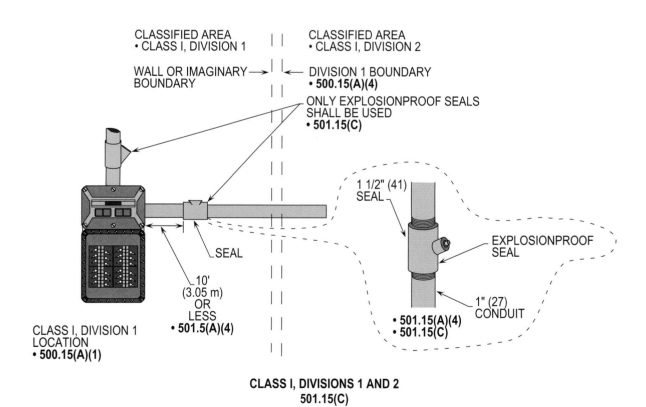

CLASS I, DIVISIONS 1 AND 2
501.15(C)

Purpose of Change: To clarify where explosion proof seals are required.

NEC Ch. 5 – Article 501
Part II – 501.15(F)(3)

 120

Type of Change	Panel Action	UL	UL 508	API 500	API 505	OSHA
New Sentence	Accept in Principle in Part	2062	-	1.1.2	1.1.2	1910.307(b)(3), Note
ROP	ROC	NFPA 70E	NFPA 70B	NFPA 79	NFPA 497	NEMA
pg. 1242 # 14-40	pg. 394 # 14-47	440.4(A), FPN	22.1	3.3.53 and A.4.1	1.1.2	-
log: 1619 CMP: 14	log: 672 Submitter: Vic Gournas			2002 NEC: 501.5(F)(3)		IEC: 60079-20

2002 NEC – ~~501.5(F)(3)~~ Canned Pumps, Process, or Service Connections, etc.

For canned pumps, process, or service connections for flow, pressure, or analysis measurement, and so forth, that depend on a single compression seal, diaphragm, or tube to prevent flammable or combustible fluids from entering the electrical raceway or cable system capable of transmitting fluids, an additional approved seal, barrier, or other means shall be provided to prevent the flammable or combustible fluid from entering the raceway or cable system capable of transmitting fluids beyond the additional devices or means, if the primary seal fails. The additional approved seal or barrier and the interconnecting enclosure shall meet the temperature and pressure conditions to which they will be subjected upon failure of the primary seal, unless other approved means are provided to accomplish ~~the~~ purpose ~~above~~. Drains, vents, or other devices shall be provided so that primary seal leakage will be obvious.

> **FPN:** See also the fine print notes to ~~501.5~~.

2005 NEC – 501.15(F)(3) Canned Pumps, Process, or Service Connections, etc.

For canned pumps, process, or service connections for flow, pressure, or analysis measurement, and so forth, that depend on a single compression seal, diaphragm, or tube to prevent flammable or combustible fluids from entering the electrical raceway or cable system capable of transmitting fluids, an additional approved seal, barrier, or other means shall be provided to prevent the flammable or combustible fluid from entering the raceway or cable system capable of transmitting fluids beyond the additional devices or means, if the primary seal fails. The additional approved seal or barrier and the interconnecting enclosure shall meet the temperature and pressure conditions to which they will be subjected upon failure of the primary seal, unless other approved means are provided to accomplish <u>this</u> purpose. Drains, vents, or other devices shall be provided so that primary seal leakage will be obvious.

> **FPN:** See also the fine print notes to <u>501.15</u>.

<u>Process-connected equipment that is listed and marked "Dual Seal" shall not require additional process sealing when used within the manufacturer's ratings.</u>

> **FPN:** <u>For construction and testing requirements for dual seal process connected equipment, refer to ISA 12.27.01, *Requirements for Process Sealing Between Electrical Systems and Potentially Flammable or Combustible Process Fluids.*</u>

Author's Substantiation. A new sentence has been added to recognize that process-connected equipment that is listed and marked "Dual Seal" shall not be required to have additional process sealing when used within the manufacturer's ratings. A new FPN has been added to reference ISA 12.27.01 for the construction and testing requirements for dual seal process connected equipment.

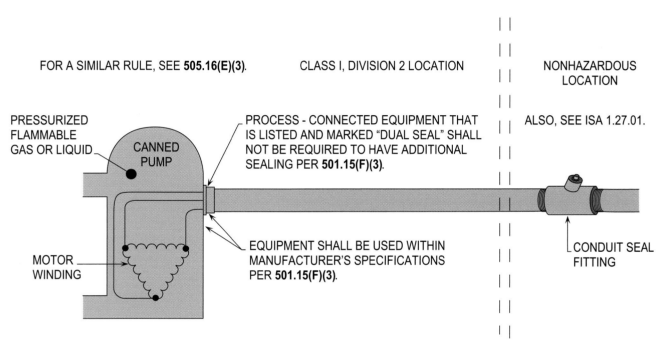

FOR A SIMILAR RULE, SEE **505.16(E)(3)**.

CLASS I, DIVISION 2 LOCATION

NONHAZARDOUS
LOCATION

PRESSURIZED
FLAMMABLE
GAS OR LIQUID

CANNED
PUMP

PROCESS - CONNECTED EQUIPMENT THAT
IS LISTED AND MARKED "DUAL SEAL" SHALL
NOT BE REQUIRED TO HAVE ADDITIONAL
SEALING PER **501.15(F)(3)**.

ALSO, SEE ISA 1.27.01.

MOTOR
WINDING

EQUIPMENT SHALL BE USED WITHIN
MANUFACTURER'S SPECIFICATIONS
PER **501.15(F)(3)**.

CONDUIT SEAL
FITTING

CLASS I, DIVISIONS 1 AND 2
501.15(F)(3)

Purpose of Change: To recognize a "dual speed" when it is used appropriately.

NEC Ch. 5 – Article 501
Part III – 501.130(B)(1)

Type of Change		Panel Action		UL	UL 508	API 500	API 505	OSHA
Revision		Accept		2062	-	1.1.2	1.1.2	1910.307(b)(3), Note
ROP		ROC		NFPA 70E	NFPA 70B	NFPA 79	NFPA 497	NEMA
pg. 1226	# 14-20	pg. 395	# 14-48	440.3(B)(1), FPN	22.1	3.3.53 and A.4.1	1.1.2	-
log: 3134	CMP: 14	log: 3439	Submitter: David Wechsler			2002 NEC: 501.9(B)(2)		IEC: 60079-20

2002 NEC – ~~501.9(B)(2) Fixed~~ Luminaires (Lighting Fixtures)

~~Luminaires (lighting fixtures) for fixed lighting shall be protected from physical damage by suitable guards or by location. Where there is danger that falling sparks or hot metal from lamps or fixtures might ignite localized concentrations of flammable vapors or gases, suitable enclosures or other effective protective means shall be provided.~~ Where lamps are of a size or type that may, under normal operating conditions, reach surface temperatures exceeding 80 percent of the ignition temperature in degrees Celsius of the gas or vapor involved, fixtures shall comply with ~~501.9(A)(1)~~ or shall be of a type that has been tested in order to determine the marked operating temperature or temperature class (T Code).

2005 NEC – 501.130(B)(1) Luminaires (Lighting Fixtures)

Where lamps are of a size or type that may, under normal operating conditions, reach surface temperatures exceeding 80 percent of the ignition temperature in degrees Celsius of the gas or vapor involved, fixtures shall comply with 501.130(A)(1) or shall be of a type that has been tested in order to determine the marked operating temperature or temperature class (T Code).

Author's Substantiation. This revision clarifies the marking and installation requirements for luminaires (lighting fixtures) installed in Class I, Division 2 locations.

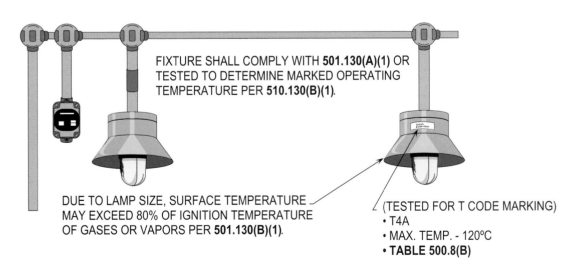

FIXTURE SHALL COMPLY WITH **501.130(A)(1)** OR TESTED TO DETERMINE MARKED OPERATING TEMPERATURE PER **510.130(B)(1)**.

DUE TO LAMP SIZE, SURFACE TEMPERATURE MAY EXCEED 80% OF IGNITION TEMPERATURE OF GASES OR VAPORS PER **501.130(B)(1)**.

(TESTED FOR T CODE MARKING)
- T4A
- MAX. TEMP. - 120°C
- **TABLE 500.8(B)**

LUMINAIRES (LIGHTING FIXTURES)
501.130(B)(1)

Purpose of Change: To outline markings and installation requirements for luminaires (lighting fixtures) installed in Class I, Division 2 locations.

NEC Ch. 5 – Article 502
Parts I through III – 502.1 through 502.150

Type of Change	Panel Action	UL	UL 508	API 500	API 505	OSHA
Renumbered	Accept	2062	-	1.1.2	1.1.2	1910.307(b)(3), Note
ROP	**ROC**	**NFPA 70E**	**NFPA 70B**	**NFPA 79**	**NFPA 499**	**NEMA**
pg. 1249 # 14-51a	pg. - # -	440.4(A), FPN	22.1	3.3.53 and A.4.1	1.1.2	-
log: CP1407 CMP: 14	log: - Submitter: CMP 14			**2002 NEC:** Article 502		**IEC:** 1241-1-1

Author's Substantiation. Article 502 has been reorganized and renumbered under the guidance of the NFPA Task Group on Usability. See cross reference table for a better understanding of these new renumbered sections of Article 502.

CROSS REFERENCE		
2005 NEC	**2002 NEC**	**SECTION HEADING**
PART I – GENERAL		
502.1	——	Scope
502.5	502.1	General
PART II – WIRING		
502.10	502.4	Wiring Methods
502.15	502.5	Sealing, Class II, Divisions 1 and 2
502.25	502.15	Uninsulated Exposed Parts, Class II, Divisions 1 and 2
502.30	502.16	Grounding and Bonding, Class II, Divisions 1 and 2
502.35	502.17	Surge Protection, Class II, Divisions 1 and 2
502.40	502.18	Multiwire Branch-Circuits
PART III – EQUIPMENT		
502.100	502.2	Transformers and Capacitors
502.105	——	
502.115	502.6	Switches, Circuit Breakers, Motor Controllers, and Fuses
502.120	502.7	Control Transformers and Resistors
502.125	502.8	Motors and Generators
502.128	502.9	Ventilating Piping
502.130	502.11	Luminaires (Lighting Fixtures)
502.135	502.10	Utilization Equipment
502.140	502.12	Flexible Cords, Class II, Divisions 1 and 2
502.145	502.13	Receptacles and Attachment Plugs
502.150	502.14	Signaling, Alarm, Remote-Control, and Communications Systems; and Meters, Instruments, and Relays

NEC Ch. 5 – Article 502
Part II – 502.10(A)(2)

 122

Type of Change		Panel Action		UL	UL 508	API 500	API 505	OSHA
Revision		Accept in Principle in Part		2062	-	1.1.2	1.1.2	1910.307(b)(3), Note
ROP		ROC		NFPA 70E	NFPA 70B	NFPA 79	NFPA 497	NEMA
pg. 1251	# 14-55	pg. 397	# 14-58	440.4(A), FPN	22.1	3.3.53 and A.4.1	1.1.2	-
log: 2690	CMP: 14	log: 661	Submitter: Phil Simmons			2002 NEC: 502.4(A)(1)(e)		IEC: 1241-1-1

2002 NEC – ~~502.4(A)(1) General~~

In Class II, Division 1 locations, the wiring methods in (a) through (e) shall be permitted.

> ~~(e)~~ Where necessary to employ flexible connections, dusttight flexible connectors, liquidtight flexible metal conduit with listed fittings, liquidtight flexible nonmetallic conduit with listed fittings, or flexible cord listed for extra-hard usage and provided with bushed fittings ~~shall be used~~. Where flexible cords are used, they shall comply with ~~502.12. Where flexible connections are subject to oil or other corrosive conditions, the insulation of the conductors shall be of a type listed for the condition or shall be protected by means of a suitable sheath.~~

2005 NEC – <u>502.10(A)(2) Flexible Connections</u>

Where necessary to employ flexible connections, <u>one or more of the following shall also be permitted:</u>

> <u>(1)</u> Dusttight flexible connectors
>
> <u>(2)</u> Liquidtight flexible metal conduit with listed fittings
>
> <u>(3)</u> Liquidtight flexible nonmetallic conduit with listed fittings
>
> <u>(4)</u> <u>Interlocked armor Type MC cable having an overall jacket of suitable polymeric material and provided with termination fittings listed for Class II, Division 1 locations.</u>
>
> <u>(5)</u> Flexible cord listed for extra-hard usage and provided with bushed fittings. Where flexible cords are used, they shall comply with <u>502.140</u>.

Author's Substantiation. This section has been reorganized to a list format which is more user friendly. A new item (4) has been added to recognize that interlocked armor Type MC cable shall be permitted to be used for flexible connections.

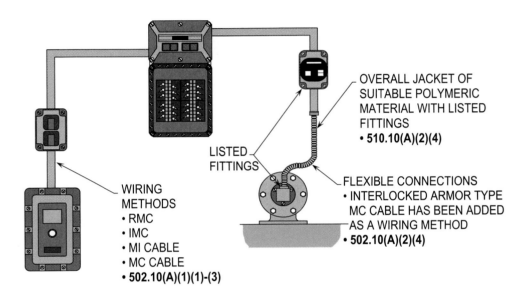

WIRING
METHODS
• RMC
• IMC
• MI CABLE
• MC CABLE
• **502.10(A)(1)(1)-(3)**

LISTED
FITTINGS

OVERALL JACKET OF
SUITABLE POLYMERIC
MATERIAL WITH LISTED
FITTINGS
• **510.10(A)(2)(4)**

FLEXIBLE CONNECTIONS
• INTERLOCKED ARMOR TYPE
MC CABLE HAS BEEN ADDED
AS A WIRING METHOD
• **502.10(A)(2)(4)**

FLEXIBLE CONNECTIONS
502.10(A)(2)(4)

Purpose of Change: To clarify that interlocked armor type MC cable can be used for flexible connections.

NEC Ch. 5 – Article 503
Parts I through III – 503.1 – 503.160

Type of Change	Panel Action	UL	UL 508	API 500	API 505	OSHA
Renumbered	Accept	2062	-	1.1.2	1.1.2	1910.307(b)(3), Note

ROP	ROC	NFPA 70E	NFPA 70B	NFPA 79	NFPA	NEMA
pg. 1255 # 14-60a	pg. - # -	440.4(A), FPN	22.1	3.3.53 and A.4.1	499	-
log: CP1408 CMP: 14	log: -	Submitter: CMP 14		2002 NEC: Article 503		IEC: 1241-1-1

Author's Substantiation. Article 503 has been reorganized and renumbered under the guidance of the NFPA Task Group on Usability. See cross reference table for a better understanding of these new renumbered sections of Article 503.

CROSS REFERENCE		
2005 NEC	**2002 NEC**	**SECTION HEADING**
PART I – GENERAL		
503.1	———	Scope
503.5	503.1	General
PART II – WIRING		
503.10	503.3	Wiring Methods
503.25	503.15	Uninsulated Exposed Parts, Class III, Divisions 1 and 2
501.30	501.16	Grounding and Bonding - Class III, Divisions 1 and 2
PART III – EQUIPMENT		
503.100	503.2	Transformers and Capacitors - Class III, Divisions 1 and 2
503.115	503.4	Switches, Circuit Breakers, Motor Controllers, and Fuses - Class III, Divisions 1 and 2
503.120	503.5	Control Transformers and Resistors - Class III, Divisions 1 and 2
503.125	503.6	Motors and Generators - Class III, Divisions 1 and 2
503.128	503.7	Ventilating Piping - Class III, Divisions 1 and 2
503.130	503.9	Luminaires (Lighting Fixtures) - Class III, Divisions 1 and 2
503.135	503.8	Utilization Equipment - Class III, Divisions 1 and 2
503.140	503.10	Flexible Cords - Class III, Divisions 1 and 2
503.145	503.11	Receptacles and Attachment Plugs - Class III, Divisions 1 and 2
503.150	503.12	Signaling, Alarm, Remote-Control, and Local Loudspeaker Intercommunications Systems - Class III, Divisions 1 and 2
503.155	503.13	Electric Cranes, Hoists, and Similar Equipment - Class III, Divisions 1 and 2
503.160	503.14	Storage Battery Charging Equipment - Class III, Divisions 1 and 2

NEC Ch. 5 – Article 503
Part II – 503.10(A)(3)

Type of Change		Panel Action		UL	UL 508	API 500	API 505	OSHA
New Subdivision		Accept		2062	-	1.1.2	1.1.2	1910.307(b)(3), Note
ROP		ROC		NFPA 70E	NFPA 70B	NFPA 79	NFPA 497	NEMA
pg. 1256	# 14-63	pg. -	# -	440.4(A), FPN	22.1	3.3.53 and A.4.1	4.1.5.2	-
log: 1474	CMP: 14	log: -	Submitter: Nicholas P. Ludlam			2002 NEC: -		IEC: 60079-20

2005 NEC – 503.10(A)(3) Nonincendive Field Wiring

Nonincendive field wiring shall be permitted using any of the wiring methods permitted for unclassified locations. Nonincendive field wiring system shall be installed in accordance with the control drawing(s). Simple apparatus, not shown on the control drawing, shall be permitted in a nonincendive field wiring circuit, provided the simple apparatus does not interconnect the nonincendive field wiring circuit to any other circuit.

> **FPN:** Simple apparatus is defined in 504.2.

Separate nonincendive field wiring circuits shall be installed in accordance with one of the following:

(1) In separate cables

(2) In multiconductor cables where the conductors of each circuit are within a grounded metal shield

(3) In multiconductor cables where the conductors of each circuit have insulation with a minimum thickness of 0.25 mm (0.01 in.)

Author's Substantiation. A new subdivision has been added to recognize nonincendive field wiring in Class III, Division 1 locations.

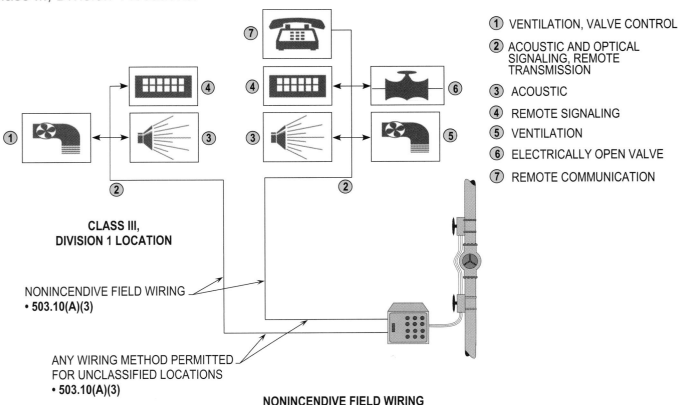

① VENTILATION, VALVE CONTROL
② ACOUSTIC AND OPTICAL SIGNALING, REMOTE TRANSMISSION
③ ACOUSTIC
④ REMOTE SIGNALING
⑤ VENTILATION
⑥ ELECTRICALLY OPEN VALVE
⑦ REMOTE COMMUNICATION

CLASS III,
DIVISION 1 LOCATION

NONINCENDIVE FIELD WIRING
• 503.10(A)(3)

ANY WIRING METHOD PERMITTED
FOR UNCLASSIFIED LOCATIONS
• 503.10(A)(3)

NONINCENDIVE FIELD WIRING
503.10(A)(3)

Purpose of Change: To add wiring methods for nonincendive field wiring in Class III, Division 1 locations.

NEC Ch. 5 – Article 504
Part – 504.10(B)

Type of Change	Panel Action	UL	UL 508	API 500	API 505	OSHA
New Paragraphs	Accept	2062	-	3.2.32	3.2.34	1910.307(b)(1)
ROP		ROC	NFPA 70E	NFPA 70B	NFPA 79	NFPA 497	NEMA
pg. 1259	# 14-67	pg. -	# -	440.2	6.7.10.2	3.3.53 and A.4.1	4.1.5.1	-
log: 1636	CMP: 14	log: -	Submitter: Nicholas P. Ludlam		2002 NEC: 504.10(B)		IEC: 60079-20

2002 NEC – 504.10(B) Location

Intrinsically safe apparatus shall be permitted to be installed in any hazardous (classified) locations for which it has been identified. General-purpose enclosures shall be permitted for intrinsically safe apparatus.

Associated apparatus shall be permitted to be installed in any hazardous (classified) location for which it has been identified or, if protected by other means, permitted by Articles 501 through 503 and Article 505.

2005 NEC – 504.10(B) Location

Intrinsically safe apparatus shall be permitted to be installed in any hazardous (classified) locations for which it has been identified. General-purpose enclosures shall be permitted for intrinsically safe apparatus.

Associated apparatus shall be permitted to be installed in any hazardous (classified) location for which it has been identified or, if protected by other means, permitted by Articles 501 through 503 and Article 505.

Simple apparatus shall be permitted to be installed in any hazardous (classified) location in which the maximum surface temperature of the simple apparatus does not exceed the ignition temperature of the flammable gases or vapors, flammable liquids, combustible dusts, or ignitible fibers or flyings present.

For simple apparatus, the maximum surface temperature can be determined from the values of the output power from the associated apparatus or apparatus to which it is connected to obtain the temperature class. The temperature class can be determined by:

(1) Reference to Table 504.10(B)

(2) Calculation using the formula:

$$T = P_o R_{th} + T_{amb}$$

Where:

T is the surface temperature

P_o is the output power marked on the associated apparatus or intrinsically safe apparatus

R_{th} is the thermal resistance of the simple apparatus

T_{amb} is the ambient temperature (normally 40°C) and reference Table 500.8(B)

In addition, components with a surface area smaller than 10 cm^2 (excluding lead wires) may be classified as T5 if their surface temperature does not exceed 150°C.

FPN: The following apparatus are examples of simple apparatus:

(1) Passive components, for example, switches, junction boxes, resistance temperature devices, and simple semiconductor devices such as LEDs

(2) Sources of generated energy, for example, thermocouples and photocells, which do not generate more than 1.5 V, 100 mA, and 25 mW

Author's Substantiation. Simple apparatus has been added to this section to recognize that such apparatus has the potential to get warmer and dissipate enough power to exceed the auto-ignition temperature of the gas or vapor present.

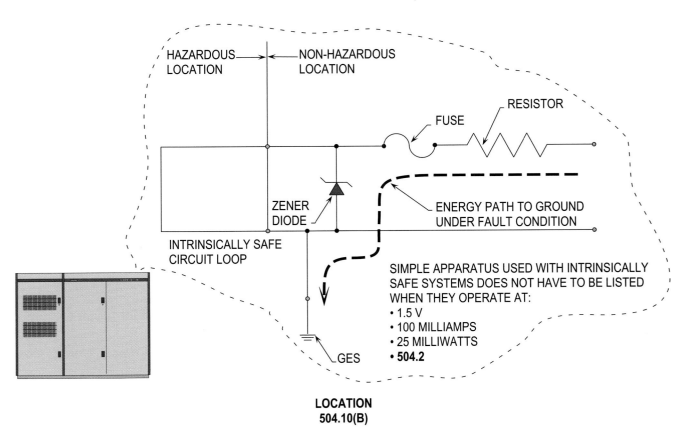

**LOCATION
504.10(B)**

Purpose of Change: To make this section easier to understand and comply with, simple apparatus was included in the list of requirements.

NEC Ch. 5 – Article 505
Part – 505.8(I)

Type of Change	Panel Action	UL	UL 508	API 500	API 505	OSHA
New Sentence	Accept in Principle	2062	-	1.1.2	1.1.2	1910.307(b)(3), Note

ROP		ROC		NFPA 70E	NFPA 70B	NFPA 79	NFPA 499	NEMA
pg. 1271	# 14-85	pg. -	# -	440.4(A), FPN	22.1	3.3.53 and A.4.1	1.1.2	-
log: 2966	CMP: 14	log: -	Submitter: David Soffrin			2002 NEC: 505.8(I)		IEC: 60079-20

2002 NEC – 505.8(I) Combustible Gas Detection System

A combustible gas detection system shall be permitted as a means of protection in industrial establishments with restricted public access and where the conditions of maintenance and supervision ensure that only qualified persons service the installation. Gas detection equipment shall be listed for detection of the specific gas or vapor to be encountered. Where such a system is installed, equipment specified in 505.8(I)(1), (2), or (3) shall be permitted.

(1) Inadequate Ventilation. In a Class I, Zone 1 location that is so classified due to inadequate ventilation, electrical equipment suitable for Class I, Zone 2 locations shall be permitted.

(2) Interior of a Building. In a building located in, or with an opening into, a Class I, Zone 2 location where the interior does not contain a source of flammable gas or vapor, electrical equipment for unclassified locations shall be permitted.

(3) Interior of a Control Panel. In the interior of a control panel containing instrumentation utilizing or measuring flammable liquids, gases, or vapors, electrical equipment suitable for Class I, Zone 2 locations shall be permitted.

> **FPN No. 1:** For further information, see ANSI/ISA-12.13.01, *Performance Requirements, Combustible Gas Detectors.*

> **FPN No. 2:** For further information, see ANSI/API RP 500, *Recommended Practice for Classification of Locations for Electrical Installations at Petroleum Facilities Classified as Class I, Division 1 or Division 2.*

> **FPN No. 3:** For further information, see ISA-RP12.13.02, *Installation, Operation, and Maintenance of Combustible Gas Detection Instruments.*

2005 NEC – 505.8(I) Combustible Gas Detection System

A combustible gas detection system shall be permitted as a means of protection in industrial establishments with restricted public access and where the conditions of maintenance and supervision ensure that only qualified persons service the installation. Gas detection equipment shall be listed for detection of the specific gas or vapor to be encountered. Where such a system is installed, equipment specified in 505.8(I)(1), (I)(2), or (I)(3) shall be permitted. The type of detection equipment, its listing, installation location(s), alarm and shutdown criteria, and calibration frequency shall be documented when combustible gas detectors are used as a protection technique.

> **FPN No. 1:** For further information, see ANSI/ISA-12.13.01, *Performance Requirements, Combustible Gas Detectors.*

> **FPN No. 2:** For further information, see ANSI/API RP 505, *Recommended Practice for Classification of Locations for Electrical Installations at Petroleum Facilities Classified as Class I, Zone 0, Zone 1, and Zone 2.*

> **FPN No. 3:** For further information, see ISA-RP12.13.02, *Installation, Operation, and Maintenance of Combustible Gas Detection Instruments.*

(1) Inadequate Ventilation. In a Class I, Zone 1 location that is so classified due to inadequate ventilation, electrical equipment suitable for Class I, Zone 2 locations shall be permitted.

(2) Interior of a Building. In a building located in, or with an opening into, a Class I, Zone 2 location where the interior does not contain a source of flammable gas or vapor, electrical equipment for unclassified locations shall be permitted.

(3) Interior of a Control Panel. In the interior of a control panel containing instrumentation utilizing or measuring flammable liquids, gases, or vapors, electrical equipment suitable for Class I, Zone 2 locations shall be permitted.

Author's Substantiation. A new sentence has been added to clarify that documentation shall be provided where combustible gas detectors are used as a protection technique. The FPN s have been relocated under the main text so as not to appear to only applying to item (3).

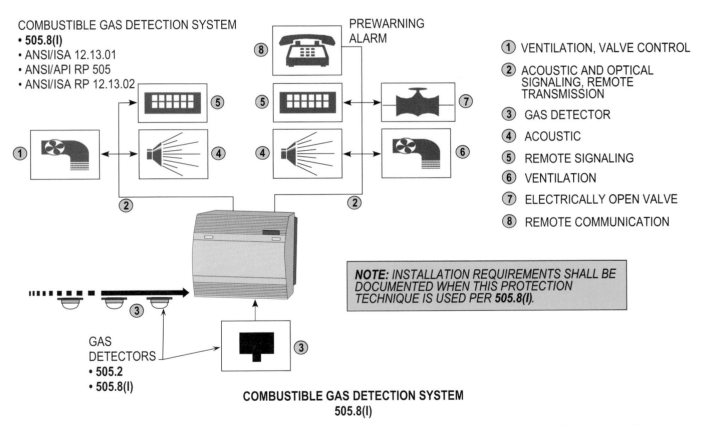

COMBUSTIBLE GAS DETECTION SYSTEM
505.8(I)

Purpose of Change: To provide guidelines to document proof of installation techniques before installing this protective scheme.

NEC Ch. 5 – Article 505
Part – 505.15(C)(1)(b)

Type of Change	Panel Action	UL	UL 508	API 500	API 505	OSHA
New Sentence	Accept in Principle	2062	-	1.1.2	1.1.2	1910.307(b)(3), FPN
ROP	**ROC**	**NFPA 70E**	**NFPA 70B**	**NFPA 79**	**NFPA 497**	**NEMA**
pg. 1280 # 14-100	pg. - # -	440.4(A), FPN	22.1	3.3.53 and A.4.1	1.1.2	-
log: 2300 CMP: 14	log: -	Submitter: H. R. Stewart		2002 NEC: 505.15(C)(1)(b)		IEC: 60079-20

2002 NEC – 505.15(C)(1) General

In Class I, Zone 2 locations, the wiring methods in (a) through (g) shall be permitted.

   (b) Types MI, MC, MV, or TC cable with termination fittings, or in cable tray systems and installed in a manner to avoid tensile stress at the termination fittings.

2005 NEC – 505.15(C)(1) General

In Class I, Zone 2 locations, the wiring methods in (C)(1)(a) through (C)(1)(g) shall be permitted.

   (b) Types MI, MC, MV, or TC cable with termination fittings, or in cable tray systems and installed in a manner to avoid tensile stress at the termination fittings. Single conductor Type MV cables shall be shielded or metallic-armored.

Author's Substantiation. A new sentence has been added to clarify that single conductor Type MV cable shall be shielded or metallic-armored to provide a ground plane that will eliminate any external electric discharge thus eliminating the ignition source and precluding any possibility of creating an explosion.

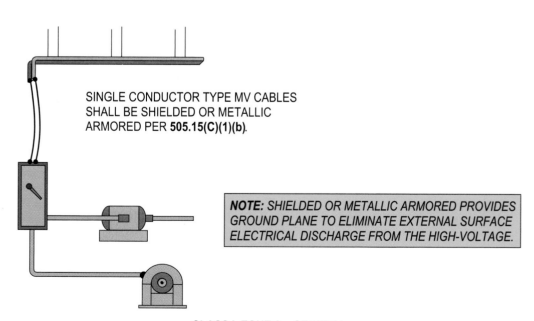

SINGLE CONDUCTOR TYPE MV CABLES SHALL BE SHIELDED OR METALLIC ARMORED PER **505.15(C)(1)(b)**.

NOTE: *SHIELDED OR METALLIC ARMORED PROVIDES GROUND PLANE TO ELIMINATE EXTERNAL SURFACE ELECTRICAL DISCHARGE FROM THE HIGH-VOLTAGE.*

CLASS I, ZONE 2 – GENERAL
505.15(C)(1)(b)

Purpose of Change: To provide requirements for shielded or metallic armored on single conductor MV cables.

NEC Ch. 5 – Article 505
Part – 505.18(A)

Type of Change		Panel Action		UL	UL 508	API 500	API 505	OSHA
New Sentence		Accept in Principle		2062	-	1.1.2	1.1.2	1910.307(b)(3), Note
ROP		ROC		NFPA 70E	NFPA 70B	NFPA 79	NFPA 499	NEMA
pg. 1268	# 14-79	pg. -	# -	440.4(A), FPN	22.1	3.3.53 and A.4.1	1.1.2	-
log: 26	CMP: 14	log: -	Submitter: Robert L. Seitz			2002 NEC: 505.18(A)		IEC: 60079-7

2002 NEC – 505.18(A) Conductors

For type of protection e, field wiring conductors shall be copper.

2005 NEC – 505.18(A) Conductors

For type of protection e, field wiring conductors shall be copper. <u>Every conductor (including spares) that enters Type e equipment shall be terminated at a Type e terminal.</u>

Author's Substantiation. A new sentence has been added to clarify the requirements for terminating conductors to Type e equipment terminals.

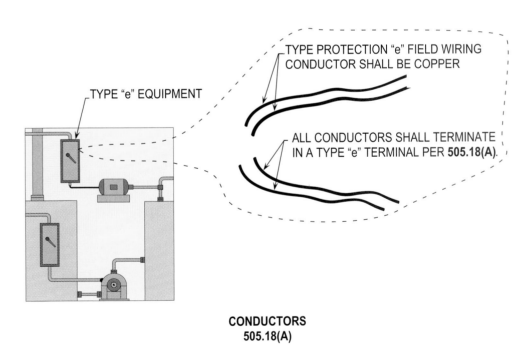

TYPE PROTECTION "e" FIELD WIRING CONDUCTOR SHALL BE COPPER

TYPE "e" EQUIPMENT

ALL CONDUCTORS SHALL TERMINATE IN A TYPE "e" TERMINAL PER **505.18(A)**.

CONDUCTORS
505.18(A)

Purpose of Change: To add requirements for terminating conductors to Type "e" equipment terminals.

NEC Ch. 5 – Article 506
Part – 506.2

124.1

Type of Change	Panel Action	UL	UL 508	API 500	API 505	OSHA
New Article	Accept in Principle	2062	-	1.1.2	1.1.2	1910.307(b)(3), Note

ROP		ROC		NFPA 70E	NFPA 70B	NFPA 79	NFPA 499	NEMA
pg. 1287	# 14-108a	pg. 407	# 14-97	440.4(A), FPN	22.1	3.3.53 and A.4.1	1.1.2	-
log: CP1409	CMP: 14	log: 1350	Submitter: CMP 14			2002 NEC: -		IEC: 1241-10

Author's Substantiation. A new Article 506 has been added to recognize dust locations using the zone concept. This section covers the definitions that apply to dust locations using the zone concept.

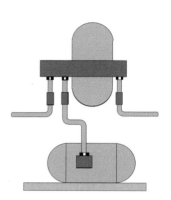

DEFINITIONS OF ZONES

Zone 20 Hazardous (Classified) Location. An area where combustible dust or ignitible fibers and flyings are present continuously or for long periods of time in quantities sufficient to be hazardous, as classified by **506.5(B)(1)**.

Zone 21 Hazardous (Classified) Location. An area where combustible dust or ignitible fibers and flyings are likely to exist occasionally under normal operation in quantities sufficient to be hazardous, as classified by **506.5(B)(2)**.

Zone 22 Hazardous (Classified) Location. An area where combustible dust or ignitible fibers and flyings are not likely to occur under normal operation in quantities sufficient to be hazardous, as classified by **506.5(B)(3)**.

DEFINITIONS
506.2

Purpose of Change: To create a new Article for dust locations using zone concepts.

NEC Ch. 5 – Article 506
Part – 506.2

124.2

Type of Change		Panel Action		UL	UL 508	API 500	API 505	OSHA
New Article		Accept in Principle		2062	-	1.1.2	1.1.2	1910.307(b)(3), Note
ROP		ROC		NFPA 70E	NFPA 70B	NFPA 79	NFPA 499	NEMA
pg. 1287	# 14-108a	pg. 407	# 14-97	440.4(A), FPN	22.1	3.3.53 and A.4.1	1.1.2	-
log: CP1409	CMP: 14	log: 1350	Submitter: CMP 14			2002 NEC: -		IEC: 1241-10

Author's Substantiation. A new Article 506 has been added to recognize dust locations using the zone concept. This section covers the definitions that apply for dust locations using the zone concept.

SCOPE - **506.1**

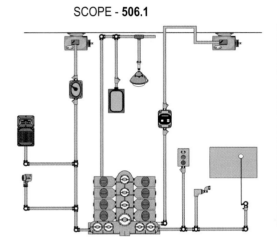

DEFINITIONS OF EQUIPMENT
ASSOCIATED NONINCENDIVE FIELD WIRING APPARATUS
DUST-IGNITIONPROOF
DUSTTIGHT
NONINCENDIVE CIRCUIT
NONINCENDIVE EQUIPMENT
NONINCENDIVE FIELD WIRING
NONINCENDIVE FIELD WIRING APPARATUS
PRESSURIZED

DEFINITIONS
506.2

Purpose of Change: To create a new Article for dust locations using zone concepts.

NEC Ch. 5 – Article 506
Part – 506.4(A) and (B)

 124.3

Type of Change		Panel Action		UL	UL 508	API 500	API 505	OSHA
New Article		Accept in Principle		2062	-	1.1.2	1.1.2	1910.307(b)(3), Note
ROP		ROC		NFPA 70E	NFPA 70B	NFPA 79	NFPA 499	NEMA
pg. 1287	# 14-108a	pg. 407	# 14-97	440.4(A), FPN	22.1	3.3.53 and A.4.1	1.1.2	-
log: CP1409	CMP: 14	log: 1350	Submitter: CMP 14			2002 NEC: -		IEC: 1241-10

2005 NEC – 506.4 General

(A) **Documentation for Industrial Occupancies.** Areas designated as hazardous (classified) locations shall be properly documented. This documentation shall be available to those authorized to design, install, inspect, maintain or operate electrical equipment.

(B) **Reference Standards.** Important information relating to topics covered in Chapter 5 are found in other publications.

FPN: It is important that the authority having jurisdiction be familiar with the recorded industrial experience as well as with standards of the National Fire Protection Association (NFPA), the ISA, International Society for Measurement and Control, and the International Electrotechnical Commission (IEC) that may be of use in the classification of various locations, the determination of adequate ventilation, and the protection against static electricity and lightning hazards.

Author's Substantiation. A new Article 506 has been added to recognize dust locations using the zone concept. This section covers the documentation required for dust locations and the reference standards.

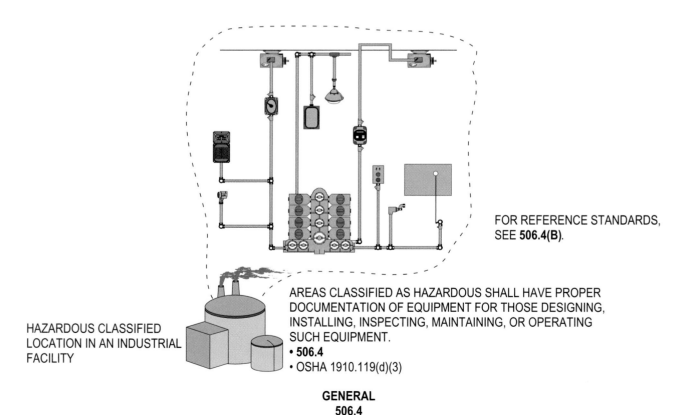

FOR REFERENCE STANDARDS, SEE **506.4(B)**.

AREAS CLASSIFIED AS HAZARDOUS SHALL HAVE PROPER DOCUMENTATION OF EQUIPMENT FOR THOSE DESIGNING, INSTALLING, INSPECTING, MAINTAINING, OR OPERATING SUCH EQUIPMENT.
• **506.4**
• OSHA 1910.119(d)(3)

HAZARDOUS CLASSIFIED LOCATION IN AN INDUSTRIAL FACILITY

**GENERAL
506.4**

Purpose of Change: To create a new Article for dust locations using the zone concept.

NEC Ch. 5 – Article 506
Part – 506.5(B)(1)

Type of Change		Panel Action		UL	UL 508	API 500	API 505	OSHA
New Article		Accept in Principle		2062	-	1.1.2	1.1.2	1910.307(b)(3), Note
ROP		ROC		NFPA 70E	NFPA 70B	NFPA 79	NFPA 499	NEMA
pg. 1287	# 14-108a	pg. 407	# 14-97	440.4(A), FPN	22.1	1.1.2	1.1.2	-
log: CP1409	CMP: 14	log: 1350	Submitter: CMP 14			2002 NEC: -		IEC: 1241-10

2005 NEC – 506.5(B)(1) Zone 20

A Zone 20 location is a location in which

   (a) Ignitible concentrations of combustible dust or ignitible fibers or flyings are present continuously.

   (b) Ignitible concentrations of combustible dust or ignitible fibers or flyings are present for long periods of time.

Author's Substantiation. A new Article 506 has been added to recognize dust locations using the zone concept. This subdivision covers the requirements for classifying a Zone 20 location.

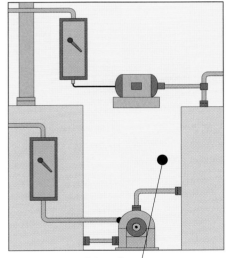

HAZARDOUS LOCATION

CLASSIFICATION OF LOCATIONS

(1) **Zone 20.** A Zone 20 location is a location where

   (a) Ignitible concentrations of combustible dust or ignitible fibers or flyings are present continuously, or

   (b) Ignitible concentrations of combustible dust or ignitible fibers or flyings are present for long periods of time.

ZONE 20
506.5(B)(1)

Purpose of Change: To create a new Article pertaining to Zones containing dust and to characterize each one.

NEC Ch. 5 – Article 506
Part – 506.5(B)(2)(a) and (b)

Type of Change		Panel Action		UL	UL 508	API 500	API 505	OSHA
New Article		Accept in Principle		2062	-	1.1.2	1.1.2	1910.307(b)(3), Note
ROP		ROC		NFPA 70E	NFPA 70B	NFPA 79	NFPA 499	NEMA
pg. 1287	# 14-108a	pg. 407	# 14-97	440.4(A), FPN	22.1	3.3.53 and A.4.1	1.1.2	-
log: CP1409	CMP: 14	log: 1350	Submitter: CMP 14			2002 NEC: -		IEC: 1241-10

2005 NEC – 506.5(B)(2) Zone 21

A Zone 21 location is a location

(a) In which ignitible concentrations of combustible dust or ignitible fibers or flyings are likely to exist occasionally under normal operating conditions; or

(b) In which ignitible concentrations of combustible dust or ignitible fiber or flyings may exist frequently because of repair or maintenance operations or because of leakage; or

Author's Substantiation. A new Article 506 has been added to recognize dust locations using the zone concept. This subdivision covers the requirements for classifying a Zone 21 location.

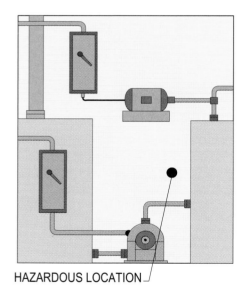

HAZARDOUS LOCATION

> **CLASSIFICATION OF LOCATIONS**
>
> **(2)   Zone 21.** A Zone 21 location is where
>
> **(a)** Ignitible concentrations of combustible dust or ignitible fibers or flyings are likely to exist occasionally under normal operation conditions, or
>
> **(b)** Ignitible concentrations of combustible dust or ignitible fibers or flyings may exist frequently because of repair or maintenance operations or because of leakage.

ZONE 21
506.5(B)(2)

Purpose of Change: To create a new Article pertaining to Zones containing dust and to characterize each one.

NEC Ch. 5 – Article 506
Part – 506.5(B)(2)(c) and (d)

 **124.6**

Type of Change	Panel Action	UL	UL 508	API 500	API 505	OSHA
New Article	Accept in Principle	2062	-	1.1.2	1.1.2	1910.307(b)(3), Note

ROP		ROC		NFPA 70E	NFPA 70B	NFPA 79	NFPA 499	NEMA
pg. 1287	# 14-108a	pg. 407	# 14-97	440.4(A), FPN	22.1	3.3.53 and A.4.1	1.1.2	-

log: CP 1409	CMP: 14	log: 1350	Submitter: CMP 14		2002 NEC: -		IEC: 1241-10

2005 NEC – <u>506.5(B)(2) Zone 21</u>

A Zone 21 location is a location

  (c) In which equipment is operated or processes are carried on, of such a nature that equipment breakdown or faulty operations could result in the release of ignitible concentrations of combustible dust, or ignitible fibers or flyings and also cause simultaneous failure of electrical equipment in a mode to cause the electrical equipment to become a source of ignition; or

  (d) That is adjacent to a Zone 20 location from which ignitible concentrations of dust or ignitible fibers or flyings could be communicated, unless communication is prevented by adequate positive pressure ventilation from a source of clean air and effective safeguards against ventilation failure are provided.

Author's Substantiation. A new Article 506 has been added to recognize dust locations using the zone concept. This subdivision covers the requirements for classifying a Zone 21 location.

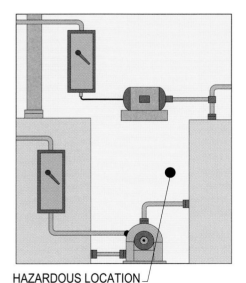

HAZARDOUS LOCATION

CLASSIFICATION OF LOCATIONS

(2) **Zone 21.** A Zone 21 location is where
  (c) Equipment is operated or processes are carried on, of such a nature that equipment breakdown or faulty operations could result in the release of ignitible concentrations of combustible dust, or ignitible fibers or flyings and also cause simultaneous failure of electrical equipment in a mode to cause the electrical equipment to become a source of ignition, or

  (d) Equipment is adjacent to a Zone 20 location from which ignitible concentrations of dust or ignitible fibers or flyings could be communicated, unless communication is prevented by adequate positive pressure ventilation from a source of clean air and effective safeguards against ventilation failure are provided.

ZONE 21
506.5(B)(2)

Purpose of Change: To create a new Article pertaining to Zones containing dust and to characterize each one.

NEC Ch. 5 – Article 506
Part – 506.5(B)(3)(a) through (c)

124.7

Type of Change	Panel Action	UL	UL 508	API 500	API 505	OSHA
New Article	Accept in Principle	2062	-	1.1.2	1.1.2	1910.307(b)(3), Note
ROP		ROC		NFPA 70E	NFPA 70B	NFPA 79	NFPA 499	NEMA

ROP		ROC		NFPA 70E	NFPA 70B	NFPA 79	NFPA 499	NEMA
pg. 1287	# 14-108a	pg. 407	# 14-97	440.4(A), FPN	22.1	3.3.53 and A.4.1	1.1.2	-
log: CP1409	CMP: 14	log: 1350	Submitter: CMP 14			2002 NEC: -		IEC: 1241-10

2005 NEC – 506.5(B)(3) Zone 22

A Zone 22 location is a location

(a) In which ignitible concentrations of combustible dust or ignitible fibers or flyings are not likely to occur in normal operation, and if they do occur, will only persist for a short period; or

(b) In which combustible dust, or fibers, or flyings are handled, processed, or used but in which the dust, fibers, or flyings are normally confined within closed containers of closed systems from which they can escape only as a result of the abnormal operation of the equipment with which the dust, or fibers, or flyings are handled, processed, or used; or

(c) That is adjacent to a Zone 21 location, from which ignitible concentrations of dust or fibers or flyings could be communicated, unless such communication is prevented by adequate positive pressure ventilation from a source of clean air and effective safeguards against ventilation failure are provided.

Author's Substantiation. A new Article 506 has been added to recognize dust locations using the zone concept. This subdivision covers the requirements for classifying a Zone 22 location.

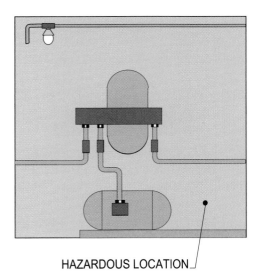

HAZARDOUS LOCATION

CLASSIFICATION OF LOCATIONS
(3) Zone 22. A Zone 22 location is where
(a) Ignitible concentrations of combustible dust or ignitible fibers or flyings are not likely to occur in normal operation, and if they do occur, will only persist for a short period, or
(b) Combustible dust, or fibers, or flyings are handled, processed, or used but in which the dust, fibers, or flyings are normally confined within closed containers of closed systems from which they can escape only as a result of the abnormal operation of the equipment with which the dust, or fibers, or flyings are handled, processed, or used, or
(c) Equipment is adjacent to a Zone 21 locations, from which ignitible concentrations of dust or fibers or flyings could be communicated, unless such communication is prevented by adequate positive pressure ventilation from a source of clean air and effective safeguards against ventilation failure are provided.

ZONE 22
506.5(B)(3)

Purpose of Change: To create a new Article pertaining to Zones containing dust and to characterize each one.

NEC Ch. 5 – Article 506
Part – 506.6(A) through (D)

 124.8

Type of Change		Panel Action		UL	UL 508	API 500	API 505	OSHA
New Article		Accept in Principle		2062	-	1.1.2	1.1.2	1910.307(b)(3), Note
ROP		ROC		NFPA 70E	NFPA 70B	NFPA 79	NFPA 499	NEMA
pg. 1287	# 14-108a	pg. 407	# 14-97	440.4(A), FPN	22.1	3.3.53 and A.4.1	1.1.2	-
log: CP1409	CMP: 14	log: 1350	Submitter: CMP 14			2002 NEC: -		IEC: 1241-10

2005 NEC – <u>506.6 Special Precaution</u>

Article 506 requires equipment construction and installation that ensures safe performance under conditions of proper use and maintenance.

> **FPN:** It is important that inspection authorities and users exercise more than ordinary care with regard to the installation and maintenance of electrical equipment in hazardous (classified) locations.

(A) Implementation of Zone Classification System. Classification of areas, engineering and design, selection of equipment and wiring methods, installation, and inspection shall be performed by qualified persons.

(B) Dual Classification. In instances of areas within the same facility classified separately, Zone 22 locations shall be permitted to abut, but not overlap, Class II or Class III, Division 2 locations. Zone 20 or Zone 21 locations shall not abut Class II or Class III, Division 1 or Division 2 locations.

(C) Reclassification Permitted. A Class II or Class III, Division 1 or Division 2 location shall be permitted to be reclassified as a Zone 20, Zone 21, or Zone 22 location, provided that all of the space that is classified because of a single combustible dust or ignitible fiber or flying source is reclassified under the requirements of this article.

(D) Simultaneous Presence of Flammable Gases and Combustible Dusts, Fibers, or Flyings. Where flammable gases or combustible dusts, fibers, or flyings are or may be present at the same time, the simultaneous presence shall be considered during the selection and installation of the electrical equipment and the wiring methods, including the determination of the safe operating temperature of the electrical equipment.

Author's Substantiation. A new Article 506 has been added to recognize dust locations using the zone concept. This section covers the special precautions associated with dust locations using the zone concept.

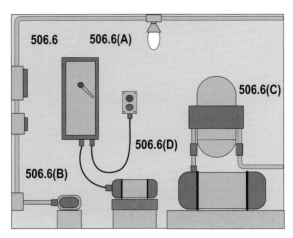

SPECIAL PRECAUTION
• 506.6
ARTICLE 506 REQUIRES EQUIPMENT CONSTRUCTION AND INSTALLATION THAT ENSURES SAFE PERFORMANCE UNDER CONDITIONS OF PROPER USE AND MAINTENANCE.

IMPLEMENTATION OF ZONE CLASSIFICATION SYSTEM
• 506.6(A)
CLASSIFICATION OF AREAS, ENGINEERING AND DESIGN, SELECTION OF EQUIPMENT AND WIRING METHODS, INSTALLATION, AND INSPECTION SHALL BE PERFORMED BY QUALIFIED PERSONS.

DUAL CLASSIFICATION
• 506.6(B)
IN INSTANCES OF AREAS WITHIN THE SAME FACILITY CLASSIFIED SEPARATELY, ZONE 22 LOCATIONS SHALL BE PERMITTED TO ABUT, BUT NOT OVERLAP, CLASS II OR III, DIVISION 2 LOCATIONS. ZONE 20 OR 21 LOCATIONS SHALL NOT ABUT CLASS II OR CLASS III, DIVISION 1 OR DIVISION 2 LOCATIONS.

RECLASSIFICATION PERMITTED
• 506.6(C)
A CLASS II OR III, DIVISION 1 OR 2 LOCATION SHALL BE PERMITTED TO BE RECLASSIFIED AS A ZONE 20, 21, OR 22 LOCATION, PROVIDED THAT ALL OF THE SPACE THAT IS CLASSIFIED BECAUSE OF A SINGLE COMBUSTIBLE DUST OR IGNITIBLE FIBER OR FLYING SOURCE IS RECLASSIFIED UNDER THE REQUIREMENTS OF THIS ARTICLE.

SIMULTANEOUS PRESENCE OF FLAMMABLE GASES AND COMBUSTIBLE DUSTS, FIBERS, OR FLYINGS
• 506.6(D)
WHERE FLAMMABLE GASES OR COMBUSTIBLE DUSTS, FIBERS, OR FLYINGS ARE OR MAY BE PRESENT AT THE SAME TIME, THE SIMULTANEOUS PRESENCE SHALL BE CONSIDERED DURING THE SELECTION AND INSTALLATION OF THE ELECTRICAL EQUIPMENT AND THE WIRING METHODS, INCLUDING THE DETERMINATION OF THE SAFE OPERATING TEMPERATURE OF THE ELECTRICAL EQUIPMENT.

SPECIAL PRECAUTIONS
506.6(A) THRU (D)

Purpose of Change: To create a new **Article 506** pertaining to dust and provide special precaution requirements for zone installations.

NEC Ch. 5 – Article 506
Part – 506.8(A) through (F)

124.9

Type of Change	Panel Action	UL	UL 508	API 500	API 505	OSHA
New Article	Accept in Principle	2062	-	1.1.2	1.1.2	1910.307(b)(3), Note
ROP		ROC		NFPA 70E	NFPA 70B	NFPA 79	NFPA 499	NEMA

ROP		ROC		NFPA 70E	NFPA 70B	NFPA 79	NFPA 499	NEMA
pg. 1287	# 14-108a	pg. 407	# 14-97	440.4(A), FPN	22.1	3.3.53 and A.4.1	1.1.2	-
log: CP1409	CMP: 14	log: 1350	Submitter: CMP 14			2002 NEC: -		IEC: 1241-10

2005 NEC – 506.8 Protection Techniques

Acceptable protection techniques for electrical and electronic equipment in hazardous (classified) locations shall be as described in 506.8(A) through 506.8(F).

(A) Dust Ignitionproof. This protection technique shall be permitted for equipment in Zone 20, Zone 21, and Zone 22 locations for which it is identified.

(B) Pressurized. This protection technique shall be permitted for equipment in Zone 21, and Zone 22 locations for which it is identified.

(C) Intrinsic Safety. This protection technique shall be permitted for equipment in Zone 20, Zone 21, and Zone 22 locations for which it is identified. Installation of intrinsically safe apparatus and wiring shall be in accordance with the requirements of Article 504.

(D) Dusttight. This protection technique shall be permitted for equipment in Zone 22 locations for which it is identified.

(E) Nonincendive Circuit. This protection technique shall be permitted for equipment in Zone 22 locations for which it is identified.

(F) Nonincendive Equipment. This protection technique shall be permitted for equipment in Zone 22 locations for which it is identified.

Author's Substantiation. A new Article 506 has been added to recognize dust locations using the zone concept. This section covers the protection techniques for electrical and electronic equipment in dust locations using the zone concept.

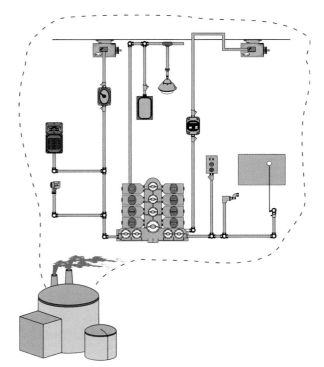

PROTECTION TECHNIQUES
(A) Dust ignitionproof can be used in Zone 20, 21 and 22 locations.
(B) Pressurized can be used in Zone 21 and 22 locations.
(C) Intrinsic safety can be used in Zone 20, 21 and 22 locations.
(D) Dusttight can be used in Zone 22 locations.
(E) Nonincendive circuit can be used in Zone 22 locations.
(F) Nonincendive equipment can be used in Zone 22 locations.

PROTECTION TECHNIQUES
506.8

Purpose of Change: To outline the zones in which each individual protection technique identified for the location can be installed.

NEC Ch. 5 – Article 506
Part – 506.9(A) through (C)

124.10

Type of Change		Panel Action		UL	UL 508	API 500	API 505	OSHA
New Article		Accept in Principle		2062	-	1.1.2	1.1.2	1910.307(b)(3), Note
ROP		ROC		NFPA 70E	NFPA 70B	NFPA 79	NFPA 499	NEMA
pg. 1287	# 14-108a	pg. 407	# 14-97	440.4(A), FPN	22.1	3.3.53 and A.4.1	1.1.2	-
log: CP1409	CMP: 14	log: 1350	Submitter: CMP 14			2002 NEC: -		IEC: 1241-10

2005 NEC – <u>506.9 Equipment Requirements</u>

(A) Suitability. Suitability of identified equipment shall be determined by one of the following:

    (1) Equipment listing or labeling

    (2) Evidence of equipment evaluation from a qualified testing laboratory or inspection agency concerned with product evaluation

    (3) Evidence acceptable to the authority having jurisdiction such as a manufacturer's self-evaluation or an owner's engineering judgment

(B) Listing

    (1) Equipment that is listed for Zone 20 shall be permitted in a Zone 21 or Zone 22 location of the same dust, or ignitible fiber, or flying. Equipment that is listed for Zone 21 may be used in a Zone 22 location of the same dust, fiber, or flying.

    (2) Equipment shall be permitted to be listed for a specific dust, or ignitible fiber or flying, or any specific combination of dusts, fibers, or flyings.

(C) Marking. Equipment identified for Class II, Division 1 or Class II, Division 2 shall, in addition to being marked in accordance with 500.8(B), be permitted to be marked with both of the following:

    (1) Zone 20, 21, or 22 (as applicable)

    (2) Temperature classification in accordance with 506.9(D)

Author's Substantiation. A new Article 506 has been added to recognize dust locations using the zone concept. This section covers the equipment requirements in dust locations using the zone concept.

OSHA RULES PER 1910.302 thru 1910.308. ALSO, SEE OSHA 1910.303(a), 1910.303(b)(1)(i) THRU (b)(vii) AND (b)(2).

NEC RULES PER **110.3(A)(1) THRU 110.3(A)(8), 110.3(B), AND 110.2**

DESIGN BY	NEC LOOP	OSHA LOOP
• NEC • UL 508 • ANSI C SERIES • NFPA 79	• **90.7** • **110.3(A); (B)** • **110.2** • **506.9**	• OSHA 1910.7 • OSHA 1910.303(a) • OSHA 1910.399

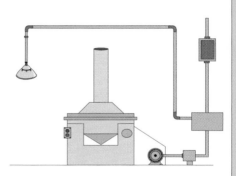

"IDENTIFIED" (AS APPLIED TO EQUIPMENT). RECOGNIZABLE AS SUITABLE FOR THE SPECIFIC PURPOSE, FUNCTION, USE, ENVIRONMENT, APPLICATION, ETC., WHERE PRESCRIBED IN A PARTICULAR CODE REQUIREMENT.

(A) SUITABILITY OF EQUIPMENT FOR A SPECIFIC PURPOSE, ENVIRONMENT OR APPLICATION MAY BE DETERMINED BY:
• EQUIPMENT LISTING OR LABELING;
• EVIDENCE OF EQUIPMENT EVALUATION FROM A QUALIFIED TESTING LABORATORY OR INSPECTION AGENCY CONCERNED WITH PRODUCT EVALUATION; OR
• EVIDENCE ACCEPTABLE TO THE AUTHORITY HAVING JURISDICTION SUCH AS A MANUFACTURER's SELF CERTIFICATION OR AN OWNER's ENGINEERING JUDGEMENT.
(SEE LISTING AND MARKING PER 506.9(B) AND (C))

EQUIPMENT REQUIREMENTS
506.9

Purpose of Change: To provide provisions for approving and accepting electrical equipment in Zone 20, 21, and 22 hazardous (classified) locations containing dust.

NEC Ch. 5 – Article 506
Part – 506.9(D) and (E)

 124.11

Type of Change		Panel Action		UL	UL 508	API 500	API 505	OSHA
New Article		Accept in Principle		2062	-	1.1.2	1.1.2	1910.307(b)(3), Note
ROP		ROC		NFPA 70E	NFPA 70B	NFPA 79	NFPA 499	NEMA
pg. 1287	# 14-108a	pg. 407	# 14-97	440.4(A), FPN	22.1	3.3.53 and A.4.1	1.1.2	-
log: CP1409	CMP: 14	log: 1350	Submitter: CMP 14			2002 NEC: -		IEC: 1241-10

2005 NEC – 506.9 Equipment Requirements

(D) **Temperature Classifications.** Equipment shall be marked to show the operating temperature referenced to a 40°C (104°F) ambient. Electrical equipment designed for use in the ambient temperature range between -20…C and +40…C shall require no additional ambient temperature marking. Electrical equipment that is designed for use in a range of ambient temperatures other than -20°C and +40°C is considered to be special; and the ambient temperature range shall then be marked on the equipment, including either the symbol Ta or Tamb together with the special range of ambient temperatures. As an example, such a marking might be $-30°C \leq Ta \leq +40°C$. Electrical equipment suitable for ambient temperatures exceeding 40°C (104°F) shall be marked with both the maximum ambient temperature and the operating temperature at that ambient temperature.

Exception No. 1: Equipment of the non-heat-producing type, such as conduit fittings, shall not be required to have a marked operating temperature.

Exception No. 2: Equipment identified for Class II, Division 1 or Class II, Division 2 locations as permitted by 506.20(B) and 506.20(C) shall be permitted to be marked in accordance with 500.6(D) and Table 500.6(D).

(E) **Threading.** All NPT threads referred to herein shall be threaded with a National (American) Standard Pipe Taper (NPT) thread that provides a taper of 1 in 16 (3/4-in taper per foot). Conduit and fittings shall be made wrenchtight to prevent sparking when the fault current flows through the conduit system, and to ensure the integrity of the conduit system. Equipment provided with threaded entries for field wiring connections shall be installed in accordance with 506.9(E)(1) or (E)(2).

(1) **Equipment Provided with Threaded Entries for NPT Threaded Conduit or Fittings.** For equipment provided with threaded entries for NPT threaded conduit or fittings, listed conduit fittings, or cable fittings shall be used.

(2) **Equipment Provided with Threaded Entries for Metric Threaded Conduit or Fittings.** For equipment with metric threaded entries, such entries shall be identified as being metric, or listed adapters to permit connection to conduit or NPT-threaded fittings shall be provided with the equipment. Adapters shall be used for connection to conduit or NPT-threaded fittings. Listed cable fittings that have metric threads shall be permitted to be used.

Author's Substantiation. A new Article 506 has been added to recognize dust locations using the zone concept. This section covers the equipment requirements in dust locations using the zone concept.

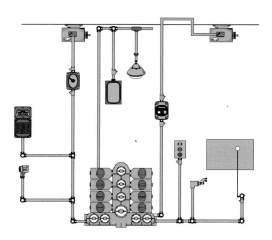

EQUIPMENT REQUIREMENTS
TEMPERATURE CLASSIFICATIONS • **506.9(D)** **Ex. 1** - NON-HEAT PRODUCING TYPE **Ex. 2** - EQUIPMENT IDENTIFIED FOR CLASS II, DIVISION 1 OR 2 LOCATIONS
THREADING • **506.9(E)**
EQUIPMENT PROVIDED WITH THREADED ENTRIES FOR NPT THREADED CONDUIT OR FITTINGS • **506.9(E)(1)**
EQUIPMENT PROVIDED WITH THREADED ENTRIES FOR METRIC THREADED CONDUIT OR FITTINGS • **506.9(E)(2)**

EQUIPMENT REQUIREMENTS
506.9

Purpose of Change: To provide requirements pertaining to temperature classifications and threading techniques in dust-related zones.

NEC Ch. 5 – Article 506
Part – 506.15(A)(1) through (A)(5)

 124.12

Type of Change	Panel Action	UL	UL 508	API 500	API 505	OSHA
New Article	Accept in Principle	2062	-	1.1.2	1.1.2	1910.307(b)(3), Note
ROP		ROC		NFPA 70E	NFPA 70B	NFPA 79	NFPA 499	NEMA

ROP		ROC		NFPA 70E	NFPA 70B	NFPA 79	NFPA 499	NEMA
pg. 1287	# 14-108a	pg. 407	# 14-97	440.4(A), FPN	22.1	3.3.53 and A.4.1	1.1.2	-
log: CP1409	CMP: 14	log: 1350	Submitter: CMP 14			2002 NEC: -		IEC: 1241-10

2005 NEC – <u>506.15 Wiring Methods</u>

Wiring methods shall maintain the integrity of the protection techniques and shall comply with 506.15(A), (B), or (C).

(A) Zone 20. In Zone 20 locations, the wiring methods in (1) through (5) shall be permitted.

(1) Threaded rigid metal conduit, or threaded steel intermediate metal conduit.

(2) Type MI cable with termination fittings listed for the location. Type MI cable shall be installed and supported in a manner to avoid tensile stress at the termination fittings.

Exception: MI cable and fittings listed for Class II, Division 1 locations are permitted to be used.

(3) In industrial establishments with limited public access, where the condition of maintenance and supervision ensure that only qualified persons service the installation, Type MC cable, listed for continuous use in Zone 20 locations, with a gas/vaportight continuous corrugated metallic sheath, and overall jacket of suitable polymeric material, separate grounding conductors in accordance with 250.122, and provided with termination fittings listed for the application, shall be permitted.

Exception: MC cable and fittings listed for Class II, Division 1 locations are permitted to be used.

(4) Fittings and boxes shall be identified for use in Zone 20 locations.

Exception: Boxes and fittings listed for Class II, Division 1 locations are permitted to be used.

(5) Where necessary to employ flexible connections, liquidtight flexible metal conduit with listed fittings, liquidtight flexible nonmetallic conduit with listed fittings, or flexible cord listed for extra-hard usage and provided with listed fittings shall be used. Where flexible cords are used, they shall also comply with 506.17. Where flexible connections are subject to oil or other corrosive conditions, the insulation of the conductors shall be of a type listed for the condition or shall be protected by means of a suitable sheath.

Exception: Flexible conduit and flexible conduit and cord fittings listed for Class II, Division 1 locations are permitted to be used.

FPN: See 506.25 for grounding requirements where flexible conduit is used.

Author's Substantiation. A new Article 506 has been added to recognize dust locations using the zone concept. This subsection covers the wiring methods permitted to be installed in Zone 20 locations.

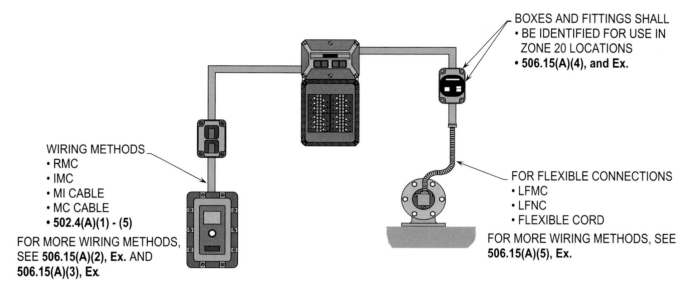

WIRING METHODS
• RMC
• IMC
• MI CABLE
• MC CABLE
• **502.4(A)(1) - (5)**
FOR MORE WIRING METHODS,
SEE **506.15(A)(2), Ex.** AND
506.15(A)(3), Ex.

BOXES AND FITTINGS SHALL
• BE IDENTIFIED FOR USE IN
ZONE 20 LOCATIONS
• **506.15(A)(4), and Ex.**

FOR FLEXIBLE CONNECTIONS
• LFMC
• LFNC
• FLEXIBLE CORD
FOR MORE WIRING METHODS, SEE
506.15(A)(5), Ex.

WIRING METHODS – ZONE 20
506.15(A)(1) THRU (A)(4)

Purpose of Change: To provide requirements in the new **Article 506** for wiring methods used in Zone 20.

NEC Ch. 5 – Article 506
Part – 506.15(B)(1) and (B)(2)

 124.13

Type of Change	Panel Action	UL	UL 508	API 500	API 505	OSHA
New Article	Accept in Principle	2062	-	1.1.2	1.1.2	1910.307(b)(3), Note
ROP		ROC		NFPA 70E	NFPA 70B	NFPA 79	NFPA 499	NEMA

ROP		ROC		NFPA 70E	NFPA 70B	NFPA 79	NFPA 499	NEMA
pg. 1287	# 14-108a	pg. 407	# 14-97	440.4(A), FPN	22.1	3.3.53 and A.4.1	1.1.2	-
log: CP1409	CMP: 14	log: 1350	Submitter: CMP 14			2002 NEC: -		IEC: 1241-10

2005 NEC – <u>506.15 Wiring Methods</u>

Wiring methods shall maintain the integrity of the protection techniques and shall comply with 506.15(A), (B), or (C).

(B) Zone 21. In Zone 21 locations, the wiring methods in (B)(1) and (B)(2) shall be permitted.

(1) All wiring methods permitted 506.15(A)

(2) Fittings and boxes that are dusttight, provided with threaded bosses for connection to conduit, in which taps, joints, or terminal connections are not made, and are not used in locations where metal dust is present, may be used

Author's Substantiation. A new Article 506 has been added to recognize dust locations using the zone concept. This subsection covers the wiring methods permitted to be installed in Zone 21 locations.

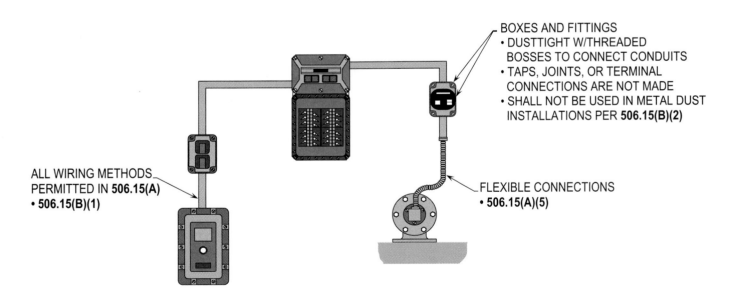

BOXES AND FITTINGS
- DUSTTIGHT W/THREADED BOSSES TO CONNECT CONDUITS
- TAPS, JOINTS, OR TERMINAL CONNECTIONS ARE NOT MADE
- SHALL NOT BE USED IN METAL DUST INSTALLATIONS PER **506.15(B)(2)**

ALL WIRING METHODS PERMITTED IN **506.15(A)**
- **506.15(B)(1)**

FLEXIBLE CONNECTIONS
- **506.15(A)(5)**

WIRING METHODS – ZONE 21
506.15(B)(1) and (B)(2)

Purpose of Change: To provide requirements in the new **Article 506** for wiring methods used in Zone 21.

NEC Ch. 5 – Article 506
Part I – 506.15(C)(1) through (C)(8)

124.14

Type of Change	Panel Action	UL	UL 508	API 500	API 505	OSHA
New Article	Accept in Principle	2062	-	1.1.2	1.1.2	1910.307(b)(3), Note
ROP	ROC	NFPA 70E	NFPA 70B	NFPA 79	NFPA 499	NEMA
pg. 1287 # 14-108a	pg. 407 # 14-97	440.4(A), FPN	22.1	3.3.53 and A.4.1	1.1.2	-
log: CP1409 CMP: 14	log: 1350 Submitter: CMP 14			2002 NEC: -		IEC: 1241-10

2005 NEC – 506.15 Wiring Methods

Wiring methods shall maintain the integrity of the protection techniques and shall comply with 506.15(A), (B), or (C).

(C) Zone 22. In Zone 22 locations, the wiring methods in (1) through (8) shall be permitted.

(1) All wiring methods permitted 502.16(B).

(2) Rigid metal conduit, intermediate metal conduit, electrical metallic tubing, dusttight wireways.

(3) Type MC or MI cable with listed termination fittings.

(4) Type PLTC in cable trays.

(5) Type ITC in cable trays.

(6) Type MC, MI, MV, or TC cable installed in ladder, ventilated trough, or ventilated channel cable trays in a single layer, with a space not less than the larger cable diameter between two adjacent cables, shall be the wiring method employed. Single conductor Type MV cables shall be shielded or metallic armored.

(7) Nonincendive field wiring shall be permitted using any of the wiring methods permitted for unclassified locations. Nonincendive field wiring systems shall be installed in accordance with the control drawing(s).

Simple apparatus, not shown on the control drawing, shall be permitted in a nonincendive field wiring circuit, provided the simple apparatus does not interconnect the nonincendive field wiring circuit to any other circuit.

FPN: *Simple apparatus is defined in 504.2.*

Separation of nonincendive field wiring circuits shall be installed in accordance with one of the following:

a. Be in separate cables

b. Be in multiconductor cables where the conductors of each circuit are within a grounded metal shield

c. Be in multiconductor cables where the conductors have insulation with a minimum thickness of 0.25 mm (0.01 in.)

(8) Boxes and fittings shall be dusttight.

Author's Substantiation. A new Article 506 has been added to recognize dust locations using the zone concept. This subsection covers the wiring methods permitted to be installed in Zone 22 locations.

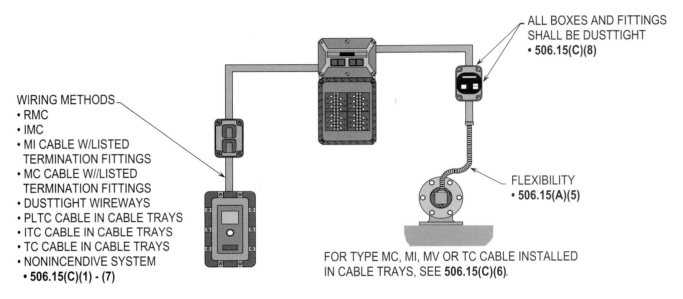

WIRING METHODS
- RMC
- IMC
- MI CABLE W/LISTED
  TERMINATION FITTINGS
- MC CABLE W//LISTED
  TERMINATION FITTINGS
- DUSTTIGHT WIREWAYS
- PLTC CABLE IN CABLE TRAYS
- ITC CABLE IN CABLE TRAYS
- TC CABLE IN CABLE TRAYS
- NONINCENDIVE SYSTEM
- **506.15(C)(1) - (7)**

ALL BOXES AND FITTINGS
SHALL BE DUSTTIGHT
- **506.15(C)(8)**

FLEXIBILITY
- **506.15(A)(5)**

FOR TYPE MC, MI, MV OR TC CABLE INSTALLED
IN CABLE TRAYS, SEE **506.15(C)(6)**.

WIRING METHODS – ZONE 22
506.15(C)(1) THRU (C)(8)

Purpose of Change: To provide requirements in the new **Article 506** for wiring methods used in Zone 22.

NEC Ch. 5 – Article 506
Part – 506.16

124.15

Type of Change	Panel Action	UL	UL 508	API 500	API 505	OSHA
New Article	Accept in Principle	2062	-	1.1.2	1.1.2	1910.307(b)(3), Note
ROP	ROC	NFPA 70E	NFPA 70B	NFPA 79	NFPA 499	NEMA
pg. 1287 # 14-108a	pg. 407 # 14-97	440.4(A), FPN	22.1	3.3.53 and A.4.1	1.1.2	-
log: CP1409 CMP: 14	log: 1350 Submitter: CMP 14			2002 NEC: -		IEC: 1241-10

2005 NEC – <u>506.16 Sealing</u>

<u>Where necessary to protect the ingress of combustible dust, or ignitible fibers, or flyings, or to maintain the type of protection, seals shall be provided. The seal shall be identified as capable of preventing the ingress of combustible dust or ignitible fibers or flyings and maintaining the type of protection but need not be explosionproof or flameproof.</u>

Author's Substantiation. A new Article 506 has been added to recognize dust locations using the zone concept. This section covers the sealing requirements for dust locations using the zone concept.

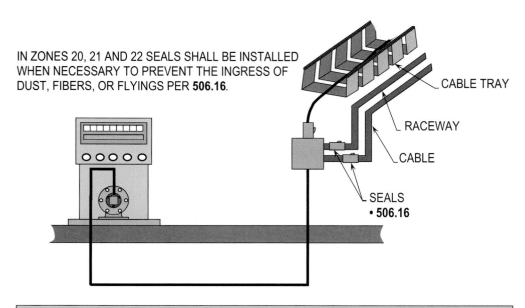

IN ZONES 20, 21 AND 22 SEALS SHALL BE INSTALLED WHEN NECESSARY TO PREVENT THE INGRESS OF DUST, FIBERS, OR FLYINGS PER **506.16**.

CABLE TRAY

RACEWAY

CABLE

SEALS
• **506.16**

NOTE: *SEALS ARE NOT REQUIRED TO BE EXPLOSIONPROOF OR FLAMEPROOF PER 501.16.*

SEALING
506.16

Purpose of Change: To require sealing to be provided where necessary due to the movement of dust, fibers, and flyings through cables and raceways to equipment.

NEC Ch. 5 – Article 506
Part – 506.17(1) through (5)

124.16

Type of Change	Panel Action	UL	UL 508	API 500	API 505	OSHA
New Article	Accept in Principle	2062	-	1.1.2	1.1.2	1910.307(b)(3), Note
ROP		ROC		NFPA 70E	NFPA 70B	NFPA 79	NFPA 499	NEMA

ROP		ROC		NFPA 70E	NFPA 70B	NFPA 79	NFPA 499	NEMA
pg. 1287	# 14-108a	pg. 407	# 14-97	440.4(A), FPN	22.1	3.3.53 and A.4.1	1.1.2	-
log: CP1409	CMP: 14	log: 1350	Submitter: CMP 14			2002 NEC: -		IEC: 1241-10

2005 NEC – 506.17 Flexible Cords

Flexible cords used in Zone 20, Zone 21, and Zone 22 locations shall comply with all of the following:

(1) Be of a type listed for extra-hard usage

(2) Contain, in addition to the conductors of the circuit, a grounding conductor in complying with 400.23

(3) Be connected to terminals or to supply conductors in an approved manner

(4) Be supported by clamps or by other suitable means in such a manner to minimize tension on the terminal connections

(5) Be provided with suitable seals to prevent the entrance of combustible dust, or ignitible fibers, or flyings where the flexible cord enters boxes or fittings

Author's Substantiation. A new Article 506 has been added to recognize dust locations using the zone concept. This section covers the requirements for flexible cords used in dust locations using the zone concept.

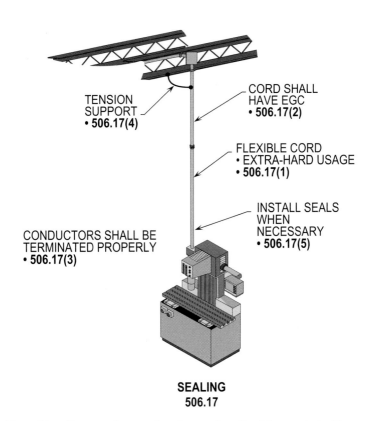

TENSION SUPPORT
• **506.17(4)**

CORD SHALL HAVE EGC
• **506.17(2)**

FLEXIBLE CORD
• EXTRA-HARD USAGE
• **506.17(1)**

INSTALL SEALS WHEN NECESSARY
• **506.17(5)**

CONDUCTORS SHALL BE TERMINATED PROPERLY
• **506.17(3)**

SEALING
506.17

Purpose of Change: To outline the requirements when using flexible cords in Zones 20, 21, and 22 of the new **Article 506**.

NEC Ch. 5 – Article 506
Part – 506.20(A) through (D)

Type of Change	Panel Action	UL	UL 508	API 500	API 505	OSHA
New Article	Accept in Principle	2062	-	1.1.2	1.1.2	1910.307(b)(3), Note
ROP		**ROC**		**NFPA 70E**	**NFPA 70B**	**NFPA 79**	**NFPA 499**	**NEMA**

ROP		ROC		NFPA 70E	NFPA 70B	NFPA 79	NFPA 499	NEMA
pg. 1287	# 14-108a	pg. 407	# 14-97	440.4(A), FPN	22.1	3.3.53 and A.4.1	1.1.2	-
log: CP1409	CMP: 14	log: 1350	Submitter: CMP 14			2002 NEC: -		IEC: 1241-10

2005 NEC – 506.20 Equipment Installation

(A) Zone 20. In Zone locations, only equipment listed and marked as suitable for the location shall be permitted.

Exception: Intrinsically safe apparatus listed for use in Class II, Division 1 locations with a suitable temperature class shall be permitted.

(B) Zone 21. In Zone 21 locations, only equipment listed and marked as suitable for the location shall be permitted.

Exception No. 1: Apparatus listed for use in Class II, Division 1 locations with a suitable temperature class shall be permitted.

Exception No. 2: Pressurized equipment identified for Class II, Division 1 shall be permitted.

(C) Zone 22. In Zone 22 locations, only equipment listed and marked as suitable for the location shall be permitted.

Exception No. 1: Apparatus listed for use in Class II, Division 1 or Class II, Division 2 locations with a suitable temperature class shall be permitted.

Exception No. 2: Pressurized equipment identified for Class II, Division 1 or Division 2 shall be permitted.

(D) Manufacturer's Instructions. Electrical equipment installed in hazardous (classified) locations shall be installed in accordance with the instructions (if any) provided by the manufacturer.

Author's Substantiation. A new Article 506 has been added to recognize dust locations using the zone concept. This section covers the equipment installation requirements for dust locations using the zone concept.

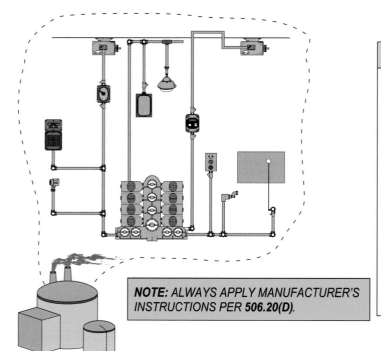

NOTE: *ALWAYS APPLY MANUFACTURER'S INSTRUCTIONS PER 506.20(D).*

EQUIPMENT INSTALLATION

ZONE 20
• **506.20(A)**
In Zone 20 locations, only equipment listed and marked as suitable for the location shall be permitted. (See Ex. for other permitted methods.)

ZONE 21
• **506.20(B)**
In Zone 21 locations, only equipment listed and marked as suitable for the location shall be permitted. (See Ex.'s 1 and 2 for other permitted methods.)

ZONE 22
• **506.20(C)**
In Zone 22 locations, only equipment listed and marked as suitable for the location shall be permitted. (See Ex. 's 1 and 2 for other permitted methods.)

EQUIPMENT INSTALLATION
506.20

Purpose of Change: To address requirements for installing equipment in Zones 20, 21, and 22 in new **Article 506**.

NEC Ch. 5 – Article 506
Part – 506.20(E)

Type of Change	Panel Action	UL	UL 508	API 500	API 505	OSHA
New Article	Accept in Principle	2062	-	1.1.2	1.1.2	1910.307(b)(3), Note
ROP	ROC	NFPA 70E	NFPA 70B	NFPA 79	NFPA 499	NEMA
pg. 1287	# 14-108a	pg. 407	# 14-97	440.4(A), FPN	22.1	3.3.53 and A.4.1	1.1.2	-
log: CP1409	CMP: 14	log: 1350	Submitter: CMP 14		2002 NEC: -		IEC: 1241-10

2005 NEC – 506.20 Equipment Installation

(E) Temperature. The temperature marking specified in 506.9(C)(2)(5) shall comply with (E)(1) or (E)(2).

(1) For combustible dusts, less than the lower of either the layer or cloud ignition temperature of the specific combustible dust. For organic dusts that may dehydrate or carbonize, the temperature marking shall not exceed the lower of either the ignition temperature or 165°C (329°F).

(2) For ignitible fibers or flyings, less than 165°C (329°F) for equipment that is not subject to overloading, or 120°C (248°F) for equipment (such as motors or power transformers) that may be overloaded.

> **FPN:** See NFPA 499-2004, *Recommended Practice for the Classification of Combustible Dusts and of Hazardous (Classified) Locations for Electrical Installations in Chemical Processing Areas*, for minimum ignition temperatures of specific dusts.

Author's Substantiation. A new Article 506 has been added to recognize dust locations using the zone concept. This section covers the equipment installation requirements for dust location using the zone concept.

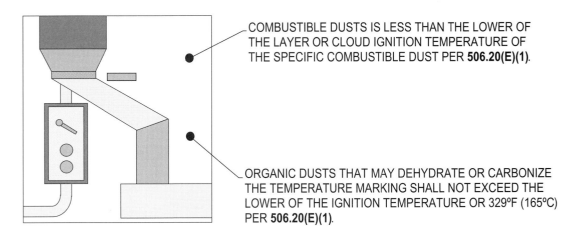

COMBUSTIBLE DUSTS IS LESS THAN THE LOWER OF THE LAYER OR CLOUD IGNITION TEMPERATURE OF THE SPECIFIC COMBUSTIBLE DUST PER **506.20(E)(1)**.

ORGANIC DUSTS THAT MAY DEHYDRATE OR CARBONIZE THE TEMPERATURE MARKING SHALL NOT EXCEED THE LOWER OF THE IGNITION TEMPERATURE OR 329°F (165°C) PER **506.20(E)(1)**.

EQUIPMENT INSTALLATION
506.20

Purpose of Change: To provide requirements for temperature ratings for combustible dust and organic dust in the new **Article 506**.

NEC Ch. 5 – Article 506
Part – 506.21

124.19

Type of Change		Panel Action		UL	UL 508	API 500	API 505	OSHA
New Article		Accept in Principle		2062	-	1.1.2	1.1.2	1910.307(b)(3), Note
ROP		ROC		NFPA 70E	NFPA 70B	NFPA 79	NFPA 499	NEMA
pg. 1287	# 14-108a	pg. 407	# 14-97	440.4(A), FPN	22.1	3.3.53 and A.4.1	1.1.2	-
log: CP1409	CMP: 14	log: 1350	Submitter: CMP 14			2002 NEC: -		IEC: 1241-10

2005 NEC – 506.21 Multiwire Branch Circuits

In Zone 20 and Zone 21 locations, a multiwire branch circuit shall not be permitted.

Exception: Where the disconnect device(s) for the circuit opens all ungrounded conductors of the multiwire circuit simultaneously.

Author's Substantiation. A new Article 506 has been added to recognize dust locations using the zone concept. This section covers the requirements for multiwire branch circuits in dust locations using the zone concept.

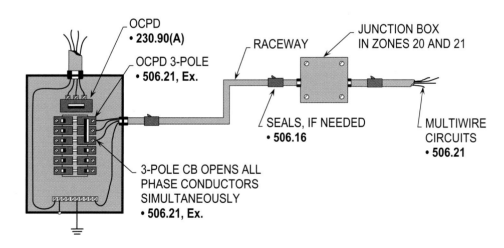

OCPD
• **230.90(A)**

OCPD 3-POLE
• **506.21, Ex.**

RACEWAY

JUNCTION BOX
IN ZONES 20 AND 21

SEALS, IF NEEDED
• **506.16**

MULTIWIRE
CIRCUITS
• **506.21**

3-POLE CB OPENS ALL
PHASE CONDUCTORS
SIMULTANEOUSLY
• **506.21, Ex.**

MULTIWIRE BRANCH-CIRCUITS
506.21

Purpose of Change: To include requirements for multiwire circuits installed in Zones 20 and 21 in the new **Article 506**.

NEC Ch. 5 – Article 506
Part – 506.25(A)

124.20

Type of Change	Panel Action	UL	UL 508	API 500	API 505	OSHA
New Article	Accept in Principle	2062	-	1.1.2	1.1.2	1910.307(b)(3), Note
ROP		ROC		NFPA 70E	NFPA 70B	NFPA 79	NFPA 499	NEMA

ROP		ROC		NFPA 70E	NFPA 70B	NFPA 79	NFPA 499	NEMA
pg. 1287	# 14-108a	pg. 407	# 14-97	440.4(A), FPN	22.1	3.3.53 and A.4.1	1.1.2	-
log: CP1409	CMP: 14	log: 1350	Submitter: CMP 14			2002 NEC: -		IEC: 1241-10

2005 NEC – 506.25 Grounding and Bonding

Grounding and bonding shall comply with Article 250 and the requirements in 506.25(A) and 506.26(B).

(A) Bonding. The locknut-bushing and double-locknut types of contacts shall not be depended on for bonding purposes, but bonding jumpers with proper fittings or other approved means of bonding shall be used. Such means of bonding shall apply to all intervening raceways, fittings, boxes, enclosures, and so forth, between Zone 20, Zone 21, and Zone 22 locations and the point of grounding for service equipment or point of grounding of a separately derived system.

Exception: The specific bonding means shall be required only to the nearest point where the grounded circuit conductor and the grounding electrode conductor are connected together on the line side of the building or structure disconnecting means as specified in 250.32(A), (B), and (C), if the branch side overcurrent protection is located on the load side of the disconnecting means.

FPN: See 250.100 for additional bonding requirements in hazardous (classified) locations.

Author's Substantiation. A new Article 506 has been added to recognize dust locations using the zone concept. This subsection covers the bonding requirements for dust locations using the zone concept.

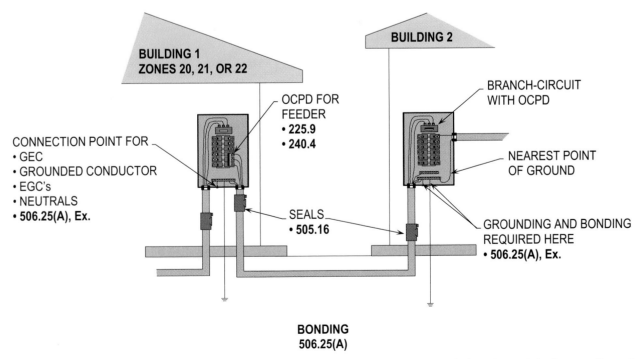

BONDING
506.25(A)

Purpose of Change: To verify that grounded system is only required to be bonded at the point of grounding where the grounded circuit conductor is connected to the grounding electrode of building 2 and not back to the grounding electrode of the first building.

NEC Ch. 5 – Article 506
Part – 506.25(B)

Type of Change	Panel Action	UL	UL 508	API 500	API 505	OSHA
New Article	Accept in Principle	2062	-	1.1.2	1.1.2	1910.307(b)(3), Note
ROP	**ROC**	**NFPA 70E**	**NFPA 70B**	**NFPA 79**	**NFPA 499**	**NEMA**
pg. 1287 # 14-108a	pg. 407 # 14-97	440.4(A), FPN	22.1	3.3.53 and A.4.1	1.1.2	-
log: CP1409 CMP: 14	log: 1350 Submitter: CMP 14			2002 NEC: -		IEC: 1241-10

2005 NEC – 506.25 Grounding and Bonding

Grounding and bonding shall comply with Article 250 and the requirements in 506.25(A) and 506.25(B).

(B) Types of Equipment Grounding Conductors. Where flexible conduit is used as permitted in 506.15, it shall be installed with internal or external bonding jumpers in parallel with each conduit and complying with 250.102.

Exception: In Zone 22 locations, the bonding jumper shall be permitted to be deleted where all of the following conditions are met:

(1) Listed liquidtight flexible metal conduit 1.8 m (6 ft) or less in length, with fittings listed for grounding, is used.

(2) Overcurrent protection in the circuit is limited to 10 amperes or less.

(3) The load is not a power utilization load.

Author's Substantiation. A new Article 506 has been added to recognize dust locations using the zone concept. This subsection covers the types of equipment grounding conductors permitted for dust locations using the zone concept.

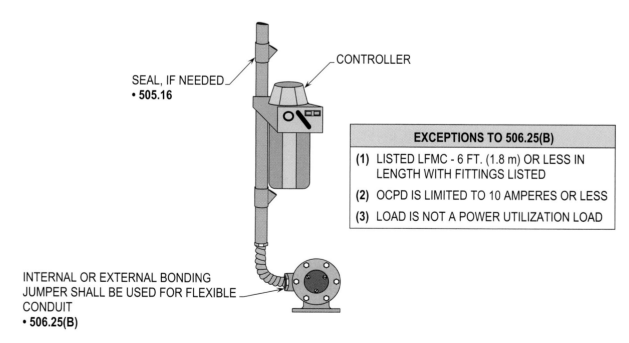

TYPES OF EQUIPMENT GROUNDING CONDUCTORS
506.25(B)

Purpose of Change: To recognize the types of equipment grounding conductors permitted for dust locations using the zone concept.

Type of Change	Panel Action	UL	UL 508	API 500	API 505	OSHA
Reorganized	Accept in Principle	-	-	-	-	1910.307(b)(3), Note
ROP	ROC	NFPA 70E	NFPA 70B	NFPA 79	NFPA	NEMA
pg. 1290 # 14-111	pg. 415 # 14-99	440.4(A), FPN	-	-	30A	-
log: 3354 CMP: 14	log: 3091 Submitter: Donald Cook			2002 NEC: 511.3(A)		IEC: 60079-19

2002 NEC – 511.3 Classification of Locations

(A) **Unclassified Locations.** Parking garages used for parking or storage ~~and where no repair work is done except exchange of parts and routine maintenance requiring no use of electrical equipment, open flame, welding, or the use of volatile flammable liquids are not classified.~~

The storage, handling, or dispensing into motor vehicles of alcohol-based windshield washer fluid in areas used for the service and repair operations of the vehicles shall not cause such areas to be classified as hazardous (classifed) locations.

FPN ~~No. 1: For further information, see NFPA 88A-1998, *Standards for Parking Structures* and NFPA 88B-1997, *Standard for Repair Garages.*~~

FPN ~~No. 2:~~ For further information, see ~~7.3.5~~ of NFPA 30A, *Code for Motor Fuel Dispensing Facilities and Repair Garages.*

2005 NEC – 511.3 Classification of Locations

(A) **Unclassified Locations**

(1) **Parking and Repair Garages.** Parking garages used for parking or storage shall be permitted to be unclassified. Repair garages shall be permitted to be unclassified when designed in accordance with 511.3(A)(2) through 511.3(A)(7).

FPN: For further information, see NFPA 88A-2002, *Standards for Parking Structures,* and NFPA 30A-2003, *Code for Motor Fuel Dispensing Facilities and Repair Garages.*

(2) **Alcohol-Based Windshield Washer Fluid.** The storage, handling, or dispensing into motor vehicles of alcohol-based windshield washer fluid in areas used for the service and repair operations of the vehicles shall not cause such areas to be classified as hazardous (classifed) locations.

FPN: For further information, see 8.3.5, Exception, of NFPA 30A-2003, *Code for Motor Fuel Dispensing Facilities and Repair Garages.*

(3) **Specific Areas Adjacent to Classified Locations.** Areas adjacent to classified locations in which flammable vapors are not likely to be released, such as stock rooms, switchboard rooms, and other similar locations, shall not be classified where mechanically ventilated at a rate of four or more air changes per hour, or designed with positive air pressure, or where effectively cut off by walls or partitions.

(4) **Pits in Lubrication or Service Room Where Class I Liquids Are Not Transferred.** Any pit, belowgrade work area, or subfloor work area that is provided with exhaust ventilation at a rate of not less than 0.3 m³/min/m² (1cfm/ft²) of floor area at all times that the building is occupied or when vehicles are parked in or over this area and where exhaust air is taken from a point within 300 mm (12 in.) of the floor of the pit, belowgrade work area, or subfloor work area is unclassified. [NFPA 30A:7.4.5.4 and Table 8.3.1]

(5) **Up to a Level of 450 mm (18 in.) Above the Floor in Lubrication or Service Rooms Where Class I Liquids Are Transferred.** For each floor, the entire area up to a level of 450 mm (18 in.) above the floor shall be considered unclassified where there is mechanical ventilation

providing a minimum of four air changes per hour or one cubic foot per minute of exchanged air for each square foot of floor area. Ventilation shall provide for air exchange across the entire floor area, and exhaust air shall be taken at a point within 0.3 m (12 in.) of the floor.

(6) Flammable Liquids Having Flashpoints Below 38°C (100°F). Where flammable liquids having a flash point below 38°C (100°F) (such as gasoline) or gaseous fuels (such as natural gas, hydrogen, or LPG) will not be transferred, such location shall be considered to be unclassified, unless the location is required to be classified in accordance with 511.3(B)(2) or (B)(4).

(7) Within 450 mm (18 in.) of the Ceiling. In major repair garages, where lighter-than-air gaseous fuels (such as natural gas or hydrogen) vehicles are repaired or stored, the area within 450 mm (18 in.) of the ceiling shall be considered unclassified where ventilation of at least 1 cfm/sq. ft. of ceiling area taken from a point within 450 mm (18 in.) of the highest point in the ceiling is provided.

FPN: For further information on the definition of *major repair garage*, see 3.3.12.1 of NFPA 30A-2003, *Code for Motor Fuel Dispensing Facilities and Repair Garages*.

Author's Substantiation. This section has been reorganized with title heads to recognize parking and repair garages, alcohol-based windshield washer fluid, specific area adjacent to classified locations, pits in lubrication or service room where Class I liquids are not transferred, up to a level of 450 mm (18 in.) above the floor in lubrication or service room where Class I liquids are transferred, flammable liquids having flashpoints below 38°C (100°F), and within 450 mm (18in.) of the ceiling for unclassified locations.

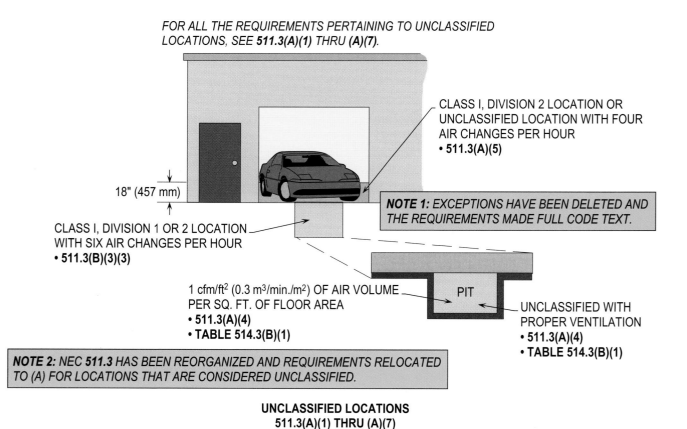

FOR ALL THE REQUIREMENTS PERTAINING TO UNCLASSIFIED LOCATIONS, SEE **511.3(A)(1)** THRU **(A)(7)**.

CLASS I, DIVISION 2 LOCATION OR UNCLASSIFIED LOCATION WITH FOUR AIR CHANGES PER HOUR
• **511.3(A)(5)**

18" (457 mm)

NOTE 1: EXCEPTIONS HAVE BEEN DELETED AND THE REQUIREMENTS MADE FULL CODE TEXT.

CLASS I, DIVISION 1 OR 2 LOCATION WITH SIX AIR CHANGES PER HOUR
• **511.3(B)(3)(3)**

1 cfm/ft² (0.3 m³/min./m²) OF AIR VOLUME PER SQ. FT. OF FLOOR AREA
• **511.3(A)(4)**
• **TABLE 514.3(B)(1)**

PIT

UNCLASSIFIED WITH PROPER VENTILATION
• **511.3(A)(4)**
• **TABLE 514.3(B)(1)**

*NOTE 2: NEC **511.3** HAS BEEN REORGANIZED AND REQUIREMENTS RELOCATED TO (A) FOR LOCATIONS THAT ARE CONSIDERED UNCLASSIFIED.*

UNCLASSIFIED LOCATIONS
511.3(A)(1) THRU (A)(7)

Purpose of Change: To reorganize and relocate unclassified requirements for locations in a commercial garage.

NEC Ch. 5 – Article 511
Part – 511.3(B)(1) through (B)(4)

Type of Change	Panel Action	UL	UL 508	API 500	API 505	OSHA
Reorganized	Accept in Principle	-	-	-	-	1910.307(b)(3), Note
ROP	**ROC**	**NFPA 70E**	**NFPA 70B**	**NFPA 79**	**NFPA**	**NEMA**
pg. 1290	# 14-111	pg. 415	# 14-99	440.4(A), FPN	-	-	30A	-
log: 3354	CMP: 14	log: 3091	Submitter: Donald Cook	2002 NEC: 511.3(B)	IEC: 60079-19

2002 NEC – 511.3(B) Classified Locations

~~Classification shall be in accordance with Article 500.~~ Areas in which flammable fuel is ~~transferred to~~ vehicle fuel tanks shall ~~also~~ conform to Article 514.

(1) **Up to a Level of 450 mm (18 in.) Above the Floor.** For each floor, the entire area up to a level of 450 mm (18 in.) above the floor shall be ~~considered to be~~ a Class I, Division 2 location.

   Exception: Where the enforcing agency determines that there is mechanical ventilation providing a minimum of four air changes per hour or one cubic foot per minute of exchanged air for each square foot of floor area. Ventilation shall provide for air exchange across the entire floor area within 0.3 m (12 in.) of the floor.

(2) **Within 457 mm (18 in.) of the Ceiling.** ~~Where compressed natural gas (CNG) vehicles are repaired or stored, the area within 457 mm (18 in.) of the ceiling shall be classified as Class I, Division 2, except where ventilation of at least four air changes per hour is provided. [NFPA 88B, 3-1.1]~~

(3) **Any Pit or Depression Below Floor Level.** Any pit or depression below floor level shall be ~~considered to be~~ a Class I, Division 1 location and shall extend up to said floor level.

   ~~Exception No. 1:~~ Any pit or depression in which six air changes per hour are exhausted ~~at~~ the floor level of the pit shall ~~be permitted to be judged by the enforcing agency to~~ be a Class I, Division 2 location.

   ~~Exception No. 2: Lubrication and service rooms without dispensing shall be classified in accordance with Table 514.3(B)(1).~~

(4) **~~Areas Adjacent to Defined Locations or with Positive-Pressure Ventilation. Areas adjacent to defined locations in which flammable vapors are not likely to be released, such as stock rooms, switchboard rooms, and other similar locations, shall not be classified where mechanically ventilated at a rate of four or more air changes per hour or where effectively cut off by walls or partitions.~~**

(5) **~~Adjacent Areas by Special Permission. Adjacent areas that by reason of ventilation, air pressure differentials, or physical spacing are such that, in the opinion of the authority enforcing this Code, no ignition hazards exists, shall be unclassified.~~**

2005 NEC – 511.3(B) Classified Locations

(1) <u>**Flammable Fuel Dispensing Areas.**</u> Areas in which flammable fuel is <u>dispensed into</u> vehicle fuel tanks shall conform to Article 514.

(2) <u>**Lubrication of Service Room Where Class I Liquids or Gaseous Fuels (Such as Natural Gas, Hydrogen, or LPG) Are Not Transferred.**</u> <u>The following spaces that are not designed in accordance with 511.3(A)(4) shall be classified as Class I, Division 2:</u>

   (1) <u>Entire area within any unventilated pit, belowgrade work area, or subfloor area.</u>

   (2) <u>Area up to 450 mm (18 in.) above any such unventilated pit, belowgrade work area, or subfloor work area and extending a distance of 900 mm (3 ft) horizontally from the edge of any such pit, belowgrade work area, or subfloor work area.</u>

(3) Lubrication or Service Room Where Class I Liquids or Gaseous Fuels (Such as Natural Gas, Hydrogen, or LPG) Are Transferred. The following spaces that are not designed in accordance with 511.3(A)(5) shall be classified as follows:

(1) Up to a Level of 450 mm (18 in.) Above the Floor. For each floor, the entire area up to a level of 450 mm (18 in.) above the floor shall be a Class I, Division 2 location.

(2) Any Unventilated Pit or Depression Below Floor Level. Any unventilated pit or depression below floor level shall be a Class I, Division 1 location and shall extend up to said floor level.

(3) Any Ventilated Pit or Depression Below Floor Level. Any ventilated pit or depression in which six air changes per hour are exhausted from a point within 300 mm (12 in.) of the floor level of the pit shall be a Class I, Division 2 location.

(4) Space Above an Unventilated Pit or Depression Below Floor Level. Above a pit, or depression below floor level, the space up to 450 mm (18 in.) above the floor or grade level and 900 mm (3 ft) horizontally from a lubrication pit shall be a Class I, Division 2 location.

(5) Dispenser for Class I liquids, Other Than Fuels. Within 900 mm (3 ft) of any fill or dispensing point, extending in all directions shall be a Class I, Division 2 location. See also 511.3(B)(1).

(4) Within 450 mm (18 in.) of the Ceiling. In major repair garages where lighter-than-air gaseous fuel (such as natural gas or hydrogen) vehicles are repaired or stored, ceiling spaces that are not designed in accordance with 511.3(A)(7) shall be classified as Class I, Division 2.

FPN: For further information on the definition of *major repair garage*, see 3.3.12.1 of NFPA 30A, 2003, *Code for Motor Fuel Dispensing Facilities and Repair Garages*.

Author's Substantiation. This section has been reorganized with title heads to recognize flammable fuel dispensing areas, lubrication or service rooms where Class I liquids or gaseous fuels (such as natural gas, hydrogen or LPG) are not transferred, lubrication or service rooms where Class I liquids or gaseous fuels (such as natural gas, hydrogen or LPG) are transferred, and within 450 mm (18 in.) of the ceiling in classified locations.

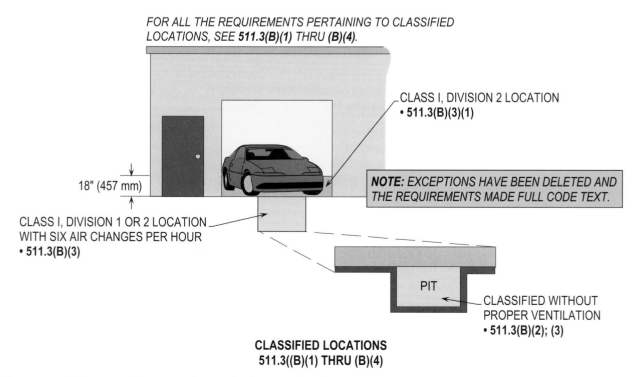

FOR ALL THE REQUIREMENTS PERTAINING TO CLASSIFIED LOCATIONS, SEE **511.3(B)(1)** THRU **(B)(4)**.

CLASS I, DIVISION 2 LOCATION
• **511.3(B)(3)(1)**

18" (457 mm)

NOTE: EXCEPTIONS HAVE BEEN DELETED AND THE REQUIREMENTS MADE FULL CODE TEXT.

CLASS I, DIVISION 1 OR 2 LOCATION WITH SIX AIR CHANGES PER HOUR
• **511.3(B)(3)**

PIT

CLASSIFIED WITHOUT PROPER VENTILATION
• **511.3(B)(2); (3)**

CLASSIFIED LOCATIONS
511.3((B)(1) THRU (B)(4)

Purpose of Change: To reorganize and relocate classified requirements for locations in a commercial garage.

NEC Ch. 5 – Article 511
Part – 511.4(A)(1), Ex.

127

Type of Change		Panel Action		UL	UL 508	API 500	API 505	OSHA
Deletion		Accept in Principle in Part		-	-	-	-	1910.307(b)(3), Note
ROP		ROC		NFPA 70E	NFPA 70B	NFPA 79	NFPA	NEMA
pg. 1293	# 14-115	pg. -	# -	440.4(A), FPN	-	-	30A	-
log: 1411	CMP: 14	log: -	Submitter: TC of Automotive and Marine Service Stations			2002 NEC: 511.4(A)(1) and Ex.		IEC: 60079-19

2002 NEC – 511.4(A) Wiring Located in Class I Locations

Within Class I locations as classified in 511.3, wiring shall conform to applicable provisions of Article 501.

(1) ~~Raceways embedded in a masonry wall or buried beneath a floor shall be considered to be within the Class I location above the floor if any connections or extensions lead into or through such areas.~~

~~**Exception:** Rigid nonmetallic conduit that complies with Article 352 shall be permitted where buried under not less than 600 mm (24 in.) of cover. Where rigid nonmetallic conduit is used, threaded rigid metal conduit or threaded steel intermediate metal conduit shall be used for the last 600 mm (24 in.) of the underground run to emergence or to the point of connection to the aboveground raceway and an equipment grounding conductor shall be included to provide electrical continuity of the raceway system and for grounding of non-current-carrying metal parts.~~

Author's Substantiation. This subsection and exception have been deleted to clarify that raceways and wiring shall conform to the applicable provision of Article 501 where installed within Class I locations.

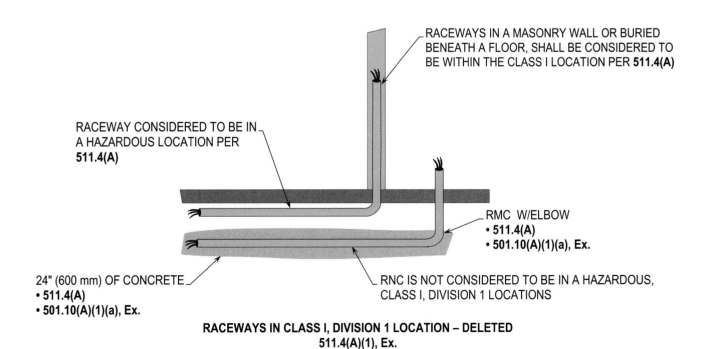

RACEWAYS IN A MASONRY WALL OR BURIED BENEATH A FLOOR, SHALL BE CONSIDERED TO BE WITHIN THE CLASS I LOCATION PER **511.4(A)**

RACEWAY CONSIDERED TO BE IN A HAZARDOUS LOCATION PER **511.4(A)**

RMC W/ELBOW
• **511.4(A)**
• **501.10(A)(1)(a), Ex.**

24" (600 mm) OF CONCRETE
• **511.4(A)**
• **501.10(A)(1)(a), Ex.**

RNC IS NOT CONSIDERED TO BE IN A HAZARDOUS, CLASS I, DIVISION 1 LOCATIONS

RACEWAYS IN CLASS I, DIVISION 1 LOCATION – DELETED
511.4(A)(1), Ex.

Purpose of Change: To delete the requirements for raceway and wiring to be considered in Class I, Division 1 locations under certain conditions of use.

NEC Ch. 5 – Article 513
Part – 513.7(A), Ex.

Type of Change		Panel Action		UL	UL 508	API 500	API 505	OSHA
Revision		Accept		2225	-	-	-	1910.307(b)(3), Note
ROP		ROC		NFPA 70E	NFPA 70B	NFPA 79	NFPA	NEMA
pg. 1299	# 14-126	pg. -	# -	440.4(A), FPN	-	-	409	-
log: 796	CMP: 14	log: -	Submitter: Dan Leaf			2002 NEC: 513.7(A), Ex.		IEC: 60079-19

2002 NEC – 513.7(A) Fixed Wiring

All fixed wiring in a hangar but not installed in a Class I location as classified in 513.3 shall be installed in metal raceways or shall be Type MI, TC, or MC cable.

> **Exception:** Wiring in unclassified locations, as ~~classified~~ in 513.3(D), shall be ~~of a~~ type recognized in Chapter 3.

2005 NEC – 513.7(A) Fixed Wiring

All fixed wiring in a hangar but not installed in a Class I location as classified in 513.3 shall be installed in metal raceways or shall be Type MI, TC, or MC cable.

> **Exception:** Wiring in unclassified locations, as <u>described</u> in 513.3(D), shall be <u>permitted to be any suitable</u> type <u>wiring method</u> recognized in Chapter 3.

Author's Substantiation. This revision clarifies the types of wiring methods that shall be permitted to be installed in unclassified locations.

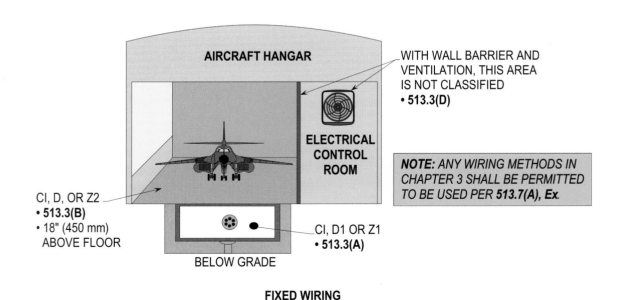

FIXED WIRING
513.7(A), Ex.

Purpose of Change: To clarify the types of wiring methods that shall be permitted to be used in unclassified locations.

NEC Ch. 5 – Article 513
Part – 513.12

128

Type of Change	Panel Action	UL	UL 508	API 500	API 505	OSHA
New Section	Accept in Principle	943	-	-	-	1910.307(b)(3), Note
ROP	ROC	NFPA 70E	NFPA 70B	NFPA 79	NFPA	NEMA
pg. 1302 # 14-132	pg. - # -	440.4(A), FPN	-	-	409	-
log: 1862 CMP: 14	log: -	Submitter: Andre R. Cartal		2002 NEC: -		IEC: 60079-19

2005 NEC – 513.12 Ground-Fault Circuit-Interrupter Protection for Personnel

All 125-volt, 50/60 Hz, single phase, 15- and 20-ampere receptacles installed in areas where electrical diagnostic equipment, electrical hand tools, or portable lighting equipment are to be used shall have ground-fault circuit-interrupter protection for personnel.

Author's Substantiation. A new section has been added to recognize ground-fault circuit-interrupter (GFCI) protection for personnel, hand tools, and equipment used in aircraft hangars. GFCI protection in the 50/60 Hz range has been recognized to avoid confusion with 400 Hz equipment commonly used on aircraft.

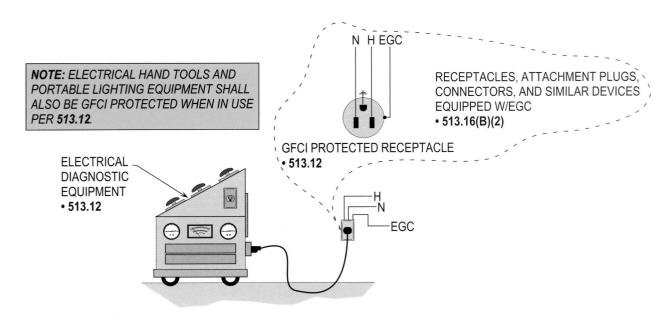

GFCI PROTECTION FOR PERSONNEL
513.12

Purpose of Change: To provide GFCI protection for personnel and tools and equipment in aircraft hangars.

NEC Ch. 5 – Article 514
Part – 514.1

Type of Change		Panel Action		UL	UL 508	API 500	API 505	OSHA
Revision		Accept in Principle		1130	-	-	-	1910.307(b)(3), Note
ROP		ROC		NFPA 70E	NFPA 70B	NFPA 79	NFPA	NEMA
pg. 1303	# 14-134	pg. -	# -	440.4(A), FPN	-	-	30A and 303	-
log: 1131	CMP: 14	log: -	Submitter: Jon Nisja			2002 NEC: 514.1		IEC: 60079-20

2002 NEC – 514.1 Scope

~~These occupancies shall include locations where gasoline or other volatile flammable liquids or liquefied flammable gases are transferred to the fuel tanks (including auxiliary fuel tanks) of self-propelled vehicles or approved containers.~~

2005 NEC – 514.1 Scope

This article shall apply to motor fuel dispensing facilities, marine/motor fuel dispensing facilities, motor fuel dispensing facilities located inside buildings, and fleet vehicle motor fuel dispensing facilities.

Author's Substantiation. This section has been revised to clarify that motor fuel dispensing facilities, marine/motor fuel dispensing facilities, motor fuel dispensing facilities located inside buildings, and fleet vehicle motor fuel dispensing facilities are covered in this article.

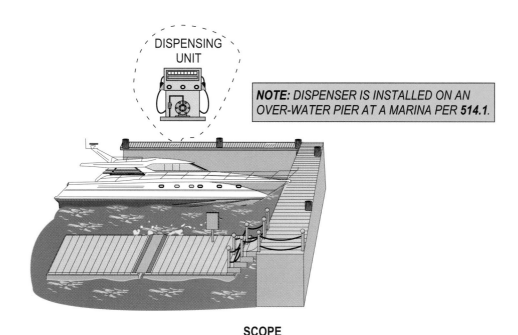

DISPENSING UNIT

NOTE: DISPENSER IS INSTALLED ON AN OVER-WATER PIER AT A MARINA PER **514.1**.

SCOPE
514.1

Purpose of Change: To provide requirements for dispensers installed on over-water piers.

NEC Ch. 5 – Article 514
Part – 514.13

Type of Change	Panel Action	UL	UL 508	API 500	API 505	OSHA
New Sentence	Accept in Principle	1130	-	-	-	1910.307(b)(3), Note
ROP		ROC		NFPA 70E	NFPA 70B	NFPA 79	NFPA	NEMA

ROP		ROC		NFPA 70E	NFPA 70B	NFPA 79	NFPA	NEMA
pg. 1308	# 14-142	pg. 418	# 14-113	440.4(A), FPN	-	-	30A	-
log: 2817	CMP: 14	log: 125	Submitter: David Shapiro			2002 NEC: 514.13		IEC: 60079-20

2002 NEC – 514.13 Provisions for Maintenance and Service of Dispensing Equipment

Each dispensing device shall be provided with a means to remove all external voltage sources, including feedback, during periods of maintenance and service of the dispensing equipment. The location of this means shall be permitted to be other than inside or adjacent to the dispensing device.

2005 NEC – 514.13 Provisions for Maintenance and Service of Dispensing Equipment

Each dispensing device shall be provided with a means to remove all external voltage sources, including feedback, during periods of maintenance and service of the dispensing equipment. The location of this means shall be permitted to be other than inside or adjacent to the dispensing device. <u>The means shall be capable of being locked in the open position.</u>

Author's Substantiation. A new sentence has been added to recognize that a means of disconnecting external voltage sources shall be capable of being locked in the open position.

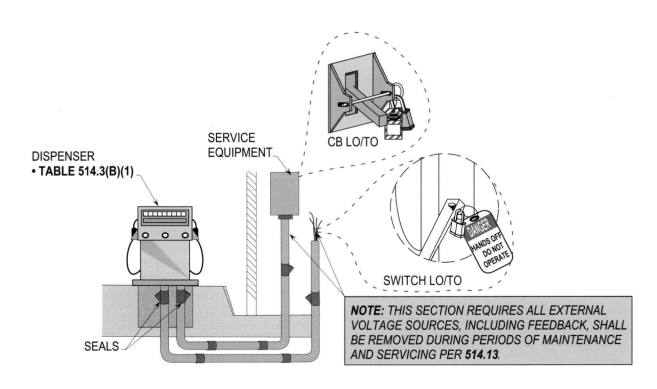

PROVISIONS FOR MAINTENANCE AND SERVICE OF DISPENSING EQUIPMENT
514.13

Purpose of Change: To require a means for locking out and tagging out hazardous energy to dispenser equipment.

NEC Ch. 5 – Article 516
Part – 516.3(B)(4)

Type of Change		Panel Action		UL	UL 508	API 500	API 505	OSHA
Revision		Accept		-	-	-	-	1910.307(b)(3), Note
ROP		ROC		NFPA 70E	NFPA 70B	NFPA 79	NFPA	NEMA
pg. 1310	# 14-145a	pg. 419	# 14-115	440.4(A), FPN	-	-	33 and 34	-
log: CP1413	CMP: 14	log: 799	Submitter: CMP 14			2002 NEC: 516.3(B)(4)		IEC: 60079

2002 NEC – ~~516.3(A)~~ Class I or Class II, Division 1 Locations

The following spaces shall be considered Class I or Class II, Division 1 locations, as applicable.

(4) For dipping and coating operations, all space within a 1.5-m (5-ft) radial distance from the vapor sources extending from these surfaces to the floor. The vapor source shall be the liquid exposed in the process and drainboard, and any dipped or coated object from which it is possible to measure vapor concentrations exceeding 25 percent of the lower flammable limit at a distance of 300 mm (1 ft), in any direction, from the object.

2005 NEC – 516.3(B) Class I or Class II, Division 1 Locations

The following spaces shall be considered Class I, Division 1, or Class I, Zone 1, or Class II, Division 1 locations, as applicable:

(4) For open dipping and coating operations, all space within a 1.5-m (5-ft) radial distance from the vapor sources extending from these surfaces to the floor. The vapor source shall be the liquid exposed in the process and drainboard, and any dipped or coated object from which it is possible to measure vapor concentrations exceeding 25 percent of the lower flammable limit at a distance of 300 mm (1 ft), in any direction, from the object.

Author's Substantiation. This revision clarifies that all space within a 1.5-m (5-ft) radial distance for open dipping and coating operations shall be considered Class I, Division 1, or Class I, Zone 1, or Class II, Division 1 locations.

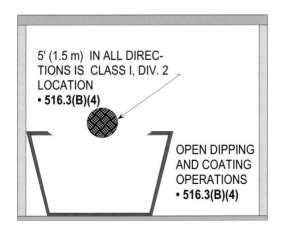

5' (1.5 m) IN ALL DIRECTIONS IS CLASS I, DIV. 2 LOCATION
• 516.3(B)(4)

OPEN DIPPING AND COATING OPERATIONS
• 516.3(B)(4)

CLASS I OR CLASS II, DIVISION 1 LOCATIONS
516.3(B)(4)

Purpose of Change: To clarify that open dipping and coating operations within a 1.5 m (5 ft) radial distance shall be considered Class I, Division 1, or Class I, Zone 1, or Class II, Division 1 locations.

NEC Ch. 5 – Article 516
Part – 516.3(B)(6)

Type of Change		Panel Action		UL	UL 508	API 500	API 505	OSHA
Revision		Accept		-	-	-	-	1910.307(b)(3), Note
ROP		ROC		NFPA 70E	NFPA 70B	NFPA 79	NFPA	NEMA
pg. 1310	# 14-145a	pg. 419	# 14-115	440.4(A), FPN	-	-	33 and 34	-
log: CP1413	CMP: 14	log: 799	Submitter: CMP 14			2002 NEC: 516.3(B)(6)		IEC: 60079

2002 NEC - ~~516.3(A)~~ Class I or Class II, Division 1 Locations

The following spaces shall be considered Class I or Class II, Division 1 locations, as applicable.

(6) ~~The interior of any enclosed dipping or coating process or apparatus. [NFPA 34, 1.6, Definitions, 4.2.1, 4.2.2, 4.3.1]~~

2005 NEC - 516.3(B) Class I or Class II, Division 1 Locations

The following spaces shall be considered Class I, Division 1, or Class I, Zone 1, or Class II, Division 1 locations, as applicable:

(6) All space in all directions outside of but within 900 mm (3 ft) of open containers, supply containers, spray gun cleaners, and solvent distillation units containing flammable liquids.

Author's Substantiation. This revision clarifies the set boundaries around containers that are filled with flammable liquids.

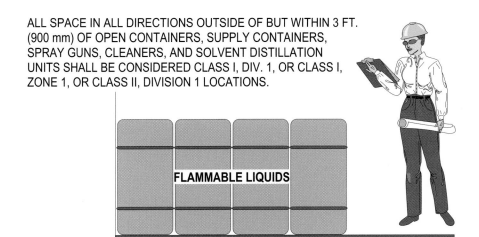

ALL SPACE IN ALL DIRECTIONS OUTSIDE OF BUT WITHIN 3 FT. (900 mm) OF OPEN CONTAINERS, SUPPLY CONTAINERS, SPRAY GUNS, CLEANERS, AND SOLVENT DISTILLATION UNITS SHALL BE CONSIDERED CLASS I, DIV. 1, OR CLASS I, ZONE 1, OR CLASS II, DIVISION 1 LOCATIONS.

FLAMMABLE LIQUIDS

**CLASS I OR CLASS II, DIVISION 1 LOCATIONS
516.3(B)(6)**

Purpose of Change: To set boundaries around containers that are filled with flammable liquids.

NEC Ch. 5 – Article 516
Part – 516.16

Type of Change	Panel Action	UL	UL 508	API 500	API 505	OSHA
Revision	Accept	467	-	1.1.2	1.1.2	1910.307(b)(3), Note
ROP	**ROC**	**NFPA 70E**	**NFPA 70B**	**NFPA 79**	**NFPA**	**NEMA**
pg. 1310	# 14-145a	pg. -	# -	440.4(A), FPN	-	-	33 and 34	-
log: CP1413	CMP: 14	log: -	Submitter: CMP 14		2002 NEC: 516.16		IEC: 60079

2002 NEC – 516.16 Grounding

All metal raceways, the metal armors or metallic sheath on cables, and all non-current-carrying metal parts of fixed or portable electrical equipment, regardless of voltage, shall be grounded as provided in Article 250. Grounding ~~in Class I and Class II locations shall comply with 501.16 and 502.16, respectively.~~

2005 NEC – 516.16 Grounding

All metal raceways, the metal armors or metallic sheath on cables, and all non-current-carrying metal parts of fixed or portable electrical equipment, regardless of voltage, shall be grounded as provided in Article 250. Grounding <u>shall comply with 501.30, 502.30, or 505.25, as applicable</u>.

Author's Substantiation. This revision clarifies that users shall comply with 501.30, 502.30, or 505.25 for additional grounding requirements.

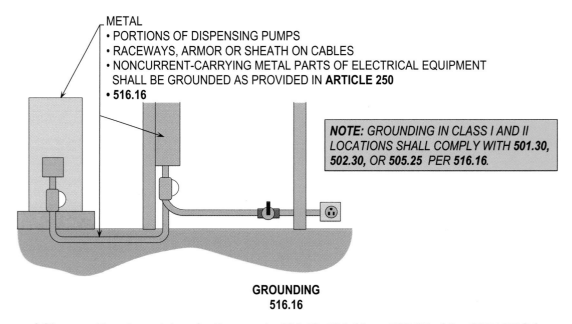

**GROUNDING
516.16**

Purpose of Change: To adequately refer the user to **501.30, 502.30,** or **505.25** of the **2005 NEC** for additional requirements.

NEC Ch. 5 – Article 517
Part II – 517.13(B)

Type of Change		Panel Action		UL	UL 508	API 500	API 505	OSHA
Revision		Accept		467	-	-	-	1910.305(a)(1)(i)
ROP		ROC		NFPA 70E	NFPA 70B	NFPA 79	NFPA	NEMA
pg. 1320	# 15-17	pg. -	# -	410.2(B)(3)	-	-	99	-
log: 2694	CMP: 15	log: -	Submitter: Phil Simmons			2002 NEC: 517.13(B)		IEC: 413.5

2002 NEC – 517.13(B) Insulated Equipment Grounding Conductor

~~In an area used for patient care,~~ the grounding terminals of all receptacles and all non-current-carrying conductive surfaces of fixed electric equipment likely to become energized that are subject to personal contact, operating at over 100 volts, shall be grounded by an insulated copper conductor. The grounding conductor shall be sized in accordance with Table 250.122 and installed in metal raceways or ~~metal-clad cables~~ with the branch-circuit conductors supplying these receptacles or fixed equipment.

2005 NEC – 517.13(B) Insulated Equipment Grounding Conductor

The grounding terminals of all receptacles and all non-current-carrying conductive surfaces of fixed electric equipment likely to become energized that are subject to personal contact, operating at over 100 volts, shall be grounded by an insulated copper conductor. The <u>equipment</u> grounding conductor shall be sized in accordance with Table 250.122 and installed in metal raceways or as a <u>part of listed cables having a metallic armor or sheath assembly</u> with the branch-circuit conductors supplying these receptacles or fixed equipment.

Author's Substantiation. This revision removes any redundant words or phrases and clarifies the grounding requirements in patient care areas.

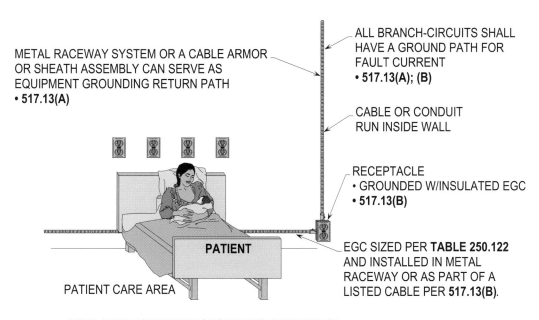

INSULATED EQUIPMENT GROUNDING CONDUCTOR
517.13(B)

Purpose of Change: To more appropriately address the grounding requirements in patient care areas.

NEC Ch. 5 – Article 517
Part II – 517.16, FPN

Type of Change		Panel Action		UL	UL 508	API 500	API 505	OSHA
New Sentence		Accept in Principle		1047	-	-	-	1910.304(f)(4)
ROP		ROC		NFPA 70E	NFPA 70B	NFPA 79	NFPA	NEMA
pg. 1323	# 15-24	pg. 422	# 15-12	420.1(A)(2)	-	-	99	-
log: 3435	CMP: 15	log: 805	Submitter: Robert Schuerger			2002 NEC: 517.16		IEC: 413.5

2002 NEC – 517.16 Receptacles with Insulated Grounding Terminals

Receptacles with insulated grounding terminals, as permitted in 250.146(D), shall be identified; such identification shall be visible after installation.

> **FPN:** Caution is important in specifying such a system with receptacles having insulated grounding terminals, since the grounding impedance is controlled only by the grounding conductors and does not benefit functionally from any parallel grounding paths.

2005 NEC – 517.16 Receptacles with Insulated Grounding Terminals

Receptacles with insulated grounding terminals, as permitted in 250.146(D), shall be identified; such identification shall be visible after installation.

> **FPN:** Caution is important in specifying such a system with receptacles having insulated grounding terminals, since the grounding impedance is controlled only by the equipment grounding conductors and does not benefit functionally from any parallel grounding paths. This type of installation is typically used where a reduction of electrical noise (electromagnetic interference) is necessary and parallel grounding paths are to be avoided.

Author's Substantiation. A new sentence has been added to clarify the benefits of using an isolation receptacle.

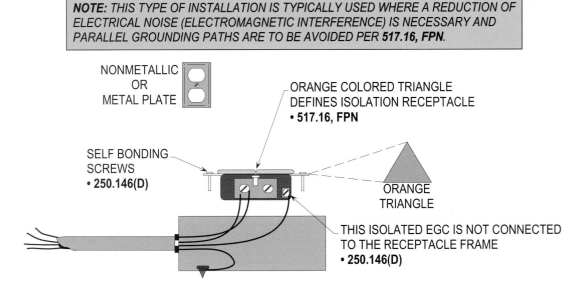

NOTE: THIS TYPE OF INSTALLATION IS TYPICALLY USED WHERE A REDUCTION OF ELECTRICAL NOISE (ELECTROMAGNETIC INTERFERENCE) IS NECESSARY AND PARALLEL GROUNDING PATHS ARE TO BE AVOIDED PER **517.16, FPN**.

NONMETALLIC OR METAL PLATE

ORANGE COLORED TRIANGLE DEFINES ISOLATION RECEPTACLE
• 517.16, FPN

SELF BONDING SCREWS
• 250.146(D)

ORANGE TRIANGLE

THIS ISOLATED EGC IS NOT CONNECTED TO THE RECEPTACLE FRAME
• 250.146(D)

RECEPTACLES WITH INSULATED GROUNDING TERMINALS
517.16, FPN

Purpose of Change: To clarify the benefits of using an isolation receptacle.

NEC Ch. 5 – Article 517
Part II – 517.17(A)

131

Type of Change	Panel Action	UL	UL 508	API 500	API 505	OSHA
New Subsection	Accept in Principle in Part	1053	-	-	-	1910.304(e)(1)(i)
ROP		ROC		NFPA 70E	NFPA 70B	NFPA 79	NFPA	NEMA

ROP		ROC		NFPA 70E	NFPA 70B	NFPA 79	NFPA	NEMA
pg. 1324	# 15-24a	pg. 423	# 15-14	410.9(A)(1)	-	-	99	AB 4
log: CP1500	CMP: 15	log: 767	Submitter: CMP 15			2002 NEC: -		IEC: 314.2

2005 NEC – 517.17(A) Applicability

The requirements of 517.17 shall apply to hospitals and other buildings (including multiple occupancy buildings) with critical care areas or utilizing electrical life support equipment, and buildings that provide the required essential utilities or services for the operation of critical care areas or electrical life support equipment.

Author's Substantiation. A new subsection has been added to clearly identify when and where a second level of ground-fault protection is needed to prevent nuisance tripping of the main.

GROUND-FAULT PROTECTION SHALL BE APPLIED TO THE FOLLOWING:
(1) HOSPITALS
(2) OTHER BUILDINGS WITH CRITICAL CARE AREAS UTILIZING ELECTRICAL SUPPORT EQUIPMENT PER **517.17(A)**.

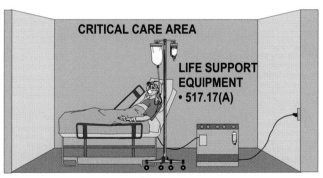

HOSPITALS AND OTHER BUILDINGS INCLUDE MULTIPLE OCCUPANCIES
• **517.17(A)**

APPLICABILITY
517.17(A)

Purpose of Change: To provide a section of applicability pertaining to ground-fault protection in **517.17(B)** through **(D)** of the **2005 NEC**.

NEC Ch. 5 – Article 517
Part II – 517.18(C)

Type of Change		Panel Action		UL	UL 508	API 500	API 505	OSHA
Revision		Accept in Principle		514(D)	-	-	-	-
ROP		ROC		NFPA 70E	NFPA 70B	NFPA 79	NFPA	NEMA
pg. 1328	# 15-33	pg. -	# -	-	-	-	99	-
log: 3403	CMP: 15	log: -	Submitter: Lee A. Mickael			2002 NEC: 517.18(C)		IEC: 413.5

2002 NEC – 517.18(C) Pediatric Locations

Receptacles located within the ~~patient care areas of pediatric wards, rooms, or areas~~ shall be listed tamper resistant or shall employ a listed tamper resistant cover.

2005 NEC – 517.18(C) Pediatric Locations

Receptacles located within the <u>rooms, bathrooms, playrooms, activity rooms, and patient care areas of pediatric wards</u> shall be listed tamper resistant or shall employ a listed tamper resistant cover.

Author's Substantiation. This revision clarifies the areas where receptacles shall be tamper resistant or shall employ a listed tamper resistant cover.

RECEPTACLES WITHIN ROOMS, BATHROOMS, PLAYROOMS, ACTIVITY
ROOMS, AND PATIENT CARE AREAS OF PEDIATRIC WARDS
• **517.18(C)**

LISTED TAMPER
RESISTANT
RECEPTACLE

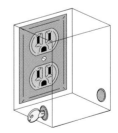

LISTED TAMPER
RESISTANT-TYPE
COVER

PEDIATRIC LOCATIONS
517.18(C)

Purpose of Change: To require all receptacles to be tamper resistant or to have tamper resistant covers in particular areas of general care.

Type of Change	Panel Action	UL	UL 508	API 500	API 505	OSHA
Revision	Accept	467	-	-	-	-

ROP		ROC		NFPA 70E	NFPA 70B	NFPA 79	NFPA	NEMA
pg. 1329	# 15-35a	pg. -	# -	-	-	-	99	-
log: CP1501	CMP: 15	log: -	-	Submitter: CMP 15		2002 NEC: 517.19(C)		IEC: 413.5

2002 NEC – 517.19(C) Patient Vicinity Grounding and Bonding (Optional)

A patient vicinity shall be permitted to have a patient equipment grounding point. The patient equipment grounding point, where supplied, shall be permitted to contain one or more ~~jacks listed for the purpose~~. An equipment bonding jumper not smaller than 10 AWG shall be used to connect the grounding terminal of all grounding-type receptacles to the patient equipment grounding point. The bonding conductor shall be permitted to be arranged centrically or looped as convenient.

2005 NEC – 517.19(C) Patient Vicinity Grounding and Bonding (Optional)

A patient vicinity shall be permitted to have a patient equipment grounding point. The patient equipment grounding point, where supplied, shall be permitted to contain one or more <u>listed grounding and bonding jacks</u>. An equipment bonding jumper not smaller than 10 AWG shall be used to connect the grounding terminal of all grounding-type receptacles to the patient equipment grounding point. The bonding conductor shall be permitted to be arranged centrically or looped as convenient.

Author's Substantiation. This revision improves the usability of this subsection and clarifies that grounding and bonding jacks shall be listed.

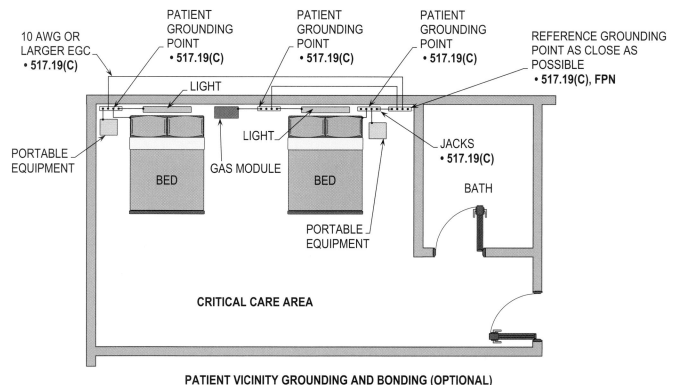

PATIENT VICINITY GROUNDING AND BONDING (OPTIONAL)
517.19(C)

Purpose of Change: To revise and delete confusing wording in an effort to make this section more user friendly.

NEC Ch. 5 – Article 517
Part II – 517.19(E)

Type of Change		Panel Action		UL	UL 508	API 500	API 505	OSHA
Revision		Accept		1047	-	-	-	-
ROP		ROC		NFPA 70E	NFPA 70B	NFPA 79	NFPA	NEMA
pg. 1329	# 15-35b	pg. -	# -	-	-	-	99	-
log: CP1502	CMP: 15	log: -	Submitter: CMP 15			2002 NEC: 517.19(E)		IEC: 413.5

2002 NEC – 517.19(E) Additional Protective Techniques in Critical Care Areas (Optional)

Isolated power systems shall be permitted to be used for critical care areas, and, if used, the isolated power system equipment shall be listed ~~for the purpose and~~ the system designed and installed ~~so that it meets the provisions of and is~~ in accordance with 517.160.

2005 NEC – 517.19(E) Additional Protective Techniques in Critical Care Areas (Optional)

Isolated power systems shall be permitted to be used for critical care areas, and, if used, the isolated power system equipment shall be listed <u>as isolated power equipment</u>. The <u>isolated power</u> system shall be designed and installed in accordance with 517.160.

Author's Substantiation. This revision improves the usability of this subsection and clarifies that the isolated power system equipment shall be listed as isolated power equipment.

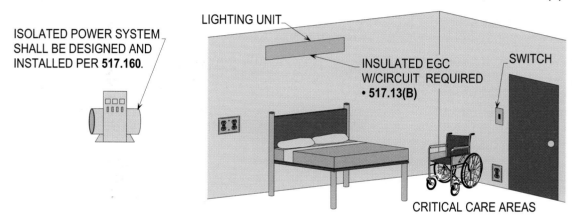

FOR A SIMILAR REQUIREMENT, SEE **517.20(B)**.

LIGHTING UNIT

ISOLATED POWER SYSTEM
SHALL BE DESIGNED AND
INSTALLED PER **517.160**.

INSULATED EGC
W/CIRCUIT REQUIRED
• **517.13(B)**

SWITCH

CRITICAL CARE AREAS

ADDITIONAL PROTECTIVE TECHNIQUES IN CRITICAL CARE AREAS (OPTIONAL)
517.19(E)

Purpose of Change: To provide requirements for installing isolated power systems in critical care areas.

NEC Ch. 5 – Article 517
Part III – 517.30(C)(3)(1) through (5)

Type of Change	Panel Action	UL	UL 508	API 500	API 505	OSHA
Revision	Accept in Principle	-	-	-	-	-
ROP		ROC		NFPA 70E	NFPA 70B	NFPA 79	NFPA	NEMA

ROP		ROC		NFPA 70E	NFPA 70B	NFPA 79	NFPA	NEMA
pg. 1333	# 15-42	pg. 425	# 15-25	-	-	-	99	-
log: 2695	CMP: 15	log: 3608	Submitter: Phil Simmons			2002 NEC: 517.30(C)(3)		IEC: 413.5

2002 NEC – 517.30(C)(3) Mechanical Protection of the Emergency System

The wiring of the emergency system of a hospital shall be mechanically protected by installation in nonflexible metal raceways, or shall be wired with Type MI cable.

Exception No. 1: Flexible power cords of appliances, or other utilization equipment, connected to the emergency system shall not be required to be enclosed in raceways.

Exception No. 2: Secondary circuits of transformer-powered communications or signaling systems shall not be required to be enclosed in raceways unless otherwise specified by Chapters 7 or 8.

Exception No. 3: Schedule 80 rigid nonmetallic conduit shall be permitted if the branch circuits do not serve patient care areas and it is not prohibited elsewhere in this Code.

Exception No. 4: Where encased in not less than 50 mm (2 in.) of concrete, Schedule 40 rigid nonmetallic conduit or electrical nonmetallic tubing shall be permitted if the branch circuits do not serve patient care areas.

Exception No. 5: Flexible metal raceways and cable assemblies shall be permitted to be used in listed prefabricated medical headwalls, listed office furnishings, or where necessary for flexible connection to equipment.

2005 NEC – 517.30(C)(3) Mechanical Protection of the Emergency System

The wiring of the emergency system in hospitals shall be mechanically protected. Where installed as branch circuits in patient care areas, the installation shall comply with the requirements of 517.13(A) and 517.13(B). The following wiring methods shall be permitted.

(1) Nonflexible metal raceways, Type MI cable, or Schedule 80 rigid nonmetallic conduit. Nonmetallic raceways shall not be used for branch circuits that supply patient care areas.

(2) Where encased in not less than 50 mm (2 in.) of concrete, Schedule 40 rigid nonmetallic conduit, flexible nonmetallic or jacketed metallic raceways, or jacketed metallic cable assemblies listed for installation in concrete. Nonmetallic raceways shall not be used for branch circuits that supply patient care areas.

(3) Listed flexible metal raceways and listed metal sheathed cable assemblies in any of the following:

a. Where used in listed prefabricated medical headwalls

b. In listed office furnishings

c. Where fished into existing walls or ceilings, not otherwise accessible and not subject to physical damage

d. Where necessary for flexible connection to equipment

(4) Flexible power cords of appliances or other utilization equipment connected to the emergency system.

(5) Secondary circuits of Class 2 or Class 3 communication or signaling systems.

Author's Substantiation. This subdivision has been reorganized with text placed into positive language and not exceptions for the permitted wiring methods to provide mechanical protection of the emergency system.

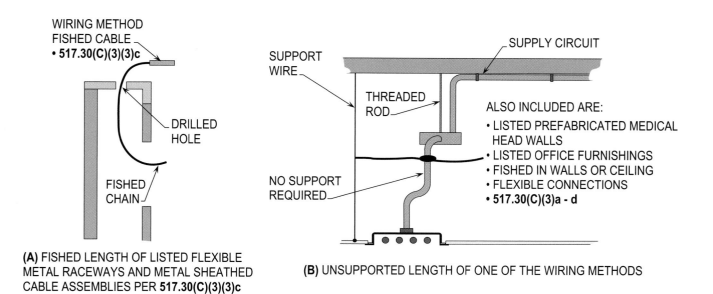

WIRING METHOD
FISHED CABLE
• **517.30(C)(3)(3)c**

DRILLED HOLE

FISHED CHAIN

(A) FISHED LENGTH OF LISTED FLEXIBLE METAL RACEWAYS AND METAL SHEATHED CABLE ASSEMBLIES PER **517.30(C)(3)(3)c**

SUPPORT WIRE

THREADED ROD

NO SUPPORT REQUIRED

SUPPLY CIRCUIT

ALSO INCLUDED ARE:
• LISTED PREFABRICATED MEDICAL HEAD WALLS
• LISTED OFFICE FURNISHINGS
• FISHED IN WALLS OR CEILING
• FLEXIBLE CONNECTIONS
• **517.30(C)(3)a - d**

(B) UNSUPPORTED LENGTH OF ONE OF THE WIRING METHODS

**MECHANICAL PROTECTION OF THE EMERGENCY SYSTEM
517.30(C)(3)**

Purpose of Change: To revise and allow listed flexible metal raceways and metal sheathed cable assemblies to protect emergency systems.

NEC Ch. 5 – Article 518
Part – 518.1

Type of Change		Panel Action		UL	UL 508	API 500	API 505	OSHA
Revision		Accept in Principle		-	-	-	-	-
ROP		ROC		NFPA 70E	NFPA 70B	NFPA 79	NFPA 101	NEMA
pg. 1341	# 15-55	pg. 427	# 15-34	-	-	-	6.1.2	-
log: 28	CMP: 15	log: 3122	Submitter: James K. Lathrop			2002 NEC: 518.1		IEC: 711

2002 NEC – 518.1 Scope

This article covers all building or portions of buildings or structures designed or intended for the ~~assembly~~ of 100 or more persons.

2005 NEC – 518.1 Scope

Except for the assembly occupancies explicitly covered by 520.1, this article covers all building or portions of buildings or structures designed or intended for the gathering together of 100 or more persons for such purposes as deliberation, worship, entertainment, eating, drinking, amusement, awaiting transportation, or similar purposes.

Author's Substantiation. This revision clarifies that multiple occupancies of 100 or more persons are considered places of assembly.

NOTE 1: ARTICLE 518 DOES NOT COVER THE ASSEMBLY OCCUPANCIES THAT ARE COVERED BY **520.1**.

NOTE 2: PLACES OF ASSEMBLY ARE CONSIDERED AS THE GATHERING TOGETHER OF 100 PERSONS OR MORE FOR SUCH PURPOSES AS DELIBERATION, WORSHIP, ENTERTAINMENT, AWAITING TRANSPORTATION, OR SIMILAR PURPOSES.

SCOPE
518.1

Purpose of Change: To provide guidelines for portions of multiple occupancies that are considered places of assembly.

NEC Ch. 5 – Article 518
Part – 518.2(B)

Type of Change		Panel Action		UL	UL 508	API 500	API 505	OSHA
New Sentence		Accept		-	-	-	-	-
ROP		ROC		NFPA 70E	NFPA 70B	NFPA 79	NFPA 101	NEMA
pg. 1348	# 15-63	pg. 428	# 15-42	-	-	-	6.1.2	-
log: 937	CMP: 15	log: 809	Submitter: Donald J. Talka			2002 NEC: 518.2(B)		IEC: 711

2002 NEC – 518.2(B) Multiple Occupancies

Occupancy of any room or space for assembly purposes by less than 100 persons in a building or other occupancy, and incidental to such other occupancy, shall be classified as part of the other occupancy and subject to the provisions applicable thereto.

2005 NEC – 518.2(B) Multiple Occupancies

Where an assembly occupancy forms a portion of a building containing other occupancies, Article 518 applies only to that portion of the building considered an assembly occupancy. Occupancy of any room or space for assembly purposes by less than 100 persons in a building or other occupancy, and incidental to such other occupancy, shall be classified as part of the other occupancy and subject to the provisions applicable thereto.

Author's Substantiation. A new sentence has been added to clarify that Article 518 applies to portions of buildings that are considered an assembly occupancy.

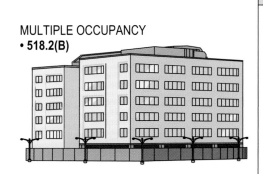

MULTIPLE OCCUPANCY
• 518.2(B)

EXAMPLES OF MULTIPLE OCCUPANCIES PER 518.2(A)	
• ARMORIES	• MORTUARY CHAPELS
• ASSEMBLY HALLS	• MULTIPURPOSE ROOMS
• AUDITORIUMS	• MUSEUMS
• BOWLING LANES	• PLACES OF AWAITING TRANSPORTATION
• CLUB ROOMS	• PLACES OF RELIGIOUS WORSHIP
• CONFERENCE ROOMS	• POOL ROOMS
• COURTROOMS	• RESTAURANTS
• DANCE HALLS	• SKATING RINKS
• DINING AND DRINKING FACILITIES	
• EXHIBITION HALLS	
• GYMNASIUMS	

NOTE: ARTICLE 518 ONLY APPLIES TO THAT PORTION OF THE BUILDING CONSIDERED AN ASSEMBLY OCCUPANCY (100 PERSONS OR MORE) PER 518.2(B).

MULTIPLE OCCUPANCIES
518.2(B)

Purpose of Change: To provide guidelines for portions of multiple occupancies that are considered places of assembly.

Type of Change		Panel Action		UL	UL 508	API 500	API 505	OSHA
New Section		Accept		50 and 1332	-	-	-	-
ROP		ROC		NFPA 70E	NFPA 70B	NFPA 79	NFPA 1	NEMA
pg. 1367	# 15-91	pg. 435	# 15-62	-	-	-	1-16.16	-
log: 3171	CMP: 15	log: 815	Submitter: TCC			2002 NEC: -		IEC: 708

2005 NEC – 525.11 Multiple Sources of Supply

Where multiple services or separately derived systems or both supply rides, attractions, and other structures, all sources of supply that serve rides, attractions, or other structures separated by less than 3.7 m (12 ft) shall be bonded to the same grounding electrode system.

Author's Substantiation. A new section has been added to recognize the grounding requirements for rides, attractions, and other structures.

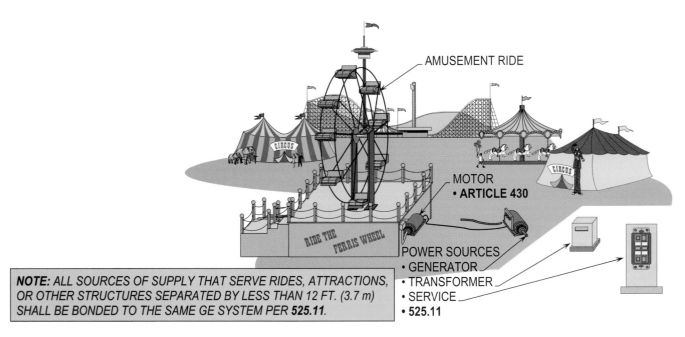

NOTE: ALL SOURCES OF SUPPLY THAT SERVE RIDES, ATTRACTIONS, OR OTHER STRUCTURES SEPARATED BY LESS THAN 12 FT. (3.7 m) SHALL BE BONDED TO THE SAME GE SYSTEM PER **525.11**.

AMUSEMENT RIDE

MOTOR
• **ARTICLE 430**

POWER SOURCES
• GENERATOR
• TRANSFORMER
• SERVICE
• **525.11**

MULTIPLE OCCUPANCIES
525.11

Purpose of Change: To clarify the bonding and grounding requirements of power sources for rides, attractions, structures, etc.

Type of Change		Panel Action		UL	UL 508	API 500	API 505	OSHA
Revision		Accept		943	-	-	-	1910.306(j)(5)
ROP		ROC		NFPA 70E	NFPA 70B	NFPA 79	NFPA	NEMA
pg. 1369	# 15-94	pg. -	# -	410.4	-	-	-	-
log: 53	CMP: 15	log: -	Submitter: TCC			2002 NEC: 525.23(A)		IEC: 708

2002 NEC – 525.23(A) General-Use 15- and 20-Ampere, 125-Volt Receptacles

~~All 125-volt, single-phase, 15- and 20-ampere receptacle outlets that are in use by personnel shall have listed ground-fault circuit-interrupter protection for personnel.~~ The ground-fault circuit interrupter shall be permitted to be an integral part of the attachment plug or located in the power-supply cord, within 300 mm (12 in.) of the attachment plug. ~~For the purposes of this section,~~ listed cord sets incorporating ground-fault circuit-interrupter protection for personnel shall be permitted. ~~Egress lighting shall not be connected to the load side terminals of a ground-fault circuit-interrupter receptacle.~~

2002 NEC – 525.23(A) <u>Where GFCI Protection Is Required</u>

The ground-fault circuit interrupter shall be permitted to be an integral part of the attachment plug or located in the power-supply cord, within 300 mm (12 in.) of the attachment plug. Listed cord sets incorporating ground-fault circuit-interrupter for personnel shall be permitted.

   (1) <u>125-volt, single-phase, 15- and 20-ampere non-locking type receptacles used for disassembly and reassembly or readily accessible to the general public.</u>

   (2) <u>Equipment that is readily accessible to the general public and supplied from a 125-volt, single-phase, 15- or 20-ampere branch circuit.</u>

<u>(B) Where GFCI Protection Is Not Required.</u>

<u>Receptacles that only facilitate quick disconnecting and reconnecting of electrical equipment shall not be required to be provided with GFCI protection. These receptacles shall be of the locking type.</u>

Author's Substantiation. This revision clarifies the types of GFCI protection that shall be permitted to be used for protecting the general public.

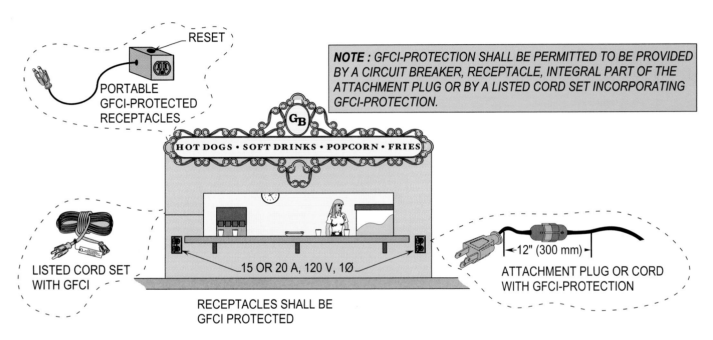

RESET

PORTABLE
GFCI-PROTECTED
RECEPTACLES

NOTE : *GFCI-PROTECTION SHALL BE PERMITTED TO BE PROVIDED BY A CIRCUIT BREAKER, RECEPTACLE, INTEGRAL PART OF THE ATTACHMENT PLUG OR BY A LISTED CORD SET INCORPORATING GFCI-PROTECTION.*

HOT DOGS · SOFT DRINKS · POPCORN · FRIES

LISTED CORD SET
WITH GFCI

15 OR 20 A, 120 V, 1Ø

◄12" (300 mm) ►

ATTACHMENT PLUG OR CORD
WITH GFCI-PROTECTION

RECEPTACLES SHALL BE
GFCI PROTECTED

WHERE GFCI PROTECTION IS REQUIRED
525.23(A)

Purpose of Change: To clarify the types of GFCI protection allowed to be used for protecting the general public.

NEC Ch. 5 – Article 547
Part – 547.2

Type of Change	Panel Action	UL	UL 508	API 500	API 505	OSHA
New Definition	Accept	-	-	-	-	-
ROP	**ROC**	**NFPA 70E**	**NFPA 70B**	**NFPA 79**	**NFPA**	**NEMA**
pg. 1392 # 19-7a	pg. - # -	-	-	-	61	-
log: CP1908 CMP: 19	log: -	Submitter: CMP 19		2002 NEC: -		IEC: 705

2005 NEC – 547.2 Definitions

Site-Isolating Device. A disconnecting means installed at the distribution point for the purposes of isolation, system maintenance, emergency disconnection, or connection of optional standby systems.

Author's Substantiation. A new definition has been added to define a site-isolating device and its purpose when used in a agricultural building.

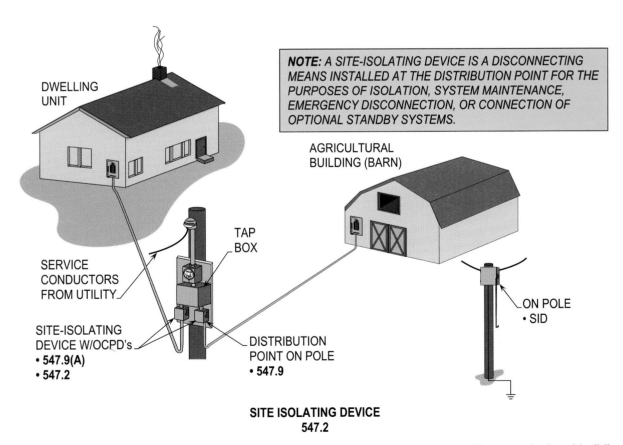

NOTE: A SITE-ISOLATING DEVICE IS A DISCONNECTING MEANS INSTALLED AT THE DISTRIBUTION POINT FOR THE PURPOSES OF ISOLATION, SYSTEM MAINTENANCE, EMERGENCY DISCONNECTION, OR CONNECTION OF OPTIONAL STANDBY SYSTEMS.

DWELLING UNIT

AGRICULTURAL BUILDING (BARN)

TAP BOX

SERVICE CONDUCTORS FROM UTILITY

ON POLE
• SID

SITE-ISOLATING DEVICE W/OCPD's
• 547.9(A)
• 547.2

DISTRIBUTION POINT ON POLE
• 547.9

SITE ISOLATING DEVICE
547.2

Purpose of Change: To define a site-isolating device and its purpose when used in an agricultural building installation.

NEC Ch. 5 – Article 547
Part – 547.5(G)

135

Type of Change		Panel Action		UL	UL 508	API 500	API 505	OSHA
New Item		Accept in Principle in Part		-	-	-	-	1910.306(j)(5)
ROP		ROC		NFPA 70E	NFPA 70B	NFPA 79	NFPA	NEMA
pg. 1394	# 19-10a	pg. 442	# 19-10	410.4	-	-	-	-
log: CP1901	CMP: 19	log: 3465	Submitter: CMP 19			2002 NEC: 547.5(G)		IEC: 705

2002 NEC – 547.5(G) Receptacles

All 125-volt, single-phase, 15- and 20-ampere general-purpose receptacles installed in the following locations shall have ground-fault circuit-interrupter protection for personnel:

(1) In areas having an equipotential plane

(2) Outdoors

(3) Damp or wet locations

2005 NEC – 547.5(G) Receptacles

All 125-volt, single-phase, 15- and 20-ampere general-purpose receptacles installed in the following locations shall have ground-fault circuit-interrupter protection for personnel:

(1) In areas having an equipotential plane

(2) Outdoors

(3) Damp or wet locations

(4) <u>Dirt confinement areas for livestock</u>

Author's Substantiation. A new item has been added to recognize that GFCI-protection for personnel shall be provided for dirt confinement areas for livestock.

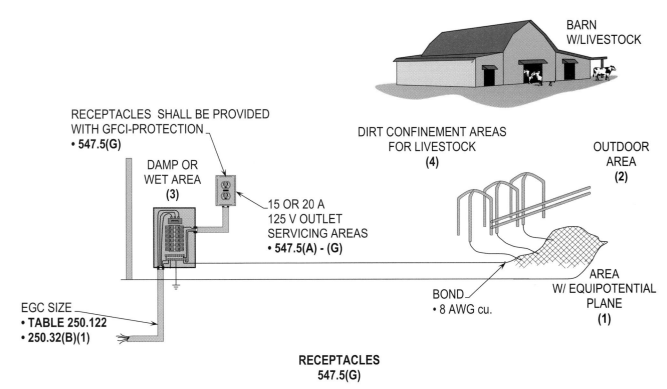

RECEPTACLES SHALL BE PROVIDED
WITH GFCI-PROTECTION
• **547.5(G)**

DAMP OR
WET AREA
(3)

15 OR 20 A
125 V OUTLET
SERVICING AREAS
• **547.5(A) - (G)**

EGC SIZE
• **TABLE 250.122**
• **250.32(B)(1)**

BARN
W/LIVESTOCK

DIRT CONFINEMENT AREAS
FOR LIVESTOCK
(4)

OUTDOOR
AREA
(2)

BOND
• 8 AWG cu.

AREA
W/ EQUIPOTENTIAL
PLANE
(1)

RECEPTACLES
547.5(G)

Purpose of Change: To recognize that GFCI-protection shall be provided for dirt confinement areas for livestock.

NEC Ch. 5 – Article 547
Part – 547.10(A)

Type of Change	Panel Action	UL	UL 508	API 500	API 505	OSHA
Revision	Accept	467	-	-	-	-
ROP	**ROC**	**NFPA 70E**	**NFPA 70B**	**NFPA 79**	**NFPA**	**NEMA**
pg. 1403 # 19-23a	pg. - # -	-	-	-	-	-
log: CP1902 CMP: 19	log: -	Submitter: CMP 19		2002 NEC: 547.10(A)		IEC: 705

2002 NEC – 547.10(A) ~~Areas Requiring Equipotential Plane~~

Equipotential planes shall be installed in all concrete floor confinement areas ~~of~~ livestock buildings ~~that contain metallic equipment that is accessible to animals and likely to become energized.~~ Outdoor confinement areas, such as feedlots, ~~shall have equipotential planes installed around~~ metallic equipment that ~~is accessible to animals and likely to~~ become energized. The equipotential plane shall encompass the area ~~around the equipment~~ where the ~~animal~~ stands while accessing the equipment.

2005 NEC – 547.10(A) <u>Where Required</u>

Equipotential planes shall be installed in all concrete floor confinement areas <u>in</u> livestock <u>buildings, and in all</u> outdoor confinement areas such as feedlots, <u>containing</u> metallic equipment that <u>may</u> become energized <u>and is accessible to livestock</u>. The equipotential plane shall encompass the area where the <u>livestock</u> stands while accessing <u>metallic</u> equipment <u>that may become energized</u>.

Author's Substantiation. This revision clarifies the conditions of where an equipotential plane is required.

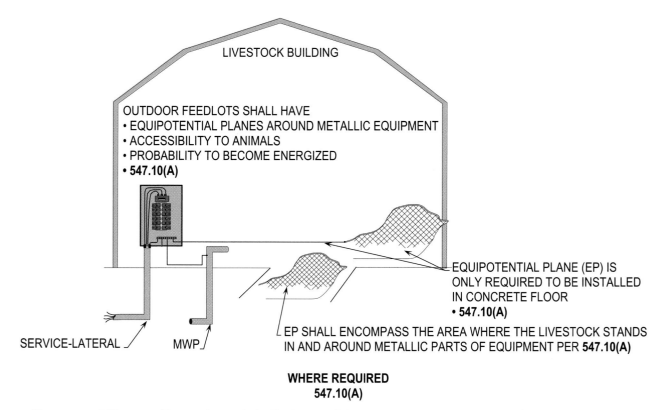

LIVESTOCK BUILDING

OUTDOOR FEEDLOTS SHALL HAVE
• EQUIPOTENTIAL PLANES AROUND METALLIC EQUIPMENT
• ACCESSIBILITY TO ANIMALS
• PROBABILITY TO BECOME ENERGIZED
• **547.10(A)**

EQUIPOTENTIAL PLANE (EP) IS ONLY REQUIRED TO BE INSTALLED IN CONCRETE FLOOR
• **547.10(A)**

EP SHALL ENCOMPASS THE AREA WHERE THE LIVESTOCK STANDS IN AND AROUND METALLIC PARTS OF EQUIPMENT PER **547.10(A)**

SERVICE-LATERAL   MWP

WHERE REQUIRED
547.10(A)

Purpose of Change: To rewrite and clarify the conditions under when an equipotential plane is required.

NEC Ch. 5 – Article 550
Part II – 550.12(C)

Type of Change	Panel Action	UL	UL 508	API 500	API 505	OSHA - HUD
New Sentence	Accept	-	-	-	-	Subpart I, 3280
ROP	**ROC**	**NFPA 70E**	**NFPA 70B**	**NFPA 79**	**NFPA**	**NEMA**
pg. 1414 # 19-41	pg. - # -	-	-	-	58	-
log: 698 CMP: 19	log: -	Submitter: Dan Leaf		2002 NEC: 550.12(C)		IEC: -

2002 NEC – 550.12(C) Laundry Area

Where a laundry area is provided, a 20-ampere branch circuit shall be provided to supply the laundry receptacle outlet(s).

2005 NEC – 550.12(C) Laundry Area

Where a laundry area is provided, a 20-ampere branch circuit shall be provided to supply the laundry receptacle outlet(s). <u>This circuit shall have no other outlets.</u>

Author's Substantiation. A new sentence has been added to clarify that no other outlets shall be installed on a laundry circuit.

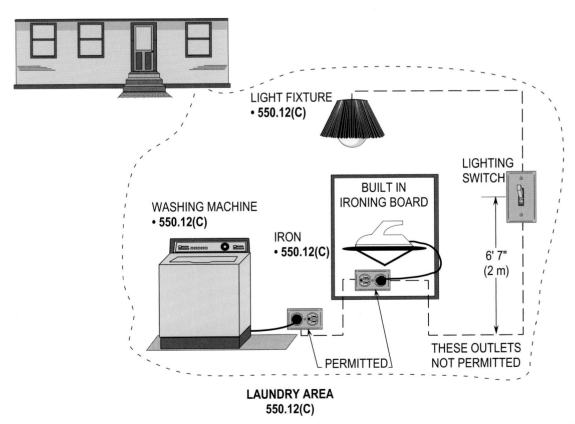

Purpose of Change: To allow no other outlets on the laundry circuit except those used for the laundry.

NEC Ch. 5 – Article 550
Part II – 550.13(F)(1)

Type of Change	Panel Action	UL	UL 508	API 500	API 505	OSHA - HUD
Revision	Accept	-	-	-	-	Subpart I, 3280
ROP	**ROC**	**NFPA 70E**	**NFPA 70B**	**NFPA 79**	**NFPA**	**NEMA**
pg. 1417 # 19-46a	pg. - # -	-	-	-	58	-
log: CP1919 CMP: 19	log: -	Submitter: CMP 19		2002 NEC: 550.13(F)		IEC: -

2002 NEC – 550.13(F) Receptacle Outlets Not Permitted

Receptacle outlets shall not be permitted in the following locations:

(1) Receptacle outlets shall not be installed ~~in or~~ within ~~reach [750 mm (30 in.)] of~~ a shower or bathtub space.

2005 NEC – 550.13(F) Receptacle Outlets Not Permitted

Receptacle outlets shall not be permitted in the following locations:

(1) Receptacle outlets shall not be installed within a bathtub or shower space.

Author's Substantiation. This revision clarifies that receptacle outlets shall not be installed within a bathtub or shower space.

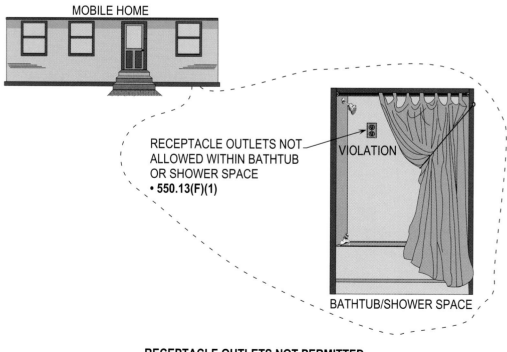

MOBILE HOME

RECEPTACLE OUTLETS NOT ALLOWED WITHIN BATHTUB OR SHOWER SPACE
• **550.13(F)(1)**

VIOLATION

BATHTUB/SHOWER SPACE

RECEPTACLE OUTLETS NOT PERMITTED
550.13(F)(1)

Purpose of Change: To clarify that receptacle outlets shall not be installed within a bathtub or shower space.

NEC Ch. 5 – Article 550
Part II – 550.15(H), Ex.

Type of Change		Panel Action		UL	UL 508	API 500	API 505	OSHA - HUD
Revision		Accept		-	-	-	-	Subpart I, 3280
ROP		ROC		NFPA 70E	NFPA 70B	NFPA 79	NFPA	NEMA
pg. 1418	# 19-49	pg. -	# -	-	-	-	58	-
log: 790	CMP: 19	log: -	Submitter: Dan Leaf			2002 NEC: -		IEC: -

2002 NEC – 550.15(H) Under-Chassis Wiring (Exposed to Weather)

Where outdoor or under-chassis line-voltage (120 volts, nominal, or higher) wiring is exposed to moisture or physical damage, it shall be protected by rigid metal conduit or intermediate metal conduit. The conductors shall be suitable for wet locations.

> **Exception:** Electrical metallic tubing or rigid nonmetallic conduit shall be permitted where closely routed against frames and equipment enclosures.

2005 NEC – 550.15(H) Under-Chassis Wiring (Exposed to Weather)

Where outdoor or under-chassis line-voltage (120 volts, nominal, or higher) wiring is exposed to moisture or physical damage, it shall be protected by rigid metal conduit or intermediate metal conduit. The conductors shall be suitable for wet locations.

> **Exception:** Type MI cable, electrical metallic tubing or rigid nonmetallic conduit shall be permitted where closely routed against frames and equipment enclosures.

Author's Substantiation. This revision clarifies that Type MI cable shall be permitted for under-chassis wiring (exposed to weather) where closely routed against frames and equipment enclosures.

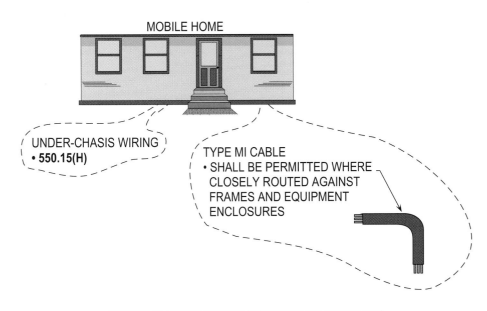

UNDER-CHASSIS WIRING (EXPOSED TO WEATHER)
550.15(H)

Purpose of Change: To allow Type MI cable as a wiring method for under-chassis of mobile homes.

NEC Ch. 5 – Article 550
Part III – 550.32(A)

Type of Change		Panel Action		UL	UL 508	API 500	API 505	OSHA - HUD
Revision		Accept in Principle		869A	-	-	-	Subpart I, 3280
ROP		ROC		NFPA 70E	NFPA 70B	NFPA 79	NFPA	NEMA
pg. 1426	# 19-62	pg. 445	# 19-32	-	-	-	58	-
log: 638	CMP: 19	log: 1850	Submitter: Dan Leaf			2002 NEC: 550.32(A)		IEC: -

2002 NEC – 550.32(A) Mobile Home Service Equipment

The mobile home service equipment shall be located adjacent to the mobile home and not mounted in or on the mobile home. The service equipment shall be located in sight from and not more than 9.0 m (30 ft) from the exterior wall of the mobile home it serves. The service equipment shall be permitted to be located elsewhere on the premises, provided that a disconnecting means suitable for service equipment is located in sight from and not more than 9.0 m (30 ft) from the exterior wall of the mobile home it serves. Grounding at the disconnecting means shall be in accordance with 250.32.

2005 NEC – 550.32(A) Mobile Home Service Equipment

The mobile home service equipment shall be located adjacent to the mobile home and not mounted in or on the mobile home. The service equipment shall be located in sight from and not more than 9.0 m (30 ft) from the exterior wall of the mobile home it serves. The service equipment shall be permitted to be located elsewhere on the premises, provided that a disconnecting means suitable for service equipment is located in sight from and not more than 9.0 m (30 ft) from the exterior wall of the mobile home it serves and is rated not less than that required for service equipment per 550.32(C). Grounding at the disconnecting means shall be in accordance with 250.32.

Author's Substantiation. This revision clarifies the minimum rating of a disconnecting means that is not service equipment.

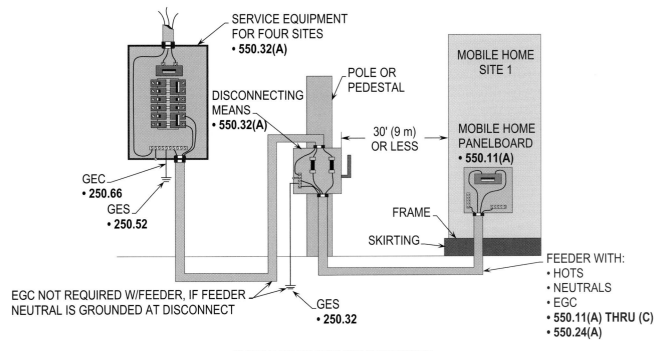

SERVICE EQUIPMENT
FOR FOUR SITES
• **550.32(A)**

POLE OR
PEDESTAL

DISCONNECTING
MEANS
• **550.32(A)**

30' (9 m)
OR LESS

MOBILE HOME
SITE 1

MOBILE HOME
PANELBOARD
• **550.11(A)**

GEC
• **250.66**

GES
• **250.52**

FRAME

SKIRTING

GES
• **250.32**

EGC NOT REQUIRED W/FEEDER, IF FEEDER
NEUTRAL IS GROUNDED AT DISCONNECT

FEEDER WITH:
• HOTS
• NEUTRALS
• EGC
• **550.11(A) THRU (C)**
• **550.24(A)**

MOBILE HOME SERVICE EQUIPMENT
550.32(A)

Purpose of Change: To require mobile home service equipment to be suitable to use as such.

NEC Ch. 5 – Article 551
Part V – 551.46(E), Ex. 3

Type of Change		Panel Action		UL	UL 508	API 500	API 505	OSHA - HUD
New Exception		Accept in Principle		869A	-	-	-	Subpart I, 3280
ROP		ROC		NFPA 70E	NFPA 70B	NFPA 79	NFPA	NEMA
pg. 1440	# 19-89	pg. -	# -	-	-	-	58	-
log: 564	CMP: 19	log: -	Submitter: Bruce A. Hopkins			2002 NEC: -		IEC: -

2005 NEC – 551.46(E) Location

The point of entrance of a power-supply assembly shall be located within 4.5 m (15 ft) of the rear, on the left (road) side or at the rear, left of the longitudinal center of the vehicle, within 450 mm (18 in.) of the outside wall.

> **Exception No. 3:** Recreational vehicles designed for transporting livestock shall be permitted to have the electrical point of entrance located on either side or the front.

Author's Substantiation. A new exception has been added to recognize that recreational vehicles for transporting livestock are often designed with left side primary entrances, thus requiring additional flexibility in the location of the power supply entrance location.

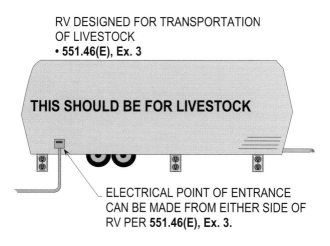

RV DESIGNED FOR TRANSPORTATION
OF LIVESTOCK
• **551.46(E), Ex. 3**

THIS SHOULD BE FOR LIVESTOCK

ELECTRICAL POINT OF ENTRANCE
CAN BE MADE FROM EITHER SIDE OF
RV PER **551.46(E), Ex. 3.**

LOCATION
551.46(E), Ex. 3

Purpose of Change: To point out that these units use traditional RV park hook-up connections.

NEC Ch. 5 – Article 553
Part II – 553.4

Type of Change		Panel Action		UL	UL 508	API 500	API 505	OSHA
Revision		Accept in Principle		869A	-	-	-	1910.304(d)
ROP		ROC		NFPA 70E	NFPA 70B	NFPA 79	NFPA	NEMA
pg. 1469	# 19-131	pg. -	# -	410.8	-	-	58	-
log: 1352	CMP: 19	log: -	Submitter: Dan Leaf			2002 NEC: 553.4		IEC: -

2002 NEC – 553.4 Location of Service Equipment

The service equipment for a floating building shall be located adjacent to, but not in or on, the building.

2005 NEC – 553.4 Location of Service Equipment

The service equipment for a floating building shall be located adjacent to, but not in or on, the building <u>or any floating structure</u>.

Author's Substantiation. This revision clarifies that service equipment shall not be located in or on any floating structure.

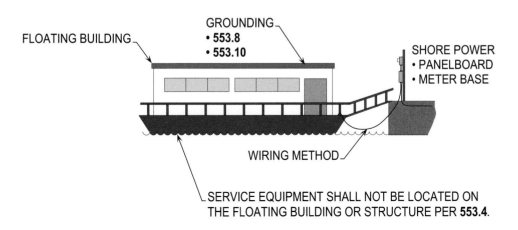

FLOATING BUILDING

GROUNDING
• **553.8**
• **553.10**

SHORE POWER
• PANELBOARD
• METER BASE

WIRING METHOD

SERVICE EQUIPMENT SHALL NOT BE LOCATED ON THE FLOATING BUILDING OR STRUCTURE PER **553.4.**

LOCATION OF SERVICE EQUIPMENT
553.4

Purpose of Change: To make it clear that service equipment shall not be installed on a floating building or structure.

NEC Ch. 5 – Article 553
Part III – 553.8(C)

Type of Change		Panel Action		UL	UL 508	API 500	API 505	OSHA
New Subsection		Accept in Principle		467	-	-	-	1910.307(b)(3), Note
ROP		ROC		NFPA 70E	NFPA 70B	NFPA 79	NFPA	NEMA
pg. 1470	# 19-132	pg. -	# -	440.4(A), FPN	-	-	-	-
log: 693	CMP: 19	log: -	Submitter: Dan Leaf			2002 NEC: -		IEC: 514.3 and 542.3

2005 NEC – 553.8(C) Identification of Equipment Grounding Conductor

The equipment grounding conductor shall be an insulated copper conductor with a continuous outer finish that is either green or green with one or more yellow stripes. For conductors larger than 6 AWG, or where multiconductor cables are used, re-identification of conductors as allowed in 250.119(A)(2)(b) and (A)(2)(c) or 250.119(B)(2) and (B)(3) shall be permitted.

Author's Substantiation. A new subsection has been added to recognize the identification requirements for equipment grounding conductors at floating buildings.

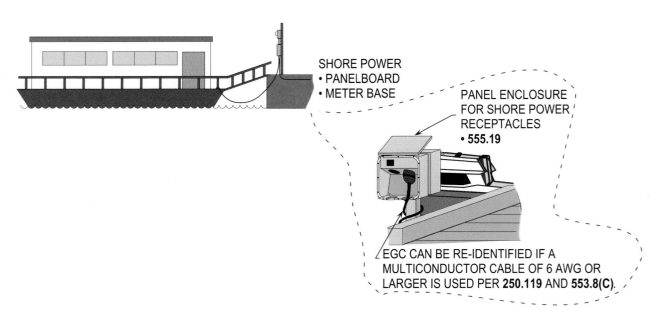

SHORE POWER
• PANELBOARD
• METER BASE

PANEL ENCLOSURE FOR SHORE POWER RECEPTACLES
• **555.19**

EGC CAN BE RE-IDENTIFIED IF A MULTICONDUCTOR CABLE OF 6 AWG OR LARGER IS USED PER **250.119** AND **553.8(C)**.

IDENTIFICATION OF EQUIPMENT GROUNDING CONDUCTOR
553.8(C)

Purpose of Change: To provide requirements for identifying the equipment grounding conductor.

NEC Ch. 5 – Article 555
Part – 555.21

Type of Change		Panel Action		UL	UL 508	API 500	API 505	OSHA
New Sentence		Accept		1130	-	-	-	1910.307(b)(3), Note
ROP		ROC		NFPA 70E	NFPA 70B	NFPA 79	NFPA	NEMA
pg. 1477	# 19-142	pg. 449	# 19-50	440.4(A), FPN	-	-	30A and 303	-
log: 1056	CMP: 19	log: 852	Submitter: Robert A. McCullough			2002 NEC: 550.21		IEC: 60079-20

2002 NEC – 555.21 ~~Gasoline~~ Dispensing Stations – Hazardous (Classified) Locations

Electrical wiring and equipment located at or serving ~~gasoline~~ dispensing stations shall comply with Article 514 in addition to the requirements of this article.

2005 NEC – 555.21 <u>Motor Fuel</u> Dispensing Stations – Hazardous (Classified) Locations

Electrical wiring and equipment located at or serving <u>motor fuel </u>dispensing stations shall comply with Article 514 in addition to the requirements of this article. <u>All electrical wiring for power and lighting shall be installed on the side of the wharf, pier, or dock opposite from the liquid piping system.</u>

> **FPN:** <u>For additional information, see NFPA 303-2000, *Fire Protection Standard for Marinas and Boatyards*, and NFPA 30A-2003, *Motor Fuel Dispensing Facilities and Repair Garages.*</u>

Author's Substantiation. A new sentence has been added to clarify that electrical wiring for power and lighting shall be installed on the load side of the wharf, pier, or dock opposite from the liquid piping system. The FPN was inadvertently deleted in the last cycle.

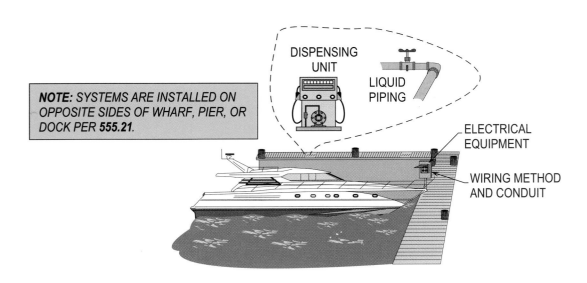

MOTOR FUEL DISPENSING SYSTEMS – HAZARDOUS (CLASSIFIED) LOCATIONS
555.21

Purpose of Change: To clarify the location of the electrical wiring for power and lighting in relationship to the liquid piping system.

NEC Ch. 5 – Article 590
Part – 590.4(B) and (C)

Type of Change		Panel Action		UL	UL 508	API 500	API 505	OSHA
Revision		Accept in Principle		1088	-	-	-	1910.305(a)(2)(c)(iii)
ROP		ROC		NFPA 70E	NFPA 70B	NFPA 79	NFPA	NEMA
pg. 1379	# 3-114	pg. 438	# 3-97	420.1(B)(2)(a) and (b)	-	-	1 and 241	-
log: 2667	CMP: 3	log: 554	Submitter: Phil Simmons			2002 NEC: 527.4(B) and (C)		IEC: 704.52

2002 NEC – ~~527.4~~ General

(B) **Feeders.** ~~Feeders shall be protected as provided in Article 240~~. They shall originate in an approved distribution center. Conductors shall be permitted within cable assemblies or within multiconductor cords or cables of a type identified in Table 400.4 for hard usage or extra-hard usage. For the purpose of this section, Type NM and Type NMC cables shall be permitted to be used in any dwelling, building, or structure without any height limitation.

> **Exception:** Single insulated conductors shall be permitted where installed for the purpose(s) specified in ~~527.3(C)~~, where accessible only to qualified persons.

(C) **Branch Circuits.** All branch circuits shall originate in an approved power outlet or panelboard. Conductors shall be permitted within cable assemblies or within multiconductor cord or cable of a type identified in Table 400.4 for hard usage or extra-hard usage. ~~All~~ conductors shall be protected ~~as provided in Article 240~~. For the purposes of this section, Type NM and Type NMC cables shall be permitted to be used in any dwelling, building, or structure without any height limitation.

> **Exception:** Branch circuits installed for the purposes specified in ~~527.3(B)~~ or ~~(C)~~ shall be permitted to be run as single insulated conductors. Where the wiring is installed in accordance with ~~527.3(B)~~, the voltage to ground shall not exceed 150 volts, the wiring shall not be subject to physical damage, and the conductors shall be supported on insulators at intervals of not more than 3.0 m (10 ft); or, for festoon lighting, the conductors shall be arranged ~~so~~ that excessive strain is not transmitted to the lampholders.

2005 NEC – <u>590.4</u> General

(B) **Feeders.** <u>Overcurrent protection shall be provided in accordance with 240.4, 240.5, 240.100, and 240.101.</u> They shall originate in an approved distribution center. Conductors shall be permitted within cable assemblies or within multiconductor cords or cables of a type identified in Table 400.4 for hard usage or extra-hard usage. For the purpose of this section, Type NM and Type NMC cables shall be permitted to be used in any dwelling, building, or structure without any height limitation <u>or limitation by building construction type and without concealment within walls, floors, or ceilings</u>.

> **Exception:** Single insulated conductors shall be permitted where installed for the purpose(s) specified in <u>590.3(C)</u>, where accessible only to qualified persons.

(C) **Branch Circuits.** All branch circuits shall originate in an approved power outlet or panelboard. Conductors shall be permitted within cable assemblies or within multiconductor cord or cable of a type identified in Table 400.4 for hard usage or extra-hard usage. Conductors shall be protected <u>from overcurrent as provided in 240.4, 240.5, and 240.100</u>. For the purposes of this section, Type NM and Type NMC cables shall be permitted to be used in any dwelling, building, or structure without any height limitation <u>or limitation by building construction type and without concealment within walls, floors, or ceilings</u>.

Exception: Branch circuits installed for the purposes specified in 590.3(B) or 590.3(C) shall be permitted to be run as single insulated conductors. Where the wiring is installed in accordance with 590.3(B), the voltage to ground shall not exceed 150 volts, the wiring shall not be subject to physical damage, and the conductors shall be supported on insulators at intervals of not more than 3.0 m (10 ft); or, for festoon lighting, the conductors shall be so arranged that excessive strain is not transmitted to the lampholders.

Author's Substantiation. This revision clarifies that Type NM and Type NMC cables shall be permitted to be used for temporary installations as feeders and branch circuits for any dwelling, building, or structure without any height limitation or limitation by building construction type and without concealment within walls, floors, or ceilings.

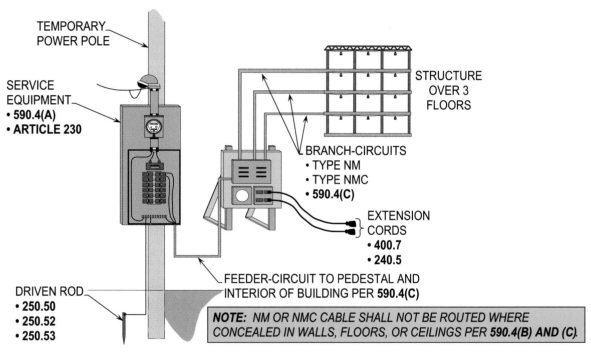

FEEDERS AND BRANCH-CIRCUITS
590.4(B); (C)

Purpose of Change: To allow type NM or NMC cable to be used as a temporary wiring method in any dwelling, building or structure under construction, regardless of their height, but such wiring method shall not be concealed.

NEC Ch. 5 – Article 590
Part – 590.4(J), Ex.

Type of Change		Panel Action		UL	UL 508	API 500	API 505	OSHA
New Exception		Accept in Principle		1088	-	-	-	1910.305(a)(2)(i)(c)
ROP		ROC		NFPA 70E	NFPA 70B	NFPA 79	NFPA	NEMA
pg. 1383	# 3-119	pg. 439	# 3-103	420.1(B)(1)(b)	-	-	1 and 241	-
log: 2512	CMP: 3	log: 3656	Submitter: Russell LeBlanc			2002 NEC: -		IEC: 704.52

2005 NEC – 590.4(J) Support

Cable assemblies and flexible cords and cables shall be supported in place at intervals that ensure that they will be protected from physical damage. Support shall be in the form of staples, cable ties, straps, or similar type fittings installed so as not to cause damage. Vegetation shall not be used for support of overhead spans of branch circuits or feeders.

> **Exception:** For holiday lighting in accordance with 590.3(B), where the conductors or cables are arranged with proper strain relief devices, tension take-up devices, or other approved means to avoid damage from the movement of the live vegetation, trees shall be permitted to be used for support of overhead spans of branch circuit conductors or cables.

Author's Substantiation. A new exception has been added to recognize that overhead spans of branch circuit conductors or cables shall be permitted to be supported by live vegetation where certain conditions are followed.

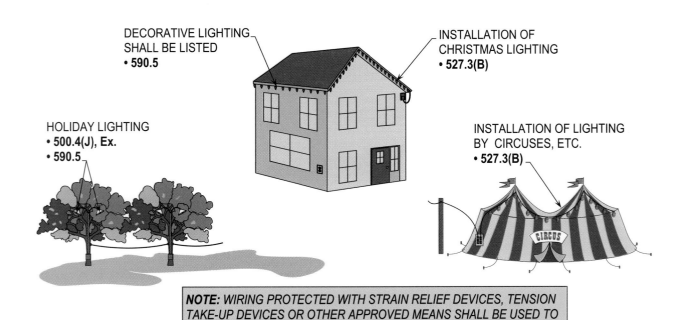

DECORATIVE LIGHTING SHALL BE LISTED
• **590.5**

INSTALLATION OF CHRISTMAS LIGHTING
• **527.3(B)**

HOLIDAY LIGHTING
• **500.4(J), Ex.**
• **590.5**

INSTALLATION OF LIGHTING BY CIRCUSES, ETC.
• **527.3(B)**

NOTE: WIRING PROTECTED WITH STRAIN RELIEF DEVICES, TENSION TAKE-UP DEVICES OR OTHER APPROVED MEANS SHALL BE USED TO AVOID PHYSICAL DAMAGE PER 590.4(J), Ex.

SUPPORT
590.4(J), Ex.

Purpose of Change: To protect the wiring run overhead and supported by live vegetation such as trees, from physical damage.

Special Equipment

Chapter 6 has always been utilized by designers, installers and inspectors when dealing with special equipment that is located and used in specific occupancies.

Special equipment, due to the specialized rules and regulations that are needed for their safe design and installation, cannot be located in **Chapter 4**, which covers Equipment for General Use. Naturally, this is due to modern technology developing highly sophisticated equipment, for which the rules in **Chapter 4** are not adequate.

If an engineer is designing an electrical system where elevators are to be installed in an office building, **Article 620** will contain the requirements necessary for such specialized equipment and its associated apparatus. Note, anytime special equipment is designed, installed and inspected, one of the 600 (series) Article numbers of **Chapter 6** will be selected, based upon the specialized equipment involved.

For example, an electrician designing an electrical circuit for a welder must consult **Article 630** and comply with the requirements pertaining to the type of welder being installed.

For those users interested in the requirements necessary to design and install disconnects for deenergizing power circuits supplying cranes, they must utilize **Article 610** to find these rules.

NEC Ch. 6 – Article 600
Part I – 600.1

Type of Change		Panel Action		UL	UL 508	API 500	API 505	OSHA
Revision		Accept		48 and 2161	-	-	-	1910.306(a)
ROP		ROC		NFPA 70E	NFPA 70B	NFPA 79	NFPA	NEMA
pg. 1482	# 18-106	pg. 449	# 18-71	430.1	-	-	-	-
log: 3137	CMP: 18	log: 837	Submitter: Stephen G. Kieffer			2002 NEC: 600.1		IEC: -

2002 NEC – 600.1 Scope

This article covers the installation of conductors and equipment for electric signs and outline lighting ~~as defined in Article 100~~.

> ~~FPN: As defined in Article 100, electric signs and outline lighting include~~ all ~~products and~~ installations ~~utilizing~~ neon tubing, such as signs, decorative elements, skeleton tubing, or art forms.

2002 NEC – 600.1 Scope

This article covers the installation of conductors and equipment for electric signs and outline lighting. All installations <u>and equipment using</u> neon tubing, such as signs, decorative elements, skeleton tubing, or art forms<u>, are covered by this article</u>.

Author's Substantiation. This revision clarifies that all installation and equipment using neon tubing is covered by this article.

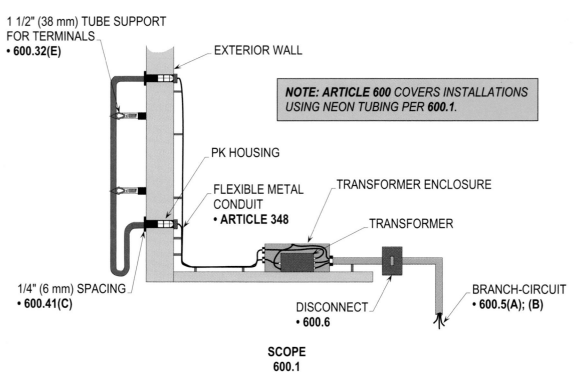

SCOPE
600.1

Purpose of Change: To make it clear that installations using neon tubing are governed in the scope section of **Article 600**.

NEC Ch. 6 – Article 600
Part I – 600.2

Type of Change		Panel Action		UL	UL 508	API 500	API 505	OSHA
New Definition		Accept in Principle in Part		48 and 2161	-	-	-	1910.306(a)
ROP		ROC		NFPA 70E	NFPA 70B	NFPA 79	NFPA	NEMA
pg. 1483	# 18-107	pg. 450	# 18-79	430.1	-	-	-	-
log: 3138	CMP: 18	log: 772	Submitter: Stephen G. Kieffer			2002 NEC: -		IEC: -

2005 NEC – 600.2 Definitions

Section Sign. A sign or outline lighting system, shipped as subassemblies, that requires field-installed wiring between the subassemblies to complete the overall sign.

Author's Substantiation. A new definition has been added to define section signs subassembled in the field to complete the overall sign.

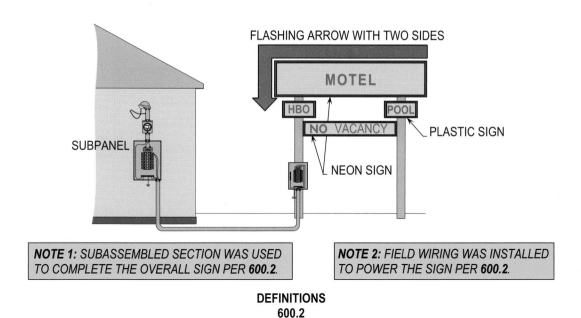

FLASHING ARROW WITH TWO SIDES

MOTEL

HBO   POOL

NO VACANCY

PLASTIC SIGN

NEON SIGN

SUBPANEL

NOTE 1: *SUBASSEMBLED SECTION WAS USED TO COMPLETE THE OVERALL SIGN PER 600.2.*

NOTE 2: *FIELD WIRING WAS INSTALLED TO POWER THE SIGN PER 600.2.*

DEFINITIONS
600.2

Purpose of Change: To recognize section signs subassembled in the field to complete the overall sign.

Type of Change	Panel Action	UL	UL 508	API 500	API 505	OSHA
New Section	Accept in Principle	-	1310 and 1585	-	-	1910.305(j)(5)(i)(D)
ROP	ROC	NFPA 70E	NFPA 70B	NFPA 79	NFPA	NEMA
pg. 1485 # 18-110	pg. - # -	450.3	-	-	-	-
log: 2887 CMP: 18	log: -	Submitter: Muhammad Khan		2002 NEC: -		IEC: -

2005 NEC – <u>600.24 Class 2 Power Sources</u>

<u>In addition to the requirements of Article 600, signs and outline lighting systems supplied by Class 2 transformers, power supplies, and power sources shall comply with 725.41.</u>

Author's Substantiation. A new section has been added to recognize that signs and outline lighting systems are now being installed with Class 2 power sources but shall comply with 725.41.

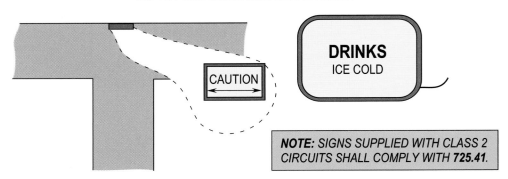

SIGNS SUPPLIED BY A CLASS 2 XFMR POWER
SUPPLY OR POWER SOURCE PER **600.24**

CAUTION

DRINKS
ICE COLD

NOTE: SIGNS SUPPLIED WITH CLASS 2
CIRCUITS SHALL COMPLY WITH **725.41.**

CLASS 2 POWER SOURCES
600.24

Purpose of Change: To clarify that signs can be supplied with a Class 2 power source where in accordance with **725.41** of the **2005 NEC**.

NEC Ch. 6 – Article 600
Part – 604.6(A)(3)

Type of Change		Panel Action		UL	UL 508	API 500	API 505	OSHA
Revision		Accept		62 and 183	-	-	-	1910.305(g)
ROP		ROC		NFPA 70E	NFPA 70B	NFPA 79	NFPA	NEMA
pg. 1497	# 19-154	pg. -	# -	410.6 and 420.7	-	-	-	-
log: 2323	CMP: 19	log: -	Submitter: Vince Baclawski			2002 NEC: 604.6(A)(3)		IEC: -

2002 NEC – 604.6(A)(3) Flexible Cord

Flexible cord suitable for hard usage, with minimum 12 AWG conductors, shall be permitted as part of a listed-made assembly not exceeding 1.8 m (6 ft) in length when making a transition between components of a manufactured wiring system and utilization equipment not permanently secured to the building structure. The cord shall be visible for its entire length and shall not be subject to strain or physical damage.

2005 NEC – 604.6(A)(3) Flexible Cord

Flexible cord suitable for hard usage, with minimum 12 AWG conductors, shall be permitted as part of a listed factory-made assembly not exceeding 1.8 m (6 ft) in length when making a transition between components of a manufactured wiring system and utilization equipment, <u>other than luminaires (fixtures),</u> not permanently secured to the building structure. The cord shall be visible for its entire length and shall not be subject to strain or physical damage.

Author's Substantiation. This revision clarifies that flexible cord shall not be limited in length to 1.8 m (6 ft) or less where connecting luminaires (fixtures).

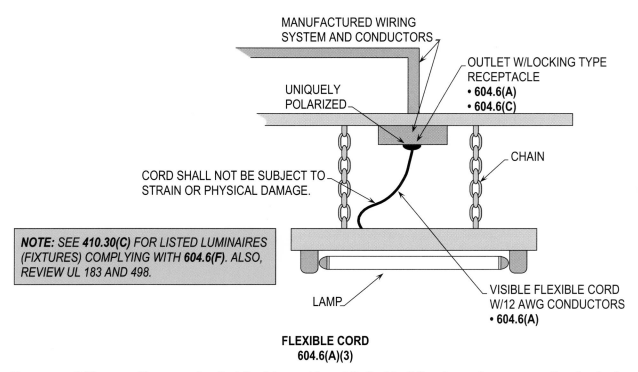

FLEXIBLE CORD
604.6(A)(3)

Purpose of Change: To recognize that flexible cord is not limited to 6 ft or less where connecting luminaires (fixtures).

NEC Ch. 6 – Article 610
Part VII – 610.61

Type of Change		Panel Action		UL	UL 508	API 500	API 505	OSHA
Revision		Accept in Principle in Part		White Book	-	-	-	1910.306(b)
ROP		ROC		NFPA 70E	NFPA 70B	NFPA 79	NFPA	NEMA
pg. 1506	# 12-15	pg. 458	# 12-10	410.10(F)(2)	-	-	-	ICS-8
log: 1340	CMP: 12	log: 3275	Submitter: Tom Young			2002 NEC: 610.61		IEC: -

2002 NEC – 610.61 Grounding

All exposed non-current-carrying metal parts of cranes, monorail hoists, hoists, and accessories, including pendant controls, shall be metallically joined together into a continuous electrical conductor so that the entire crane or hoist will be grounded in accordance with Article 250. Moving parts, other the removable accessories or attachments, that have metal-to-metal bearing surfaces shall be considered to be electrically connected to each other through the bearing surfaces for grounding purposes. The trolley frame and bridge frame shall be considered as electrically grounded through the bridge and trolley wheels and ~~their~~ respective tracks ~~unless local conditions, such as paint or other insulating material, prevent reliable metal-to-metal contact. In this case,~~ a separate bonding conductor shall be provided.

2005 NEC – 610.61 Grounding

All exposed non-current-carrying metal parts of cranes, monorail hoists, hoists, and accessories, including pendant controls, shall be metallically joined together into a continuous electrical conductor so that the entire crane or hoist will be grounded in accordance with Article 250. Moving parts, other the removable accessories or attachments that have metal-to-metal bearing surfaces, shall be considered to be electrically connected to each other through the bearing surfaces for grounding purposes. The trolley frame and bridge frame shall <u>not</u> be considered as electrically grounded through the bridge and trolley wheels and <u>its</u> respective tracks. A separate bonding conductor shall be provided.

Author's Substantiation. This revision clarifies that the trolley frame and bridge frame shall not be considered as electrically grounded through the bridge and trolley wheels and its respective tracks.

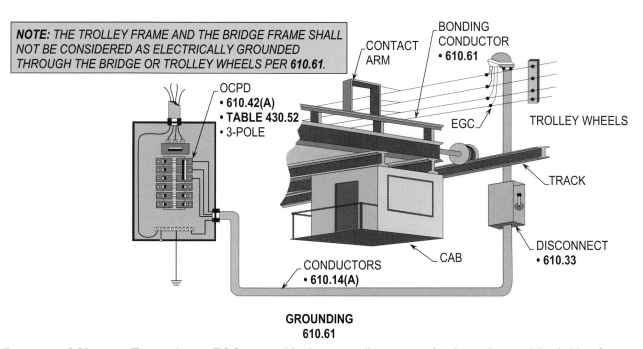

NOTE: THE TROLLEY FRAME AND THE BRIDGE FRAME SHALL NOT BE CONSIDERED AS ELECTRICALLY GROUNDED THROUGH THE BRIDGE OR TROLLEY WHEELS PER **610.61**.

CONTACT ARM

BONDING CONDUCTOR
• **610.61**

OCPD
• **610.42(A)**
• **TABLE 430.52**
• 3-POLE

EGC

TROLLEY WHEELS

TRACK

DISCONNECT
• **610.33**

CAB

CONDUCTORS
• **610.14(A)**

GROUNDING
610.61

Purpose of Change: To require an EGC to provide the grounding means for the trolley and the bridge frame.

NEC Ch. 6 – Article 620
Part I – 620.2

Type of Change	Panel Action	UL	UL 508	API 500	API 505	OSHA
New Definitions	Accept in Principle	104	Ch. 6	-	-	1910.306(c)(1)
ROP		ROC	NFPA 70E	NFPA 70B	NFPA 79	NFPA	NEMA
pg. 1507	# 12-16	pg. 459	# 12-11	430.2	-	Ch. 9, 10, and 12	-	-
log: 340	CMP: 12	log: 429	Submitter: Andy Juhasz		2002 NEC: -		IEC: -

2005 NEC – 620.2 Definitions

Control Room (for Elevator, Dumbwaiter). An enclosed control space outside the hoistway, intended for full bodily entry, that contains the elevator motor controller. The room could also contain electrical and/or mechanical equipment used directly in connection with the elevator or dumbwaiter but not the electric driving machine or the hydraulic machine.

Control Space (for Elevator, Dumbwaiter). A space inside or outside the hoistway, intended to be accessed with or without full bodily entry, that contains the elevator motor controller. This space could also contain electrical and/or mechanical equipment used directly in connection with the elevator or dumbwaiter but not the electric driving machine or the hydraulic machine.

Machine Room (for Elevator, Dumbwaiter). An enclosed machinery space outside the hoistway, intended for full bodily entry, that contains the electrical driving machine or the hydraulic machine. The room could also contain electrical and/or mechanical equipment used directly in connection with the elevator or dumbwaiter.

Machinery Space (for Elevator, Dumbwaiter). A space inside or outside the hoistway, intended to be accessed with or without full bodily entry, that contains elevator or dumbwaiter mechanical equipment, and could also contain electrical equipment used directly in connection with the elevator or dumbwaiter. This space could also contain the electric driving machine or the hydraulic machine.

Author's Substantiation. New definitions have been added to recognize the terms for control room (for elevator, dumbwaiter), control space (for elevator, dumbwaiter), machine room (for elevator, dumbwaiter), and machinery space (for elevator, dumbwaiter) to coordinate their usage with the proposed revisions to A17.1, *Safety Code for Elevators and Escalators*.

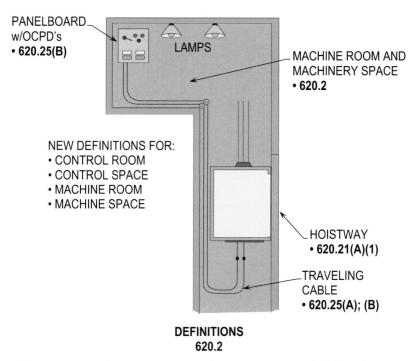

PANELBOARD
w/OCPD's
• **620.25(B)**

LAMPS

MACHINE ROOM AND
MACHINERY SPACE
• **620.2**

NEW DEFINITIONS FOR:
• CONTROL ROOM
• CONTROL SPACE
• MACHINE ROOM
• MACHINE SPACE

HOISTWAY
• **620.21(A)(1)**

TRAVELING
CABLE
• **620.25(A); (B)**

**DEFINITIONS
620.2**

Purpose of Change: To add new definitions that were introduced in the 2002 update and coordinate their use in various other sections.

NEC Ch. 6 – Article 620
Part III – 620.22(A)

Type of Change	Panel Action	UL	UL 508	API 500	API 505	OSHA
New Sentence	Accept in Principle	104	Ch. 6	-	-	1910.306(c)(1)
ROP		ROC		NFPA 70E	NFPA 70B	NFPA 79	NFPA	NEMA

ROP		ROC		NFPA 70E	NFPA 70B	NFPA 79	NFPA	NEMA
pg. 1510	# 12-23	pg. -	# -	430.2	-	Ch. 9, 10, and 12	-	-
log: 2418	CMP: 12	log: -	Submitter: William Davis			2002 NEC: 620.22(A)		IEC: -

2002 NEC – 620.22(A) Car Light Source

A separate branch circuit shall supply the car lights, receptacle(s), auxiliary lighting power source, and ventilation on each elevator car. The overcurrent device protecting the branch circuit shall be located in the elevator machine room or control room/machinery space or control space.

2005 NEC – 620.22(A) Car Light Source

A separate branch circuit shall supply the car lights, receptacle(s), auxiliary lighting power source, and ventilation on each elevator car. The overcurrent device protecting the branch circuit shall be located in the elevator machine room or control room/machinery space or control space.

<u>Required lighting shall not be connected to the load side of a ground-fault circuit interrupter.</u>

Author's Substantiation. A new sentence has been added to clarify that car lighting shall not be connected to the load side of a ground-fault circuit interrupter.

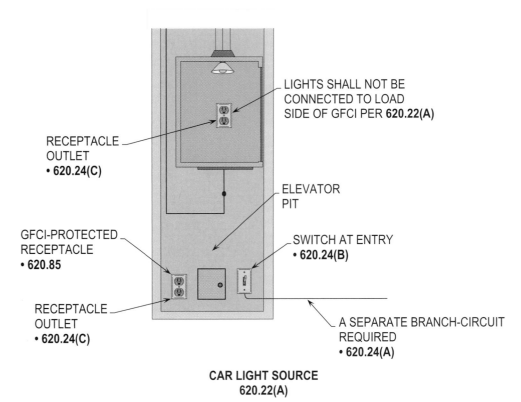

RECEPTACLE OUTLET
• 620.24(C)

LIGHTS SHALL NOT BE CONNECTED TO LOAD SIDE OF GFCI PER 620.22(A)

ELEVATOR PIT

GFCI-PROTECTED RECEPTACLE
• 620.85

SWITCH AT ENTRY
• 620.24(B)

RECEPTACLE OUTLET
• 620.24(C)

A SEPARATE BRANCH-CIRCUIT REQUIRED
• 620.24(A)

CAR LIGHT SOURCE
620.22(A)

Purpose of Change: To clarify that car lighting shall not be connected to the load side of the GFCI.

NEC Ch. 6 – Article 625
Part IV – 625.26

Type of Change		Panel Action		UL	UL 508	API 500	API 505	OSHA
New Section		Accept in Principle		583 and 1564	-	-	-	-
ROP		ROC		NFPA 70E	NFPA 70B	NFPA 79	NFPA	NEMA
pg. 1517	# 12-35	pg. 460	# 12-20	-	-	-	-	PE 5
log: 2639	CMP: 12	log: 636	Submitter: Frank C. Lambert			2002 NEC: -		IEC: -

2005 NEC – <u>625.26 Interactive Systems</u>

<u>Electric vehicle supply equipment and other parts of a system, either on-board or off-board the vehicle, that are identified for and intended to be interconnected to a vehicle and also serve as an optional standby system or an electric power production source or provide for bi-directional power feed shall be listed as suitable for that purpose. When used as an optional standby system, the requirements of Article 702 shall apply, and when used as an electric power production source, the requirements of Article 705 shall apply.</u>

Author's Substantiation. A new section has been added to recognize interactive systems that are either on-board or off-board the vehicle.

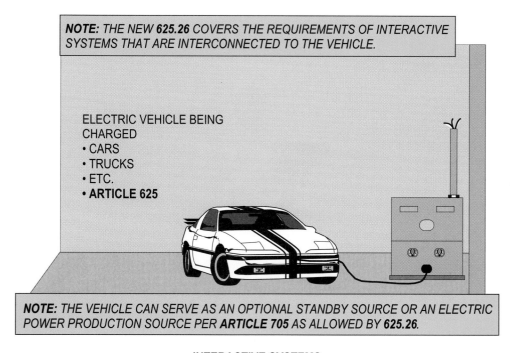

NOTE: THE NEW 625.26 COVERS THE REQUIREMENTS OF INTERACTIVE SYSTEMS THAT ARE INTERCONNECTED TO THE VEHICLE.

ELECTRIC VEHICLE BEING CHARGED
• CARS
• TRUCKS
• ETC.
• **ARTICLE 625**

NOTE: THE VEHICLE CAN SERVE AS AN OPTIONAL STANDBY SOURCE OR AN ELECTRIC POWER PRODUCTION SOURCE PER ARTICLE 705 AS ALLOWED BY 625.26.

INTERACTIVE SYSTEMS
625.26

Purpose of Change: To add requirements for interactive systems either on-board or off-board the vehicle.

NEC Ch. 6 – Article 630
Part I – 630.1

Type of Change		Panel Action		UL	UL 508	API 500	API 505	OSHA
Revision		Accept in Principle in Part		-	-	-	-	-
ROP		ROC		NFPA 70E	NFPA 70B	NFPA 79	NFPA	NEMA
pg. 1518	# 12-36	pg. -	# -	-	-	-	-	-
log: 2318	CMP: 12	log: -	Submitter: Vince Baclawski			2002 NEC: 630.1		IEC: -

2002 NEC – 630.1 Scope

This article covers electric arc welding, resistance welding ~~apparatus~~, and other similar welding equipment that is connected to an electric supply system.

2005 NEC – 630.1 Scope

This article covers <u>apparatus for</u> electric arc welding, resistance welding, <u>plasma cutting,</u> and other similar welding <u>and cutting process</u> equipment that is connected to an electric supply system.

Author's Substantiation. This revision recognizes plasma cutting as a cutting process.

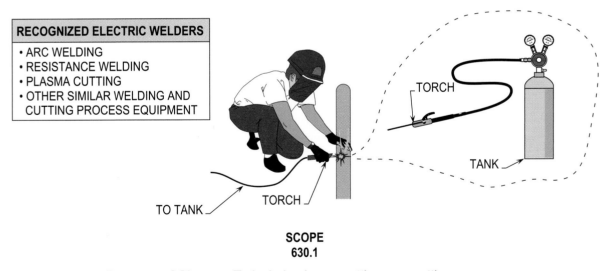

RECOGNIZED ELECTRIC WELDERS
• ARC WELDING
• RESISTANCE WELDING
• PLASMA CUTTING
• OTHER SIMILAR WELDING AND CUTTING PROCESS EQUIPMENT

TORCH
TANK
TO TANK
TORCH

**SCOPE
630.1**

Purpose of Change: To include plasma cutting as a cutting process.

NEC Ch. 6 – Article 630
Part II – 630.11(B)

Type of Change		Panel Action		UL	UL 508	API 500	API 505	OSHA
Revision		Accept		551	-	-	-	1910.306(d)
ROP		ROC		NFPA 70E	NFPA 70B	NFPA 79	NFPA	NEMA
pg. 1519	# 12-39	pg. -	# -	430.4	-	-	-	EW 1
log: 3243	CMP: 12	log: -	Submitter: Paul Dobrowsky			2002 NEC: 630.11(B)		IEC: -

2002 NEC – 630.11(B) Group of Welders

Conductor ampacity shall be based on the individual currents determined in 630.11(A) as the sum of 100 percent of the two largest welders, plus 85 percent of the third largest welder, plus 70 percent of the fourth largest welder, plus 60 percent of all remaining welders.

2002 NEC – 630.11(B) Group of Welders

Minimum conductor ampacity shall be based on the individual currents determined in 630.11(A) as the sum of 100 percent of the two largest welders, plus 85 percent of the third largest welder, plus 70 percent of the fourth largest welder, plus 60 percent of all remaining welders.

Author's Substantiation. This revision clarifies that the minimum ampacity shall be used to determine the load and select the size conductors.

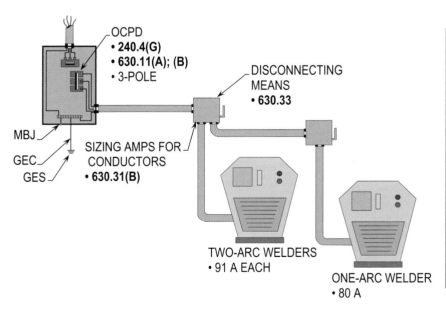

GROUP OF WELDERS
630.11(B)

Purpose of Change: To revise section to clarify that it is the minimum ampacity used to determine load and select conductors.

NEC Ch. 6 – Article 640
Part I – 640.6

Type of Change		Panel Action		UL	UL 508	API 500	API 505	OSHA
Revision		Accept in Principle		1270	-	-	-	-
ROP		ROC		NFPA 70E	NFPA 70B	NFPA 79	NFPA	NEMA
pg. 1524	# 12-47	pg. 462	# 12-29	-	-	-	-	-
log: 854	CMP: 12	log: 3130	Submitter: James E. Brunssen			2002 NEC: 640.6		IEC: -

2002 NEC – 640.6 Mechanical Execution of Work

Equipment and cabling shall be installed in a neat and workmanlike manner. Cables installed exposed on the surface of ceiling and sidewalls shall be supported by the structural components of the building structure in such a manner that the cable will not be damaged by normal building use. Such cables shall be attached to structural components by straps, staples, hangers, or similar fittings designed and installed so as not to damage the cable. The installation shall also conform with 300.4(D).

2005 NEC – 640.6 Mechanical Execution of Work

Equipment and cabling shall be installed in a neat and workmanlike manner. Cables installed exposed on the surface of ceilings and sidewalls shall be supported in such a manner that the cables will not be damaged by normal building use. Such cables shall be supported by straps, staples, hangers, or similar fittings designed and installed so as not to damage the cable. The installation shall also conform to 300.4(D) and 300.11.

Author's Substantiation. This revision clarifies the supporting means for cables used for audio signal processing, amplification, and reproduction equipment.

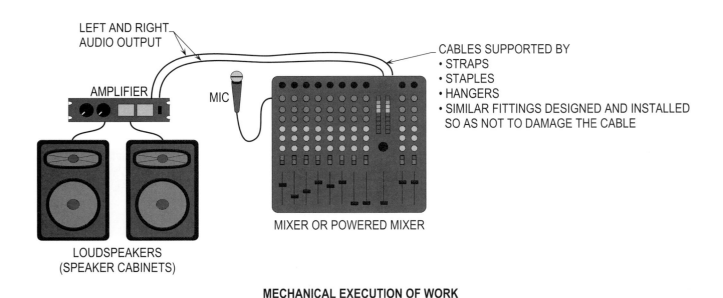

MECHANICAL EXECUTION OF WORK
640.6

Purpose of Change: To recognize supporting means of cables to the exposed surface of ceilings and sidewalls.

NEC Ch. 6 – Article 645
Part – 645.17

Type of Change		Panel Action		UL	UL 508	API 500	API 505	OSHA
New Section		Accept		75 and 1950	32	-	-	1910.306(e)
ROP		ROC		NFPA 70E	NFPA 70B	NFPA 79	NFPA	NEMA
pg. 1067	# 9-110	pg. 325	# 12-4	340 and 430.5	21.4.3	7.2.12	-	-
log: 3519	CMP: 9	log: 607	Submitter: Stephen W. McCluer			2002 NEC: -		IEC: 707

2005 NEC – 645.17 Power Distribution Units

Power distribution units that are used for information technology equipment shall be permitted to have multiple panelboards within a single cabinet, provided that each panelboard has no more than 42 overcurrent devices and the power distribution unit is utilization equipment listed for information technology application.

Author's Substantiation. A new section has been added to provide requirements when using panelboards in power distribution units.

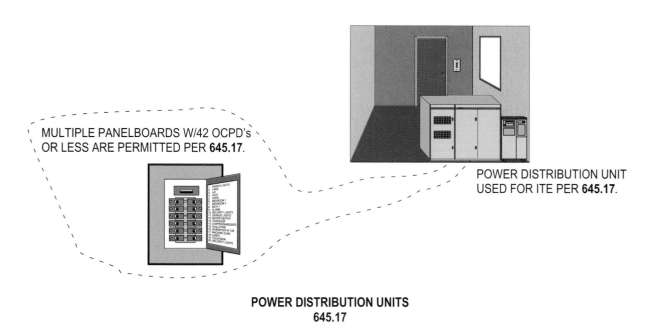

MULTIPLE PANELBOARDS W/42 OCPD's OR LESS ARE PERMITTED PER **645.17**.

POWER DISTRIBUTION UNIT USED FOR ITE PER **645.17**.

POWER DISTRIBUTION UNITS
645.17

Purpose of Change: To outline requirements when using panelboards in power distribution units.

NEC Ch. 6 – Article 647
Part – 647.4(A)

Type of Change	Panel Action	UL	UL 508	API 500	API 505	OSHA
Revision	Accept in Principle	75	-	-	-	-
ROP		ROC		NFPA 70E	NFPA 70B	NFPA 79	NFPA	NEMA

ROP		ROC		NFPA 70E	NFPA 70B	NFPA 79	NFPA	NEMA
pg. 1536	# 12-65	pg. -	# -	340 and 430.5	Ch. 12	Ch. 11	-	-
log: 3030	CMP: 12	log: -		Submitter: Todd Lottmann		2002 NEC: 647.4(A)		IEC: 707

2002 NEC – 647.4(A) Panelboards and Overcurrent Protection

Use of standard single-phase panelboards and distribution equipment with a higher voltage rating shall be permitted. The system shall be clearly marked on the face of the panel or on the inside of the panel doors. Common-trip two-pole circuit breakers that are identified for ~~operation~~ at the system voltage shall be provided for both ungrounded conductors in all feeders and branch circuits.

2005 NEC – 647.4(A) Panelboards and Overcurrent Protection

Use of standard single-phase panelboards and distribution equipment with a higher voltage rating shall be permitted. The system shall be clearly marked on the face of the panel or on the inside of the panel doors. Common trip two-pole circuit breakers <u>or a combination two-pole fused disconnecting means</u> that are identified for <u>use</u> at the system voltage shall be provided for both ungrounded conductors in all feeders and branch circuits. <u>Branch circuits and feeders shall be provided with a means to simultaneously disconnect all ungrounded conductors.</u>

Author's Substantiation. This revision clarifies that common trip two-pole circuit breakers or combination two-pole fused disconnecting means shall be permitted for disconnecting simultaneously all ungrounded conductors.

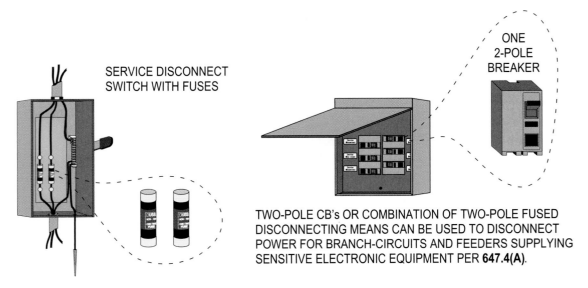

SERVICE DISCONNECT
SWITCH WITH FUSES

ONE
2-POLE
BREAKER

TWO-POLE CB's OR COMBINATION OF TWO-POLE FUSED
DISCONNECTING MEANS CAN BE USED TO DISCONNECT
POWER FOR BRANCH-CIRCUITS AND FEEDERS SUPPLYING
SENSITIVE ELECTRONIC EQUIPMENT PER **647.4(A)**.

**PANELBOARDS AND OVERCURRENT PROTECTION
647.4(A)**

Purpose of Change: To allow common trip two-pole CB's or combination two-pole fused disconnecting means for disconnecting simultaneously all ungrounded conductors.

NEC Ch. 6 – Article 670
Part – 670.3(A)(1) through (A)(5)

Type of Change		Panel Action		UL	UL 508	API 500	API 505	OSHA
Revision		Accept		White Book	67	-	-	-
ROP		ROC		NFPA 70E	NFPA 70B	NFPA 79	NFPA	NEMA
pg. 1546	# 12-82	pg. -	# -	400.11	6.2	17.4	-	ICS 6
log: 3047	CMP: 12	log: -	Submitter: Todd F. Lottmann			2002 NEC: 670.3(A)		IEC: 60204-1

2002 NEC – 670.3(A) Permanent Nameplate

A permanent nameplate ~~that lists supply voltage, phase, frequency, full-load current, the maximum ampere rating of the short-circuit and ground-fault protective device, ampere rating of largest motor or load, short-circuit interrupting rating of the machine overcurrent-protective device, if furnished, and diagram number~~ shall be attached to the control equipment enclosure or machine ~~where~~ plainly visible after installation.

2005 NEC – 670.3(A) Permanent Nameplate

A permanent nameplate shall be attached to the control equipment enclosure or machine <u>and shall be</u> plainly visible after installation. <u>The nameplate shall include the following information:</u>

(1) <u>Supply voltage, phase, frequency, and full-load current</u>

(2) <u>Maximum ampere rating of the short-circuit and ground-fault protective device</u>

(3) <u>Ampere rating of largest motor, from the motor nameplate, or load</u>

(4) <u>Short circuit current rating of the machine industrial control panel based on one of the following:</u>

    a. <u>Short circuit current rating of a listed and labeled machine control enclosure or assembly</u>

    b. <u>Short circuit current rating established utilizing an approved method</u>

        FPN: <u>UL 508A-2001, Supplement SB, is an example of an approved method.</u>

(5) <u>Electrical diagram number(s) or the number of the index to the electrical drawings</u>

Author's Substantiation. The text has been placed into a list for usability of the marking requirements.

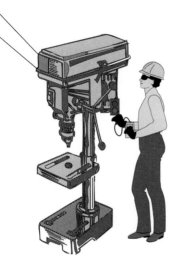

MACHINE NAMEPLATE DATA

(1) Supply voltage, phase, frequency, and full-load current

(2) Maximum ampere rating of the short-circuit and ground-fault protective device

(3) Ampere rating of largest motor, from the motor nameplate, or load

(4) Short circuit current rating of the machine industrial control panel based on one of the following:
   a. Short circuit current rating of a listed and labeled machine-control enclosure or assembly
   b. Short circuit current rating established utilizing an approved method

(5) Electrical diagram number(s) or the number of the index to the electrical drawings

MACHINE NAMEPLATE DATA
670.3(A)(1) THRU (A)(5)

Purpose of Change: To number the text into a list for increasing usability of the marking requirements.

NEC Ch. 6 – Article 670
Part – 670.4(B) and (C)

Type of Change		Panel Action		UL	UL 508	API 500	API 505	OSHA
Revision		Accept		363	16	-	-	1910.305(j)(4)(ii)
ROP		ROC		NFPA 70E	NFPA 70B	NFPA 79	NFPA	NEMA
pg. 1547	# 12-83	pg. -	# -	420.10(E)(2)	11.5	5.5	-	ICS 6
log: 3048	CMP: 12	log: -	Submitter: Todd F. Lottmann			2002 NEC: -		IEC: 60204-1

2002 NEC – 670.4(B) Overcurrent Protection

A machine shall be considered as an individual unit and therefore shall be provided with a disconnecting means. The disconnecting means shall be permitted to be supplied by branch circuits protected by either fuses or circuit breakers. The disconnecting means shall not be required to incorporate overcurrent protection. Where furnished as part of the machine, overcurrent protection shall consist of a single circuit breaker or set of fuses, the machine shall bear the marking required in 670.3, and the supply conductors shall be considered either as feeders or taps as covered by 240.21.

The rating or setting of the overcurrent protective device for the circuit supplying the machine shall not be greater than the sum of the largest rating or setting of the branch-circuit short-circuit and ground-fault protective device provided with the machine, plus 125 percent of the full-load current rating of all resistance heating loads, plus the sum of the full-load currents of all other motors and apparatus that could be in operation at the same time.

> **Exception:** Where one or more instantaneous trip circuit breakers or motor short-circuit protectors are used for motor branch-circuit short-circuit and ground-fault protection as permitted by 430.52(C), the procedure specified above for determining the maximum rating of the protective device for the circuit supplying the machine shall apply with the following provision: For the purpose of the calculation, each instantaneous trip circuit breaker or motor short-circuit protector shall be assumed to have a rating not exceeding the maximum percentage of motor full-load current permitted by Table 430.52 for the type of machine supply circuit protective device employed.

Where no branch-circuit short-circuit and ground-fault protective device is provided with the machine, the rating or setting of the overcurrent protective device shall be based on 430.52 and 430.53, as applicable.

2005 NEC – 670.4(B) Disconnecting Means

A machine shall be considered as an individual unit and therefore shall be provided with disconnecting means. The disconnecting means shall be permitted to be supplied by branch circuits protected by either fuses or circuit breakers. The disconnecting means shall not be required to incorporate overcurrent protection.

> **Exception:** Where one or more instantaneous trip circuit breakers or motor short-circuit protectors are used for motor branch-circuit short-circuit and ground-fault protection as permitted by 430.52(C), the procedure specified above for determining the maximum rating of the protective device for the circuit supplying the machine shall apply with the following provision: For the purpose of the calculation, each instantaneous trip circuit breaker or motor short-circuit protector shall be assumed to have a rating not exceeding the maximum percentage of motor full-load current permitted by Table 430.52 for the type of machine supply circuit protective device employed.

Where no branch-circuit short-circuit and ground-fault protective device is provided with the machine, the rating or setting of the overcurrent protective device shall be based on 430.52 and 430.53, as applicable.

670.4(C) Overcurrent Protection

Where furnished as part of the machine, overcurrent protection <u>of each supply circuit</u> shall consist of a single circuit breaker or set of fuses, the machine shall bear the marking required in 670.3 and the supply conductors shall be considered either as feeders or taps as covered by 240.21.

The rating or setting of the overcurrent protective device for the circuit supplying the machine shall not be greater than the sum of the largest rating or setting of the branch-circuit short-circuit and ground-fault protective device provided with the machine, plus 125 percent of the full-load current rating of all resistance heating loads, plus the sum of the full-load currents of all other motors and apparatus that could be in operation at the same time.

Author's Substantiation. This revision separates the requirements into two subsections for determining the requirements for the disconnecting means and overcurrent protection for industrial machinery.

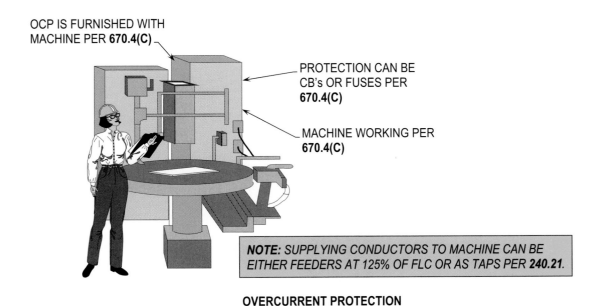

OCP IS FURNISHED WITH MACHINE PER **670.4(C)**

PROTECTION CAN BE CB's OR FUSES PER **670.4(C)**

MACHINE WORKING PER **670.4(C)**

NOTE: SUPPLYING CONDUCTORS TO MACHINE CAN BE EITHER FEEDERS AT 125% OF FLC OR AS TAPS PER **240.21**.

OVERCURRENT PROTECTION
670.4(C)

Purpose of Change: To revise title heads and separate the two topics in existing **670.4(B)** into **670.4(C)** with requirements for overcurrent protection.

NEC Ch. 6 – Article 670
Part – 670.5

Type of Change		Panel Action		UL	UL 508	API 500	API 505	OSHA
Deletion		Accept in Principle		-	-	-	-	Table S-1
ROP		ROC		NFPA 70E	NFPA 70B	NFPA 79	NFPA	NEMA
pg. 1547	# 12-85	pg. -	# -	Table 400.15(A)(1)	-	-	-	-
log: 3261	CMP: 12	log: -	Submitter: Gary J. Locke			2002 NEC: 670.5		IEC: 60204-1

2002 NEC – ~~670.5 Clearance~~

~~Where the conditions of maintenance and supervision ensure that only qualified persons service the installation, the dimensions of the working space in the direction of access to live parts operating at not over 150 volts line-to-line or line-to-ground that are likely to require examination, adjustment, servicing, or maintenance while energized shall be a minimum of 750 mm (2 1/2 ft). Where controls are enclosed in cabinets, the door(s) shall open at least 90 degrees or be removable.~~

Author's Substantiation. This section has been deleted because the text is redundant with that of NFPA 79, *Electrical Standard for Industrial Machinery*.

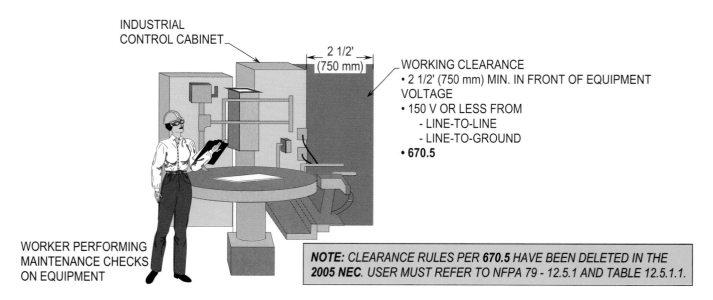

INDUSTRIAL CONTROL CABINET

2 1/2' (750 mm)

WORKING CLEARANCE
• 2 1/2' (750 mm) MIN. IN FRONT OF EQUIPMENT VOLTAGE
• 150 V OR LESS FROM
  - LINE-TO-LINE
  - LINE-TO-GROUND
• 670.5

WORKER PERFORMING MAINTENANCE CHECKS ON EQUIPMENT

NOTE: *CLEARANCE RULES PER* **670.5** *HAVE BEEN DELETED IN THE* **2005 NEC***. USER MUST REFER TO NFPA 79 - 12.5.1 AND TABLE 12.5.1.1.*

CLEARANCE (SECTION DELETED)
670.5

Purpose of Change: To delete clearance requirements because clearance requirements for industrial machinery are outlined in NFPA 79.

NEC Ch. 6 – Article 680
Part I – 680.8

Type of Change		Panel Action		UL	UL 508	API 500	API 505	OSHA
New Text		Accept in Principle		676A	-	-	-	1910.306(j)
ROP		ROC		NFPA 70E	NFPA 70B	NFPA 79	NESC	NEMA
pg. 1554	# 17-61	pg. 467	# 17-102	430.10	-	-	234 and Table 234-3	-
log: 1584	CMP: 17	log: 1872	Submitter: Michael J. Johnston			2002 NEC: 680.8		IEC: 702

2002 NEC – 680.8 Overhead Conductor Clearances

2005 NEC – 680.8 Overhead Conductor Clearances

Overhead conductors shall meet the clearance requirements in this section. Where a minimum clearance from the water level is given, the measurement shall be taken from the maximum water level of the specified body of water.

Author's Substantiation. New explanatory text has been added to this section before applying the clearance requirements for overhead conductors.

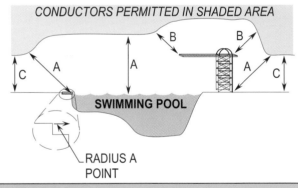

CONDUCTORS PERMITTED IN SHADED AREA

SWIMMING POOL

RADIUS A POINT

NOTE: *WHERE A MINIMUM CLEARANCE FROM THE WATER LEVEL IS GIVEN, SUCH MEASUREMENT SHALL BE TAKEN FROM THE MAXIMUM WATER LEVEL OF THE SPECIFIED BODY OF WATER PER 680.8.*

Clearances required above pools, hot tubs, and spas, based upon a voltage of 750 volts or less to ground.

A - Clearances in any direction from the water level, edge of water surface, base of diving platform or permanently anchored raft.

B - Clearances in any direction to diving platform or tower.

C - Horizontal limit of clearance measured from inside wall of the pool.

Note: Local codes and ordinances may not permit conductors to pass over pools, adjacent equipment or fixtures.

Requirement	Messenger cable
A	22.5 ft. (6.9 m)
B	14.5 ft. (4.4 m)
C	This limit extends to the outer edge of the structures listed in A and B of this Table but not less than 10 ft. (3 m).

OVERHEAD CONDUCTOR CLEARANCES
680.8

Purpose of Change: To add explanatory text before jumping into clearance requirements in **680.8(A)**, **(B)**, and **(C)** and applying the various clearance measurements based on the maximum water level.

NEC Ch. 6 – Article 680
Part II – 680.23(B)(6)

Type of Change		Panel Action		UL	UL 508	API 500	API 505	OSHA
New Subdivision		Accept in Principle		676A	-	-	-	1910.306(j)
ROP		ROC		NFPA 70E	NFPA 70B	NFPA 79	NFPA	NEMA
pg. 1570	# 17-98	pg. 475	# 17-139	430.10(E)	-	-	-	-
log: 3273	CMP: 17	log: 1881	Submitter: Leonard F. Devine, Jr.			2002 NEC: -		IEC: 702

2005 NEC – <u>680.23(B)(6) Servicing</u>

<u>All luminaires shall be removable from the water for relamping or normal maintenance. Luminaires shall be installed in such a manner that personnel can reach the luminaire for relamping, maintenance, or inspection while on the deck or equivalently dry location.</u>

Author's Substantiation. A new subdivision has been added to recognize the servicing requirements for wet-niche luminaires (fixtures) installed in and under the water of the pool.

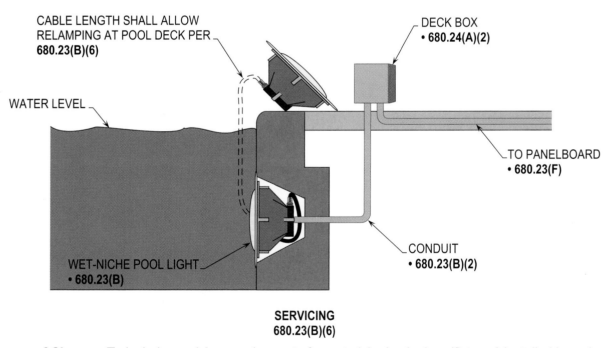

CABLE LENGTH SHALL ALLOW RELAMPING AT POOL DECK PER 680.23(B)(6)

DECK BOX
• 680.24(A)(2)

WATER LEVEL

TO PANELBOARD
• 680.23(F)

WET-NICHE POOL LIGHT
• 680.23(B)

CONDUIT
• 680.23(B)(2)

SERVICING
680.23(B)(6)

Purpose of Change: To include servicing requirements for wet-niche luminaires (fixtures) installed in and under the water of the pool.

NEC Ch. 6 – Article 680
Part II – 680.26(C)(3)a through c

140

Type of Change		Panel Action	UL	UL 508	API 500	API 505	OSHA
Revision		Accept in Principle in Part	467	-	-	-	1910.306(j)
ROP		ROC	NFPA 70E	NFPA 70B	NFPA 79	NFPA	NEMA
pg. 1582	# 17-122	pg. 481 # 17-174	430.10	-	-	-	-
log: 1699	CMP: 17	log: 3385 Submitter: Gregory L. Olsen			2002 NEC: 680.26(C)		IEC: 702

2002 NEC – 680.26(C) ~~Common~~ Bonding Grid

The parts specified in 680.26(B) shall be connected to a ~~common~~ bonding grid with a solid copper conductor, insulated, covered, or bare, not smaller than 8 AWG. Connection shall be made by exothermic welding or by pressure connectors or clamps that are labeled as being suitable for the purpose and are of stainless steel, brass, copper, or copper alloy. The ~~common~~ bonding grid shall be permitted to be any of the following:

(3) ~~A solid copper conductor, insulated, covered, or bare, not smaller than 8 AWG~~

2005 NEC – 680.26(C) <u>Equipotential</u> Bonding Grid

The parts specified in 680.26(B) shall be connected to an <u>equipotential</u> bonding grid with a solid copper conductor, insulated, covered, or bare, not smaller than 8 AWG <u>or rigid metal conduit of brass or other identified corrosion-resistant metal conduit</u>. Connection shall be made by exothermic welding or by <u>listed</u> pressure connectors or clamps that are labeled as being suitable for the purpose and are of stainless steel, brass, copper, or copper alloy. The <u>equipotential common</u> bonding grid shall <u>extend under paved walking surfaces for 1 m (3 ft) horizontally beyond the inside walls of the pool and shall</u> be permitted to be any of the following:

(3) <u>Alternate Means. This system shall be permitted to be constructed as specified in (a) through (c):</u>

    a. <u>Materials and Connections. The grid shall be constructed of minimum 8 AWG bare solid copper conductors. Conductors shall be bonded to each other at all points of crossing. Connections shall be made as required by 680.26(D).</u>

    b. <u>Grid Structure. The equipotential bonding grid shall cover the contour of the pool and the pool deck extending 1 m (3 ft) horizontally from the inside walls of the pool. The equipotential boding grid shall be arranged in a 300 mm (12 in.) by 300 mm (12 in.) network of conductors in a uniformly spaced perpendicular grid pattern with tolerance of 100 mm (4 in.).</u>

    c. <u>Securing. The below-grade grid shall be secured within or under the pool and deck media.</u>

Author's Substantiation. This subdivision has been revised to clarify the material and connections, grid structure, and securing for the equipotential bonding grid.

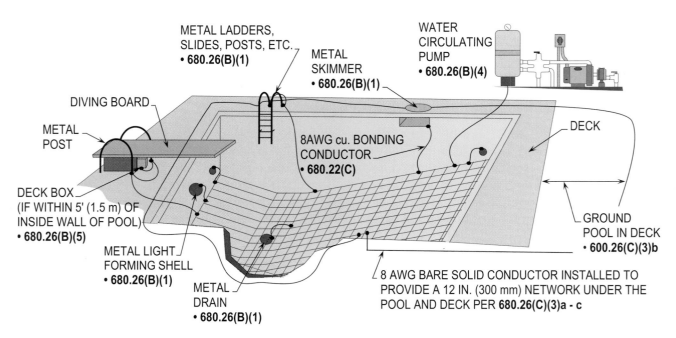

METAL LADDERS,
SLIDES, POSTS, ETC.
• **680.26(B)(1)**

METAL
SKIMMER
• **680.26(B)(1)**

WATER
CIRCULATING
PUMP
• **680.26(B)(4)**

DIVING BOARD

METAL
POST

8AWG cu. BONDING
CONDUCTOR
• **680.22(C)**

DECK

DECK BOX
(IF WITHIN 5' (1.5 m) OF
INSIDE WALL OF POOL)
• **680.26(B)(5)**

GROUND
POOL IN DECK
• **600.26(C)(3)b**

METAL LIGHT
FORMING SHELL
• **680.26(B)(1)**

METAL
DRAIN
• **680.26(B)(1)**

8 AWG BARE SOLID CONDUCTOR INSTALLED TO
PROVIDE A 12 IN. (300 mm) NETWORK UNDER THE
POOL AND DECK PER **680.26(C)(3)a - c**

EQUIPOTENTIAL BONDING GRID
680.26(C)(3)a THRU c

Purpose of Change: To revise 680.26(C) and add "a. Materials and Connections", "b. Grid Structure", and "c. Securing" with language that electricians can relate to.

NEC Ch. 6 – Article 600
Part III – 680.32

 141

Type of Change		Panel Action		UL	UL 508	API 500	API 505	OSHA
Revision		Accept		White Book	-	-	-	1910.306(j)
ROP		ROC		NFPA 70E	NFPA 70B	NFPA 79	NFPA	NEMA
pg. 1588	# 17-132a	pg. 483	# 17-177a	430.10(B)(3)	-	-	-	WD 1
log: CP1701	CMP: 17	log: CC1700	Submitter: CMP 17			2002 NEC: -		IEC: 702

2002 NEC – 680.32 Ground-Fault Circuit Interrupters Required

All electrical equipment, including power-supply cords, used with storable pools shall be protected by ground-fault circuit interrupters.

> **FPN:** For flexible cord usage, see 400.4.

2005 NEC – 680.32 Ground-Fault Circuit Interrupters Required

All electrical equipment, including power-supply cords, used with storable pools shall be protected by ground-fault circuit interrupters.

<u>All 125-volt receptacles located within 6.0 m (20 ft) of the inside walls of a storable pool shall be protected by a ground-fault circuit interrupter. In determining these dimensions, the distance to be measured shall be the shortest path the supply cord of an appliance connected to the receptacle would follow without piercing a floor, wall, ceiling, doorway with hinged or sliding door, window opening, or other effective permanent barrier.</u>

> **FPN:** For flexible cord usage, see 400.4.

Author's Substantiation. This revision clarifies that all 125-volt receptacles located with 6.0 m (20 ft) of the inside walls of a storable pool shall be protected by a ground-fault circuit interrupter.

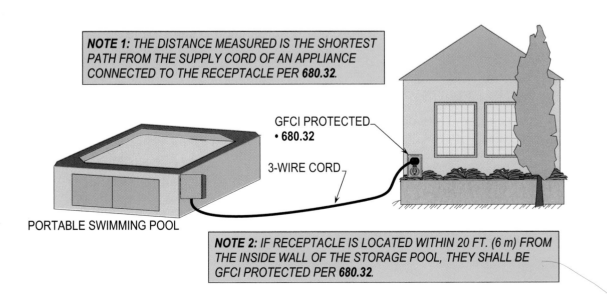

NOTE 1: THE DISTANCE MEASURED IS THE SHORTEST PATH FROM THE SUPPLY CORD OF AN APPLIANCE CONNECTED TO THE RECEPTACLE PER 680.32.

GFCI PROTECTED
• 680.32

3-WIRE CORD

PORTABLE SWIMMING POOL

NOTE 2: IF RECEPTACLE IS LOCATED WITHIN 20 FT. (6 m) FROM THE INSIDE WALL OF THE STORAGE POOL, THEY SHALL BE GFCI PROTECTED PER 680.32.

GROUND-FAULT CIRCUIT INTERRUPTER REQUIRED
680.32

Purpose of Change: To require storage pools to be GFCI protected, if receptacle is located within 20 ft. (6 m).

NEC Ch. 6 – Article 682
Part I – 682.2

 **142.1**

Type of Change		Panel Action		UL	UL 508	API 500	API 505	OSHA
New Article		Accept		-	-	-	-	-
ROP		ROC		NFPA 70E	NFPA 70B	NFPA 79	NFPA	NEMA
pg. 1599	# 17-154	pg. 484	# 17-184	-	-	-	-	-
log: 1625	CMP: 17	log: 829	Submitter: Neil F. LaBrake, Jr.			2002 NEC: -		IEC: -

2005 NEC – 682.2 Definitions

Artificially Made Bodies of Water. Bodies of water that have been constructed or modified to fit some decorative or commercial purpose such as, but not limited to, aeration ponds, fish farm ponds, storm retention basins, treatment ponds, irrigation (channel) facilities. Water depths may vary seasonally or be controlled.

Electrical Datum Plane. The electrical datum plane as used in this article is defined as follows:

(1) In land areas subject to tidal fluctuation, the electrical datum plane is a horizontal plane 600 mm (2 ft) above the highest tide level for the area occurring under normal circumstances, that is, highest high tide.

(2) In land areas not subject to tidal fluctuation, the electrical datum plane is a horizontal plane 600 mm (2 ft) above the highest water level for the area occurring under normal circumstances.

(3) In land areas subject to flooding, the electrical datum plane based on (1) or (2) above is a horizontal plane 600 mm (2 ft) above the point identified as the prevailing high water mark or an equivalent benchmark based on seasonal or storm-driven flooding from the authority having jurisdiction.

(4) The electrical datum plane for floating structures and landing stages that are (1) installed to permit rise and fall response to water level, without lateral movement, and (2) that are so equipped that they can rise to the datum plane established for (1) or (2) above, is a horizontal plane 750 mm (30 in.) above the water level at the floating structure or landing stage and a minimum of 300 mm (12 in.) above the level of the deck.

Equipotential Plane. An area where wire mesh or other conductive elements are on, embedded in, or placed under the walk surface within 75 mm (3 in.), bonded to all metal structures and fixed nonelectrical equipment that may become energized, and connected to the electrical grounding system to prevent a difference in voltage from developing within the plane.

Natural Bodies of Water. Bodies of water such as lakes, streams, ponds, rivers, and other naturally occurring bodies of water, which may vary in depth throughout the year.

Shoreline. The farthest extent of standing water under the applicable conditions that determine the electrical datum plane for the specified body of water.

Author's Substantiation. New definitions have been added to recognize artificially made bodies of water, electrical datum plane, equipotential plane, natural bodies of water, and shoreline. A new article has been added to recognize the installation of electrical wiring and equipment in and adjacent to natural and artificially made bodies of water.

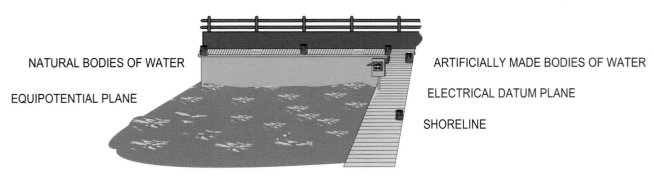

NATURAL BODIES OF WATER

ARTIFICIALLY MADE BODIES OF WATER

EQUIPOTENTIAL PLANE

ELECTRICAL DATUM PLANE

SHORELINE

PART I - DEFINITIONS
682.2

Purpose of Change: To add definitions to be used when applying requirements in this new Article.

NEC Ch. 6 – Article 682
Part II – 682.10 through 682.15

 142.2

Type of Change		Panel Action		UL	UL 508	API 500	API 505	OSHA
New Article		Accept		-	-	-	-	-
ROP		ROC		NFPA 70E	NFPA 70B	NFPA 79	NFPA	NEMA
pg. 1599	# 17-154	pg. 484	# 17-184	-	-	-	-	-
log: 1625	CMP: 17	log: 829	Submitter: Neil F. LaBrake, Jr.			2002 NEC: -		IEC: -

2005 NEC – Part II - Installation

682.10 Electrical Equipment and Transformers. Electrical equipment and transformers, including their enclosures, shall be specifically approved for the intended location. No portion of an enclosure for electrical equipment not identified for operation while submerged shall be located below the electrical datum plane.

682.11 Location of Service Equipment. On land, the service equipment for floating structures and submersible electrical equipment shall be located no closer than 1.5 m (5 ft) horizontally from the shoreline and live parts elevated a minimum of 300 mm (12 in.) above the electrical datum plane. Service equipment shall disconnect when the water level reaches the height of the established electrical datum plane.

682.12 Electrical Connections. All electrical connections not intended for operation while submerged shall be located at least 300 mm (12 in.) above the deck of a floating or fixed structure, but not below the electrical datum plane.

682.13. Wiring Methods and Installation. Wiring methods and installations of Chapter 3 and Articles 553, 555, and 590 shall be permitted where identified for use in wet locations.

682.14 Disconnecting Means for Floating Structures or Submersible Electrical Equipment.

    (A) Type. The disconnecting means shall be permitted to consist of a circuit breaker, switch, or both and shall be properly identified as to which structure or equipment it controls.

    (B) Location. The disconnecting means shall be readily accessible on land and shall be located in the supply circuit ahead of the structure or the equipment connection. The disconnecting means shall be located not more than 750 mm (30 in.) from the structure or equipment connection. The disconnecting means shall be within sight of, but not closer than, 1.5 m (5 ft) horizontally from the edge of the shoreline and live parts elevated a minimum of 300 mm (12 in.) above the electrical datum plane.

682.15. Ground Fault Circuit Interrupter (GFCI) Protection. Fifteen- and 20-ampere single-phase, 125-volt through 250-volt receptacles installed outdoors and in or on floating buildings or structures within the electrical datum plane area that are used for storage, maintenance, or repair where portable electric hand tools, electrical diagnostic equipment, or portable lighting equipment are to be used shall be provided with GFCI protection. The GFCI protection device shall be located not less than 300 mm (12 in.) above the established electrical datum plane.

Author's Substantiation. Part II covers the installation of electrical equipment and transformers, location of service equipment, electrical connections, and wiring methods and installations in and adjacent to natural and artificially made bodies of water. A new article has been added to recognize the installation of electrical wiring and equipment in and adjacent to natural and artificially made bodies of water.

ELECTRICAL EQUIPMENT
AND TRANSFORMERS
• **682.10**

LOCATION OF SERVICE
EQUIPMENT
• **682.11**

ELECTRICAL
CONNECTIONS
• **682.12**

DISCONNECTION MEANS FOR
FLOATING STRUCTURES OR
SUBMERSIBLE ELECTRICAL
EQUIPMENT
• **682.14**

GFCI PROTECTION
• **682.15**

WIRING METHODS AND INSTALLATION
• **680.13**

PART II - INSTALLATION
682.10 THRU 15

Purpose of Change: To add installation requirements in **Part II** of **Article 682**.

NEC Ch. 6 – Article 682
Part III – 682.30 through 682.33

 142.3

Type of Change		Panel Action		UL	UL 508	API 500	API 505	OSHA
New Article		Accept		-	-	-	-	-
ROP		ROC		NFPA 70E	NFPA 70B	NFPA 79	NFPA	NEMA
pg. 1599	# 17-154	pg. 484	# 17-184	-	-	-	-	-
log: 1625	CMP: 17	log: 829	Submitter: Neil F. LaBrake, Jr.			2002 NEC: -		IEC: -

2005 NEC – Part III – Grounding and Bonding

682.30 Grounding. Wiring and equipment within the scope of this article shall be grounded as specified in Articles 250, 553, and 555 and with the requirements in Part III.

682.31 Equipment Grounding Conductors

(A) Type. Equipment grounding conductors shall be insulated copper conductors sized in accordance with 250.122 but not smaller than 12 AWG.

(B) Feeders. Where a feeder supplies a remote panel board, an insulated equipment grounded conductor shall extend from a grounding terminal in the service to a grounding terminal and busbar in the remote panel board.

(C) Branch Circuits. The insulated equipment grounding conductor for branch circuits shall terminate at a grounding terminal in a remote panel board or the grounding terminal in the main service equipment.

(D) Cord-and-Plug-Connected Appliances. Where required to be grounded, cord-and-plug-connected appliances shall be grounded by means of an equipment grounding conductor in the cord and a grounding-type attachment plug.

682.32 Bonding of Non-Current-Carrying Metal Parts. All metal parts in contact with the water, all metal piping, tanks, and all non-current-carrying metal parts that may become energized shall be bonded to the grounding bus in the panelboard.

682.33 Equipotential Planes and Bonding of Equipment Planes. An equipotential plane shall be installed where required in this section to mitigate step and touch voltages at electrical equipment.

(A) Areas Requiring Equipotential Planes. Equipotential planes shall be installed adjacent to all outdoor service equipment or disconnecting means that control equipment in or on water, that have a metallic enclosure and controls accessible to personnel, and are likely to become energized. The equipotential plane shall encompass the area around the equipment and shall extend from the area directly below the equipment out not less than 900 mm (36 in.) in all directions from which a person would be able to stand and come in contact with the equipment.

(B) Areas Not Requiring Equipotential Planes. Equipotential planes shall not be required for the controlled equipment supplied by the service equipment or disconnecting means. All circuits rated not more than 60 amperes at 120 through 250 volts, single phase, shall have GFCI protection.

(C) Bonding. Equipotential planes shall be bonded to the electrical grounding system. The bonding conductor shall be solid copper, insulated, covered or bare, and not smaller than 8 AWG. Connections shall be made by exothermic welding or by listed pressure connectors or clamps that are labeled as being suitable for the purpose and are of stainless steel, brass, copper, or copper alloy.

Author's Substantiation. Part II covers the grounding and bonding requirements for electrical wiring and equipment in and adjacent to natural and artificially made bodies of water. A new article has been added to recognize the installation of electrical wiring and equipment in and adjacent to natural and artificially made bodies of water.

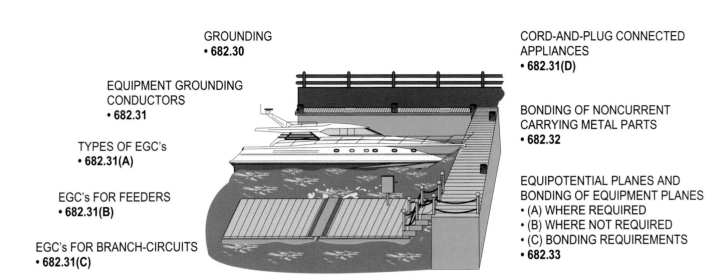

GROUNDING
• **682.30**

EQUIPMENT GROUNDING
CONDUCTORS
• **682.31**

TYPES OF EGC's
• **682.31(A)**

EGC's FOR FEEDERS
• **682.31(B)**

EGC's FOR BRANCH-CIRCUITS
• **682.31(C)**

CORD-AND-PLUG CONNECTED
APPLIANCES
• **682.31(D)**

BONDING OF NONCURRENT
CARRYING METAL PARTS
• **682.32**

EQUIPOTENTIAL PLANES AND
BONDING OF EQUIPMENT PLANES
• (A) WHERE REQUIRED
• (B) WHERE NOT REQUIRED
• (C) BONDING REQUIREMENTS
• **682.33**

**PART III - GROUNDING AND BONDING
682.30 THRU 33**

Purpose of Change: To include requirements for grounding and bonding in the new **Article 682**.

Type of Change	Panel Action	UL	UL 508	API 500	API 505	OSHA
Revision	Accept in Part	-	-	-	-	-
ROP	**ROC**	**NFPA 70E**	**NFPA 70B**	**NFPA 79**	**NFPA**	**NEMA**
pg. 1602 # 12-89	pg. 486 # 12-49	-	-	-	-	-
log: 3468 CMP: 12	log: 1252 Submitter: Charles M. Trout			2002 NEC: 685.1(2)		IEC: -

2002 NEC – 685.1 Scope

This article covers integrated electrical systems, other than unit equipment, in which orderly shutdown is necessary to ensure safe operation. An *integrated electrical system* as used in this article is a unitized segment of an industrial wiring system where all of the following conditions are met:

(1) An orderly shutdown is required to minimize personnel hazard and equipment damage.

(2) The conditions of maintenance and supervision ensure that qualified persons service the system.

(3) Effective safeguards, acceptable to the authority having jurisdiction, are established and maintained.

2005 NEC – 685.1 Scope

This article covers integrated electrical systems, other than unit equipment, in which orderly shutdown is necessary to ensure safe operation. An *integrated electrical system* as used in this article is a unitized segment of an industrial wiring system where all of the following conditions are met:

(1) An orderly shutdown is required to minimize personnel hazard and equipment damage.

(2) The conditions of maintenance and supervision ensure that qualified persons service the system. The name(s) of the qualified person(s) shall be kept in a permanent record at the office of the establishment in charge of the completed installation.

A person designated as a qualified person shall possess the skills and knowledge related to the construction and operation of the electrical equipment and installation and shall have received documented safety training on the hazards involved. Documentation of their qualifications shall be on file with the office of the establishment in charge of the completed installation.

(3) Effective safeguards acceptable to the authority having jurisdiction are established and maintained.

Author's Substantiation. This revision clarifies the prescriptive requirements of qualified persons who maintain integrated electrical systems.

INTEGRATED ELECTRICAL SYSTEMS

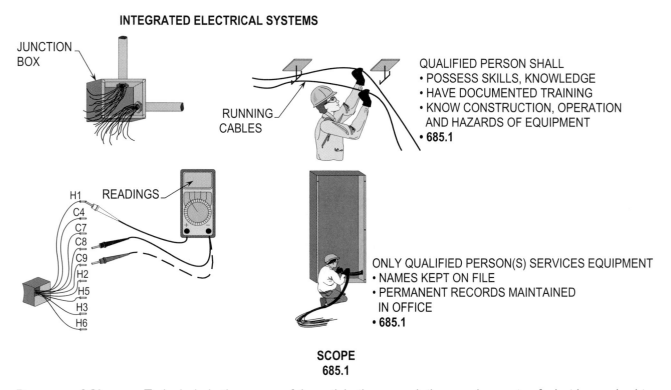

JUNCTION BOX

RUNNING CABLES

QUALIFIED PERSON SHALL
• POSSESS SKILLS, KNOWLEDGE
• HAVE DOCUMENTED TRAINING
• KNOW CONSTRUCTION, OPERATION
  AND HAZARDS OF EQUIPMENT
• **685.1**

READINGS

H1
C4
C7
C8
C9
H2
H5
H3
H6

ONLY QUALIFIED PERSON(S) SERVICES EQUIPMENT
• NAMES KEPT ON FILE
• PERMANENT RECORDS MAINTAINED
  IN OFFICE
• **685.1**

**SCOPE
685.1**

Purpose of Change: To include in the scope of the article the prescriptive requirements of what is required to be a qualified person.

NEC Ch. 6 – Article 690
Part I – 690.2

Type of Change		Panel Action		UL	UL 508	API 500	API 505	OSHA
New Definition		Accept		-	-	-	-	-
ROP		ROC		NFPA 70E	NFPA 70B	NFPA 79	NFPA	NEMA
pg. 1604	# 13-23	pg. 487	# 13-9	-	-	-	-	-
log: 2504	CMP: 13	log: 1047	Submitter: John C. Wiles, Jr.			2002 NEC:		IEC: -

2005 – 690.2 Definitions

Building Integrated Photovoltaics. Photovoltaic cells, devices, modules, or modular materials that are integrated into the outer surface or structure of a building and serve as the outer protective surface of that building.

Author's Substantiation. A new definition has been added to define a type of photovoltaic product being installed throughout the country.

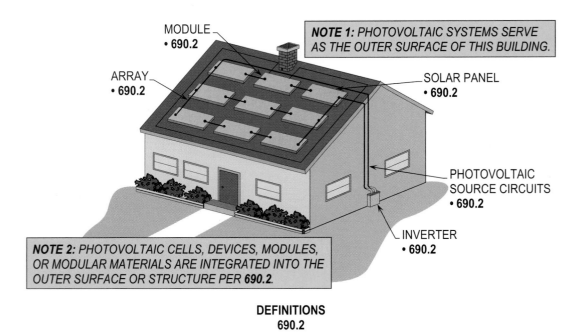

MODULE • 690.2

NOTE 1: PHOTOVOLTAIC SYSTEMS SERVE AS THE OUTER SURFACE OF THIS BUILDING.

ARRAY • 690.2

SOLAR PANEL • 690.2

PHOTOVOLTAIC SOURCE CIRCUITS • 690.2

INVERTER • 690.2

NOTE 2: PHOTOVOLTAIC CELLS, DEVICES, MODULES, OR MODULAR MATERIALS ARE INTEGRATED INTO THE OUTER SURFACE OR STRUCTURE PER 690.2.

DEFINITIONS
690.2

Purpose of Change: To include the definition "Building Integrated Photovoltaics" as a type of product being installed in buildings today.

NEC Ch. 6 – Article 692
Part I – 692.1

Type of Change		Panel Action		UL	UL 508	API 500	API 505	OSHA
New Sentence		Accept		-	-	-	-	-
ROP		ROC		NFPA 70E	NFPA 70B	NFPA 79	NFPA	NEMA
pg. 1630	# 13-67	pg. -	# -	-	-	-	-	-
log: 3379	CMP: 13	log: -	Submitter: Kenneth Krastims			2002 NEC: 692.1		IEC: -

2002 NEC – 692.1 Scope

This article identifies the requirements for the installation of fuel cell power systems, which may be stand-alone or interactive with other electrical power production sources and may be with or without electrical energy storage such as batteries.

2005 NEC – 692.1 Scope

This article identifies the requirements for the installation of fuel cell power systems, which may be stand-alone or interactive with other electrical power production sources and may be with or without electrical energy storage such as batteries. <u>These systems may have ac or dc output for utilization.</u>

Author's Substantiation. A new sentence has been added to clarify that ac or dc output may be used for utilization to fuel cell systems.

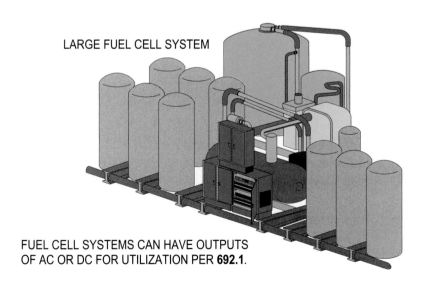

LARGE FUEL CELL SYSTEM

FUEL CELL SYSTEMS CAN HAVE OUTPUTS
OF AC OR DC FOR UTILIZATION PER **692.1**.

**SCOPE
692.1**

Purpose of Change: To clarify that output power can be AC or DC for the purpose of utilization.

NEC Ch. 6 – Article 695
Part – 695.2

Type of Change		Panel Action		UL	UL 508	API 500	API 505	OSHA
Revision		Accept		1247	-	-	-	-
ROP		**ROC**		**NFPA 70E**	**NFPA 70B**	**NFPA 79**	**NFPA**	**NEMA**
pg. 1638	# 13-79	pg. 496	# 13-48	-	-	-	-	ICS 15 and MG 1
log: 553	CMP: 13	log: 2400	Submitter: TC on Fire Pumps			2002 NEC: 695.2		IEC: -

2002 NEC – 695.2 Definitions

Fault Tolerant External Control ~~Conductors~~. Those control ~~conductors~~ entering ~~and/~~or leaving the fire pump controller enclosure, which if broken, disconnected, or shorted will not prevent the controller from starting the fire pump from all other internal or external means and may cause the controller to start the pump under these conditions.

2005 NEC – 695.2 Definitions

Fault Tolerant External Control <u>Circuits</u>. Those control <u>circuits either</u> entering or leaving the fire pump controller enclosure, which if broken, disconnected, or shorted will not prevent the controller from starting the fire pump from all other internal or external means and may cause the controller to start the pump under these conditions.

Author's Substantiation. This revision clarifies that *fault tolerant* applies to the entire circuit, not just the conductors.

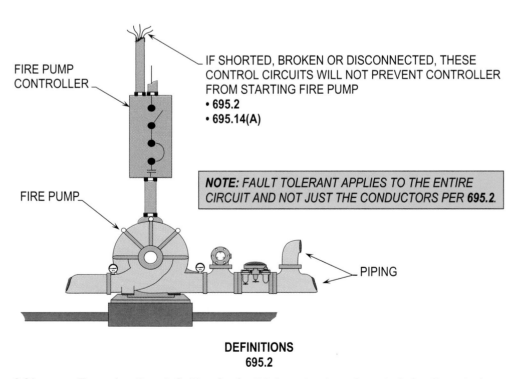

FIRE PUMP CONTROLLER

IF SHORTED, BROKEN OR DISCONNECTED, THESE CONTROL CIRCUITS WILL NOT PREVENT CONTROLLER FROM STARTING FIRE PUMP
• **695.2**
• **695.14(A)**

FIRE PUMP

*NOTE: FAULT TOLERANT APPLIES TO THE ENTIRE CIRCUIT AND NOT JUST THE CONDUCTORS PER **695.2**.*

PIPING

DEFINITIONS
695.2

Purpose of Change: To revise the definition for fault tolerant external control circuits entering and leaving the fire pump controller enclosure.

NEC Ch. 6 – Article 695
Part – 695.6(D)

Type of Change		Panel Action		UL	UL 508	API 500	API 505	OSHA
Revision		Accept		1247	-	-	-	-
ROP		ROC		NFPA 70E	NFPA 70B	NFPA 79	NFPA	NEMA
pg. 1650	# 13-102	pg. 499	# 13-64	-	-	-	-	ICS 15 and MG 1
log: 1618	CMP: 13	log: 23	Submitter: R. Schneider			2002 NEC: 695.6(D)		IEC: -

2002 NEC – 695.6(D) Overload Protection

Power circuits shall not have automatic protection against overloads. ~~Except as provided in 695.5(C)(2),~~ branch-circuit and feeder conductors shall be protected against short circuit only. Where a tap is made to supply a fire pump, ~~and~~ the wiring ~~is run~~ in accordance with 230.6~~,~~ the applicable distance and size restrictions in 240.21 shall not apply.

2005 NEC – 695.6(D) Overload Protection

Power circuits shall not have automatic protection against overloads. Branch-circuit and feeder conductors shall be protected against short circuit only. Where a tap is made to supply a fire pump, the wiring <u>shall be treated as service conductors</u> in accordance with 230.6. The applicable distance and size restrictions in 240.21 shall not apply.

Author's Substantiation. This revision clarifies when overload, short-circuit, and ground-fault protection of circuits is required and not required.

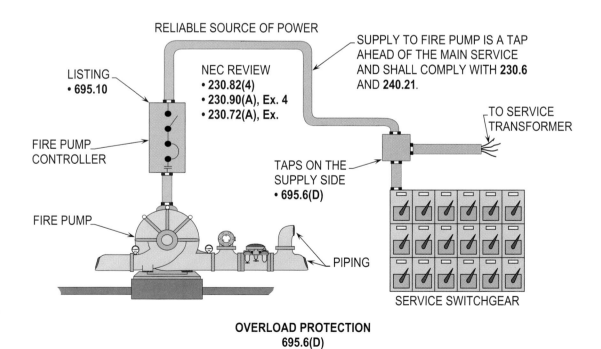

OVERLOAD PROTECTION
695.6(D)

Purpose of Change: To clarify when overload, short-circuit, and ground-fault protection of circuits is and is not required.

Special Conditions

The requirements of **Chapter 7** have always been used to design and install the electrical elements that are related to emergency required power sources. Emergency systems and their power sources are mandated by codes other than the NEC. The NEC establishes the provisions for the installation of emergency systems, once they are required.

Emergency systems must operate automatically and supply illumination to designated areas and power to designated equipment, and their wiring must not be run with the normal power system. Legally required standby systems must also operate automatically, but unlike emergency systems, their wiring can be run with the wiring of the normal power system. Operation standby systems are not required to operate automatically, and their wiring can also be run with the wiring of the normal power system.

Articles and Sections of the 700 series govern wiring methods and equipment related to special conditions and use.

NEC Ch. 7 – Article 700
Part I – 700.5(B)

Type of Change		Panel Action		UL	UL 508	API 500	API 505	OSHA
Revision		Accept		1004	-	-	-	-
ROP		ROC		NFPA 70E	NFPA 70B	NFPA 79	NFPA	NEMA
pg. 1652	# 13-109	pg. -	# -	-	6.4.4.6(2)		-	MG 1
log: 1019	CMP: 13	log: -	Submitter: Lawrence A. Bey			2002 NEC: 700.5(B)		IEC: 56

2002 NEC – 700.5(B) Selective Load Pickup, Load Shedding, and Peak Load Shaving

The alternate power source shall be permitted to supply emergency, legally required standby, and optional standby system loads where automatic selective load pickup and load shedding is provided as needed to ensure adequate power to (1) the emergency circuits, (2) the legally required standby circuits, and (3) the optional standby circuits, in that order of priority. The alternate power source shall be permitted to be used for peak load shaving, provided ~~the above~~ conditions are met.

2005 NEC – 700.5(B) Selective Load Pickup, Load Shedding, and Peak Load Shaving

The alternate power source shall be permitted to supply emergency, legally required standby, and optional standby system loads where <u>the source has adequate capacity or where</u> automatic selective load pickup and load shedding is provided as needed to ensure adequate power to (1) the emergency circuits, (2) the legally required standby circuits, and (3) the optional standby circuits, in that order of priority. The alternate power source shall be permitted to be used for peak load shaving, provided <u>these</u> conditions are met.

Author's Substantiation. This revision clarifies that load shed and add is not required if the source has the capacity to supply all the connected loads.

NOTE: *GENERATOR HAS CAPACITY TO SUPPLY (1) EMERGENCY CIRCUITS, (2) LEGALLY REQUIRED STANDBY CIRCUITS, (3) OPTIONAL STANDBY CIRCUITS IN THIS ORDER OF PRIORITY PER 700.5(B).*

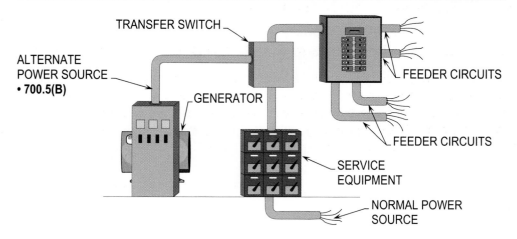

**SELECTIVE LOAD PICKUP, LOAD SHEDDING, AND PEAK LOAD SHAVING
700.5(B)**

Purpose of Change: To clarify that load shed and add is not required if the source has the capacity to supply all the connected loads.

NEC Ch. 7 – Article 700
Part II – 700.9(D)(1)(1)

Type of Change		Panel Action	UL	UL 508	API 500	API 505	OSHA
Revision		Accept	1004	-	-	-	-
ROP		ROC	NFPA 70E	NFPA 70B	NFPA 79	NFPA	NEMA
pg. 1656	# 13-116	pg. -	# -	-	6.4.4.6(2)	-	-	MG 1
log: 1791	CMP: 13	log: -	Submitter: Barry F. O'Connell		2002 NEC: 700.9(D)(1)		IEC: 56

2002 NEC – 700.9(D)(1) Feeder-Circuit Wiring

Feeder-circuit wiring shall meet one of the following conditions:

(1) Be installed ~~with buildings~~ that are fully protected by an approved automatic fire suppression system

2005 NEC – 700.9(D)(1) Feeder-Circuit Wiring

Feeder-circuit wiring shall meet one of the following conditions:

(1) Be installed <u>in spaces or areas</u> that are fully protected by an approved automatic fire suppression system

Author's Substantiation. This revision clarifies that feeder-circuit wiring shall be installed in spaces or areas that are fully protected by an approved automatic fire suppression system.

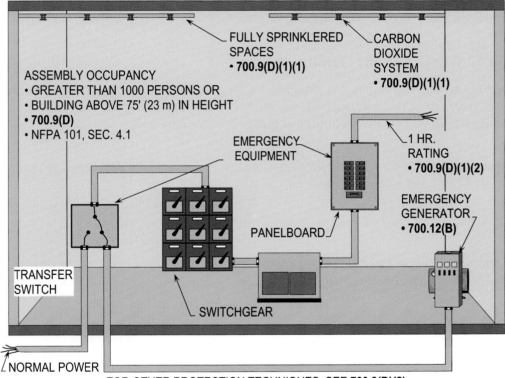

NOTE: *FEEDER-CIRCUIT WIRING SHALL BE INSTALLED IN SPACES OR AREAS THAT ARE FULLY PROTECTED BY AN APPROVED AUTOMATIC FIRE SUPPRESSION SYSTEM.*

FULLY SPRINKLERED SPACES
• **700.9(D)(1)(1)**

CARBON DIOXIDE SYSTEM
• **700.9(D)(1)(1)**

ASSEMBLY OCCUPANCY
• GREATER THAN 1000 PERSONS OR
• BUILDING ABOVE 75' (23 m) IN HEIGHT
• **700.9(D)**
• NFPA 101, SEC. 4.1

EMERGENCY EQUIPMENT

1 HR. RATING
• **700.9(D)(1)(2)**

EMERGENCY GENERATOR
• **700.12(B)**

PANELBOARD

TRANSFER SWITCH

SWITCHGEAR

NORMAL POWER

FOR OTHER PROTECTION TECHNIQUES, SEE **700.9(D)(2).**

FEEDER-CIRCUIT WIRING
700.9(D)(1)(1)

Purpose of Change: To provide rules for the protection of feeder-circuit wiring by deleting and adding words to clarify the intent.

144

Type of Change		Panel Action		UL	UL 508	API 500	API 505	OSHA
Revision		Accept in Principle		1004	-	-	-	-
ROP		ROC		NFPA 70E	NFPA 70B	NFPA 79	NFPA	NEMA
pg. 1659	# 13-124	pg. 501	# 13-78	-	6.4.4.6(2)	-	-	MG 1
log: 1283	CMP: 13	log: 640	Submitter: David H. Kendall			2002 NEC: 700.12(B)(6)		IEC: 56

2002 NEC – 700.12(B)(6) Outdoor Generator Sets

Where an outdoor housed generator set is equipped with a readily accessible disconnecting means located within sight of the building or structure supplied, an additional disconnecting means shall not be required where ungrounded conductors pass through the building or structure.

2005 NEC – 700.12(B)(6) Outdoor Generator Sets

Where an outdoor housed generator set is equipped with a readily accessible disconnecting means located within sight of the building or structure supplied, an additional disconnecting means shall not be required where ungrounded conductors <u>serve or</u> pass through the building or structure.

Author's Substantiation. This revision clarifies the requirements for disconnecting conductors entering or passing through buildings or structures.

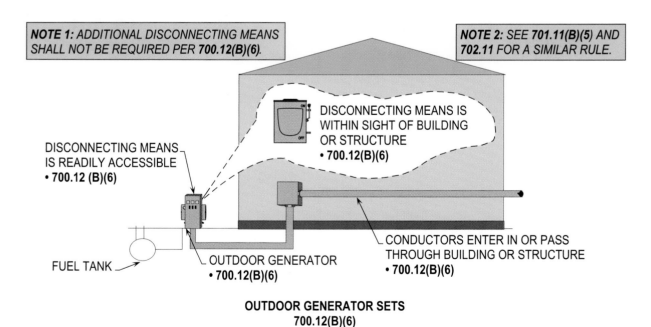

OUTDOOR GENERATOR SETS
700.12(B)(6)

Purpose of Change: To provide requirements for disconnecting conductors entering or passing through buildings or structures.

NEC Ch. 7 – Article 700
Part III – 700.12(D)

Type of Change		Panel Action		UL	UL 508	API 500	API 505	OSHA
Revision		Accept		1004	-	-	-	-
ROP		ROC		NFPA 70E	NFPA 70B	NFPA 79	NFPA	NEMA
pg. 1660	# 13-129	pg. -	# -	-	6.4.4.6(2)	-	-	MG 1
log: 665	CMP: 13	log: -	Submitter: Dan Leaf			2002 NEC: 700.12(D)		IEC: 56

2002 NEC – 700.12(D) Separate Service

Where acceptable to the authority having jurisdiction as suitable for use as an emergency source, ~~a second~~ service shall be permitted. This service shall be in accordance with Article 230~~, with~~ separate service drop or lateral~~, widely separated~~ electrically and physically from ~~the normal service~~ to minimize the possibility of simultaneous interruption of supply.

2005 NEC – 700.12(D) Separate Service

Where acceptable to the authority having jurisdiction as suitable for use as an emergency source <u>of power,</u> <u>an additional</u> service shall be permitted. This service shall be in accordance with <u>the applicable provisions of</u> Article 230 <u>and the following additional requirements:</u>

(1) Separate service drop or <u>service</u> lateral

(2) <u>Service conductors sufficiently remote</u> electrically and physically from <u>any other service conductors</u> to minimize the possibility of simultaneous interruption of supply

Author's Substantiation. This revision clarifies the specific requirements for a separate service drop or lateral to supply emergency equipment.

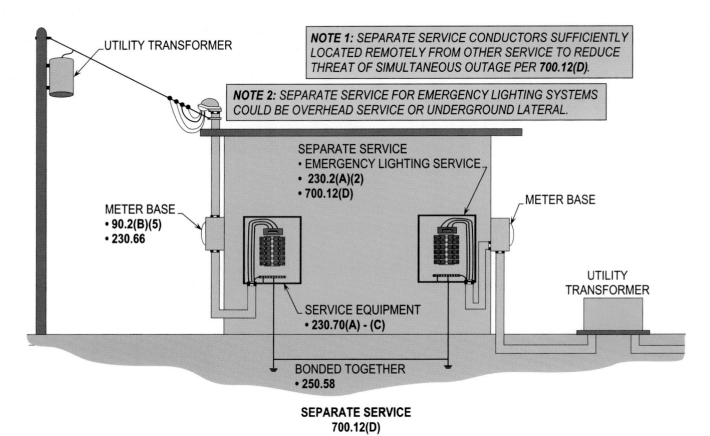

NOTE 1: *SEPARATE SERVICE CONDUCTORS SUFFICIENTLY LOCATED REMOTELY FROM OTHER SERVICE TO REDUCE THREAT OF SIMULTANEOUS OUTAGE PER **700.12(D)**.*

NOTE 2: *SEPARATE SERVICE FOR EMERGENCY LIGHTING SYSTEMS COULD BE OVERHEAD SERVICE OR UNDERGROUND LATERAL.*

UTILITY TRANSFORMER

SEPARATE SERVICE
• EMERGENCY LIGHTING SERVICE
• 230.2(A)(2)
• 700.12(D)

METER BASE
• 90.2(B)(5)
• 230.66

METER BASE

UTILITY TRANSFORMER

SERVICE EQUIPMENT
• 230.70(A) - (C)

BONDED TOGETHER
• 250.58

SEPARATE SERVICE
700.12(D)

Purpose of Change: To provide specific requirements for a separate service drop or lateral to supply emergency equipment.

NEC Ch. 7 – Article 700
Part III – 700.12(E)

Type of Change		Panel Action		UL	UL 508	API 500	API 505	OSHA
New Subsection		Accept in Principle in Part		1004	-	-	-	-
ROP		ROC		NFPA 70E	NFPA 70B	NFPA 79	NFPA	NEMA
pg. 1661	**#** 13-128	**pg.** -	**#** -	-	6.4.4.6(2)	-	-	MG 1
log: 171	**CMP:** 13	**log:** -	Submitter: Frank Seiler			2002 NEC: -		IEC: 56

2005 NEC – 700.12(E) Fuel Cell System

Fuel cell systems used as a source of power for emergency systems shall be of suitable rating and capacity to supply and maintain the total load for not less than 2 hours of full-demand operation.

Installation of a fuel cell system shall meet the requirements of Parts II through VIII of Article 692.

Where a single fuel cell system serves as the normal supply for the building or group of buildings concerned, it shall not serve as the sole source of power for the emergency standby system.

Author's Substantiation. A new subsection has been added to recognize fuel cell systems as a source of power for emergency systems.

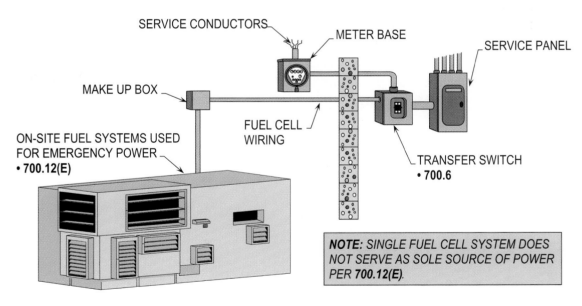

FUEL CELL SYSTEM
700.12(E)

Purpose of Change: To include a fuel cell system as a source of power for emergency systems.

NEC Ch. 7 – Article 701
Part I – 701.6

Type of Change	Panel Action	UL	UL 508	API 500	API 505	OSHA
Revision	Accept in Principle in Part	1004	-	-	-	-
ROP	**ROC**	**NFPA 70E**	**NFPA 70B**	**NFPA 79**	**NFPA**	**NEMA**
pg. 1665 # 13-137	pg. - # -	-	B.1.21	-	-	MG 1
log: 1020 CMP: 13	log: -	Submitter: Lawrence A. Bey		2002 NEC: 701.6		IEC: 56

2002 NEC – 701.6 Capacity and Rating

A legally required standby system shall have adequate capacity and rating for the supply of all equipment intended to be operated at one time. Legally required standby system equipment shall be suitable for the maximum available fault current at its terminals.

The alternate power source shall be permitted to supply legally required standby and optional standby system loads where automatic selective load pickup and load shedding is provided ~~as needed~~ to ensure adequate power to the legally required standby circuits.

2005 NEC – 701.6 Capacity and Rating

A legally required standby system shall have adequate capacity and rating for the supply of all equipment intended to be operated at one time. Legally required standby system equipment shall be suitable for the maximum available fault current at its terminals.

The legally required standby alternate power source shall be permitted to supply both legally required standby and optional standby system loads under either of the following conditions:

(1) Where the alternate source has adequate capacity to handle all connected loads

(2) Where automatic selective load pickup and load shedding is provided that will ensure adequate power to the legally required standby circuits

Author's Substantiation. This revision clarifies that alternate power sources have the capability to supply the connected loads.

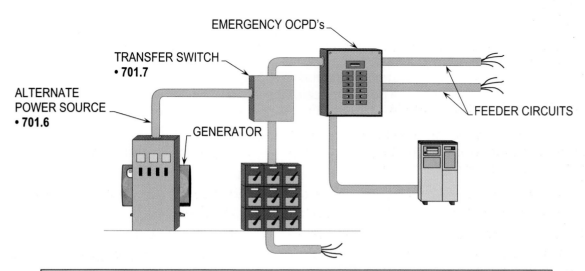

EMERGENCY OCPD's

TRANSFER SWITCH
• **701.7**

ALTERNATE
POWER SOURCE
• **701.6**

GENERATOR

FEEDER CIRCUITS

NOTE: *GENERATOR HAS ADEQUATE CAPACITY TO SUPPLY ALL CONNECTED LOADS. IF AUTOMATIC SELECTIVE LOAD PICKUP AND LOAD SHEDDING IS PROVIDED, GENERATOR HAS ADEQUATE POWER FOR THE LEGALLY REQUIRED STANDBY CIRCUITS PER **701.6(1) AND (2)**.*

CAPACITY AND RATING
701.6

Purpose of Change: To ensure the alternate power source has the capability to supply the connected loads.

NEC Ch. 7 – Article 702
Part I – 702.2

Type of Change		Panel Action		UL	UL 508	API 500	API 505	OSHA
Revision		Accept		1004	-	-	-	-
ROP		ROC		NFPA 70E	NFPA 70B	NFPA 79	NFPA	NEMA
pg. 1669	# 13-146	pg. -	# -	-	21.4.3	-	-	MG 1
log: 666	CMP: 13	log: -	Submitter: Dan Leaf			2002 NEC: 702.2		IEC: -

2002 NEC – 702.2 Definition

Optional Standby Systems. Those systems intended to ~~protect~~ public or private facilities or property where life safety does not depend on the performance of the system. Optional standby systems are intended to supply on-site generated power to selected loads either automatically or manually.

2005 NEC – 702.2 Definition

Optional Standby Systems. Those systems intended <u>to supply power</u> to public or private facilities or property where life safety does not depend on the performance of the system. Optional standby systems are intended to supply on-site generated power to selected loads either automatically or manually.

Author's Substantiation. This revision clarifies that optional standby systems are intended to supply systems.

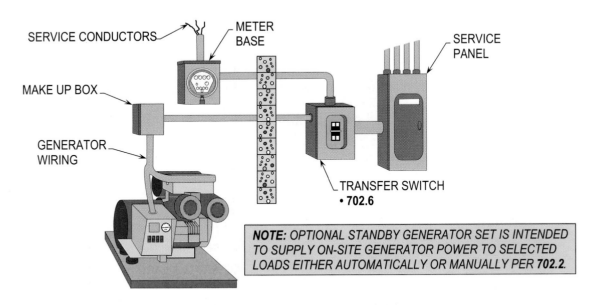

NOTE: OPTIONAL STANDBY GENERATOR SET IS INTENDED TO SUPPLY ON-SITE GENERATOR POWER TO SELECTED LOADS EITHER AUTOMATICALLY OR MANUALLY PER **702.2**.

DEFINITION - OPTIONAL STANDBY SYSTEMS
702.2

Purpose of Change: To clarify that optional standby systems are intended to supply systems.

Type of Change		Panel Action		UL	UL 508	API 500	API 505	OSHA
New Exception		Accept in Principle in Part		1004	-	-	-	-
ROP		ROC		NFPA 70E	NFPA 70B	NFPA 79	NFPA	NEMA
pg. 1670	# 13-148	pg. 504	# 13-101	-	21.4.3	-	-	MG 1
log: 2924	CMP: 13	log: 1103	Submitter: Dorothy Kellogg			2002 NEC: -		IEC: -

2005 NEC – 702.6 Transfer Equipment

Transfer equipment shall be suitable for the intended use and designed and installed so as to prevent the inadvertent interconnection of normal and alternate sources of supply in any operation of the transfer equipment. Transfer equipment and electric power production systems installed to permit operation in parallel with the normal source shall meet the requirements of Article 705.

Transfer equipment, located on the load side of branch circuit protection, shall be permitted to contain supplementary overcurrent protection having an interrupting rating sufficient for the available fault current that the generator can deliver. The supplementary overcurrent protection devices shall be part of a listed transfer equipment.

Transfer equipment shall be required for all standby systems subject to the provisions of this article and for which an electric-utility supply is either the normal or standby source.

> **Exception:** Temporary connection of a portable generator without transfer equipment shall be permitted where conditions of maintenance and supervision ensure that only qualified persons service the installation and where the normal supply is physically isolated by a lockable disconnect means or by disconnection of the normal supply conductors.

Author's Substantiation. A new exception has been added to include requirements for conditions where a transfer switch is not required.

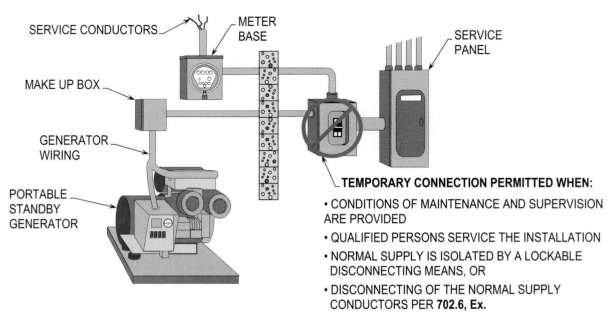

SERVICE CONDUCTORS

METER BASE

SERVICE PANEL

MAKE UP BOX

GENERATOR WIRING

PORTABLE STANDBY GENERATOR

TEMPORARY CONNECTION PERMITTED WHEN:

• CONDITIONS OF MAINTENANCE AND SUPERVISION ARE PROVIDED

• QUALIFIED PERSONS SERVICE THE INSTALLATION

• NORMAL SUPPLY IS ISOLATED BY A LOCKABLE DISCONNECTING MEANS, OR

• DISCONNECTING OF THE NORMAL SUPPLY CONDUCTORS PER **702.6, Ex.**

TRANSFER EQUIPMENT
702.6

Purpose of Change: To include requirements for conditions where a transfer switch is not required.

NEC Ch. 7 – Article 702
Part I – 702.7(2), Ex.

Type of Change	Panel Action	UL	UL 508	API 500	API 505	OSHA
New Exception	Accept	1004	-	-	-	-
ROP		ROC		NFPA 70E	NFPA 70B	NFPA 79	NFPA	NEMA

ROP		ROC		NFPA 70E	NFPA 70B	NFPA 79	NFPA	NEMA
pg. 1671	# 13-149	pg. -	# -	-	21.4.3	-	-	MG 1
log: 997	CMP: 13	log: -	Submitter: Noel Williams			2002 NEC: -		IEC: -

2002 NEC – 702.7 Signals

Audible and visual signal devices shall be provided, where practicable, for the following purposes.

(1) **Derangement.** To indicate derangement of the optional standby source.

(2) **Carrying Load.** To indicate that the optional standby source is carrying load.

2005 NEC – 702.7 Signals

Audible and visual signal devices shall be provided, where practicable, for the following purposes.

(1) **Derangement.** To indicate derangement of the optional standby source.

(2) **Carrying Load.** To indicate that the optional standby source is carrying load.

Exception: Signals shall not be required for portable standby power sources.

Author's Substantiation. A new exception has been added to clarify that portable optional standby power sources shall not be required to receive signals.

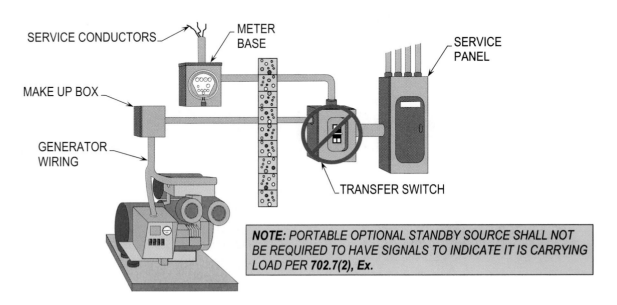

NOTE: PORTABLE OPTIONAL STANDBY SOURCE SHALL NOT BE REQUIRED TO HAVE SIGNALS TO INDICATE IT IS CARRYING LOAD PER 702.7(2), Ex.

SIGNALS
702.7(2), Ex.

Purpose of Change: To verify that portable optional standby power sources shall not be required to receive signals.

NEC Ch. 7 – Article 725
Part III – 725.42

Type of Change		Panel Action		UL	UL 508	API 500	API 505	OSHA
Revision		Accept		1310 and 1585	-	-	-	1910.308(c)
ROP		ROC		NFPA 70E	NFPA 70B	NFPA 79	NFPA	NEMA
pg. 1692	# 3-157	pg. 550	# 3-181	-	21.4	6.3	-	SB 3
log: 1010	CMP: 3	log: 898	Submitter: Noel Williams			2002 NEC: 725.42		IEC: SELF; PELV; FELV

2002 NEC – 725.42 Circuit Marking

The equipment shall be durably marked where plainly visible to indicate each circuit that is a Class 2 or Class 3 circuit.

2005 NEC – 725.42 Circuit Marking

The equipment <u>supplying the circuits</u> shall be durably marked where plainly visible to indicate each circuit that is a Class 2 or Class 3 circuit.

Author's Substantiation. This revision clarifies that the equipment supplying the circuits shall be durably marked.

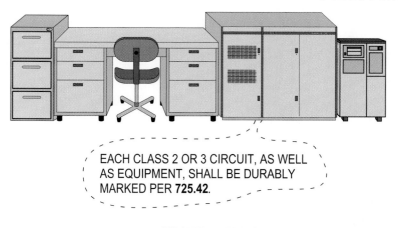

EQUIPMENT SUPPLYING CIRCUITS SHALL BE DURABLE AND PLAINLY MARKED AND VISIBLE TO INDICATE EACH CIRCUIT THAT IS CLASS 2 OR 3.

EACH CLASS 2 OR 3 CIRCUIT, AS WELL AS EQUIPMENT, SHALL BE DURABLY MARKED PER **725.42**.

CIRCUIT MARKING
725.42

Purpose of Change: To provide requirements for marking Class 2 and Class 3 circuits.

NEC Ch. 7 – Article 725
Part III – 725.56(F)

146

Type of Change		Panel Action		UL	UL 508	API 500	API 505	OSHA
New Subsection		Accept		1310 and 1585	-	-	-	1910.308(c)
ROP		ROC		NFPA 70E	NFPA 70B	NFPA 79	NFPA	NEMA
pg. 1695	# 3-162a	pg. 550	# 3-183	-	21.4	6.3	-	SB 3
log: CP 305	CMP: 3	log: 1425	Submitter: CMP 3			2002 NEC: -		IEC: SELF; PELV; FELV

2005 NEC – <u>725.56(F) Class 2 or Class 3 Conductors or Cables and Audio System Circuits</u>

<u>Audio system circuits described in 640.9(C), and installed using Class 2 or Class 3 wiring methods in compliance with Sections 725.54 and 725.61, shall not be permitted to be installed in the same cable or raceway with Class 2 or Class 3 conductors or cables.</u>

Author's Substantiation. A new subsection has been added to recognize that audio system circuits shall not be installed in the same cable or raceway with Class 2 or Class 3 conductors or cables.

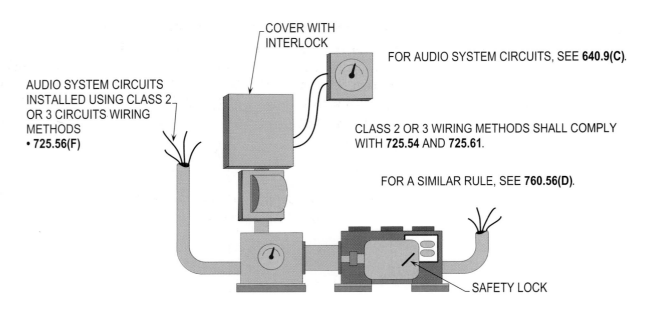

COVER WITH INTERLOCK

FOR AUDIO SYSTEM CIRCUITS, SEE **640.9(C)**.

AUDIO SYSTEM CIRCUITS INSTALLED USING CLASS 2 OR 3 CIRCUITS WIRING METHODS
• **725.56(F)**

CLASS 2 OR 3 WIRING METHODS SHALL COMPLY WITH **725.54** AND **725.61**.

FOR A SIMILAR RULE, SEE **760.56(D)**.

SAFETY LOCK

**CLASS 2 OR CLASS 3 CONDUCTORS OR CABLES AND AUDIO SYSTEMS CIRCUITS
725.56(F)**

Purpose of Change: To clarify that audio system circuits shall not be installed in the same raceway with Class 2 or 3 circuits.

Type of Change	Panel Action	UL	UL 508	API 500	API 505	OSHA
Reorganized	Accept in Principle	1424	-	-	-	1910.308(d)

ROP		ROC		NFPA 70E	NFPA 70B	NFPA 79	NFPA	NEMA
pg. 1762	# 3-223	pg. -	# -	450.4	-	-	-	SB 3
log: 122	CMP: 3	log: -	Submitter: Stanley D. Kahn			2002 NEC: Article 760		IEC: SELV; PELV; FELV

Author's Substantiation. This article has been reorganized to relocate appropriate parts and sections for usability of the NEC.

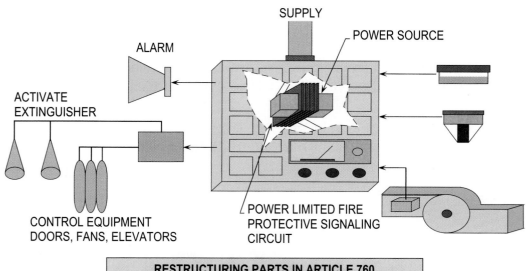

SUPPLY

POWER SOURCE

ALARM

ACTIVATE
EXTINGUISHER

CONTROL EQUIPMENT
DOORS, FANS, ELEVATORS

POWER LIMITED FIRE
PROTECTIVE SIGNALING
CIRCUIT

RESTRUCTURING PARTS IN ARTICLE 760 FIRE ALARM SYSTEMS
I. GENERAL
II. NON-POWER-LIMITED FIRE ALARM (NPLFA) CIRCUITS
III. POWER-LIMITED FIRE ALARM (PLFA) CIRCUITS
IV. LISTING REQUIREMENTS

FIRE ALARM SYSTEMS
ARTICLE 760 – PARTS I THRU IV

Purpose of Change: To recognize the requirements and relocate into the appropriate part and sections for better usability of the NEC.

NEC Ch. 7 – Article 760
Part III – 760.41

Type of Change		Panel Action		UL	UL 508	API 500	API 505	OSHA
Revision		Accept		1424	-	-	-	1910.308(d)
ROP		ROC		NFPA 70E	NFPA 70B	NFPA 79	NFPA	NEMA
pg. 1785	# 3-256	pg. 661	# 3-590	450.4	-	-	-	SB 3
log: 1597	CMP: 3	log: 1338	Submitter: Joseph A. Ross			2002 NEC: 760.41		IEC: SELV; PELV; FELV

2002 NEC – 760.41 Power Sources for PLFA Circuits

The power source for a power-limited fire alarm circuit shall be as specified in 760.41(A), (B), or (C). These circuits shall not be supplied through ground-fault circuit interrupters.

2005 NEC – 760.41 Power Sources for PLFA Circuits

The power source for a power-limited fire alarm circuit shall be as specified in 760.41(A), (B), or (C). These circuits shall not be supplied through ground-fault circuit interrupters <u>or arc-fault circuit interrupters</u>.

Author's Substantiation. This revision clarifies that fire alarm circuits shall not be supplied through ground-fault circuit interrupters or arc-fault circuit interrupters.

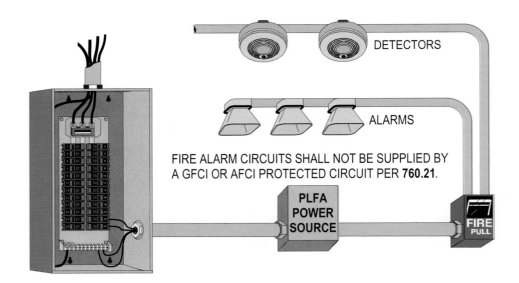

FIRE ALARM CIRCUITS SHALL NOT BE SUPPLIED BY A GFCI OR AFCI PROTECTED CIRCUIT PER **760.21**.

POWER SOURCES FOR PLFA CIRCUITS
760.41

Purpose of Change: To clarify that fire alarm circuits shall not be supplied by GFCI or AFCI protected circuits.

NEC Ch. 7 – Article 760
Part III – 760.42

Type of Change		Panel Action		UL	UL 508	API 500	API 505	OSHA
Revision		Accept		1424	-	-	-	1910.308(d)
ROP		ROC		NFPA 70E	NFPA 70B	NFPA 79	NFPA	NEMA
pg. 1786	# 3-258	pg. 662	# 3-594	450.4	-	-	-	SB 3
log: 1009	CMP: 3	log: 902	Submitter: Noel Williams			2002 NEC: 760.42		IEC: SELV; PELV; FELV

2002 NEC – 760.42 Circuit Marking

The equipment shall be durably marked where plainly visible to indicate each circuit that is a power-limited fire alarm circuit.

2005 NEC – 760.42 Circuit Marking

The equipment <u>supplying PLFA circuits</u> shall be durably marked where plainly visible to indicate each circuit that is a power-limited fire alarm circuit.

Author's Substantiation. This revision clarifies that equipment supplying PLFA circuits shall be durably marked.

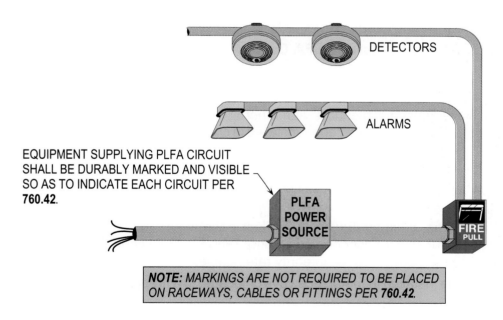

EQUIPMENT SUPPLYING PLFA CIRCUIT SHALL BE DURABLY MARKED AND VISIBLE SO AS TO INDICATE EACH CIRCUIT PER **760.42**.

DETECTORS

ALARMS

PLFA POWER SOURCE

FIRE PULL

NOTE: MARKINGS ARE NOT REQUIRED TO BE PLACED ON RACEWAYS, CABLES OR FITTINGS PER **760.42**.

CIRCUIT MARKING
760.42

Purpose of Change: To clarify such marking shall be done at the equipment where the circuitry originates.

NEC Ch. 7 – Article 760
Part III – 760.61(A) and (B)

Type of Change		Panel Action		UL	UL 508	API 500	API 505	OSHA
Revision		Accept in Principle		1424	-	-	-	1910.308(d)
ROP		ROC		NFPA 70E	NFPA 70B	NFPA 79	NFPA	NEMA
pg. 1798	# 3-273	pg. 682	# 3-680	450.4	-	-	-	SB 3
log: 2433	CMP: 3	log: 1813	Submitter: Larry Neibauer			2002 NEC: 760.61(A) and (B)		IEC: SELV; PELV; FELV

2002 NEC – 760.61 Applications of Listed PLFA Cables

PLFA cables shall comply with the requirements described in either 760.61(A), (B), or (C) or where cable substitutions are made as shown in 760.61(D).

   (A) Plenum. Cables installed in ducts, plenums, and other spaces used for environmental air shall be Type FPLP. Abandoned cables shall not be permitted to remain. Types FPLP, FPLR, and FPL cables installed in compliance with 300.22 shall be permitted.

   (B) Riser. Cables installed in risers shall be as described in either (1), (2), or (3):

   (1) Cables installed in vertical runs and penetrating more than one floor, or cables installed in vertical runs in a shaft, shall be Type FPLR. Floor penetrations requiring Type FPLR shall contain only cables suitable for riser or plenum use. Abandoned cables shall not be permitted to remain.

   (2) Other cables shall be installed in metal raceways or located in a fireproof shaft having firestops at each floor.

   (3) Type FPL cable shall be permitted in one- and two-family dwellings.

2005 NEC – 760.61 Applications of Listed PLFA Cables

PLFA cables shall comply with the requirements described in either 760.61(A), (B), or (C) or where cable substitutions are made as shown in 760.61(D).

   (A) Plenum. Cables installed in ducts, plenums, and other spaces used for environmental air shall be Type FPLP. Types FPLP, FPLR, and FPL cables installed in compliance with 300.22 shall be permitted. Type FPLP-CI cable shall be permitted to be installed to provide a 2-hour circuit integrity rated cable.

   (B) Riser. Cables installed in risers shall be as described in either (1), (2), or (3):

   (1) Cables installed in vertical runs and penetrating more than one floor, or cables installed in vertical runs in a shaft, shall be Type FPLR. Floor penetrations requiring Type FPLR shall contain only cables suitable for riser or plenum use. Type FPLR-CI cable shall be permitted to be installed to provide a 2-hour circuit integrity rated cable.

   (2) Other cables shall be installed in metal raceways or located in a fireproof shaft having firestops at each floor.

   (3) Type FPL cable shall be permitted in one- and two-family dwellings.

Author's Substantiation. This revision provides requirements for installing Type FPLR-CI cable in plenums and risers.

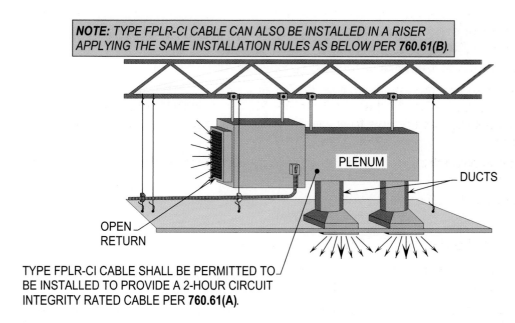

NOTE: *TYPE FPLR-CI CABLE CAN ALSO BE INSTALLED IN A RISER APPLYING THE SAME INSTALLATION RULES AS BELOW PER **760.61(B)**.*

PLENUM

DUCTS

OPEN RETURN

TYPE FPLR-CI CABLE SHALL BE PERMITTED TO BE INSTALLED TO PROVIDE A 2-HOUR CIRCUIT INTEGRITY RATED CABLE PER **760.61(A)**.

PLENUM AND RISER
760.61(A) AND (B)

Purpose of Change: To clarify such marking must be done at the equipment where the circuitry originates.

NEC Ch. 7 – Article 770
Parts I through IV – 770.1 through 770.182

Type of Change		Panel Action		UL	UL 508	API 500	API 505	OSHA
Reorganized		Accept in Principle		910	-	-	-	1910.331(a)(4)
ROP		ROC		NFPA 70E	NFPA 70B	NFPA 79	NFPA	NEMA
pg. 1829	# 16-5	pg. 706	# 16-9	-	-	-	-	-
log: 123	CMP: 16	log: 996	Submitter: Stanley D. Kahn			2002 NEC: Article 770		IEC: -

Author's Substantiation. This article has been reorganized to relocate appropriate parts and sections for usability of the NEC.

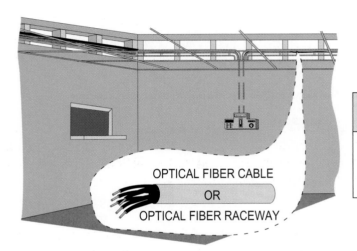

**RESTRUCTURING PARTS IN ARTICLE 770
OPTICAL FIBER CABLES**

I. GENERAL
II. PROTECTION
III. CABLES WITHIN BUILDINGS
IV. LISTING REQUIREMENTS

OPTICAL FIBER CABLE
OR
OPTICAL FIBER RACEWAY

**RESTRUCTURING
ARTICLE 770 – PARTS I THRU IV**

Purpose of Change: To reorganize the requirements and relocate into the appropriate part and sections for better usability.

Communications Systems

Chapter 8 of the NEC covers communications systems. Articles and Sections of this chapter stand alone and are independent of the other chapters, except in specific cases where they are referenced. Communications systems include such systems as telephone, telegraph, radio, alarm, etc.

The rules and regulations in this Article are mainly applied to those systems that are connected to a central station and operate as elements of such systems. When communications are involved, one of the Articles in the 800 series must be selected, based upon the system utilized.

Chapter 8 also includes **Article 810,** which deals with radio and television equipment. The requirements in **Article 810** pertain to radio and television receiving equipment and amateur radio transmitting and receiving equipment, but not equipment and antenna use for coupling carrier current to power line conductors.

Article 820 of **Chapter 8** covers coaxial cables distribution of radio frequency signals typically employed in community antenna television (CATV) systems.

When communications are involved, one of the Articles in the 800 series must be selected, based upon the system utilized. The 810 series of **Chapter 8** is used for radio and television equipment, and **Article 820** is used with various Sections for CATV-related equipment and cables.

NEC Ch. 8 – Article 800
Parts I through V – 800.1 through 800.182

148

Type of Change		Panel Action		UL	UL 508	API 500	API 505	OSHA
Reorganized		Accept in Principle		4978	-	-	-	1910.308(e)
ROP		ROC		NFPA 70E	NFPA 70B	NFPA 79	NFPA	NEMA
pg. 1893	# 16-70	pg. -	# -	450.5	-	-	-	SB 3
log: 124	CMP: 16	log: -		Submitter: Stanley D. Kahn		2002 NEC: Article 800		IEC: 443

Author's Substantiation. This article has been reorganized to relocate appropriate parts and sections for usability of the NEC.

CROSS REFERENCE		
2005 NEC	**2002 NEC**	**SECTION HEADING**
PART I – GENERAL		
800.1	800.1	Scope
800.2	800.2	Definitions
800.3	——	Other Articles
800.3(A)	800.3	Hybrid Power and Communications Cables
800.3(B)	800.8	Hazardous (Classified) Locations
800.3(C)	800.52(B)	Spread of Fire or Products of Combustion
800.3(D)	——	Equipment in Other Space Used for Environmental Air
800.18	800.4	Installation of Equipment
800.21	800.5	Access to Electrical Equipment Behind Panels Designed to Allow Access
800.24	800.6	Mechanical Execution of Work
PART II – WIRES AND CABLES OUTSIDE AND ENTERING BUILDINGS		
800.44	800.10	Overhead Communications Wires and Cables
800.44(A)	800.10(A)	On Poles and In-Span
800.44(B)	800.10(B)	Above Roofs
800.47	800.11	Underground Circuits Entering Buildings
800.47(A)	800.11(A)	With Electric Light or Power Conductors
800.47(B)	800.11(B)	Underground Block Distribution
800.50	800.12	Circuits Requiring Primary Protectors
800.50(A)	800.12(A)	Insulation, Wires, and Cables
800.50(B)	800.12(B)	On Buildings
800.50(C)	800.12(C)	Entering Buildings
800.53	800.13	Lightning Conductors
PART III – PROTECTION		
800.90	800.30	Protective Devices
800.90(A)	800.30(A)	Application
800.90(B)	800.30(B)	Location
800.90(C)	800.30(C)	Hazardous (Classified) Locations
800.90(D)	800.32	Secondary Protectors
800.93	800.33	Cable Grounding

CROSS REFERENCE		
2005 NEC	**2002 NEC**	**SECTION HEADING**
PART IV – GROUNDING METHODS		
800.100	800.40	Cable and Primary Protection Grounding
800.100(A)	800.40(A)	Grounding Conductor
800.100(B)	800.40(B)	Electrode
800.100(C)	800.40(C)	Electrode Connection
800.100(D)	800.40(D)	Bonding of Electrodes
800.106	800.41	Primary Protector Grounding and Bonding at Mobile Homes
800.106(A)	800.41(A)	Grounding
800.106(B)	800.41(B)	Bonding
PART V – COMMUNICATIONS WIRES AND CABLES WITHIN BUILDINGS		
800.110	800.48	Raceways for Communications Wires and Cables
800.113	800.50	Installation and Marking of Communications Wires and Cables
800.133	800.52	Installation of Communications Wires, Cables, and Equipment
800.133(A)	800.52(A)	Separation from Other Conductors
800.133(B)	800.52(D)	Cable Trays
800.133(C)	800.52(E)	Support of Conductors
800.133(D)	800.52(C)	Wiring in Ducts for Dust, Loose Stock, or Vapor Removal
800.154	800.53	Applications of Listed Communications Wires and Cables and Communications Raceways
800.154(A)	800.53(A)	Plenum
800.154(B)	800.53(B)	Riser
800.154(C)	800.53(C)	Distributing Frames and Cross-Connect Arrays
800.154(D)	800.53(D)	Cable Trays
800.154(E)	800.53(E)	Other Wiring Within Buildings
800.154(F)	800.53(F)	Hybrid Power and Communications Cable
800.154(G)	800.53(G)	Cable Substitutions
PART VI – LISTING REQUIREMENTS		
800.170	800.4	Equipment
800.170(A)	800.31	Primary Protectors
800.170(B)	800.32	Secondary Protectors
800.173	——	Drop Wire and Cable
800.179	800.51	Communications Wires and Cables
800.179(A)	800.51(A)	Type CMP
800.179(B)	800.51(B)	Type CMR
800.179(C)	800.51(C)	Type CMG
800.179(D)	800.51(D)	Type CM
800.179(E)	800.51(E)	Type CMX
800.179(F)	800.51(F)	Type CMUC Undercarpet Wire and Cable
800.179(G)	800.51(G)	Multipurpose (MP) Cables
800.179(H)	——	Communications Circuit Integrity (CI) Cable
800.179(I)	800.51(H)	Communications Wires
800.179(J)	800.51(I)	Hybrid Power and Communications Cable
800.182	——	Communications Raceways
800.182(A)	800.51(J)	Plenum Communications Raceways
800.182(B)	800.51(K)	Riser Communications Raceways
800.182(C)	800.51(L)	General-Purpose Communications Raceways

NEC Ch 8 – Article 800
Part I – 800.2

Type of Change		Panel Action		UL	UL 508	API 500	API 505	OSHA
New Definitions		Accept in Principle		4978	-	-	-	1910.308(e)
ROP		ROC		NFPA 70E	NFPA 70B	NFPA 79	NFPA	NEMA
pg. 1895	# 16-74	pg. -	# -	450.5	-	-	-	SB 3
log: 3235	CMP: 16	log: -	Submitter: Edward Walton			2002 NEC: -		IEC: 443

2005 NEC – 800.2 Definitions

Air Duct. A conduit or passageway for conveying air to or from heating, cooling, air conditioning, or ventilating equipment, but not including the plenum. [NFPA 97:1.2.6]

Communications Circuit Integrity (CI) cable. Cable used in communications systems to ensure continued operation of critical circuits during a specified time under fire conditions.

Communications Equipment. The electronic equipment that performs the telecommunications operations for the transmission of audio, video, and data, and including power equipment (e.g., dc converters, inverters and batteries) and technical support equipment (e.g., computers).

Author's Substantiation. New definitions have added to define an air duct, communications circuit integrity (CI) cable, and communications equipment.

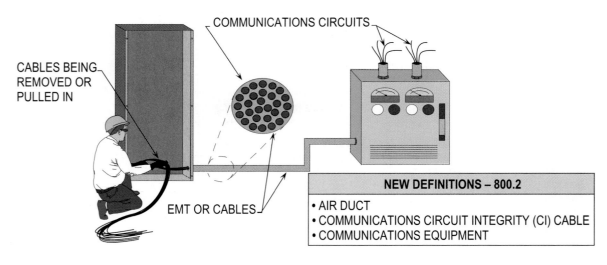

DEFINITIONS
800.2

Purpose of Change: To add definitions that are used in this Article in everyday installations of communications circuits.

NEC Ch. 8 – Article 800
Part IV – 800.100(A)(4), FPN

Type of Change	Panel Action	UL	UL 508	API 500	API 505	OSHA
New FPN	Accept	4978	-	-	-	1910.308(e)
ROP	**ROC**	**NFPA 70E**	**NFPA 70B**	**NFPA 79**	**NFPA**	**NEMA**
pg. 1907 # 16-96	pg. 866 # 16-389	450.5	-	-	-	SB 3
log: 846 CMP: 16	log: -	Submitter: James E. Brunssen		2002 NEC: -		IEC: 443

2005 NEC – 800.100(A)(4) Length

The primary protector grounding conductor shall be as short as practicable. In one- and two-family dwellings, the primary protector grounding conductor shall be as short as practicable, not to exceed 6.0 m (20 ft) in length.

> **FPN:** Similar grounding conductor length limitations applied at apartment buildings and commercial buildings help to reduce voltages that may be developed between the building's power and communications systems during lightning events.

Author's Substantiation. A new FPN has been added to recognize that apartment buildings and commercial buildings do not have restrictions on the length of the primary protector grounding conductor.

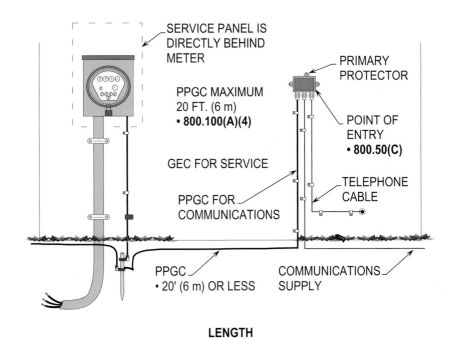

LENGTH
800.100(A)(4), FPN

Purpose of Change: The new FPN points out that similar grounding techniques using the 20 ft. (6 m) or less rule at apartment and commercial buildings is a good practice.

NEC Ch. 8 – Article 800
Part V – 800.133(A)(1)(c), Ex. 1

Type of Change		Panel Action		UL	UL 508	API 500	API 505	OSHA
Revision		Accept		4978	-	-	-	1910.308(e)
ROP		ROC		NFPA 70E	NFPA 70B	NFPA 79	NFPA	NEMA
pg. 1954	# 16-133	pg. -	# -	450.5	-	-	-	SB 3
log: 1290	CMP: 16	log: -	Submitter: David H. Kendall			2002 NEC: 800.133(A)(1)(c), Ex. 1		IEC: 443

2002 NEC – ~~800.52(A)(1)(c)~~ Electric Light, Power, Class 1, Non-Power-Limited Fire Alarm, and Medium Power Network-Powered Broadband Communications Circuits in Raceways, Compartments, and Boxes

Communications conductors shall not be placed in any raceway, compartment, outlet box, junction box, or similar fitting with conductors of electric light, power, Class 1, non-power-limited fire alarm or medium power network-powered broadband communications circuits.

> **Exception No. 1:** Where all of the conductors of electric light, power, Class 1, non-power-limited fire alarm, and medium power network-powered broadband communications circuits are separated from all of the conductors of communications circuits by a barrier.

2005 NEC – <u>800.133(A)(1)(c)</u> Electric Light, Power, Class 1, Non-Power-Limited Fire Alarm, and Medium Power Network-Powered Broadband Communications Circuits in Raceways, Compartments, and Boxes

Communications conductors shall not be placed in any raceway, compartment, outlet box, junction box, or similar fitting with conductors of electric light, power, Class 1, non-power-limited fire alarm, or medium power network-powered broadband communications circuits.

> **Exception No. 1:** Where all of the conductors of electric light, power, Class 1, non-power-limited fire alarm, and medium power network-powered broadband communications circuits are separated from all of the conductors of communications circuits by a <u>permanent</u> barrier <u>or listed divider</u>.

Author's Substantiation. This revision clarifies that a permanent barrier or listed dividers shall be permitted to be installed to separate fire alarm circuits.

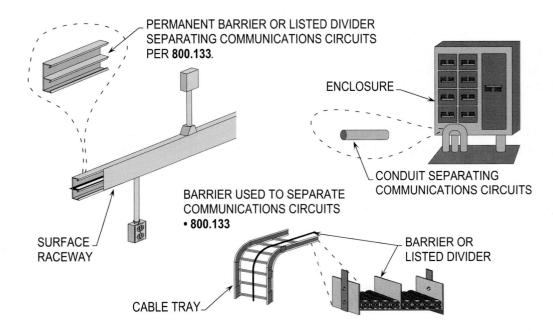

PERMANENT BARRIER OR LISTED DIVIDER
SEPARATING COMMUNICATIONS CIRCUITS
PER **800.133**.

ENCLOSURE

CONDUIT SEPARATING
COMMUNICATIONS CIRCUITS

BARRIER USED TO SEPARATE
COMMUNICATIONS CIRCUITS
• **800.133**

SURFACE
RACEWAY

BARRIER OR
LISTED DIVIDER

CABLE TRAY

**COMMUNICATIONS CIRCUITS IN CABLES, COMPARTMENTS, ENCLOSURES, OUTLET BOXES OR RACEWAYS
800.133(A)(1)(c), Ex. 1**

Purpose of Change: To clarify that certain types of permanent barriers or listed dividers can serve as a separate fire alarm circuits.

NEC Ch. 8 – Article 800
Part V – Figure 800.154

Type of Change		Panel Action		UL	UL 508	API 500	API 505	OSHA
Revision		Accept in Principle		4978	-	-	-	1910.308(e)
ROP		**ROC**		**NFPA 70E**	**NFPA 70B**	**NFPA 79**	**NFPA**	**NEMA**
pg. 1958	# 16-139	pg. -	# -	450.5	-	-	-	SB 3
log: 2575	CMP: 16	log: -	Submitter: Sanford Egesdal			2002 NEC: Figure 800.53		IEC: 443

Author's Substantiation. This figure has been revised by deleting the references to cables marked MPP, MPR, and MPG-MP.

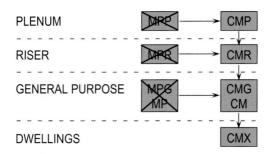

X = CABLES DELETED IN 2005 NEC

TYPE CM — COMMUNICATIONS CABLES
~~TYPE MP — MULTIPURPOSE CABLE~~

A → B CABLE A SHALL BE PERMITTED TO BE USED IN PLACE OF CABLE B

CABLE SUBSTITUTION
FIGURE 800.154

Purpose of Change: To delete reference to cables marked MPP, MPR, and MPG/MP in **Table 800.154** and **Figure 800.154** of the **2005 NEC**.

NEC Ch. 8 – Article 820
Parts I through VI – 820.1 through 820.182

149

Type of Change	Panel Action	UL	UL 508	API 500	API 505	OSHA
Reorganized	Accept in Principle	1410	-	-	-	-

ROP		ROC		NFPA 70E	NFPA 70B	NFPA 79	NFPA	NEMA
pg. 1969	# 16-154	pg. -	# -	-	-	-	-	PP 1
log: 125	CMP: 16	log: -	Submitter: Stanley D. Kahn			2002 NEC: Article 820		IEC: 443 and 528

Author's Substantiation. This article has been reorganized to relocate appropriate parts and sections for usability of the NEC.

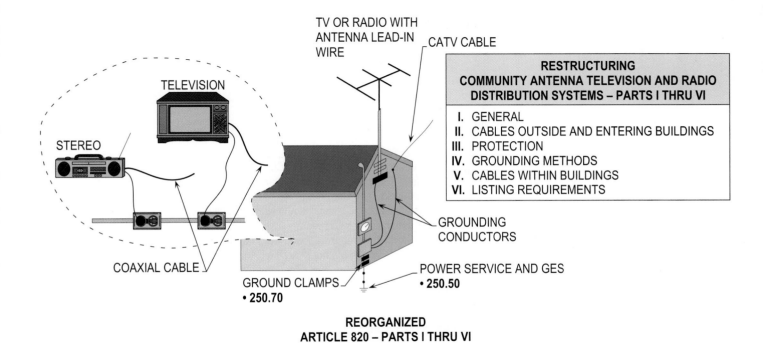

TV OR RADIO WITH ANTENNA LEAD-IN WIRE

CATV CABLE

TELEVISION

STEREO

RESTRUCTURING
COMMUNITY ANTENNA TELEVISION AND RADIO
DISTRIBUTION SYSTEMS – PARTS I THRU VI

I. GENERAL
II. CABLES OUTSIDE AND ENTERING BUILDINGS
III. PROTECTION
IV. GROUNDING METHODS
V. CABLES WITHIN BUILDINGS
VI. LISTING REQUIREMENTS

GROUNDING CONDUCTORS

COAXIAL CABLE

GROUND CLAMPS
• 250.70

POWER SERVICE AND GES
• 250.50

REORGANIZED
ARTICLE 820 – PARTS I THRU VI

Purpose of Change: To include a new **Part VI** Titled "Listing Requirements" to correlate with **Articles 800, 810, and 830** of the **2005 NEC**.

NEC Ch. 8 – Article 820
Parts I through VI – 820.1 through 820.182

149

Type of Change	Panel Action	UL	UL 508	API 500	API 505	OSHA
Reorganized	Accept in Principle	1410	-	-	-	-

ROP		ROC		NFPA 70E	NFPA 70B	NFPA 79	NFPA	NEMA
pg. 1969	# 16-154	pg. -	# -	-	-	-	-	PP 1
log: 125	CMP: 16	log:	-	Submitter: Stanley D. Kahn		2002 NEC: Article 820		IEC: 443 adn 528

Author's Substantiation. This article has been reorganized to relocate appropriate parts and sections for usability of the NEC.

CROSS REFERENCE		
2005 NEC	**2002 NEC**	**SECTION HEADING**
PART I – GENERAL		
820.1	820.1	Scope
820.2	820.2	Definitions
820.3	820.3	Other Articles
820.3(A)	820.3(A)	Spread of Fire or Products of Combustion
820.3(B)	820.3(B)	Ducts, Plenums, and Other Air-Handling Spaces
820.3(C)	820.3(C)	Installation and Use
820.3(D)	820.3(D)	Installations of Conductive and Nonconductive Optical Fiber Cables
820.3(E)	820.3(E)	Communications Circuits
820.3(F)	820.3(F)	Network-Powered Broadband Communications Systems
820.3(G)	820.3(G)	Alternate Wiring Methods
820.15	820.4	Energy Limitations
820.21	820.5	Access to Electrical Equipment Behind Panels Designed to Allow Access
820.24	820.6	Mechanical Execution of Work
PART II – CABLES OUTSIDE AND ENTERING BUILDINGS		
820.44	820.10	Overhead Cables
820.44(A)	820.10(A)	On Poles
820.44(B)	820.10(B)	Lead-in Clearance
820.44(C)	820.10(C)	On Masts
820.44(D)	820.10(D)	Above Roofs
820.44(E)	820.10(E)	Between Buildings
820.44(F)	820.10(F)	On Buildings
820.47	820.11	Underground Circuits Entering Buildings
820.47(A)	820.11(A)	Underground Systems
820.47(B)	820.11(B)	Direct-Buried Cables and Raceways
PART III – PROTECTION		
820.93	820.33	Grounding of Outer Conductive Shield of a Coaxial Cable
820.93(A)	820.33(A)	Shield Grounding
820.93(B)	820.33(B)	Shield Protection Devices

CROSS REFERENCE		
2005 NEC	**2002 NEC**	**SECTION HEADING**
PART IV – GROUNDING METHODS		
820.100	820.40	Cable Grounding
820.100(A)	820.40(A)	Grounding Conductor
820.100(B)	820.40(B)	Electrode
820.100(C)	820.40(C)	Electrode Connection
820.100(D)	820.40(D)	Bonding of Electrodes
820.103	820.41	Equipment Grounding
820.106	820.42	Bonding and Grounding at Mobile Homes
820.106(A)	820.42(A)	Grounding
820.106(B)	820.42(B)	Bonding
PART V – CABLES WITHIN BUILDINGS		
820.110	——	Raceways for Coaxial Cables
820.113	820.50	Installation and Marking of Coaxial Cables
820.133	820.52	Installation of Cables and Equipment
820.133(A)	820.52(A)	Separation from Other Conductors
820.133(B)	820.52(C)	Hybrid Power and Coaxial Cabling
820.133(C)	820.52(D)	Support of Cables
820.154	820.53	Applications of Listed CATV Cables and CATV Raceways
820.154(A)	820.53(A)	Plenums
820.154(B)	820.53(B)	Riser
820.154(C)	820.53(C)	Cable Trays
820.154(D)	820.53(D)	Other Wiring Within Buildings
PART VI – LISTING REQUIREMENTS		
820.179	820.51	Coaxial Cables
820.179(A)	820.51(A)	Type CATVP
820.179(B)	820.51(B)	Type CATVR
820.179(C)	820.51(C)	Type CATV
820.179(D)	820.15(D)	Type CATVX
820.182	——	CATV Raceways
820.182(A)	——	Plenum CATV Raceways
820.182(B)	——	Riser CATV Raceways
820.182(C)	——	General-Purpose CATV Raceways

NEC Ch. 8 – Article 830
Parts I through VI – 830.1 through 830.179

Type of Change		Panel Action		UL	UL 508	API 500	API 505	OSHA
Reorganized		Accept in Principle		1581	-	-	-	-
ROP		ROC		NFPA 70E	NFPA 70B	NFPA 79	NFPA	NEMA
pg. 2019	# 16-205	pg. -	# -	450.5(D)(4)	-	-	-	PP 1
log: 126	CMP: 16	log: -	Submitter: Stanley D. Kahn			2002 NEC: Article 830		IEC: 443 and 528

Author's Substantiation. This article has been reorganized to relocate appropriate parts and sections for usability of the NEC.

CROSS REFERENCE		
2005 NEC	**2002 NEC**	**SECTION HEADING**
PART I – GENERAL		
830.1	830.1	Scope
830.2	830.2	Definitions
830.3	830.3	Other Articles
830.3(A)	830.3(A)	Spread of Fire or Products of Combustion
830.3(B)	830.3(B)	Ducts, Plenums, and Other Air-Handling Spaces
830.3(C)	———	Equipment in Other Space Used for Environmental Air
830.3(D)	830.3(D)	Output Circuits
830.3(E)	830.9	Hazardous (Classified) Locations
830.15	830.4	Power Limitations
830.21	830.6	Access to Electrical Equipment Behind Panels Designed to Allow Access
830.24	830.7	Mechanical Execution of Work
PART II – CABLES OUTSIDE AND ENTERING BUILDINGS		
830.40	830.10	Entrance Cables
830.40(A)	830.10(A)	Medium Power Circuits
830.40(B)	830.10(B)	Low-Power Circuits
830.44	830.11	Aerial Cables
830.44(A)	830.11(A)	On Poles
830.44(B)	830.11(B)	Climbing Space
830.44(C)	830.11(C)	Lead-In Clearance
830.44(D)	830.11(D)	Clearance from Ground
830.44(E)	830.11(E)	Over Pools
830.44(F)	830.11(F)	Above Roofs
830.44(G)	830.11(G)	Final Spans
830.44(H)	830.11(H)	Between Buildings
830.44(I)	830.11(I)	On Buildings
830.47	830.12	Underground Circuits Entering Buildings
830.47(A)	830.12(A)	Underground Systems
830.47(B)	830.12(B)	Direct-Buried Cables and Raceways
830.47(C)	830.12(C)	Mechanical Protection
830.47(D)	8310.12(D)	Pools

CROSS REFERENCE		
2005 NEC	**2002 NEC**	**SECTION HEADING**
PART III – PROTECTION		
830.90	830.30	Primary Electrical Protection
830.90(A)	830.30(A)	Application
830.90(B)	830.30(B)	Location
830.90(C)	830.30(C)	Hazardous (Classified) Locations
830.93	830.33	Grounding or Interruption of Metallic Members of Network-Powered Broadband Communications Cables
PART IV – GROUNDING METHODS		
830.100	830.40	Cable, Network Interface Unit, and Primary Protector Grounding
830.100(A)	830.40(A)	Grounding Conductor
830.100(B)	830.40(B)	Electrode
830.100(C)	830.40(C)	Electrode Connection
830.100(D)	830.40(D)	Bonding of Electrodes
830.106	830.42	Bonding and Grounding at Mobile Homes
830.106(A)	830.42(A)	Grounding
830.106(B)	830.42(B)	Bonding
PART V – WIRING METHODS WITHIN BUILDINGS		
830.133	830.58	Installation of Network-Powered Broadband Communications Cables and Equipment
830.133(A)	830.58(A)	Separation of Conductors
830.133(B)	830.58(D)	Support of Conductors
830.133(C)	830.58(E)	Cable Substitutions
830.133(D)	——	Installation and Use
830.151	830.54	Medium Power Network-Powered Broadband Communications System Wiring Methods
830.151(A)	830.54(A)	Ducts, Plenums, and Other Air-Handling Spaces
830.151(B)	830.54(B)	Riser
830.151(C)	830.54(C)	Other Wiring
830.154	830.55	Low-Power Network-Powered Broadband Communications System Wiring Methods
830.154(A)	830.55(A)	In Buildings
830.154(B)	830.55(B)	Ducts, Plenums, and Other Air-Handling Spaces
830.154(C)	830.55(C)	Riser
830.154(D)	830.55(D)	Other Wiring
830.157	830.56	Protection Against Physical Damage
830.160	830.57	Bends
PART VI – LISTING REQUIREMENTS		
830.179	——	Network-Powered Broadband Communications Equipment and Cables
830.179(A)	——	Listing and Marking

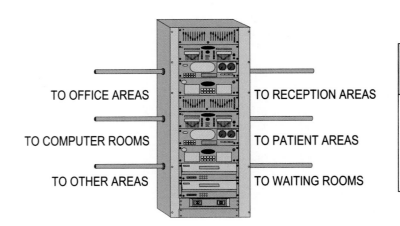

TO OFFICE AREAS

TO COMPUTER ROOMS

TO OTHER AREAS

TO RECEPTION AREAS

TO PATIENT AREAS

TO WAITING ROOMS

**RESTRUCTURING
NETWORK-POWERED BROADBAND
COMMUNICATIONS SYSTEMS**

- GENERAL
- CABLES OUTSIDE AND ENTERING BUILDINGS
- PROTECTION
- GROUNDING METHODS
- WIRING METHODS WITHIN BUILDINGS
- LISTING REQUIREMENTS

**REORGANIZED
ARTICLE 830 – PARTS I THRU VI**

Purpose of Change: To include a new **Part VI** Titled "Listing Requirements" to correlate with **Articles 800, 810,** and **820** of the **2005 NEC**.

NEC Ch. 9 – Table 2

150

Type of Change		Panel Action		UL	UL 508	API 500	API 505	OSHA
Relocated		Accept		6	-	-	-	-
ROP		ROC		NFPA 70E	NFPA 70B	NFPA 79	NFPA	NEMA
pg. 875	# 8-24a	pg. 1021	# 8-202	-	29.3.2	14.5.3	-	C 80.2
log: CP800	CMP: 8	log: 2039	Submitter: -			2002 NEC: Table 344.24		IEC: 521

Author's Substantiation. This table has been relocated so all raceway articles may be referred to this table for conduit or tubing radius bends.

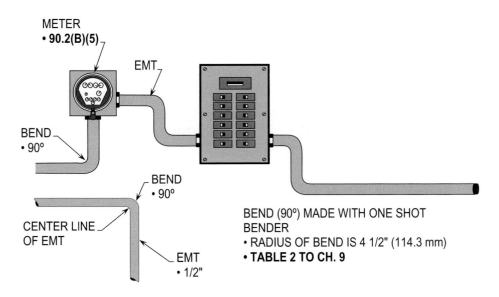

METER
• 90.2(B)(5)

EMT

BEND
• 90°

BEND
• 90°

CENTER LINE OF EMT

EMT
• 1/2"

BEND (90°) MADE WITH ONE SHOT BENDER
• RADIUS OF BEND IS 4 1/2" (114.3 mm)
• **TABLE 2 TO CH. 9**

RADIUS OF CONDUIT AND TUBING BENDS
TABLE 2, CHAPTER 9

Purpose of Change: To relocate "Table 344.24" to "Table 2" in Chapter 9 because all raceway articles can refer to this table for conduit or tubing radius bends.

NEC Ch. 9 – Table 12(B)

Type of Change		Panel Action		UL	UL 508	API 500	API 505	OSHA
Revision		Accept		-	-	-	-	-
ROP		ROC		NFPA 70E	NFPA 70B	NFPA 79	NFPA	NEMA
pg. 2042	# 3-301	pg. 1022	# 3-805	-	-	-	-	-
log: 2322	CMP: 3	log: 1821	Submitter: Vince Baclawski			2002 NEC: Ch. 9, Table 12(B)		IEC: -

Author's Substantiation. The column for over "100 and through 250" has been deleted from this table due to this column applying to equipment (smoke detectors) that have been out of production for over 20 years.

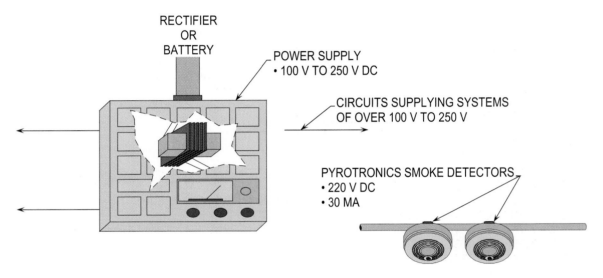

PLFA DIRECT-CURRENT POWER SOURCE LIMITATIONS
TABLE 12(B) TO CHAPTER 9

Purpose of Change: To delete "over 100 volts to 250 volts" because the equipment (smoke detector) that this voltage was to accommodate has been out of production for about 20 years.

Annex G
Article 80

Type of Change		Panel Action		UL	UL 508	API 500	API 505	OSHA
-		-		-	-	-	-	-
ROP		ROC		NFPA 70E	NFPA 70B	NFPA 79	NFPA	NEMA
pg. -	# -	pg. -	# -	-	-	-	-	-
log: -	CMP: -	log: -	Submitter: -			2002 NEC: -		IEC: -

Author's Substantiation. Article 80 has been relocated as a new annex, Annex G, because this article in the 2002 NEC was not part of the NEC unless Article 80 was adopted by federal, state, or local municipalities.

ARTICLE 80 – OUTLINED
- **80.1** SCOPE
- **80.2** DEFINITIONS
- **80.3** PURPOSE
- **80.5** ADOPTION
- **80.7** TITLE
- **80.9** APPLICATION
- **80.11** OCCUPANCY OF BUILDING OR STRUCTURE
- **80.13** AUTHORITY
- **80.15** ELECTRICAL BOARD
- **80.17** RECORDS AND REPORTS
- **80.19** PERMITS AND APPROVALS
- **80.21** PLANS REVIEW
- **80.23** NOTICE OF VIOLATIONS, PENALTIES
- **80.25** CONNECTION TO ELECTRICITY SUPPLY
- **80.27** INSPECTOR's QUALIFICATIONS
- **80.29** LIABILITY FOR DAMAGES
- **80.31** VALIDITY
- **80.33** REPEAL OF CONFLICTING ACTS
- **80.35** EFFECTIVE DATE

APPLYING EXISTING **ARTICLE 80** IN NEW **ANNEX G**

AHJ
- **90.4**
- **90.7**
- **90.2(C)**

KEY POINT: *SIMILAR RULES AND REGULATIONS CAN BE FOUND IN EXISTING **NFPA 70A**.*

ADMINISTRATION AND ENFORCEMENT
ARTICLE 80 (ANNEX G)

Purpose of Change: Because **Article 80** was not part of the NEC unless adopted by federal, state, or local authorities, it was relocated as a new **Annex G**.

Reference Standards

API RP 500 - Recommended Practice for Classification of Locations for Electrical Installations at Petroleum Facilities Classified as Class I, Division 1 and Division 2

API RP 505 - Recommended Practice for Classification of Locations for Electrical Installations at Petroleum Facilities Classified as Class I, Zone 0, Zone 1 and Zone 2

APR RP 540 - Recommended Practice for Electrical Installations in Petroleum Processing Plants

IEC 79 - Electrical Apparatus for Explosive Gas Atmospheres

NFPA 20 - Standard for the Installation of Stationary Pumps for Fire Protection

NFPA 30B - Code for the Manufacture and Storage of Aerosol Products

NFPA 33 - Standard for Spray Application Using Flammable or Combustible Materials

NFPA 34 - Standard for Dipping and Coating Processes Using Flammable or Combustible Liquids

NFPA 58 - Liquefied Petroleum Gas Code

NFPA 70B - Recommended Practice for Electrical Equipment Maintenance

NFPA 70E - Standard for Electrical Safety in the Workplace

NFPA 72® - National Fire Alarm Code®

NFPA 79 - Electrical Standard for Industrial Machinery

NFPA 88A - Standard for Parking Structures

NFPA 99 - Standard for Health Care Facilities

NFPA 110 - Standard for Emergency and Standby Power Systems

NFPA 251 - Standard Methods of Tests of Fire Endurance of Building Construction and Materials

NFPA 302 - Fire Protection Standard for Pleasure and Commercial Motor Craft

NFPA 303 - Fire Protection Standard for Marinas and Boatyards

NFPA 325 - Guide to Fire Hazard Properties of Flammable Liquids, Gases and Volatile Solids
(now found only in *Fire Protection Guide to Hazardous Materials*, published by NFPA)

NFPA 407 - Standard for Aircraft Fuel Servicing

NFPA 409 - Standard on Aircraft Hangars

NFPA 410 - Standard on Aircraft Maintenance

NFPA 496 - Standard for Purged and Pressurized Enclosures for Electrical Equipment

NFPA 497 - Recommended Practice for the Classification of Flammable Liquids, Gases, or Vapors and of Hazardous (Classified) Locations for Electrical Installations in Chemical Process Areas

NFPA 499 - Recommended Practice for the Classification of Combustible Dusts and of Hazardous (Classified) Locations for Electrical Installations in Chemical Process Areas

NFPA 505 - Fire Safety Standard for Powered Industrial Trucks Including Type Designations, Areas of Use, Conversions, Maintenance, and Operation

Questions

Section	Answer	

Section   **Answer**

_____   T   F   **1.** Where a negative impact on safety would result, hard conversion shall be used.

_____   T   F   **2.** Electrical installation in a vault, room, or closet or in an area surrounded by a wall, screen, or fence, access to which is controlled a lock(s) or other approved means, shall be considered to be accessible to qualified persons only.

_____   T   F   **3.** An insulated grounded conductor larger than 6 AWG shall be permitted to be identified at time of installation, by a distinctive gray marking at its terminations.

_____   T   F   **4.** Where the premises wiring system has branch circuits supplied from more than one nominal voltage system, each ungrounded conductor of a branch-circuit, where accessible, shall not be required to be identified by system.

_____   T   F   **5.** All 125-volt, single-phase, 15- and 20-ampere receptacles installed in commercial kitchens shall have GFCI-protection.

_____   T   F   **6.** All 125-volt, single-phase, 15- and 20-ampere receptacles installed in outdoor public spaces are not required to have GFCI-protection.

_____   T   F   **7.** The arc-fault circuit interrupter shall be installed within 10 ft. (3 m) of the branch circuit overcurrent device as measured along the branch circuit conductors.

_____   T   F   **8.** A receptacle shall not be required to be mounted in the wall or partition where it is installed on the side or face of the basin cabinet not more than 12 in. (300 mm) below the countertop.

_____   T   F   **9.** A receptacle outlet shall be required at one- and two-family dwellings for the service of evaporative coolers.

_____   T   F   **10.** Where the premises wiring system has feeders supplied from more than one nominal system voltage, each ungrounded conductor of a feeder, where accessible, shall be identified by system.

_____   T   F   **11.** Where raceway-type masts are used, all raceway fittings shall be identified for used with masts.

_____   T   F   **12.** Cable trays used to support service-entrance conductors shall contain only service-entrance conductors.

_____   T   F   **13.** In guest suites of hotels and motels, overcurrent devices shall be permitted to be located in bathrooms.

Questions

Section	Answer	

Section   **Answer**

_____   T   F   **14.** The earth shall be considered as an effective ground-fault current path.

_____   T   F   **15.** For a grounded system, a spliced main bonding jumper shall be permitted to be used to connect the equipment grounding conductor(s) and the service disconnect enclosure to the grounded conductor within the enclosure for each service disconnect.

_____   T   F   **16.** Where more than one separately derived system is installed, it shall be permissible to connect a tap from each separately derived system to a common grounding electrode conductor.

_____   T   F   **17.** Exposed structural metal that is interconnected to form a metal building frame and is not intentionally grounded and is likely to become energized shall be bonded.

_____   T   F   **18.** Equipment grounding conductors larger than 6 AWG shall be marked in conduit bodies that contain no splices or unused hubs.

_____   T   F   **19.** Equipment grounding conductors run with feeder taps shall not be smaller than shown in Table 250.122 based on the rating of the overcurrent device ahead of the feeder but shall not be required to be larger than the tap conductors.

_____   T   F   **20.** A listed and marked steel plate less than 1/16 in. (1.6 mm) shall be permitted to protect cables run through studs.

_____   T   F   **21.** Short sections of raceways used to contain conductors for protection from physical damage shall be installed complete between outlet, junction, or splicing points.

_____   T   F   **22.** Each current-carrying conductor of a paralleled set of conductors shall be counted as a current-carrying conductor.

_____   T   F   **23.** For enclosures in wet locations, raceways or cables entering above the level of uninsulated live parts shall use fittings approved for wet locations.

_____   T   F   **24.** A looped, unbroken conductor not less than twice the minimum length required for free conductors in 300.14 shall be counted once.

_____   T   F   **25.** Handhole enclosure covers shall have an identifying mark or logo that prominently identifies the function of the enclosure.

_____   T   F   **26.** Type AC cable fittings shall not be permitted to be used as means of cable support.

_____   T   F   **27.** Where Type UF cable is installed as a substitute wiring method for NM cable, the conductor insulation shall be rated 90ßC.

Questions

Section   **Answer**

_____   T   F   **28.** In industrial establishments where the conditions of maintenance and supervision ensure that only qualified persons service the installation, flexible cords and cables shall be permitted to be installed in aboveground raceways that are not longer than 50 ft. (15 m) to protect the flexible cord and cable from physical damage.

_____   T   F   **29.** Nonmetallic poles shall not be permitted to support luminaires (lighting fixtures).

_____   T   F   **30.** Where no equipment grounding conductor exists at the outlet, replacement luminaires (lighting fixtures) that are GFCI protected shall not be required to be connected to an equipment grounding conductor.

_____   T   F   **31.** A disconnecting shall be installed for luminaires (fixtures) installed in hazardous (classified) locations.

_____   T   F   **32.** Cord-and-plug-connected vending machines not incorporating integral GFCI protection shall be connected to a GFCI protected outlet.

_____   T   F   **33.** The provision for locking or adding a lock to the disconnecting means shall be installed on or at the switch or circuit breaker used as the disconnecting means and shall remain place with or without the lock installed.

_____   T   F   **34.** System isolation equipment shall not be required to be listed for disconnecting purposes.

_____   T   F   **35.** Combustible gas detection systems shall be documented when used as a protection technique in hazardous (classified) locations.

_____   T   F   **36.** Equipment that depends on a single compression seal, diaphragm, or tube to prevent flammable or combustible fluids from entering the equipment shall be identified for a Class I, Division 2 location even if installed in an unclassified location.

_____   T   F   **37.** Process-connected equipment that is listed and marked Dual Seal shall have additional process sealing in hazardous (classified) locations.

_____   T   F   **38.** In Zones 20, 21, and 22, seals shall not be required to be installed to protect the ingress of dust, fibers, or flyings.

_____   T   F   **39.** Parking garages used for parking or storage shall be classified.

_____   T   F   **40.** All 125-volt, 50/60 Hz, single-phase, 15- and 20-ampere receptacles installed in areas where electrical diagnostic equipment is to be used shall have GFCI protection.

e

Questions

Section   **Answer**

_____   T   F   **41.** In critical care areas, the patient equipment grounding point, where supplied, where supplied, shall be permitted to contain one or more approved grounding and bonding jacks.

_____   T   F   **42.** Where multiple services or separately derived systems or both supply rides, attractions, and other structures, all sources of supply that serve rides, attractions, or other structures separated by less than 12 ft. (3.7 m) shall be bonded to the same grounding electrode system.

_____   T   F   **43.** An equipotential bonding grid for a swimming pool shall extend under paved walking surfaces for 3 ft. (1 m) horizontally beyond the inside walls of the pool.

_____   T   F   **44.** All 125-volt receptacles located with 20 ft. (6 m) of the inside walls of a storable pool shall be protected by a GFCI.

_____   T   F   **45.** On land, the service equipment for floating structures and submersible electrical equipment shall be located no closer than 10 ft. (3 m) horizontally from the shore line.

_____   T   F   **46.** Equipment grounding conductors shall be insulated copper conductors and not smaller than 14 AWG where installed around artificially made bodies of water.

_____   T   F   **47.** Documentation shall be kept on file for qualified persons that maintain integrated electrical systems.

_____   T   F   **48.** Where an outdoor housed generator set is equipped with a readily accessible disconnecting means located within sight of the building or structure supplied, an additional disconnecting means shall be required where ungrounded conductors serve or pass through the building or structure.

_____   T   F   **49.** Temporary connection of a portable generator without transfer equipment shall be permitted where conditions of maintenance and supervision ensure that only qualified persons service the installation.

_____   T   F   **50.** Audio system circuits shall not be installed in the same raceway with Class 2 or Class 3 circuits.

Answers

1.	F	90.9(C)(4)	26.	F	320.30(D)(3)
2.	T	110.31	27.	T	340.112
3.	T	200.6(B)(3)	28.	T	400.14
4.	F	210.5(C)	29.	F	410.15(B)
5.	T	210.8(B)(2)	30.	T	410.18(B), Ex. 2
6.	F	210.8(B)(4)	31.	F	410.73(G), Ex. 1
7.	F	210.12(B), Ex.	32.	T	422.51
8.	T	210.15(D), Ex.	33.	T	430.102(B), Ex.
9.	F	210.63, Ex.	34.	F	430.109(A)(7)
10.	T	215.12(C)	35.	T	500.7(K)
11.	T	225.17	36.	T	500.8(A)(4)
12.	T	230.44, Ex.	37.	F	501.15(F)(3), FPN
13.	F	240.24(E)	38.	F	506.16
14.	F	250.4(A)(5)	39.	F	511.3(A)
15.	F	250.24(B)	40.	T	513.12
16.	T	250.30(A)(4)	41.	F	517.19(C)
17.	T	250.104(C)	42.	T	525.11
18.	F	250.119(A), Ex.	43.	T	680.26(C)
19.	T	250.122(G)	44.	T	680.32
20.	T	300.4(A)(1), Ex. 2	45.	F	682.11
21.	F	300.18(A), Ex.	46.	F	682.31(A)
22.	T	310.15(B)(2)(a)	47.	T	685.1(2)
23.	F	312.2(A)	48.	F	700.12(B)(6)
24.	F	314.16(B)(1)	49.	T	702.6, Ex.
25.	T	314.30(D)	50.	T	725.56(F)